Advances in Experimental Medicine and Biology

Volume 1117

More information about this series at http://www.springer.com/series/5584

Katsumi Matsuzaki
Editor

Antimicrobial Peptides

Basics for Clinical Application

Springer

Editor
Katsumi Matsuzaki
Graduate School of Pharmaceutical Sciences
Kyoto University
Sakyo-ku, Kyoto, Japan

ISSN 0065-2598 ISSN 2214-8019 (electronic)
Advances in Experimental Medicine and Biology
ISBN 978-981-13-3587-7 ISBN 978-981-13-3588-4 (eBook)
https://doi.org/10.1007/978-981-13-3588-4

Library of Congress Control Number: 2019933895

This Springer imprint is published by the registered company Springer Nature Singapore Pte Ltd.
The registered company address is: 152 Beach Road, #21-01/04 Gateway East, Singapore 189721, Singapore

Preface

Antimicrobial peptides (AMPs) discovered about 30 years ago are responsible for part of the innate immunity of animals and plants, and almost 3000 peptides have been reported so far. Previously, AMPs were thought to possess only antimicrobial activity, but subsequent research revealed that they also exhibit various properties including immune-modulating, anticancer, antibiofilm, and cell-penetrating activities.

The emergence of multidrug-resistant bacteria and new pathogens is a threat to human health. Therefore, the development of novel antimicrobial agents is a pressing need. AMPs have been considered promising candidates for new therapeutics to combat this problem because their antimicrobial spectra are broad and the development of bacterial resistance against them is difficult compared to conventional antibiotics.

This book of 15 chapters gives an overview of AMPs, how they work, what activities they have other than antimicrobial activity, how to design and discover them, what cautions should be taken into consideration before commercialization, and examples of clinical applications. Each chapter is written by leading scientists in that field. I appreciate their contributions very much.

Now that commercial development of AMPs has been reignited, it is a good time to publish this book to celebrate the 30-year anniversary of the discovery of AMPs. I hope this book will provide scientists in both academia and industry with the basic and comprehensive knowledge needed to develop AMPs of clinical use.

Kyoto, Japan — Katsumi Matsuzaki

Contents

Part I

Introduction

1 Antimicrobial Peptides of Multicellular Organisms: My Perspective

Michael Zasloff

Antimicrobial peptides of multicellular organisms were first characterized in the 1980s by investigators who felt that known systems of immunity could not explain what they observed: the resistance to bacterial infection of a *Cecropia* moth pupa lacking antibodies or lymphocytes (cecropins (Steiner 1981)), the potent microbicidal activity of neutrophils from a rabbit (defensins (Selsted et al. 1985)), and the healing of a wound on the skin of the African clawed frog without infection in a non-sterile aquarium (magainins (Zasloff 1987)). Since then AMPs have been discovered in diverse species of fungi, plants, and animals (Seshadri Sundararajan et al. 2012; Fan et al. 2016; Waghu et al. 2016; Wang et al. 2016). It is likely that we will discover that every multicellular organism expresses antimicrobial peptides as a key element of their immune system. Why are antimicrobial peptides so popular in Nature?

The basic design of an AMP is exceedingly simple: a short peptide synthesized from the 20 usual amino acids that can organize into a secondary structure in which hydrophilic and hydrophobic amino acids are segregated spatially and in which the peptide has a net cationic charge. Peptides of this type have the remarkable property of being amphiphilic: they are soluble in aqueous environments but also can partition into lipid environments, such as membranes (Zasloff 2002).

What makes them antimicrobial is the surprising difference in the structure of the membranes that surround microbes (most species of bacteria and some species of fungi and protozoa) and the cells of multicellular plants and animals. For reasons that remain unclear, the membranes that surround many species of microbe display negatively charged phospholipid headgroups (i.e., phosphatidylglycerol, cardiolipin) to their outside world. In contrast, the membranes of the cells that comprise the multicellular organism, in general, place zwitterionic phospholipids (i.e., phosphatidylcholine) on the outer leaflet of their plasma membrane and segregate phospholipids with anionic headgroups (i.e., phosphatidylserine, phosphatidylinositol polyphosphates) on the inner face of the plasma membrane. AMPs are attracted electrostatically to an accessible membrane with a strong negative surface charge, a feature that characterizes microbes. However, this selectivity is only relative, in the sense that AMPs will, due to their fundamental detergent-like properties, eventually damage the membranes of their host's cells if the concentrations are increased sufficiently. This becomes an issue in the development of AMPs as anti-infective therapeutics.

In the case of Gram-negative bacteria, which are enclosed by an outer membrane in addition to

M. Zasloff (✉)
MedStar Georgetown Transplant Institute,
Georgetown University School of Medicine,
Washington, DC, USA
e-mail: maz5@georgetown.edu

K. Matsuzaki (ed.), *Antimicrobial Peptides*, Advances in Experimental Medicine and Biology 1117,
https://doi.org/10.1007/978-981-13-3588-4_1

an inner membrane surrounding the cytoplasm, AMPs initially interact electrostatically with the anionic lipopolysaccharide outer membrane, disrupt its structure, and gain access to the inner. In the case of Gram-positive bacteria, AMPs are attracted electrostatically to the anionic teichoic acids that decorate the cellular proteoglycan envelope, accumulate on the surface, and then diffuse inward onto the inner cytoplasmic membrane. Following the initial interaction with the inner membrane, AMPs organize into simple polymeric structures. In most cases the function of the membrane is irreversibly disturbed, leading to loss of cellular contents, dissipation of membrane potential, and rapid cellular death (Brogden 2005). Certain AMPs exhibit an affinity for a specific membrane lipid, such as lipids I and II in the case of Gram-positive bacteria, and inhibit proteoglycan synthesis (Schneider et al. 2010), or galactosylceramide, in the case of fungi, and induce apoptosis (Thevissen et al. 2004). In addition, some AMPs flip from the outer to the inner face of the plasma membrane and then into the cytoplasm where they disturb the health of the cell by binding electrostatically to a vital target (Park et al. 2000).

By virtue of their secondary structure and their targeting of the microbial membrane through electrostatic interaction, AMPs generally exhibit effective microbicidal concentrations in the low micromolar range. To achieve these concentrations, AMPs are generally directed to provide antimicrobial defense in close proximity to the cell from which they are secreted. On dry skin surfaces, they create an antimicrobial barrier that can help shape the epidermal microbiome. On wet mucosal surfaces, they are secreted from the epithelium into the thin aqueous biofilm in direct contact with the epithelium and underneath the overriding mucous layer in contact with the lumen or cavity. In this setting, AMPs serve to prevent microbes from gaining access to the epithelial layer, killing them rapidly should they penetrate the physical barrier posed by the mucous layer. AMPs that are produced by hematopoietic cells are at the service of the particular cell in which they are carried, which either releases them at a site of infection or directs them into phagolysosomes in which microbes have been sequestered. Neuropeptides produced by the nervous system are amphiphilic, since they, like AMPs, are secreted from nerve endings and then captured onto membrane receptors of neighboring cells. It isn't surprising then that certain neuropeptides exhibit antimicrobial activity, and that by this means, the nervous system can provide immune support in certain contexts (Brogden et al. 2005; Augustin et al. 2017).

Because there are no specific sequence constraints on the design of an AMP, other than the requirement to adopt an amphiphilic secondary structure, an enormous diversity of AMPs is seen across Nature. That being said, a single amino acid substitution in an AMP, while not profoundly altering the biophysical properties of the molecule, can dramatically alter its antimicrobial spectrum (Seshadri Sundararajan et al. 2012; Fan et al. 2016; Waghu et al. 2016; Wang et al. 2016). In most instances, the mechanism underlying the change in specificity is not understood and likely involves specific interactions between the AMP and the microbial membrane that cannot be deciphered using current biophysical techniques. As a consequence when organisms find themselves facing microbes against which their existing AMPs are inadequate, effective AMPs can often evolve within the threatened population through simple amino acid substitutions.

The simple mechanism by which AMPs kill their targets has another striking benefit: It is very difficult for a microbe to evade the attack by AMPs on its membrane. Such a mutation would require a change in lipid composition or organization. Since the AMP is constructed from ordinary amino acids with no specific primary sequence signature, the microbe would have a tough task of protecting itself with a protease that could cleave the AMP, but not any of its own proteins. Furthermore, all multicellular organisms deploy a cocktail of AMPs in each physiological context, tuned to cover the full spectrum of microbes likely to be encountered in the particular niche.

One should appreciate that a healthy multicellular organism presents a variety of ecological niches to the microbes in its environment. These niches have relatively stable characteristics, maintained by active homeostatic physiological mechanisms, and within an environment certain microbes will gain a foothold over others. For example, in the absence of an effective antimicrobial defense, *Pseudomonas aeruginosa* will become the dominant Gram-negative bacterial species in the moist mucus-lined bronchial tubes of the human airway, and it is against this organism that the AMPs normally secreted from the bronchial epithelium of a healthy person are directed. This defense fails in cystic fibrosis due to a failure of the epithelium to maintain a pH in which the cocktail of AMPs can work effectively (Pezzulo et al. 2012). Pseudomonas then invades the bronchial epithelium, provoking the body to direct all of its backup inflammatory immune defenses to prevent further tissue invasion. A profound destructive inflammatory process unfolds. In the human ileum, the most distal segment of the small intestine, thousands of species of bacteria are effectively contained in the lumen, separated from the interior wall of the bowel by a single-celled epithelium. AMPs secreted from the Paneth cells and from the enterocytes, along with mucous secreted from goblet cells, deter the luminal microbes from attaching to the epithelium. In Crohn's disease Paneth cells fail to function properly, weakening the antimicrobial barrier and permitting certain microorganisms to establish a foothold within the fluid layer covering the epithelium, eventually invading the layer and provoking an inflammatory process as described in cystic fibrosis (Wehkamp et al. 2005).

Finally, a word about the development of AMPs as therapeutics. Because of their simple design, many AMPs have been synthesized differing in sequence, length, and amino acid composition. Molecules that exhibit amphiphilic properties and antimicrobial activity but are not themselves peptides have been designed. Very few of these molecules have been effective as systemically administered anti-infectives. AMPs evolved, in most multicellular organisms, to defend a microenvironment and not to be delivered into the systemic circulation. The organism can direct a high concentration of a cocktail of AMPs at the microbe, and the AMPs released can be tuned to exhibit selectivity for the microbes likely to be encountered in that niche as well as maintain an AMP concentration that will not damage the host's cells within that niche. From this perspective, the development of AMP therapeutics should be informed by the physiological context of the infection to be treated.

The discovery of AMPs has led to a deeper appreciation of the importance of innate immunity as an arm of vertebrate defense and provided insight into the immune defenses of the vast majority of plants and animals that now exist. Much remains to be explored.

References

Augustin R et al (2017) A secreted antibacterial neuropeptide shapes the microbiome of Hydra. Nat Commun 8(1):698

Brogden KA (2005) Antimicrobial peptides: pore formers or metabolic inhibitors in bacteria? Nat Rev Microbiol 3(3):238–250

Brogden KA et al (2005) The nervous system and innate immunity: the neuropeptide connection. Nat Immunol 6(6):558–564

Fan L et al (2016) DRAMP: a comprehensive data repository of antimicrobial peptides. Sci Rep 6:24482

Park CB et al (2000) Structure-activity analysis of buforin II, a histone H2A-derived antimicrobial peptide: the proline hinge is responsible for the cell-penetrating ability of buforin II. Proc Natl Acad Sci U S A 97(15):8245–8250

Pezzulo AA et al (2012) Reduced airway surface pH impairs bacterial killing in the porcine cystic fibrosis lung. Nature 487(7405):109–113

Schneider T et al (2010) Plectasin, a fungal defensin, targets the bacterial cell wall precursor Lipid II. Science 328(5982):1168–1172

Selsted ME et al (1985) Primary structures of six antimicrobial peptides of rabbit peritoneal neutrophils. J Biol Chem 260(8):4579–4584

Seshadri Sundararajan V et al (2012) DAMPD: a manually curated antimicrobial peptide database. Nucleic Acids Res 40(Database issue):D1108–D1112

Steiner H et al (1981) Sequence and specificity of two antibacterial proteins involved in insect immunity. Nature 292(5820):246–248

Thevissen K et al (2004) Defensins from insects and plants interact with fungal glucosylceramides. J Biol Chem 279(6):3900–3905

Waghu FH et al (2016) CAMPR3: a database on sequences, structures and signatures of antimicrobial peptides. Nucleic Acids Res 44(D1):D1094–D1097

Wang G et al (2016) APD3: the antimicrobial peptide database as a tool for research and education. Nucleic Acids Res 44(D1):D1087–D1093
Wehkamp J et al (2005) Reduced Paneth cell alpha-defensins in ileal Crohn's disease. Proc Natl Acad Sci U S A 102(50):18129–18134
Zasloff M (1987) Magainins, a class of antimicrobial peptides from Xenopus skin: isolation, characterization of two active forms, and partial cDNA sequence of a precursor. Proc Natl Acad Sci U S A 84(15):5449–5453
Zasloff M (2002) Antimicrobial peptides of multicellular organisms. Nature 415(6870):389–395

Part II

Mechanisms of Antimicrobial Action

Membrane Permeabilization Mechanisms

2

Katsumi Matsuzaki

Abstract

Many antimicrobial peptides are considered to kill microbes by permeabilizing cell membranes. This chapter summarizes the driving force of peptide binding to membranes; various mechanisms of lipid bilayer permeabilization including the barrel-stave, toroidal pore, and carpet models; and modes of permeabilization of bacterial and mammalian membranes.

Keywords

Membrane permeabilization · Membrane binding · Membrane curvature · Barrel-stave model · Toroidal pore model · Carpet model

2.1 Introduction

Membrane permeabilization as a mechanism for bacterial killing has already been suggested in a paper on the discovery of magainins from the African clawed frog *Xenopus laevis*, an archetypical antimicrobial peptide (AMP) isolated from a vertebrate for the first time (Zasloff 1987). As shown in Fig. 2.1, the addition of magainin 2 to *E. coli* cells induced an efflux of intracellular K^+ ions and concomitantly a decrease in cell viability in minutes (Matsuzaki et al. 1997a). Rapid killing is an important self-defense action of AMPs. Observations that enantiomeric peptides composed of all D-amino acids are equipotent to the parent L-peptides indicate that proteins requiring chiral recognition such as receptors and enzymes are not involved in the membrane permeabilization process (Bessalle et al. 1990; Wade et al. 1990). Furthermore, magainins induce the leakage of water-soluble dyes entrapped in artificial lipid vesicles (Matsuzaki et al. 1989, 1991a), suggesting that the lipid matrix of membranes is a target of the peptides. Many AMPs such as tachyplesin I (Matsuzaki et al. 1991b) and LL-37 (Lee et al. 2011) also permeabilize membranes.

K. Matsuzaki (✉)
Graduate School of Pharmaceutical Sciences, Kyoto University, Sakyo-ku, Kyoto, Japan
e-mail: mkatsumi@pharm.kyoto-u.ac.jp

2.2 Membrane Binding

The first step in the membrane permeabilization process is the binding of peptides to membranes. AMPs are generally polycationic and amphipathic. In many cases, linear peptides take unordered structures in an aqueous solution, whereas they are conformed to be amphipathic secondary structures, typically α-helices or β-strands, upon membrane binding (Matsuzaki et al. 1989, 1991a). Such amphipathic structures fit the membrane–water interface. Even conformationally restricted cyclic peptides also change their structures to accommodate themselves to membrane environments (Imura et al. 2007).

K. Matsuzaki (ed.), *Antimicrobial Peptides*, Advances in Experimental Medicine and Biology 1117,
https://doi.org/10.1007/978-981-13-3588-4_2

The main driving forces for membrane binding are electrostatic attraction and hydrophobic interaction (Matsuzaki et al. 1995a, 2009). Positively charged peptides preferentially interact with negatively charged membranes. Bacterial membranes are rich in acidic phospholipids (phosphatidylglycerol and cardiolipin). Furthermore, cell walls containing lipopolysaccharides (LPS) and peptidoglycans are also negatively charged. In contrast, mammalian cell membranes are less negatively charged. Acidic phospholipids, such as phosphatidylserine, are essentially sequestered on the cytoplasmic face of the membrane, although cancer cells show an increased exposure of phosphatidylserine (Utsugi et al. 1991). However, gangliosides containing sialic acid residues have some negative charges on the cell surface, which are a target of AMPs (Miyazaki et al. 2012). Interestingly, acidic phospholipids and gangliosides interact differently with AMPs (Fig. 2.2). Cationic peptides specifically bind to gangliosides containing anionic sialic acid residues so that the charge is neutralized. The binding is described by the Langmuir-type equation. Fluorescent resonance transfer experiments clearly revealed that AMPs preferentially interacted with monosialoganglioside GM1 compared with phosphatidylcholine in a GM1/phosphatidylcholine mixed bilayer mimicking mammalian cell membranes. In contrast, peptides equally interact with anionic phosphatidylglycerol and zwitterionic phosphatidylcholine in a phosphatidylglycerol/phosphatidylcholine mixed membrane, a model for bacterial cell membranes (Miyazaki et al. 2012). The interaction is theoretically explained by a combination of the Gouy-Chapman theory (electrostatic concentration immediately above the membrane surface) and a partition equilibrium (Wenk and Seelig 1998; Wieprecht et al. 1999).

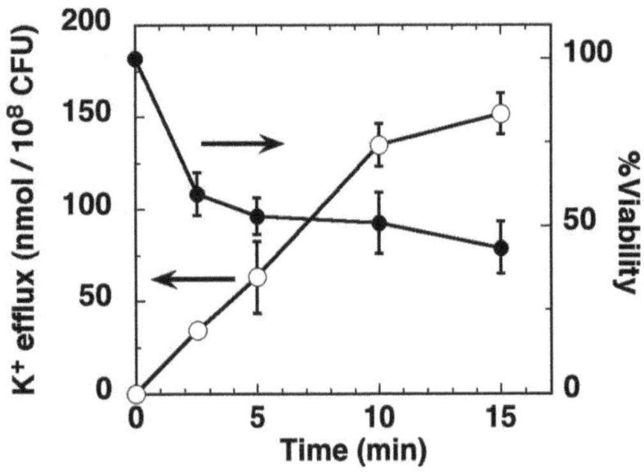

Fig. 2.1 Correlation between membrane permeabilization and bacterial death. Magainin 2 (50 nmol) was added to *E. coli* cells (5×10^8 CFU/mL). K^+ efflux (open circles, left axis) and the percent cell viability (closed circles, right axis) are plotted as a function of time

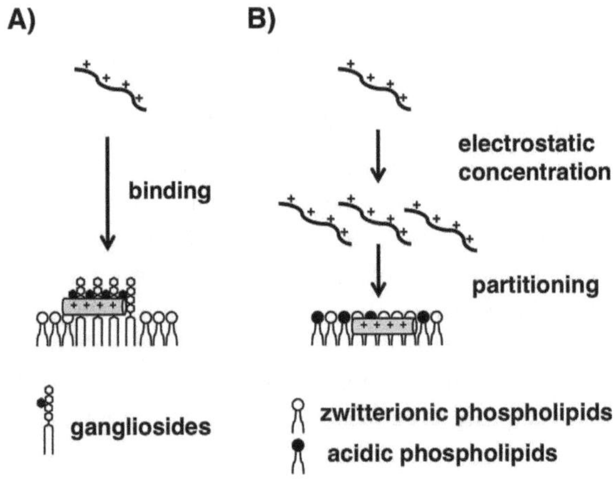

Fig. 2.2 Different binding modes of AMPs to (**a**) mammalian and (**b**) bacterial model membranes. (**a**) Cationic AMPs specifically bind to gangliosides containing anionic sialic acid residues. The binding is described by the Langmuir-type equation. (**b**) In contrast, peptides are electrostatically concentrated immediately above the membrane surface according to the Gouy-Chapman theory and then partitioned into the membrane. There is no specific interaction between AMPs and acidic phospholipids

2.3 Permeabilization of Model Membranes

Membrane-bound AMPs change membrane structures and organizations leading to membrane permeabilization. Using model membranes such as liposomes, several mechanisms so far proposed for this can be roughly classified into two categories, i.e., membrane curvature modulation and phase separation. The recently proposed latter mechanism includes clustering of acidic lipids by cationic AMPs and is described in Chap. 5. The "barrel-stave channel," "toroidal pore," and original "carpet" mechanisms are categorized as the former mechanism (Fig. 2.3). Refer to Chap. 3 and a review by Huang (2006) for theoretical treatments.

Amphipathic secondary structures on the surface of membranes can modulate membrane curvature. The peptides expand the interfacial region, making a void in the hydrocarbon region of the membrane. Consequently, membrane-

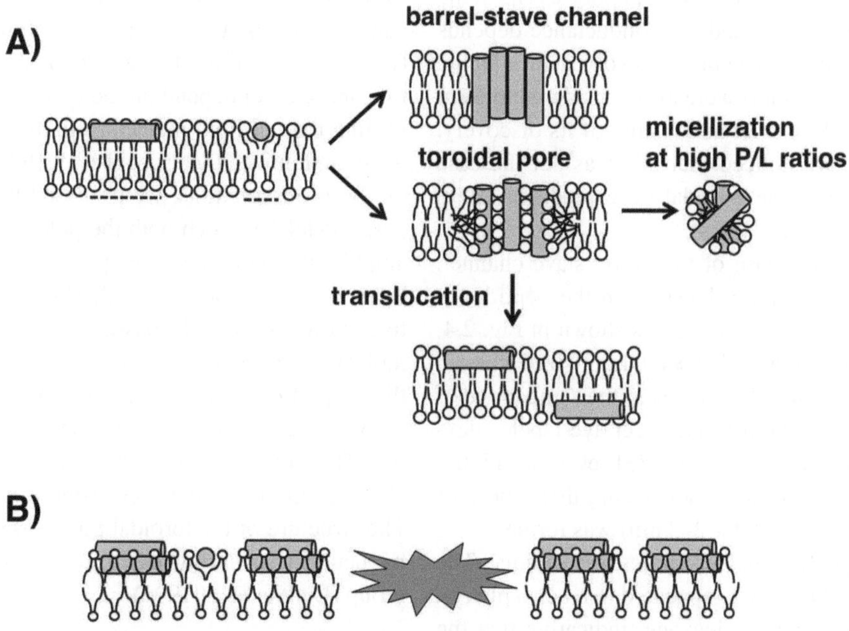

Fig. 2.3 Major mechanisms of membrane permeabilization induced by AMPs. (**a**) AMPs form amphipathic secondary structures (α-helix in this case) on the surface of membranes. The peptides expand the interfacial region making a void in the hydrocarbon region of the membrane. Consequently, membrane-thinning and positive curvature (concave) strain are induced (broken line). At threshold peptide-to-lipid ratios, typically below ~1:100, either the "barrel-stave channel" or the "toroidal pore" is formed. The former is solely composed of peptides, making a water-filled channel. The ionic current is discrete and the conductance depends on the number of peptides involved in a single channel. A typical example of this class of peptide is the peptaibol alamethicin. However, most AMPs form a toroidal pore, in which both the polar faces of the amphiphilic structures and the polar headgroups of lipids constitute the pore wall. This unique structure allows not only the passage of ions and small molecules through the pore but also the rapid flip–flop of lipids along the pore wall. The pore is not stable (the lifetime is typically ms), and upon its disintegration a fraction of peptide molecules translocates across bilayers. At much higher peptide-to-lipid ratios, membranes are solubilized into micelles. (**b**) When membranes have a negative-curvature (convex) tendency, it counteracts the positive curvature induced by the peptides, stabilizing the peptide–lipid system and allowing accumulation of large amounts of peptides. Eventually (e.g., peptide-to-lipid ratios above 1:10), membrane disruption occurs. This mechanism corresponds to the original "carpet model." It should be noted that the modified carpet model also includes toroidal pore formation, although its scientific validity needs careful consideration. Similar phenomena can happen in zero-curvature bilayers with peptides having a large hydrophobic surface capable of expanding the hydrocarbon core of the membrane. Thus, the mechanism of membrane permeabilization is not unique to peptides but also depends on the physicochemical properties of membranes

thinning and positive curvature (concave) strain are induced (Fig. 2.3a left). At threshold peptide-to-lipid ratios, typically below ~1:100, the surface-lying peptides are cooperatively inserted into the membrane, forming a water-filled pore (Fig. 2.3a right). Peptaibols, peptides containing Aib (aminoisobutyric acid) residues with the C-terminal alcohol, have been known to form a barrel-stave channel (Sansom 1991). These peptides are electrostatically almost neutral. A typical example is the antibiotic alamethicin produced by the fungus *Trichoderma viride*. The channel is solely composed of helical peptides. The ionic current is discrete and the conductance depends on the number of peptides involved in a single channel. Magainins were also considered to form this type of channel at the time of its discovery. However, we noticed that this was not the case and proposed the toroidal pore model in 1996 (Matsuzaki et al. 1996a).

First, in the case of the barrel-stave channel, the size of the pore depends on the peptide-to-lipid molar ratio. However, as shown in Fig. 2.4, the small I^- ion and the medium-sized calcein dye leaked out of liposomes in the same range of peptide-to-lipid ratio, and larger dye-labeled dextran molecules were retained even at higher peptide-to-lipid ratios, suggesting that a pore of defined size (diameter 2–3 nm) was formed.

Second, the leakage kinetics were unique. The percent leakage value appeared to reach a plateau instead of complete leakage, indicating that the number of pores decreased with increasing time (Fig. 2.5). We hypothesized that pores were unstable and a fraction of peptide molecules was translocated into the inner leaflet. Thus, the peptide density in the outer leaflet decreased, decelerating pore formation. Indeed, the translocation was coupled to the leakage. Kinetic analysis suggested that a pore is composed of ~5 magainin molecules, which appeared to be too small to allow the passage of calcein (Matsuzaki et al. 1995b). Thus, we hypothesized that lipid molecules were also involved in the pore structure and examined the flip–flop of lipids. The flop was again coupled to leakage and translocation (Fig. 2.5). The flip rate was identical to the flop rate and did not depend on the type of lipid, suggesting that all lipid molecules in the membrane were randomized (Matsuzaki et al. 1996a). Based on these observations, we proposed the toroidal pore model, in which both the polar faces of the amphiphilic helices and the polar headgroups of lipids constitute the pore wall. This unique structure allows not only the passage of ions and small molecules through the pore but also the rapid flip–flop of lipids along the pore wall. Both types of events seem to contribute to bactericidal activity. The flop of phosphatidylserine was also observed in mammalian cells (Imura et al. 2008). The structure of the toroidal pore was confirmed by neutron scattering experiments by Huang's group (Ludtke et al. 1996).

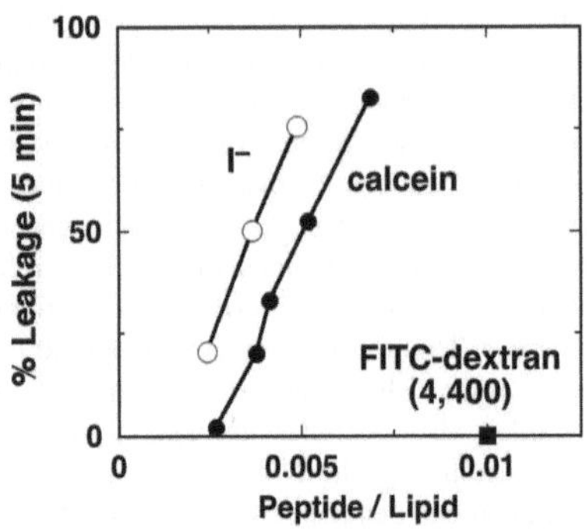

Fig. 2.4 Estimation of the pore size formed by magainin 2. The percent leakage values are plotted as a function of the peptide-to-lipid ratio for I^- ions (open circles), calcein (MW 623, closed circles), and FITC-dextran (MW 4400, closed squares)

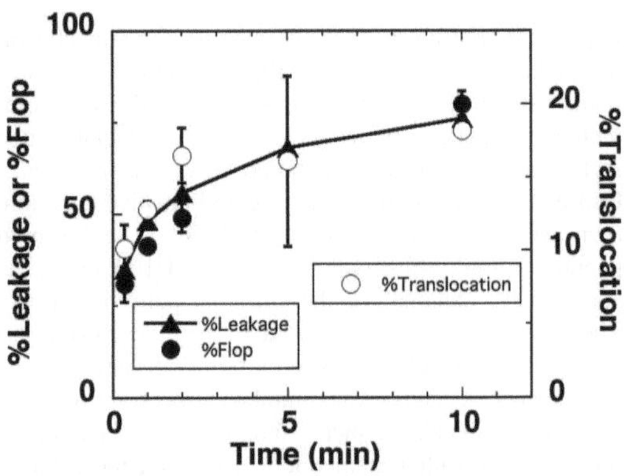

Fig. 2.5 Coupling between leakage, peptide translocation, and lipid flop. The percent leakage of calcein (solid line), percent peptide translocation (open circles), and percent flop of NBD-phosphatidylethanolamine (closed circles) are plotted as a function of time

Not only magainins but also other AMPs and membrane-acting peptides form toroidal pores, including helical PGLa (Matsuzaki et al. 1998a), mastoparan X (Matsuzaki et al. 1996b), melittin (Yang et al. 2001), buforin 2 (Kobayashi et al. 2004), and cyclic β-sheet tachyplesin I (Imura et al. 1768). The pore size depends on the type of peptide. The two substitutions F5Y and F16W for magainin 2 enlarged the pore size to 4–7 nm (Hara et al. 2001). The bacteriocin lacticin Q also forms a large toroidal pore of similar size (Yoneyama et al. 2009).

The mode of dye leakage is classified into an all-or-none or graded mode. In the former, some vesicles are empty, whereas the rest are intact. In the latter, all vesicles partially lose their contents. Although these modes are sometimes considered to originate from different leakage mechanisms, earlier theoretical studies showed that the pore lifetime τ determines the mode of leakage (Schwarz and Arbuzova 1995; Schwarz and Robert 1992). The all-or-none and the graded mode correspond to $\tau >> \tau_0$ (the time necessary for a $1/e$ reduction in the intravesicular dye concentration) and $\tau << \tau_0$, respectively.

The pore lifetime is governed by electrostatics within the pore and is therefore modulated by both peptide charge and lipid composition because the toroidal pore is composed of peptides and lipids. An increase in peptide positive charge destabilizes the pore because of enhanced electrostatic repulsion between closely spaced peptides, facilitating peptide translocation (Matsuzaki et al. 1997b). Note that translocation occurs only upon the disintegration of the pore. For example, an increase in the positive charge of magainin from +4 to +6 reduced the τ value from $9\tau_0$ to $0.1\tau_0$. Buforin 2 with +6 charges is translocated across liposomal (Kobayashi et al. 2000) and bacterial (Park et al. 1998) membranes without permeabilizing them. Similarly, an increase in acidic lipid content stabilizes the pore by reducing electrostatic repulsion between peptides in a pore (Kobayashi et al. 2004). Dye leakage and lipid flip–flop are observed in this case.

Another well-known mechanism of membrane permeabilization is the original "carpet model," in which peptides cover the membrane surface like a carpet and disrupt the bilayer organization (Shai 1995) (Fig. 2.3b). This model was proposed following studies on the interaction between cecropins and dermaseptins with phosphatidylserine-containing bilayers. We found that even magainins do not form toroidal pores, but accumulate on the membrane surface and eventually disrupt it at peptide-to-lipid ratios above 1:10 in membranes containing phosphatidylserine (Fig. 2.6), phosphatidic acid, or cardiolipin (Matsuzaki et al. 1998b). These lipids form hexagonal-II phases under charge-neutralizing conditions and thus tend to induce negative curvature strain on the membrane, counteracting the positive curvature strain imposed by magainins and allowing accumulation of huge amounts of peptides on the membrane. Thus, the mechanism of membrane permeabilization is not unique to peptides, but also depends on the physicochemical properties of membranes. It should be noted that the current modified carpet model also includes toroidal pore formation (Oren and Shai 1998), although its scientific validity needs careful consideration. Recently, the detergent-like model (Bechinger and Lohner 2006) and the interfacial activity model (Wimley 2010) have been proposed for the mechanism of AMPs. These more comprehensive models are similar to the curvature modulation mechanism described above in that the incorporation of amphipathic structures in lipid bilayers modifies the bilayer organization.

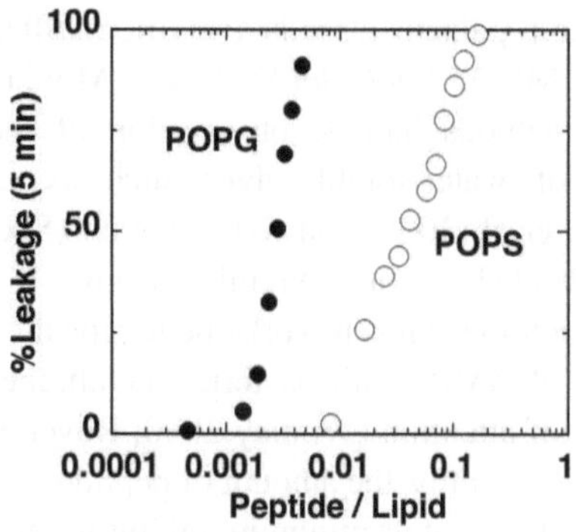

Fig. 2.6 Dependence of magainin 2 leakage activity on lipid species. The percent leakage values are plotted as a function of the peptide-to-lipid ratios in a logarithmic scale for palmitoyloleoylphosphatidylglycerol (POPG, closed circles) and palmitoyloleoylphosphatidylserine (POPS, open circles)

2.4 Permeabilization of Bacterial Membranes

AMPs have been proposed to disrupt and permeabilize the outer membrane of Gram-negative bacteria by the "self-promoted pathway" (Hancock and Chapple 1999). Cationic peptides compete with Mg^{2+} ions that bridge between adjacent phosphates of LPS. Magainin forms a helix upon binding to the lipid A moiety of LPS (Matsuzaki et al. 1999) and induces blebs on the outer membrane, while the presence of Mg^{2+} ions inhibits its antimicrobial activity (Matsuzaki et al. 1997a). The permeabilization of the outer membrane can be detected by the permeation of impermeable substances such as nonionic detergents (Matsuzaki et al. 1997a) and 1-*N*-phenylnaphthylamine (NPN) (Zhang et al. 1999). Recently, a method based on the leakage of green fluorescent protein (GFP) expressed in the periplasmic space was developed (Sochacki et al. 2011). Note that the diameter of the protein (4.5 nm) is larger than that of the magainin toroidal pore (2–3 nm).

The permeabilization of cytoplasmic membranes can be monitored by several techniques. The efflux of intracellular K^+ ions is detected by a K^+-selective electrode (Matsuzaki et al. 1997a). Potential sensitive dyes, such as $diSC_35$, detect the dissipation of transmembrane potential (Patrzykat et al. 2002). The hydrolysis of *o*-nitrophenyl-β-D-galactoside (ONPG) by cytoplasmic β-galactosidase is also often utilized for *E. coli* ML-35 (*lacI*⁻, *lacY*⁻, *lacZ*⁺). More convenient methods include observation of the cell entry of water-soluble dyes such as calcein (Imura et al. 2008) and Sytox Green (Sochacki et al. 2011) by fluorescent microscopy.

Which mechanism works better for the interaction of AMPs with bacterial membranes is a matter of attention (Wimley 2010; Roversi et al. 2014). Generally, the amount of peptide bound to bacterial cells at its minimum inhibitory or bactericidal concentration (MIC or MBC) is considered. However, it should be noted that MIC or MBC is the concentration needed to inhibit the growth of or to kill the most resistant population of bacteria and is thus too strong for the other more susceptible populations. The experimental conditions in Fig. 2.1 may be appropriate for discussion because about 50% of bacteria are killed. Assuming that (1) peptide molecules are intact and completely bound to both leaflets of the outer and inner membranes, (2) the binding to other bacterial components is negligible, and (3) the number of lipids per cell is 4×10^7 (Roversi et al. 2014), the peptide-to-lipid ratio in the inner membrane is ~1. To examine the direct interaction of magainin 2 with inner membranes, we used *E. coli* spheroplasts lacking outer membranes (Matsuzaki et al. 1997a). The peptide lysed spheroplasts in a peptide-to-lipid ratio range of 0.1–1, in accordance with the rough estimate above. The lipid composition of the spheroplasts was 68% phosphatidylethenolamine/24% phosphatidylglycerol/8% cardiolipin. Asymmetry of lipid distribution in the inner membrane has not yet been observed (Furse and Scott 2016). Nevertheless, even if the inner leaflet is composed of phosphatidylethenolamine alone, the outer leaflet is composed of ~50% phosphatidylglycerol and ~50% lipids with a negative-curvature tendency (cardiolipin and phosphatidylethenolamine). Thus, the original carpet mechanism is more likely, although further studies are needed to confirm this. Transmission electron microscopy revealed that LL-37 caused local perturbations and breaks along *P. aeruginosa* cell membranes (Andersson et al. 2004). However, the mechanism of membrane permeabilization may depend on the type of AMPs, as well as bacterial strains. Magainin appears to form a toroidal pore against the Gram-positive *B. megaterium* because the peptide forms membrane lesions of definite size (~2.8 nm, <6.6 nm) (Imura et al. 2008).

2.5 Permeabilization of Mammalian Cell Membranes

Relatively little is known about how AMPs permeabilize and are translocated across mammalian cell membranes. Our earlier work suggested that prototypical membrane-permeabilizing magainin 2 and intracellular-targeting buforin 2 interact with human cells differently (Takeshima et al. 2003). Dye-labeled magainin was translocated across HeLa cells via both energy-dependent and energy-independent mechanisms, and the internalization was accompanied by cytotoxicity. In contrast, dye-labeled buforin penetrates cells in an energy-independent fashion and exerts little toxicity. Zanetti's group reported that the Pro-rich peptide Bac7 (1–35) is taken up by murine and human cells through a nontoxic energy- and temperature-dependent process (Tomasinsig et al. 2006). Details on so-called cell-penetrating peptide (CPP) are described in Chap. 7.

2.6 Conclusion

Thirty years of extensive studies, both experimental and theoretical, have revealed several basic mechanisms for the permeabilization of lipid bilayers by AMPs, as described above. It depends on both peptide sequences and lipid compositions. However, the molecular details are difficult to understand because initial membrane permeabilization is often dynamic and transient in nature, whereas spectroscopic methods usually detect equilibrium states after the permeabilization ceases. In contrast to model membrane studies, little is known on the permeabilization mechanisms of bacterial and mammalian cells, which remain subjects of further study.

References

Andersson E, Rydengard V, Sonesson A, Morgelin M, Bjorck L, Schmidtchen A (2004) Antimicrobial activities of heparin-binding peptides. Eur J Biochem 271:1219–1226

Bechinger B, Lohner K (2006) Detergent-like actions of linear amphipathic cationic antimicrobial peptides. Biochim Biophys Acta 1758:1529–1539

Bessalle R, Kapitkovsky A, Gorea A, Shalit I, Fridkin M (1990) All-D-magainin: chirality, antimicrobial activity and proteolytic resistance. FEBS Lett 274:151–155

Furse S, Scott DJ (2016) Three-dimensional distribution of phospholipids in gram negative bacteria. Biochemistry 55:4742–4747

Hancock REW, Chapple DS (1999) Peptide antibiotics. Antimicrob Agents Chemother 43:1317–1323

Hara T, Kodama H, Kondo M, Wakamatsu K, Takeda A, Tachi T et al (2001) Effect of peptide dimerization on pore formation: antiparallel disulfide-dimerized magainin 2 analog. Biopolymers 58:437–446

Huang HW (2006) Molecular mechanism of antimicrobial peptides: the origin of cooperativity. Biochim Biophys Acta 1758:1292–1302

Imura Y, Nishida M, Ogawa Y, Takakura Y, Matsuzaki K (2007) Action mechanism of tachyplesin I and effects of PEGylation. Biochim Biophys Acta 1768:1160–1169

Imura Y, Choda N, Matsuzaki K (2008) Magainin 2 in action: distinct modes of membrane permeabilization in living bacterial and mammalian cells. Biophys J 95:5757–5765

Kobayashi S, Takeshima K, Park CB, Kim SC, Matsuzaki K (2000) Interactions of the novel antimicrobial peptide buforin 2 with lipid bilayers: proline as a translocation promoting factor. Biochemistry 39:8648–8654

Kobayashi S, Chikushi A, Tougu S, Imura Y, Nishida M, Yano Y et al (2004) Membrane translocation mechanism of the antimicrobial peptide buforin 2. Biochemistry 43:15610–15616

Lee C-C, Sun Y, Qian S, Huang HW (2011) Transmembrane pores formed by human antimicrobial peptide LL-37. Biophys J 100:1688–1696

Ludtke SJ, He K, Heller WT, Harroun TA, Yang L, Huang HW (1996) Membrane pores induced by magainin. Biochemistry 35:13723–13728

Matsuzaki K (2009) Control of cell selectivity of antimicrobial peptides. Biochim Biophys Acta 1788:1687–1692

Matsuzaki K, Harada M, Handa T, Funakoshi S, Fujii N, Yajima H et al (1989) Magainin 1-induced leakage of entrapped calcein out of negatively-charged lipid vesicles. Biochim Biophys Acta 981:130–134

Matsuzaki K, Harada M, Funakoshi S, Fujii N, Miyajima K (1991a) Physicochemical determinants for the interactions of magainins 1 and 2 with acidic lipid bilayers. Biochim Biophys Acta 1063:162–170

Matsuzaki K, Fukui M, Fujii N, Miyajima K (1991b) Interactions of an antimicrobial peptide, tachyplesin I, with lipid membranes. Biochim Biophys Acta 1070:259–264

Matsuzaki K, Sugishita K, Fujii N, Miyajima K (1995a) Molecular basis for membrane selectivity of an antimicrobial peptide, magainin 2. Biochemistry 34:3423–3429

Matsuzaki K, Murase O, Miyajima K (1995b) Kinetics of pore formation induced by an antimicrobial peptide, magainin 2. Biochemistry 34:12553–12559

Matsuzaki K, Murase O, Fujii N, Miyajima K (1996a) An antimicrobial peptide, magainin 2, induced rapid flip-flop of phospholipids coupled with pore formation and peptide translocation. Biochemistry 35:11361–11368

Matsuzaki K, Yoneyama S, Murase O, Miyajima K (1996b) Transbilayer transport of ions and lipids coupled with mastoparan X translocation. Biochemistry 35:8450–8456

Matsuzaki K, Sugishita K, Harada M, Fujii N, Miyajima K (1997a) Interactions of an antimicrobial peptide, magainin 2, with outer and inner membranes of gram-negative bacteria. Biochim Biophys Acta 1327:119–130

Matsuzaki K, Nakamura A, Murase O, Sugishita K, Fujii N, Miyajima K (1997b) Modulation of magainin 2–lipid bilayer interactions by peptide charge. Biochemistry 36:2104–2111

Matsuzaki K, Mitani Y, Akada K, Murase O, Yoneyama S, Zasloff M et al (1998a) Mechanism of synergism between antimicrobial peptides magainin 2 and PGLa. Biochemistry 37:15144–15153

Matsuzaki K, Sugishita K, Ishibe N, Ueha M, Nakata S, Miyajima K et al (1998b) Relationship of membrane curvature to the formation of pores by magainin. Biochemistry 37:11856–11863

Matsuzaki K, Sugishita K, Miyajima K (1999) Interactions of an antimicrobial peptide, magainin 2 with lipopolysaccharide-containing liposomes as a model for outer membranes of gram-negative bacteria. FEBS Lett 449:221–224

Miyazaki Y, Aoki M, Yano Y, Matsuzaki K (2012) Interaction of antimicrobial peptide magainin 2 with gangliosides as a target for human cell binding. Biochemistry 51:10229–10235

Oren Z, Shai Y (1998) Mode of action of linear amphipathic alpha-helical antimicrobial peptides. Biopolymers 47:451–463

Park CB, Kim HS, Kim SC (1998) Mechanism of action of the antimicrobial peptide buforin II: buforin II kills microorganisms by penetrating the cell membrane and inhibiting cellular functions. Biochem Biophys Res Commun 244:253–257

Patrzykat A, Friedrich CL, Zhang L, Mendoza V, Hancock RE (2002) Sublethal concentrations of pleurocidin-derived antimicrobial peptides inhibit macromolecular synthesis in *Escherichia coli*. Antimicrob Agents Chemother 46:605–614

Roversi D, Luca V, Aureli S, Park Y, Mangoni ML, Stella L (2014) How many antimicrobial peptide molecules kill a bacterium? The case of PMAP-23. ACS Chem Biol 9:2003–2007

Sansom MSP (1991) The biophysics of peptide models of ion channels. Prog Biophys Mol Biol 55:139–235

Schwarz G, Arbuzova A (1995) Pore kinetics reflected in the dequenching of a lipid vesicle entrapped fluorescent dye. Biochim Biophys Acta 1239:51–57

Schwarz G, Robert CH (1992) Kinetics of pore-mediated release of marker molecules from liposomes or cells. Biophys Chem 42:291–296

Shai Y (1995) Molecular recognition between membrane-spanning polypeptides. Trends Biol Sci 20:460–465

Sochacki KA, Barns KJ, Bucki R, Weisshaar JC (2011) Real-time attack on single *Escherichia coli* cells by the human antimicrobial peptide LL-37. Proc Natl Acad Sci U S A 108:E77–E81

Takeshima K, Chikushi A, Lee K-K, Yonehara S, Matsuzaki K (2003) Translocation of analogues of the antimicrobial peptides magainin and buforin across human cell membranes. J Biol Chem 278:1310–1315

Tomasinsig L, Skerlavaj B, Papo N, Giabbai B, Shai Y, Zanetti M (2006) Mechanistic and functional studies of the interaction of a proline-rich antimicrobial peptide with mammalian cells. J Biol Chem 281:383–391

Utsugi T, Schroit AJ, Connor J, Bucana CD, Fidler IJ (1991) Elevated expression of phosphatidylserine in the outer membrane leaflet of human tumor cells and recognition by activated human blood monocytes. Cancer Res 51:3062–3066

Wade D, Boman A, Wåhlin B, Drain CM, Andreu D, Boman HG et al (1990) All-D amino acid-containing channel forming antibiotic peptides. Proc Natl Acad Sci U S A 87:4761–4765

Wenk MR, Seelig J (1998) Magainin 2 amide interaction with lipid membranes: calorimetric detection of peptide binding and pore formation. Biochemistry 37:3909–3916

Wieprecht T, Beyermann M, Seelig J (1999) Binding of antibacterial magainin peptides to electrically neutral membranes: thermodynamics and structure. Biochemistry 38:10377–10387

Wimley WC (2010) Describing the mechanism of antimicrobial peptide action with the interfacial activity model. ACS Chem Biol 5:905–917

Yang L, Harroun TA, Weiss TM, Ding L, Huang HW (2001) Barrel-stave model or toroidal model? A case study on melittin pores. Biophys J 81:1475–1485

Yoneyama F, Imura Y, Ohno K, Zendo T, Nakayama J, Matsuzaki K et al (2009) Peptide-lipid huge toroidal pore, a new antimicrobial mechanism mediated by a lactococcal bacteriocin, lacticin Q. Antimicrob Agents Chemother 53:3211–3217

Zasloff M (1987) Magainins, a class of antimicrobial peptides from *Xenopus* skin: isolation, characterization of two active forms, and partial cDNA sequence of a precursor. Proc Natl Acad Sci U S A 84:5449–5453

Zhang L, Benz R, Hancock REW (1999) Influence of proline residues on the antibacterial and synergistic activities of α-helical peptides. Biochemistry 38:8102–8111

3 Elementary Processes and Mechanisms of Interactions of Antimicrobial Peptides with Membranes—Single Giant Unilamellar Vesicle Studies—

Moynul Hasan and Masahito Yamazaki

Abstract

To elucidate the mechanisms of action of antimicrobial peptides (AMPs) and to develop de novo designed peptides with activities similar to those of AMPs, it is essential to elucidate the detailed processes of AMP interactions with plasma membranes of bacterial and fungal cells and model membranes (lipid bilayers). In this mini-review, we summarize the present state of knowledge of the interactions of AMPs with lipid vesicles obtained using the single giant unilamellar vesicle (GUV) method. Currently, three modes of action of AMPs on GUVs have been defined. The elementary processes of interactions of AMPs with lipid vesicles revealed by the single GUV method, and the advantages of this technique, are described and discussed. For example, the single GUV method can be used to determine rate constants of AMP-induced pore formation or local rupture and membrane permeation of internal contents through the pore or the local rupture, the transbilayer movement of lipids, and the relationship between the location of AMPs and pore formation. Effects of membrane tension and of asymmetric lipid packing in the bilayer on AMP-induced pore formation also are described. On the basis of these data, we discuss the present state of understanding of the interaction of AMPs with lipid bilayers and future prospects for AMP studies.

Keywords

Antimicrobial peptides · Giant unilamellar vesicle · Pore formation · Local rupture · Translocation across membranes · Lipid bilayers · Elementary process · Rate constant · Membrane tension

M. Hasan
Integrated Bioscience Section, Graduate School of Science and Technology, Shizuoka University, Shizuoka, Japan

M. Yamazaki (✉)
Integrated Bioscience Section, Graduate School of Science and Technology, Shizuoka University, Shizuoka, Japan

Nanomaterials Research Division, Research Institute of Electronics, Shizuoka University, Shizuoka, Japan

Department of Physics, Graduate School of Science, Shizuoka University, Shizuoka, Japan
e-mail: yamazaki.masahito@shizuoka.ac.jp

3.1 Introduction

Antimicrobial peptides (AMPs) and antimicrobial proteins with bactericidal and fungicidal activities are produced by various organisms (e.g., animals including humans, plants, and insects) to defend themselves against microbes. Most AMPs target bacterial and fungal plasma membranes and induce damage in the membranes,

K. Matsuzaki (ed.), *Antimicrobial Peptides*, Advances in Experimental Medicine and Biology 1117,
https://doi.org/10.1007/978-981-13-3588-4_3

including pore formation that increases membrane permeability (Zasloff 2002; Melo et al. 2009; Hwang and Vogel 1998; Propheter et al. 2017). Some AMP-induced damage to the plasma membrane is considered to reflect the interaction of these AMPs with the lipid membrane regions of bacterial and fungal plasma membranes, as indicated by the observation that synthetic AMP analogues composed of all-D amino acids have the same antibacterial activity as wild-type AMPs composed of all-L amino acids (Wade et al. 1990; Wakabayashi et al. 1999).

It has been established that the physicochemical properties of lipid bilayer vesicles are similar to those of native plasma membranes, and lipid bilayers are therefore excellent models of plasma membranes (Lipowsky and Sackmann 1995; Baumgart et al. 2003; McLaughlin and Murray 2005; Bigay and Antonny 2012). To investigate the interaction of AMPs with the lipid membrane regions of microbial plasma membranes, various kinds of vesicles composed of lipid bilayers have been extensively used. To date, most research on the interactions of AMPs with lipid bilayers has been performed using suspensions of large unilamellar vesicles (LUVs) with a diameter of 50 nm–1 μm (100–300 nm for most uses) (i.e., the LUV suspension method). Given their small size, we cannot observe the structure and the physical quantities of each LUV in buffer; hence, using the LUV suspension method, we can obtain only ensemble averages of physical quantities of many LUVs in a suspension, as measured by various biophysical and physicochemical techniques. Therefore, the LUV suspension technique does not permit observation of each elementary process in the interaction of peptides such as AMPs with lipid bilayers (Yamazaki 2008; Islam et al. 2018). In contrast, the shape and physical properties of each giant unilamellar vesicle (GUV) with a diameter larger than 1 μm (10–40 μm for most uses) in buffer can be observed in real time using various optical microscopy techniques. Using this advantage, GUVs have been employed to investigate shape changes, the elastic modulus of lipid bilayers, phase separation, and tension-induced pore formation (Sandre et al. 1999; Rawictz et al. 2000; Baumgart et al. 2003; Tanaka et al. 2004).

We recently developed the single GUV method to investigate interactions of compounds such as peptides/proteins with lipid bilayers (Tamba and Yamazaki 2005, 2009; Yamazaki 2008; Tamba et al. 2010; Islam et al. 2014a, b, 2018). In this method, performed under various optical microscopes, a compound solution is delivered continuously to the vicinity of a single GUV through a micropipette to induce the interaction of the compound with the single GUV. During the interaction, changes in the structure and physical properties of the single GUV are measured as a function of time and spatial coordinates using fluorescence microscopy coupled with an electron-multiplying charge-coupled device (EM-CCD) camera and confocal laser scanning microscopy (CLSM). We perform the same experiment using many "single GUVs" under the same conditions and then analyze the results statistically (Yamazaki 2008; Islam et al. 2018). The single GUV method enables us to distinguish the elementary processes of individual events occurring in a single GUV during the interaction of an AMP with the GUV, and statistical analysis permits us to derive the rate constants of each elementary process.

In this mini-review, we focus on research on the interactions of AMPs with lipid vesicles using the single GUV method. We review what kinds of new information on the interaction of AMPs with lipid bilayers have been obtained using the single GUV method, compared with the information obtained using the LUV suspension method. Here, as the examples of AMPs, we use magainin 2, lactoferricin B (LfcinB), LfcinB (4–9) (which has the amino acid sequence RRWQWR, making this AMP one of the shortest versions of LfcinB), and transportan 10 (TP10). Among these examples, TP10 is not a native AMP because this molecule is a synthetic, cell-penetrating peptide (CPP). However, TP10 has been shown to exhibit strong antimicrobial activity while being able to enter eukaryotic cells without affecting viability (Nekhotiaeva et al. 2004). Hence, we can reasonably infer that TP10 has an activity similar to AMPs, given that TP10 induces preferential damage in microbes.

3.2 Mode of Action of AMPs

To date, three distinct modes of action (MoAs) of AMPs on lipid bilayers have been observed using the single GUV method: (A) small pore formation, (B) local rupture and complete rupture (or burst), and (C) no damage to the bilayers. In the type A MoA, AMPs induce small pores in lipid bilayers; these pores cannot be observed using optical microscopy but can be detected by the membrane permeation (i.e., leakage) of water-soluble fluorescent probes from the lumen of single GUVs to the outside. After the leakage, the same spherical GUVs without any detectable breaks remain, and their sizes do not change significantly. First, we consider the results for magainin 2. The magainin 2-induced leakage of internal contents has been extensively investigated using the LUV suspension method (Matsuzaki et al. 1995, 1998; Gregory et al. 2009). Those studies indicated that magainin 2 induces leakage of water-soluble fluorescent probes from the lumen of LUVs, suggesting that magainin 2 induces pores in lipid membranes. However, there are several factors (such as pore formation, membrane fusion, large change in local curvature, rupture, and fragmentation of the liposomes) that induce leakage of the internal contents from vesicle lumens, and it is difficult to clearly identify the cause of leakage using the LUV suspension method (Yamazaki 2008). Figure 3.1a shows a result for the magainin 2-induced leakage of a water-soluble fluorescent probe, calcein, from single GUVs composed of dioleoylphosphatidylglycerol (DOPG; hereafter PG) and dioleoylphosphatidylcholine (DOPC; hereafter PC) (Tamba and Yamazaki 2005, 2009). The fluorescence intensity (FI) of the GUV lumen in the fluorescence microscopic images (Fig. 3.1a (II)) was proportional to calcein concentration in the lumen. The time course of the change in FI indicated that the calcein began to leak from the lumen suddenly after an extended time of interaction, and this leakage continued gradually (Fig. 3.1b). During this leakage, local disruption, rupture of the GUVs, changes in local curvature of the membrane, membrane fusion, and vesicle fission did not occur, indicating that magainin 2 formed pores in the membrane through which the membrane permeation of calcein occurred. Hence, the MoA of magainin 2 is type A. Similarly, lysenin, a pore-forming toxin (PFT) protein, induced gradual membrane permeation of calcein from GUVs of lipid bilayers composed of sphingomyelin (SM), PC, and cholesterol (chol) (Alam et al. 2012); the MoA of lysenin also is type A.

In the type B MoA, AMPs induce local rupture or complete rupture of single GUVs, through which fluorescent probes rapidly leak away. After the leakage, the size of GUVs decreases a little, and sometimes several partially thickened membranes appear at the rim of the GUVs (corresponding to the GUV membranes). First, we consider the results for LfcinB. Figure 3.2a shows the result of LfcinB-induced leakage of calcein from single PG/PC-GUVs (Moniruzzaman et al. 2015). The calcein began to leak from the lumen suddenly after an extended time of the interaction, and the leakage then was completed within several seconds (Fig. 3.2a, c), after which the spherical GUV structure remained, albeit with a decreased diameter. These observations indicate the occurrence of a local rupture or a large pore in the GUV membrane, and hence, LfcinB belongs to type B. In the case of the interaction of TP10 with single GUVs, a transient local rupture was sometimes observed, although the rate of leakage was slower than that observed with LfcinB (Islam et al. 2014a). On the other hand, epigallocatechin gallate (EGCg), a tea catechin (antimicrobial substance), induced rapid leakage of calcein from single PC-GUVs within a few seconds, and after this leakage, the spherical GUVs were converted to single smaller lumps composed of lipid and EGCg (Tamba et al. 2007). During the conversion, we observed a large pore in a GUV membrane using optical microscopy. In this case, no spherical GUVs remained after the leakage, and therefore, we can infer that EGCg induces rupture of GUVs.

In the type C MoA, AMPs do not induce any significant damage in lipid bilayers, and hence no leakage of fluorescent probes occurs; however, AMPs translocate across lipid bilayers to

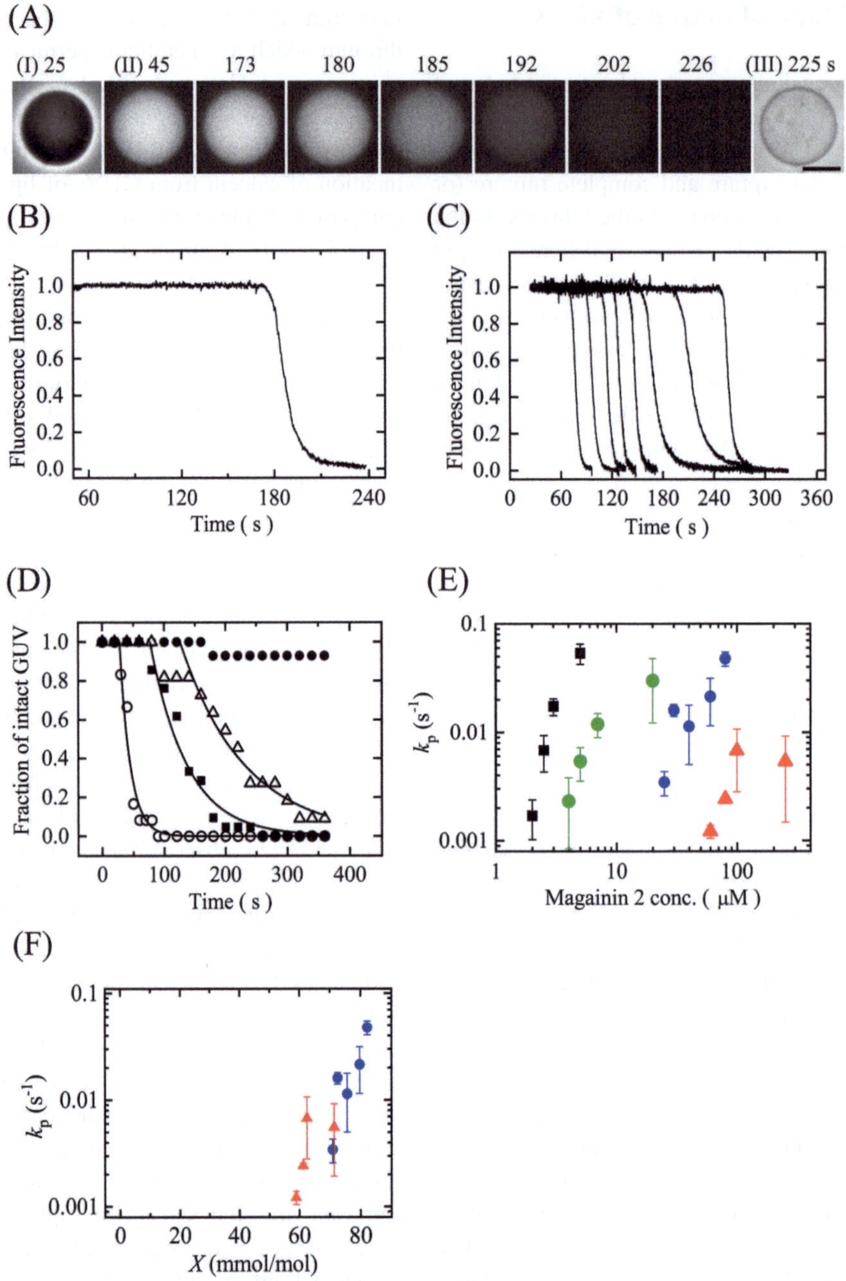

Fig. 3.1 Magainin 2-induced membrane permeation of calcein from single GUVs. Buffer A (10 mM PIPES, pH 7.0, 150 mM NaCl, 1 mM EGTA) was used at 25 °C. (**a**) Fluorescence images (II) show the time course of the FI due to calcein in a PG/PC (6/4)-GUV lumen during the interaction with 3 μM magainin 2. The numbers above each image show the time in seconds after the magainin 2 addition was started. Also shown are phase-contrast images of the GUV (I, III). The bar corresponds to 10 μm. (**b**) Time course of the change in the normalized FI of the GUV shown in (**a**). (**c**) Other examples of the time course of the change in FI of single GUVs under the same conditions as in (**a**). (**d**) Time course of P_{intact} of PG/PC (6/4)-GUV in the presence of various concentrations of magainin 2: (○) 5, (■) 3, (Δ) 2.5, and (●) 1 μM. The solid lines represent the best fit curves of Eq. 3.1. (**e**) Dependence of the rate constant of pore formation, k_P, on magainin 2 concentration in buffer, *C*. (■), PG/PC (6/4)-; (green ●), PG/PC (5/5)-; (blue ●), PG/PC (4/6)-; and (red ▲), PG/PC (3/7)-GUVs. (**f**) Dependence of k_P on *X* (**a–e** and **f** are reprinted from Tamba and Yamazaki 2009; Karal et al. 2015, respectively, with permission from the American Chemical Society)

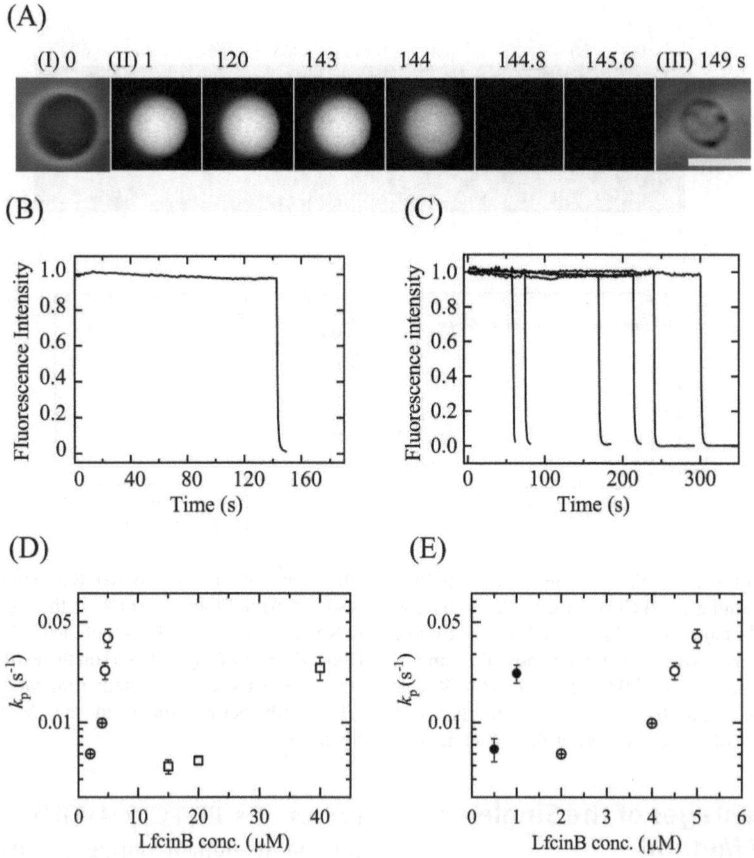

Fig. 3.2 LfcinB-induced membrane permeation of calcein from single GUVs. Buffer A was used at 25 °C. (**a**) Fluorescence images (II) show the time course of the FI due to calcein in a PG/PC (1/1)-GUV lumen during the interaction with 2.0 μM LfcinB. The numbers above each image show the time in seconds after the LfcinB addition was started. Also shown are phase-contrast images of the GUV (I, III). The bar corresponds to 20 μm. (**b**) Time course of the change in the normalized FI of the GUV shown in (**a**). (**c**) Other examples of the time course of the change in the FI of single GUVs. (**d**) The LfcinB concentration dependence of the rate constant of pore formation, k_P, in PG/PC (1/4)-GUV (□) and PG/PC (1/1)-GUV (○). (E) The LfcinB concentration dependence of the k_P in PG/PC (1/1)-GUV in buffer A without NaCl (10 mM PIPES, pH 7.0, 1.0 mM EGTA) (●) and in buffer A containing 150 mM NaCl (○) (**a–e** are reprinted from Moniruzzaman et al. 2015 with permission from the American Chemical Society)

enter the vesicle lumen. Here, we consider the results for LfcinB (4-9). The interaction of LfcinB (4-9) with single PG/PC-GUVs did not induce any leakage of calcein or AF647 (another water-soluble fluorescent probe). Nonetheless, Rh-LfcinB (4-9), which is LfcinB (4-9) labeled with a fluorescent probe (lissamine rhodamine B red), was translocated across a GUV membrane and entered the GUV lumen without leakage of AF647 (Fig. 3.3a) (Moniruzzaman et al. 2017). Moreover, Rh-LfcinB (4-9) entered the cytoplasm of *E. coli* cells without leakage of calcein. These data indicated that the antimicrobial activity of LfcinB (4-9) is not the result of damage to plasma membranes. Since AMPs with a type C MoA exhibit cell penetration activity, they are similar to cell-penetrating peptides (CPPs).

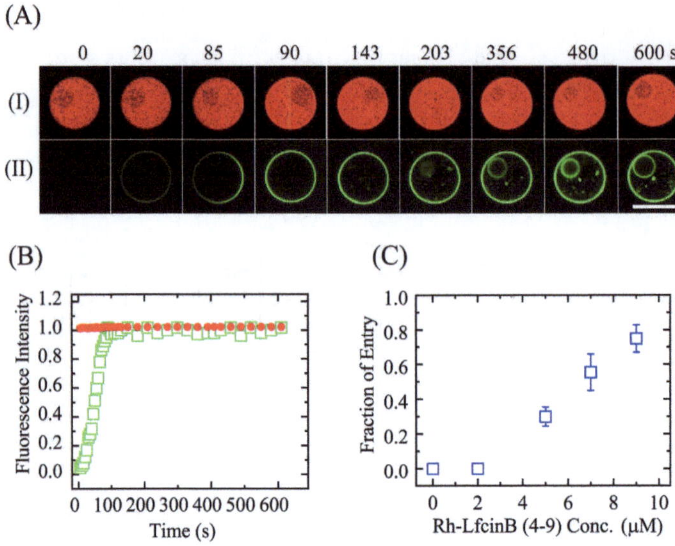

Fig. 3.3 Entry of Rh-LfcinB (4-9) into single PG/PC (1/1)-GUVs containing small GUVs. Buffer A was used at 25 °C. (**a**) CLSM images of (I) AF647 and (II) Rh-LfcinB (4-9). The numbers above each image show the time in seconds after the addition of 5.0 μM Rh-LfcinB (4-9) was started. The bar is 20 μm. (**b**) Time course of the change in the normalized FI of the GUV during the interaction of Rh-LfcinB (4-9) shown in (**a**). Red and green points correspond to the FI due to AF647 in the GUV lumen and due to Rh-LfcinB (4-9) in the rim of the GUV, respectively. (**c**) Dependence of P_{entry} (10 min) on the Rh-LfcinB (4-9) concentration (**a**–**c** are reprinted from Moniruzzaman et al. 2017 with permission from the American Chemical Society)

3.3 Advantages of the Single GUV Method

3.3.1 Separation of Elementary Processes and Determination of the Rate Constant of Each Elementary Process

AMPs of types A and B induce membrane permeation of the internal contents via a mechanism that consists of at least two elementary processes. The first process is the AMP-induced damage in the membrane such as pore formation, local rupture, and burst; the second process is membrane permeation through the damage in the membrane. One of the largest disadvantages of the LUV suspension method is the inability to distinguish these two steps (Yamazaki 2008). On the other hand, the single GUV method enables us to separate these elementary processes of AMP-induced pore formation in the lipid bilayers and (local) rupture of GUVs.

First, we show the results for magainin 2. Figure 3.1a (II) shows fluorescence microscopic images of a PG/PC (6/4)-GUV containing calcein in its lumen during the interaction with magainin 2. The time course of the FI of the GUV lumen due to calcein (Fig. 3.1b) provides two pieces of information: one is the time when pore formation starts in the membrane (based on the time at which the FI starts to decrease), and the other is the characteristics of membrane permeation of the fluorescent probe such as its rate constant (based on the time course of the decrease in FI after pore formation). If we perform the same experiments using multiple individual "single GUVs," we can obtain more information. In the case of magainin 2, the membrane permeation of calcein from the single GUVs began stochastically (Fig. 3.1c), indicating that pore formation started stochastically in the lipid bilayers. Membrane permeation in each GUV was complete within a short time (~30 s under these conditions) after the start of membrane permeation (Fig. 3.1c). These results indicated that pore formation in the membrane is the rate-determining step in the total reaction of the magainin 2-induced leakage, because the other elementary process (i.e., membrane permeation through the pore) is completed rapidly. This may correspond to the concept of the "all-or-none" mode of leakage in the LUV suspension method (Gregory et al. 2009). To analyze this type of stochastic phenomenon, the concept of a probability can be used; in this case, the probability is assessed as $P_{\text{intact}}(t)$, the fraction (among all examined GUVs) of intact GUVs where no membrane permeation of fluorescent probes occurs (Tamba and Yamazaki 2005, 2009). $P_{\text{intact}}(t)$ decreased monotonically over time. If we apply an irreversible two-state transition model from the intact state to the pore state (i.e., the state of the GUV with pores), we can obtain the theoretical equation of the time course of $P_{\text{intact}}(t)$ as follows:

$$P_{\text{intact}}(t) = \exp\{-k_P(t - t_{\text{eq}})\} \quad (3.1)$$

where k_P is the rate constant of the two-state transition, which can be considered as the rate constant of magainin 2-induced pore formation at the initial state, and t_{eq} is a fitting parameter and implies the time required to achieve the binding equilibrium of magainin 2 from aqueous solution to the GUV membrane. All the curves of the time courses of $P_{\text{intact}}(t)$ were well fit to Eq. 3.1, providing the values of k_P (Fig. 3.1d). The rate constant k_P increased with an increase in magainin 2 concentration (Fig. 3.1e). Similarly, for AMPs with a type B MoA, we can obtain the rate constant of the two-state transition from the intact state to the rupture state (i.e., the rate constant of local rupture or complete rupture) from analysis of the time course of $P_{\text{intact}}(t)$ using Eq. 3.1 (Moniruzzaman et al. 2015; Tamba et al. 2007). These rate constants also increased with increasing LfcinB and EGCg concentrations.

On the other hand, if we analyze the time course of the decrease in FI of the GUV lumen due to fluorescent probe after pore formation, we can obtain information on the AMP-induced membrane permeation such as the rate constant k_{mp} (Tamba and Yamazaki 2005; Tamba et al. 2010). The value of k_{mp} is determined as follows:

$$I_N(t) = I(t)/I(0) = \exp(-k_{\text{mp}}(t)t) \quad (3.2)$$

where $I_N(t)$ is the normalized FI of the GUV lumen, $I(t)$ is the FI of the GUV lumen at time t after initiation of membrane permeation, and $I(0)$ is $I(t)$ at $t = 0$. If we plot the log of $I_N(t)$ as a function of time, the $k_{\text{mp}}(t)$ can be obtained quantitatively from the slope of the curve. The value of k_{mp} depends on the radius of the GUVs (r), but the membrane permeability coefficient, $P(t)$, which is derived from $k_{\text{mp}}(t)$ (i.e., $P(t) = r\,k_{\text{mp}}(t)/3$), does not depend on the value of r (Alam et al. 2012).

3.3.2 Rate Constant of Pore Formation and Local Rupture

The rate constants of pore formation or local rupture greatly depend not only on AMP concentration in buffer but also on lipid composition and buffer conditions. As the concentration of the negatively charged lipid (such as PG) in the bilayers (i.e., the surface charge density of the membrane) increases, the k_P greatly increases (Figs. 3.1e and 3.2d). Moreover, as the concentration of the salt (e.g., NaCl) decreases, the k_P also greatly increases (Fig. 3.2e) (Tamba and Yamazaki 2009; Moniruzzaman et al. 2015). It is well recognized that electrostatic interactions due to the surface charges of lipid membranes in buffer increase with an increase in surface charge density or with a decrease in salt concentration (Israelachvili 1992). Therefore, these results clearly indicate that for magainin 2 and LfcinB, the values of k_P increase with an increase in electrostatic interactions.

Using the binding constant of magainin 2 to lipid bilayers, we can convert the magainin 2 concentration in the buffer, C, to the magainin 2 surface concentration in the membrane, X (mol/mol; the molar ratio of magainin 2 bound to the membrane interface to lipid in the outer monolayer of a GUV). Figure 3.1f shows the relationship between k_P and X. The dependences of k_P on X for GUVs with two different lipid compositions are largely superimposable, indicating that X determines the rate constant of magainin 2-induced pore formation (Tamba and Yamazaki 2009).

3.3.3 Rate Constant of Membrane Permeation

Generally, two factors (the size and the number of pores) determine the rate constant of membrane permeation of fluorescent probes through pores, k_{mp}. If we use various fluorescent probes of different sizes (Stokes-Einstein radius, R_{SE}) for the AMP-induced membrane permeation experiment, we can determine the size of the AMP-induced pore and obtain information on the temporal change in pore size and the number of pores by analyzing the time course of k_{mp} (Tamba et al. 2010). In the case of magainin 2-induced pore formation in PG/PC (4/6)-GUVs, rapid membrane permeation of larger fluorescent probes [such as Texas-Red dextran 3000 (TRD-3k) (R_{SE} = 1.4 nm) and Texas-Red dextran 10,000 (TRD-10k) (R_{SE} = 2.7 nm)] occurred at the initial time, but the rate of membrane permeation subsequently decreased. Figure 3.4a shows more quantitative data for TRD-3k and TRD-10k. Using the two linear regions of these curves, we can obtain the values of k_{mp} at the initial stage $\left(k_{mp}^{initial}\right)$ and the final steady stage $\left(k_{mp}^{steady}\right)$. The value of $k_{mp}^{initial}$ is 20–40 times greater than that of k_{mp}^{steady} under these conditions. In contrast, larger fluorescent probes [such as TRD-40k (R_{SE} = 5.0 nm) and FITC-BSA (R_{SE} = 3.6 nm)] leaked only at the initial stage of magainin 2-induced pore formation. Based on these results, we can reasonably conclude that magainin 2 initially induces a large, transient pore in lipid bilayers; within several minutes, the radius of the pore then decreases to a stable smaller size at the final steady stage. The result of no membrane permeation of Alexa-Fluor trypsin inhibitor (AF-SBTI) at the final stage with 7 μM magainin 2 indicates that the radius of the pore at the final stage is smaller than 2.8 nm (R_{SE} of AF-SBTI) but is larger than 1.4 nm (R_{SE} of TRD-3k). This value agrees with the 1.9 nm radius of magainin 2-induced pores in multilayer membranes at equilibrium, as determined by neutron in-plane scattering (Ludtke et al. 1996). These data provide the first information concerning the kinetic pathway of AMP-induced pore formation in lipid bilayers (i.e., evolution of pores). Subsequently, other examples of the change in the size of peptide-/protein-induced pores have been found using the GUV suspension method (Fuertes et al. 2010; Bleicken et al. 2013).

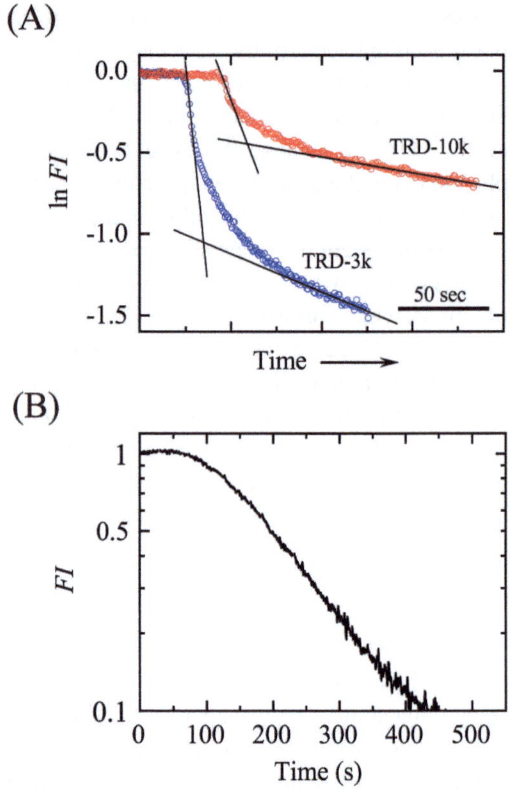

Fig. 3.4 Time course of membrane permeation. (**a**) Magainin 2-induced membrane permeation of fluorescent probes from a PG/PC (1/1)-GUV. Time course of the logarithm of the normalized FI of the GUV containing TRD-3 k or TRD-10 k during the interaction with 4 μM magainin 2 in buffer A at 25 °C. (**b**) Lysenin-induced membrane permeation of calcein from a SM/PC/chol (42/30/28)-GUV. Time course of the normalized FI of the GUV (log-10 scale) during the interaction with 40 ng/mL lysenin in PBS buffer at 37 °C (**a, b** are reprinted from Tamba et al. 2010; Alam et al. 2012, respectively, with permission from the American Chemical Society)

Contrast these results with those obtained with lysenin-induced membrane permeation of a small fluorescent probe, calcein (R_{SE} = 0.7 nm), in single SM/PC/chol (42/30/28)-GUVs. After membrane permeation was initiated, the k_{mp} gradually increased with time before reaching a steady, maximum value, which persisted for an extended time (e.g., 280 s) (Fig. 3.4b) (Alam et al. 2012). This result indicated that after lysenin-induced pore formation began, the value of P (t) increased with time before reaching a steady, maximum value of P^s, which was maintained for an extended time. Based on the experimental results that lysenin molecules form a specific oligomer in SM/PC/chol bilayers, we infer that these oligomers form pores with the same diameter. Therefore, the result of Fig. 3.4b indicated that the number of pores increased with time to reach the steady, maximum number.

3.3.4 Entry of AMPs Without Pore Formation

To detect the entry of AMPs into a GUV lumen with high sensitivity, we can use the single GUV method for CPPs. In this instance, we used CLSM to investigate the interaction of fluorescent probe-labeled AMPs with single GUVs containing smaller GUVs or LUVs and water-soluble fluorescent probes such as AF647 in their lumen (Islam et al. 2014a, 2018; Moghal et al. 2018). The binding of the fluorescent probe-labeled AMPs with bilayers of the small GUVs and LUVs increases their FI, and hence, we can detect the entry of the AMPs into the GUV lumen with higher sensitivity. The comparison between the FI of the GUV lumen due to water-soluble fluorescent probe and that due to fluorescent probe-labeled AMPs provides information on the relationship between the entry of AMPs and pore formation. Figure 3.3 shows a result for the interaction of Rh-LfcinB (4-9) with single PG/PC (1/1)-GUVs (Moniruzzaman et al. 2017). The FI of the GUV lumen due to AF647 remained constant (Fig. 3.3a (I) and red line in Fig. 3.3b) for 10 min, indicating no pore formation. On the other hand, the FI of small vesicles in the GUV lumen was observed after 143 s (Fig. 3.3a (II)), and the FI of the GUV membrane (i.e., the rim intensity) due to Rh-LfcinB (4-9) increased rapidly with time to reach a steady value at 100 s (Fig. 3.3b). These results indicated that Rh-LfcinB (4-9) first bound to the GUV membrane, then translocated across the bilayer, and finally entered into the GUV lumen without pore formation.

At present, the rate of entry of AMPs into the vesicle lumen is estimated as P_{entry} (t), the fraction of GUVs into which AMPs enter before a specific time t. Figure 3.3c indicates that P_{entry} (10 min) increased with an increase in Rh-LfcinB (4-9) concentration. It is also possible to estimate the rate constant of the binding of AMPs to the monolayer and the rate constant of unbinding by analyzing the rim intensity (Islam et al. 2014a, 2018).

3.3.5 AMP-Induced Transbilayer Movement of Lipids

The studies using the LUV suspension method suggested that the interactions of AMPs and venom peptides with lipid vesicles yielded an increase in k_{FF}, the rate constant of the transbilayer movement (i.e., flip-flop) of lipids (Matsuzaki et al. 1996; Müller et al. 2000). Matsuzaki et al. (1996) indicated a strong correlation between the magainin 2-induced leakage of internal contents and the flip-flop of lipids, suggesting the toroidal model of magainin 2-induced pore formation. Since the peptide-induced transbilayer movement of lipids is composed of several elementary steps, it was difficult to determine directly the main process for growth of k_{FF} in the framework of the LUV suspension method. On the other hand, the use of the single GUV method to measure the k_{FF} has an advantage (Hasan et al. 2018a, b). In this method, we use single GUVs with asymmetric lipid compositions in two monolayers, where a low concentration of fluorescent probe-labeled lipid [such as 18:1-NBD-lyso-PE (hereafter NBD-LPE)] exists only in their inner leaflet (e.g., PG/PC/NBD-LPE (40/59/1; inner leaflet)-PG/PC (40/59; outer leaflet)-GUVs). The analysis of the time course of the decrease in rim intensity due to NBD-LPE (measured using CLSM) provides the value of k_{FF}, because the rate of transfer of NBD-lipids from the outer leaflet to aqueous solution is large, and hence the rate-determining step of the decrease in rim intensity is the transbilayer movement of lipids from the inner leaflet to the outer one (Hasan et al. 2018b). In the interaction of magainin 2 with single GUVs with this asymmetric composition (Fig. 3.5), rim intensity decreased rapidly after the start of magainin

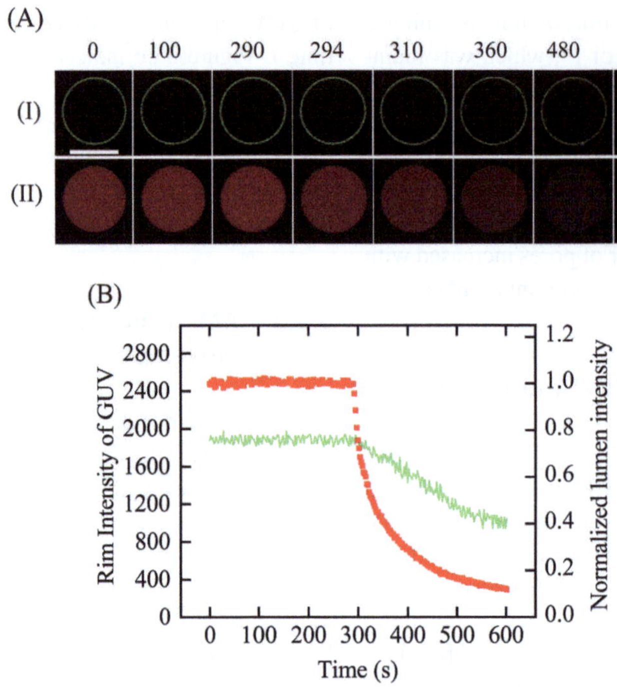

Fig. 3.5 Relationship between the transbilayer movement of lipids and magainin 2-induced pore formation. (**a**) Relationship between the rim intensity of the GUVs due to NBD-LPE and the FI of the GUV lumen due to AF647 during the interaction of 31 μM magainin 2 with PG/PC/NBD-LPE (40/59/1; inner)-PG/PC (40/59; outer)-GUVs. CLSM images of (I) NBD-LPE and (II) AF647. The numbers above each image show the time in seconds after magainin 2 addition was started. The bar corresponds to 20 μm. (**b**) Time course of the change in FI of the GUV shown in (**a**). Red and green points correspond to the normalized FI of AF647 in the GUV lumen and the FI of NBD-LPE in the rim of the GUV, respectively

2-induced membrane permeation of AF647, indicating that the transbilayer movement of NBD-LPE occurred rapidly after pore formation, whereas before pore formation, the k_{FF} value was similar to that in the absence of magainin 2. This result clearly indicated that the occurrence of pores induced by magainin 2 increased the k_{FF} of NBD-LPE, whereas the binding of magainin 2 only to the membrane did not increase the k_{FF} value. The k_{FF} after pore formation was much larger than that in the normal lipid bilayer, indicating that the mechanism of the transbilayer movement of lipids after pore formation is different from that of normal transbilayer movement. If we assume that the structure of the magainin 2-induced pore at the initial stage is a toroidal pore, we can reasonably explain this great increase in k_{FF}. In a toroidal pore, the outer and inner monolayers bend and merge with each other in a toroidal fashion to form a pore in which the inner wall is composed of peptides and lipid head groups (Matsuzaki et al. 1996; Ludtke et al. 1996; Yang et al. 2000; Qian et al. 2008). Therefore, lipid molecules can rapidly diffuse along this integrated lipid monolayer from the inner leaflet to the outer one through the wall of the pore, thereby inducing rapid transbilayer movement.

3.4 Relationship Between the Location of AMPs and Pore Formation

To elucidate the mechanism of AMP-induced pore formation, it is important to know the location of peptides in a lipid bilayer during pore formation. To elucidate the relationship between the

AMP-induced pore formation and the location of AMP in a GUV, interactions of fluorescent probe-labeled AMPs (e.g., carboxyfluorescein (CF)-labeled AMP) with single GUVs containing AF647 in their lumen have been investigated using CLSM. At present, we have defined two kinds of relationships: (A) asymmetric transmembrane distribution of AMPs before pore formation (e.g., as with magainin 2) and (B) symmetric transmembrane distribution of AMPs before pore formation (e.g., as with TP10).

First, we consider the case of magainin 2. Figure 3.6a shows the time course of FI in the interaction of CF-magainin 2/magainin 2 with a single PG/PC (4/6)-GUV containing AF647 in its lumen (Karal et al. 2015). The FI of the GUV lumen due to AF647 started to decrease at 149 s, indicating that the membrane permeation of AF647 from the GUV started through a magainin 2-induced pore. On the other hand, the rim intensity due to CF-magainin 2 rapidly increased to a steady value, I_1, at t = 50 s, remained constant for a long time (~90 s) until the intensity started to increase at a time similar to that of the start of membrane permeation of AF647, and then rapidly reached a final steady value, I_2. The average value of I_2/I_1 is approximately 2.0. This result indicates that the steady binding of magainin 2 to the outer monolayer of the GUV membrane is attained at ~50 s and the surface concentration of magainin 2 in the outer monolayer, X, remains constant until just before pore formation. Therefore, we can reasonably infer that the asymmetric transmembrane distribution of magainin 2 (i.e., localization of magainin 2 only in the outer monolayer) induces pore formation.

In contrast, the result for TP10 showed a different pattern. Figure 3.6b shows the time course of FI in the interaction of CF-TP10 with a single PG/PC-GUV containing AF647 in its lumen (Islam et al. 2014a). The rim intensity due to CF-TP10 gradually increased and reached a final, steady value at 125 s (Fig. 3.6b). On the other hand, the FI of the GUV lumen due to AF647 started to decrease at 212 s, indicating that the membrane permeation of AF647 started through CF-TP10-induced pores. After pore formation, the rim intensity did not change. This result indicated that CF-TP10 translocates from the outer leaflet to the inner one from the beginning of the interaction, without pore formation; the steady binding of TP10 in both leaflets is achieved rapidly, and then after an extended time, pore formation occurs. Therefore, the symmetric transmembrane distribution of CF-TP10 (i.e., localization of CF-TP10 in both monolayers) induces pore formation. The AMPs with a type C MoA (e.g., LfcinB (4-9)) also can translocate across lipid bilayers without pore formation, but the symmetric distribution of these AMPs does not induce pore formation (Moniruzzaman et al. 2017).

The different relationship between the AMP-induced pore formation and the location of AMP in the membrane is due to the activity of translocation of AMPs across the lipid bilayer. This activity is more important for the CPPs, and various research have been performed on this topic (Madani et al. 2011; Islam et al. 2018). However, at present, the molecular mechanism of the translocation of peptides across the lipid bilayer is not clear, although several mechanisms have been proposed (Islam et al. 2018). The characteristics of peptides (e.g., hydrophobicity and amino acid sequence) required for translocation activity without pore formation also are not clear. Notably, the translocation activity depends not only on peptide properties but also on lipid bilayer properties. The presence of a high concentration of cholesterol in the bilayer greatly suppressed the translocation of CF-TP10 from the outer leaflet to the inner one, but higher concentrations of CF-TP10 induced pore formation in the same bilayer, through which CF-TP10 rapidly translocated across the bilayer (Islam et al. 2017). Therefore, the relationship between the CF-TP10-induced pore formation and the location of CF-TP10 in the bilayers containing high concentrations of cholesterol was almost the same as that of magainin 2 in PG/PC (4/6) membrane (Fig. 3.6a).

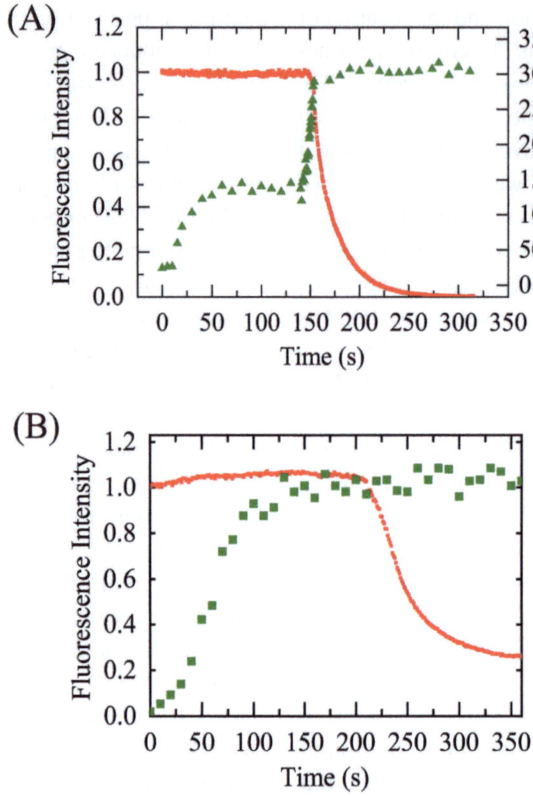

Fig. 3.6 Relationship between the location of AMPs and pore formation. (**a**) Membrane permeation of AF647 and location of CF-magainin 2 in a PG/PC (4/6)-GUV. Time course of the change in FI due to AF647 and CF-magainin 2 of the GUV interacting with 31 μM CF-magainin 2/ magainin 2 is shown. The red squares and green triangles correspond to the normalized FI of AF647 inside the GUV (left axis) and the FI of CF-magainin 2 in the rim of the GUV (right axis), respectively. (**b**) Membrane permeation of AF647 and location of CF-TP10 in a PG/PC (2/8)-GUV. Time course of the change in FI due to AF647 and CF-TP10 of the GUV interacting with 1.9 μM CF-TP10 is shown. The red and green points correspond to the normalized FI of AF647 inside the GUV and of CF-TP10 in the rim of the GUV, respectively (**a, b** are reprinted from Karal et al. 2015; Islam et al. 2014a, respectively, with permission from the American Chemical Society)

3.5 Effect of Membrane Tension (or Stretching of Membranes) on AMP-Induced Pore Formation

3.5.1 AMP-Induced Stretching of Lipid Bilayers and Membrane Tension

Generally, there is a possibility that the interaction with peptides/proteins such as AMPs affects the mechanical properties of plasma membranes and lipid bilayers. To measure the change in area of a GUV, we can use the micropipette aspiration method (Rawictz et al. 2000), where the GUV is fixed at the tip of the micropipette by a small aspiration pressure, producing a small tension (σ ~ 0.5 mN/m) in the GUV membrane. In the case of magainin 2, the fractional change in the area of a GUV membrane (δ) increased rapidly with time after the interaction of magainin 2 with the GUV started, reaching a steady value of δ (δ^{ST}) within a short time (~30 s). The δ^{ST} increased with an increase in magainin 2 concentration in the buffer, C (Fig. 3.7a) (Karal et al. 2015). After conversion of C to the surface concentration of magainin 2, X (mol/mol), the result indicated that $\delta^{ST} \propto X$ (Fig. 3.7b). The combination of Figs. 3.1f and 3.7b indicated that the rate constant of magainin 2-induced pore formation (k_p) had a positive correlation with δ^{ST} (Fig. 3.7c). At a high concentration of magainin 2, during the steady

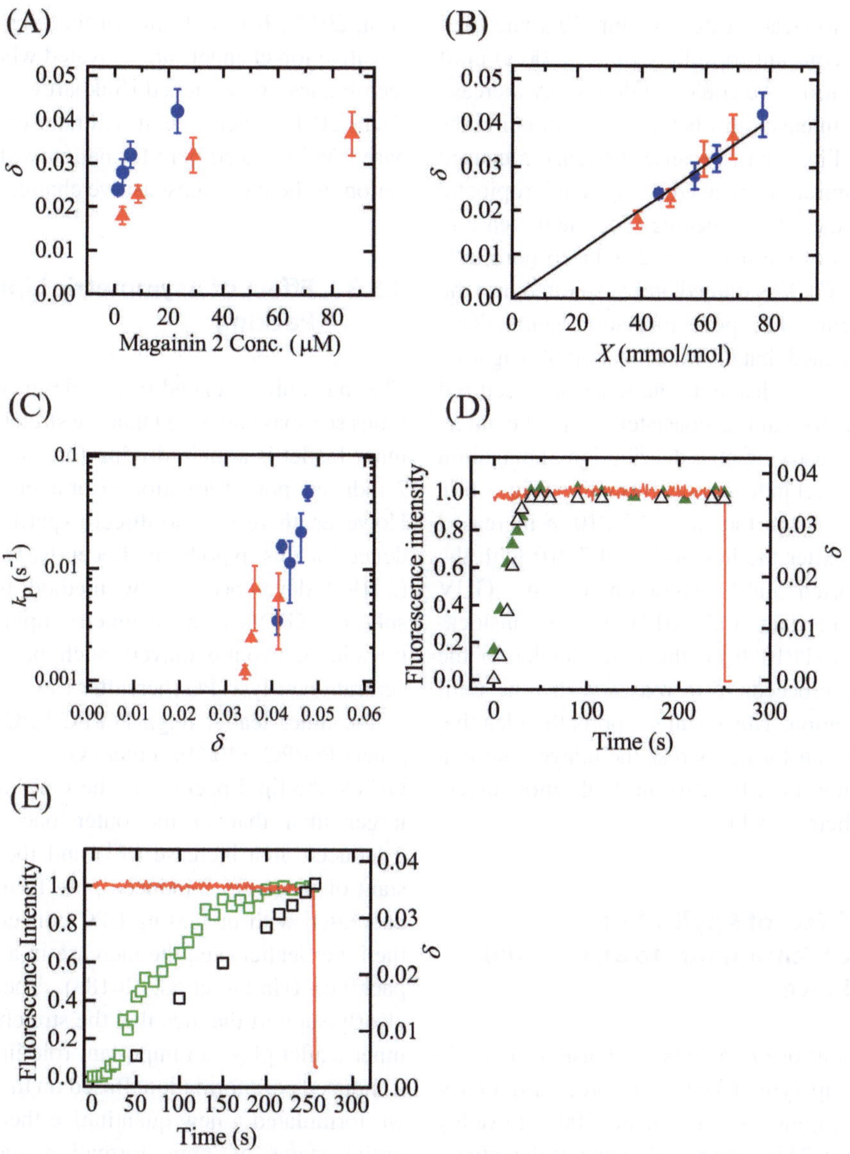

Fig. 3.7 Effect of binding of AMP to a GUV on the membrane area. (**a**) The dependence of the fractional change in area of single GUVs, δ, on the magainin 2 concentration in buffer A. (**b**) Dependence of δ on the magainin 2 surface concentration, X (mmol/mol). (●) PG/PC (4/6)- and (▲) PG/PC (3/7)-GUV. (**c**) Dependence of k_P on δ for magainin 2. (**d**) Time course of the FI of a PG/PC (4/6)-GUV and δ interacting with 15 μM CF-magainin 2/magainin 2 at σ = 0.5 mN/m. Red line and green solid triangles represent the normalized FI of AF647 in the GUV lumen and of CF-magainin 2 at its rim, respectively. Open triangles represent δ (right axis). (**e**) Time course of the FI of a PG/PC (2/8)-GUV and δ interacting with 0.3 μM CF-TP10 at σ = 1.0 mN/m. Red line and green squares correspond to FI of AF647 in the GUV lumen and of CF-TP10 at the rim of the GUV, respectively. Open squares represent δ (**a**, **b**, **d**, and **e** are reprinted from Karal et al. 2015; Islam et al. 2017, respectively, with permission from the American Chemical Society)

state of stretching (i.e., $\delta = \delta^{ST}$), aspiration of GUV (which is induced by pore formation in the lipid bilayer) occurred stochastically. Figure 3.7d shows the result of the simultaneous measurement of the magainin 2-induced change in the area, the rim intensity due to CF-magainin 2, and the membrane permeation of AF647 using CLSM. Both the rim intensity and δ rapidly increased to steady values within 50 s and then remained constant for a long time (~200 s) until pore formation. We could not detect any increase in the rim intensity just before pore formation (as shown in Fig. 3.6a), because the tension caused by the aspiration of a GUV by a micropipette increases the rate of opening (i.e., the increase in the radius) of the magainin 2-induced pore, and hence, the GUV is immediately aspirated into the micropipette after pore formation starts. This result indicated that the translocation of magainin 2 from the outer leaflet to the inner one occurred after pore formation, consistent with the inference that the asymmetric distribution of magainin 2 in the bilayer induces pore formation (Fig. 3.6a).

In contrast, in the case of TP10, δ increased with time after the interaction of TP10 with the GUV started until aspiration of the GUV (Fig. 3.7e) (Islam et al. 2017). As the translocation of CF-TP10 from the outer leaflet to the inner one proceeds, δ increases with time until pore formation. This result supports the idea that CF-TP10 translocates across the bilayer and that the binding of CF-TP10 in both monolayers induces their stretching.

3.5.2 Effect of Application of Tension Due to an External Force

We can control the membrane tension of single GUVs by applying an external force with micropipette aspiration (Evans et al. 2003; Levadny et al. 2013). This aspect is also one of the advantages of experimentation using single GUVs. Magainin 2-induced pore formation is affected greatly by membrane tension due to an external force; the rate constant of magainin 2-induced pore formation, k_p, increased with an increase in membrane tension (Karal et al. 2015). This result, as well as the result that the k_p increased with an increase in magainin 2-induced stretching of the bilayer, indicated that the magainin 2-induced pore is a stretch-activated pore (Karal et al. 2015). The membrane tension also increased the rate of translocation of CF-TP10 across the bilayer and that of CF-TP10-induced pore formation (Islam et al. 2017). It is well known that the mechanosensitive ion channels are activated when plasma membranes are stretched (Sukharev et al. 1994; Sachs 2010). Therefore, it will be useful to compare AMP-induced pore formation with the activation of the mechanosensitive channels.

3.5.3 Effect of Asymmetric Lipid Packing

The magainin 2-related results described in previous sections suggested that the stretching of the inner leaflet is a main driving force of magainin 2-induced pore formation (Karal et al. 2015). However, there was no direct experimental evidence for this hypothesis. Recently, Hasan et al. (2018a) developed a new method to prepare spherical GUVs with asymmetric lipid compositions in the two monolayers, such that a low concentration of lyso-PC (hereafter LPC) exists only in the inner leaflet (e.g., PG/PC/LPC (40/59/1; inner)-PG/PC (40/59; outer)-GUVs). In these GUVs, the lipid packing in the inner leaflet was larger than that in the outer one. Magainin 2-induced area increase (δ^{ST}) and the rate constant of magainin 2-induced pore formation (k_p) decreased with increasing LPC concentration in the inner leaflet (i.e., the increase in asymmetric packing) (Hasan et al. 2018a). These results clearly support the idea that the stretching of the inner leaflet plays an important role in magainin 2-induced pore formation. Based on these results, we formulated a new quantitative theory on the initial stage of pore formation induced by magainin 2 (Hasan et al. 2018a). A theoretical equation of k_p as a function of magainin 2 surface concentration (X) provided a good fit to the experimental results (Fig. 3.1f). This theory can explain the effect on k_p of the LPC concentration in the inner leaflet (Hasan et al. 2018a).

3.6 Concluding Remarks

In this mini-review, we summarized the present state of knowledge on the interactions of AMPs with lipid vesicles that were obtained using the single GUV method. AMPs exhibit three MoAs on lipid bilayers. For a limited number of AMPs, we have obtained detailed information on the elementary processes of these MoAs, indicating that various processes are mediated by AMPs. On the basis of this information, we have proposed a mechanism for the initial stage of magainin 2-induced pore formation (Hasan et al. 2018a). However, we do not know the mechanism of the AMP-induced local rupture (type B MoA) and that of the entry of AMPs into the vesicle lumen (or bacterial cytoplasm) without pore formation (type C MoA). If we could examine other AMPs using single GUV methods, we might identify other kinds of MoAs and other elementary processes within the present MoAs. Further investigations of the MoAs of AMPs and the elementary processes of interaction of AMPs with lipid bilayers will be essential.

Even in the case of AMPs (such as magainin 2 and LfcinB) that attack the lipid membrane regions of bacterial and fungal plasma membranes, other factors of these microbial cells may affect the interaction of AMPs with the lipid membrane regions of the plasma membranes. However, we believe that these factors are additional to the fundamental characteristics of the interactions between AMPs and lipid bilayers. A close comparison between the results of interactions of AMPs with microbial cells and those of GUVs of lipid bilayers is expected to provide deeper insight into the elementary processes of action of AMPs and the mechanism of their antimicrobial activity.

References

Alam JM, Kobayashi T, Yamazaki M (2012) The single giant unilamellar vesicle method reveals lysenin induced pore formation in lipid membranes containing sphingomyelin. Biochemistry 51:5160–5172

Baumgart T, Hess ST, Webb WW (2003) Imaging coexisting fluid domains in biomembrane models coupling curvature and line tension. Nature 425:821–824

Bigay J, Antonny B (2012) Curvature, lipid packing, and electrostatics of membrane organelles: defining cellular territories in determining specificity. Dev Cell 23:886–895

Bleicken S, Landeta O, Landajuela L, Basañez G, García-Sáez AJ (2013) Proapoptotic Bax and Bak proteins form stable protein-permeable pores of tunable size. J Biol Chem 2013(288):33241–33252

Evans E, Heinrich V, Ludwig F, Rawicz W (2003) Dynamic tension spectroscopy and strength of biomembranes. Biophys J 85:2342–2350

Fuertes G, Garcia-Sáez A, Esteban-Martin S, Giménez D, Sánchez-Muñoz OL, Schwille P, Salgado J (2010) Pores formed by Baxα5 relax to a smaller size and keep at equilibrium. Biophys J 99:2917–2925

Gregory SM, Pokorny A, Almeida PFF (2009) Magainin 2 revisited: a test of the quantitative model for the all-or-none permeabilization of phospholipid vesicles. Biophys J 96:116–131

Hasan M, Karal MAS, Levadnyy V, Yamazaki M (2018a) Mechanism of initial stage of pore formation induced by antimicrobial peptide magainin 2. Langmuir 34:3349–3362

Hasan M, Saha SK, Yamazaki M (2018b) Effect of membrane tension on transbilayer movement of lipids. J Chem Phys 148:245101

Hwang PM, Vogel HJ (1998) Structure-function relationships of antimicrobial peptides. Biochem Cell Biol 76:235–246

Islam MZ, Ariyama H, Alam JM, Yamazaki M (2014a) Entry of cell-penetrating peptide transportan 10 into a single vesicle by translocating across lipid membrane and its induced pores. Biochemistry 53:386–396

Islam MZ, Alam JM, Tamba Y, Karal MAS, Yamazaki M (2014b) The single GUV method for revealing the functions of antimicrobial, pore-forming toxin, and cell-penetrating peptides or proteins. Phys Chem Chem Phys 16:15752–15767

Islam MZ, Sharmin S, Levadnyy V, Shibly SUA, Yamazaki M (2017) Effects of mechanical properties of lipid bilayers on entry of cell-penetrating peptides into single vesicles. Langmuir 33:2433–2443

Islam MZ, Sharmin S, Moniruzzaman M, Yamazaki M (2018) Elementary processes for the entry of cell-penetrating peptides into lipid bilayer vesicles and bacterial cells. Appl Microbiol Biotechnol 102:3879–3892

Israelachvili JN (1992) Intermolecular & surface forces, 2nd edn. Academic, New York

Karal MAS, Alam JM, Takahashi T, Levadny V, Yamazaki M (2015) Stretch-activated pore of antimicrobial peptide magainin 2. Langmuir 31:3391–3401

Levadny V, Tsuboi T, Belaya M, Yamazaki M (2013) Rate constant of tension-induced pore formation in lipid membranes. Langmuir 29:3848–3852

Lipowsky R, Sackmann E (eds) (1995) Structure and dynamics of membranes. Elsevier Science BV, Amsterdam

Ludtke SJ, He K, Heller KH, Harroun TA, Yang L, Huang HW (1996) Membrane pores induced by magainin. Biochemistry 35:13723–13728

Madani F, Lindberg S, Langel Ű, Futaki S, Gräslund A (2011) Mechanisms of cellular uptake of cell-penetrating peptides. J Biophys 2011:414729

Matsuzaki K, Murase K, Fujii N, Miyajima K (1995) Translocation of a channel-forming antimicrobial peptide, magainin 2, across lipid bilayers by forming a pore. Biochemistry 34:6521–6526

Matsuzaki K, Murase O, Fujii N, Miyajima K (1996) An antimicrobial peptide, magainin 2, induced rapid flip-flop of phospholipids coupled with pore formation and peptide translocation. Biochemistry 35:11361–11368

Matsuzaki K, Sugishita K, Ishibe N, Ueha M, Nakata S, Miyajima K, Epand RM (1998) Relationship of membrane curvature to the formation of pores by magainin 2. Biochemistry 37:11856–11863

McLaughlin S, Murray D (2005) Plasma membrane phosphoinositide organization by protein electrostatics. Nature 438:605–611

Melo MN, Ferre R, Castanho ARB (2009) Antimicrobial peptides: linking partition, activity and high membrane-bound concentrations. Nat Rev Microbiol 8:1–5

Moghal MMR, Islam MZ, Sharmin S, Levadnyy V, Moniruzzaman M, Yamazaki M (2018) Continuous detection of entry of cell-penetrating peptide transportan 10 into single vesicles. Chem Phys Lipids 212:120–129

Moniruzzaman M, Alam JM, Dohra H, Yamazaki M (2015) Antimicrobial peptide lactoferricin B-induced rapid leakage of internal contents from single giant unilamellar vesicles. Biochemistry 54:5802–5814

Moniruzzaman M, Islam MZ, Sharmin S, Dohra H, Yamazaki M (2017) Entry of a six-residue antimicrobial peptide derived from lactoferricin B into single vesicles and *Escherichia coli* cells without damaging their membranes. Biochemistry 56:4419–4431

Müller P, Schiller S, Wieprecht T, Dathe M, Herrmann A (2000) Continuous measurement of rapid transbilayer movement of a pyrene-labeled phospholipid analogue. Chem Phys Lipids 106:89–99

Nekhotiaeva N, Elmquist A, Rajarao GK, Hällbrink M, Langel C, Good L (2004) Cell entry and antimicrobial properties of eukaryotic cell-penetrating peptides. FASEB J 18:394–396

Propheter DC, Chara AL, Harris TA, Ruhn KA, Hooper LV (2017) Resistin-like molecule β is a bactericidal protein that promotes spatial segregation of the microbiota and the colonic epithelium. Proc Natl Acad Sci U S A 114:11027–11033

Qian S, Wang W, Yang L, Huang HW (2008) Structure of transmembrane pore induced by Bax-derived peptide: evidence for lipidic pores. Proc Natl Acad Sci U S A 105:17379–17383

Rawictz W, Olbrich KC, McIntosh T, Needham D, Evans E (2000) Effect of chain length and unsaturation on elasticity of lipid bilayers. Biophys J 79:328–339

Sachs F (2010) Stretch-activated ion channels: what are they? Physiology 25:50–56

Sandre O, Moreaux L, Brochard-Wyard F (1999) Dynamics of transient pores in stretched vesicles. Proc Natl Acad Sci U S A 96:10591–10596

Sukharev SI, Blount P, Martinac B, Blattner FR, Kung C (1994) A large-conductance mechanosensitive channel in *E. coli* encoded by mscL alone. Nature 368:265–268

Tamba Y, Yamazaki M (2005) Single giant unilamellar vesicle method reveals effect of antimicrobial peptide, magainin 2, on membrane permeability. Biochemistry 44:15823–15833

Tamba Y, Yamazaki M (2009) Magainin 2-induced pore formation in membrane depends on its concentration in membrane interface. J Phys Chem B 113:4846–4852

Tamba Y, Ohba S, Kubota M, Yoshioka H, Yoshioka H, Yamazaki M (2007) Single GUV method reveals interaction of tea catechin (-)-epigallocatechin gallate with lipid membranes. Biophys J 92:3178–3194

Tamba Y, Ariyama H, Levadny V, Yamazaki M (2010) Kinetic pathway of antimicrobial peptide magainin 2-induced pore formation in lipid membranes. J Phys Chem B 114:12018–12026

Tanaka T, Sano R, Yamashita Y, Yamazaki M (2004) Shape changes and vesicle fission of giant unilamellar vesicles of liquid-ordered phase membrane induced by lysophosphatidylcholine. Langmuir 20:9526–9534

Wade D, Boman A, Wahlin B, Drain CM, Andreu A, Boman HG, Merrifield RB (1990) All-D amino acid-containing channel-forming antibiotic peptides. Proc Natl Acad Sci U S A 87:4761–4765

Wakabayashi H, Matsumoto H, Hashimoto K, Teraguchi S, Takase M, Hayasawa H (1999) N-acylated and D enantiomer derivatives of a nonamer core peptide of lactoferricin B showing improved antimicrobial activity. Antimicrob Agents Chemother 43:1267–1269

Yamazaki M (2008) The single GUV method to reveal elementary processes of leakage of internal contents from liposomes induced by antimicrobial substances. Adv Planar Lipid Bilayers Liposomes 7:121–142

Yang LT, Weiss M, Lehrer RI, Huang HW (2000) Crystallization of antimicrobial pores in membranes: magainin and protegrin. Biophys J 79:2002–2009

Zasloff M (2002) Antimicrobial peptides of multicellular organisms. Nature 415:389–395

4 The Mechanisms of Action of Cationic Antimicrobial Peptides Refined by Novel Concepts from Biophysical Investigations

Christopher Aisenbrey, Arnaud Marquette, and Burkhard Bechinger

Abstract

Even 30 years after the discovery of magainins, biophysical and structural investigations on how these peptides interact with membranes can still bear surprises and add new interesting detail to how these peptides exert their antimicrobial action. Early on, using oriented solid-state NMR spectroscopy, it was found that the amphipathic helices formed by magainins are active when being oriented parallel to the membrane surface. More recent investigations indicate that this in-planar alignment is also found when PGLa and magainin in combination exert synergistic pore-forming activities, where studies on the mechanism of synergistic interaction are ongoing. In a related manner, the investigation of dimeric antimicrobial peptide sequences has become an interesting topic of research which bears promise to refine our views how antimicrobial action occurs. The molecular shape concept has been introduced to explain the effects of lipids and peptides on membrane morphology, locally and globally, and in particular of cationic amphipathic helices that partition into the membrane interface. This concept has been extended in this review to include more recent ideas on soft membranes that can adapt to external stimuli including membrane-disruptive molecules. In this manner, the lipids can change their shape in the presence of low peptide concentrations, thereby maintaining the bilayer properties. At higher peptide concentrations, phase transitions occur which lead to the formation of pores and membrane lytic processes. In the context of the molecular shape concept, the properties of lipopeptides, including surfactins, are shortly presented, and comparisons with the hydrophobic alamethicin sequence are made.

Keywords

Magainin · PGLa · Cecropin · LL37 · Surfactin · Alamethicin · Membrane topology · Membrane pore · Membrane macroscopic phase · SMART model · Carpet model · Toroidal pore model · Peptide-lipid interactions · Molecular shape concept

C. Aisenbrey · A. Marquette
Université de Strasbourg/CNRS, UMR7177, Institut de Chimie, Strasbourg, France

B. Bechinger (✉)
Université de Strasbourg/CNRS, UMR7177, Institut de Chimie, Strasbourg, France

Faculté de chimie, Institut le Bel, Strasbourg, France
e-mail: bechinge@unistra.fr

K. Matsuzaki (ed.), *Antimicrobial Peptides*, Advances in Experimental Medicine and Biology 1117,
https://doi.org/10.1007/978-981-13-3588-4_4

Abbreviations

Aib	α-Aminobutyric acid
AMP	Antimicrobial peptide
ATR FTIR	Attenuated total reflection Fourier transform infrared
CD	Circular dichroism
CL	Cardiolipin
CMC	Critical micelle concentration
DLPC	1,2-Lauroyl-*sn*-glycero-3-phosphocholine
DMPC	1,2-Dimyristoyl-*sn*-glycero-3-zhosphocholine
DMPG	1,2-Dimyristoyl-*sn*-glycero-3-phospho-(1′-*rac*-glycerol)
DOPC	1,2-Dioleoyl-*sn*-glycero-3-phosphocholine
DOPG	1,2-Dioleoyl-*sn*-glycero-3-phospho-(1′-*rac*-glycerol)
DPC	Dodecylphosphocholine
GUV	Giant unilamellar vesicle
ITC	Isothermal titration calorimetry
LUV	Large unilamellar vesicle
MD	Molecular dynamics
MIC	Minimal inhibitory concentration
NMR	Nuclear magnetic resonance
PC	Phosphatidylcholine
PE	Phosphatidylethanolamine
PG	Phosphatidylglycerol
POPC	1-Palmitoyl-2-oleoyl-*sn*-glycero-3-phosphocholine
POPE	1-Palmitoyl-2-oleoyl-*sn*-glycero-3-phosphoethanolamine
POPG	1-Palmitoyl-2-oleoyl-*sn*-glycero-3-phospho-(1′-*rac*-glycerol)
POPS	1-Palmitoyl-2-oleoyl-*sn*-glycero-3-phosphoserine
SMART	Soft Membranes Adapt and Respond, also Transiently

4.1 Introduction

The innate immune system of higher organisms provides a first line of defense against a multitude of pathogenic microorganisms, where the release of antimicrobial peptides (AMP) is a powerful and fast response to fight bacterial and fungal infections (Boman 1995; Zasloff 2002). A multitude of AMPs belonging to different classes have been detected in the plant and animal kingdom, including in the human body (Agerberth et al. 1995). These complement a multitude of antibiotic peptides produced by microorganisms that have early on been identified and investigated (Leitgeb et al. 2007; Rautenbach et al. 2016a, b). The corresponding databases encompass thousands of sequences (Pirtskhalava et al. 2016; Wang et al. 2016), and early on several of them have been investigated by cell biological and biophysical approaches in order to reveal their mechanisms of action (Sansom 1991; Bechinger 1997). Magainins from the African clawed frog were among the first of those peptides found in the animal kingdom, where the antimicrobial activity has been described about 30 years ago (Zasloff 1987), thereby adding a new exciting perspective to earlier publications describing peptides from toads and frogs (Kiss and Michl 1962; Giovannini et al. 1987).

The very first attempts to develop these peptides into commercial products were unsuccessful which resulted in a temporary decline in interest on this topic around the turn of the millennium. However, the rapid increase in multiresistant pathogens (Chang et al. 2003) has again stimulated research activities on AMPs because they promise to be paradigms for new classes of antibiotics with mechanisms of action that are less prone to be neutralized by microbial resistance (Zasloff 2002). To protect such polypeptides from proteases, special formulations inside nanostructures, attachment to surfaces, or the use of unnatural amino acids is explored (Yang et al. 2014; Yuksel and Karakecili 2014; Reijmar et al. 2016). Furthermore, mimetics in the shape of small molecules or foldamers have also been developed (Violette et al. 2006; Arnusch et al. 2012; Ghosh and Haldar 2015; Ghosh et al. 2018).

Magainin and other antimicrobial peptides discussed in this paper are thought to interfere with the barrier function of bacterial membranes rather than specific membrane receptors (Bechinger 2015). Whereas molecules whose interactions are with proteinaceous receptors can be made inefficient by one or a few changes in amino acid sequence, polypeptides that act by disrupting the lipid bilayer physicochemical properties are less likely to become inactivated by resistance (Rollins-Smith et al. 2002). Indeed the amphipathic nature of AMPs has been found essential and can be achieved by helical (Sansom 1991; Bechinger 1997), cyclic (Cao et al. 2018; Laurencin et al. 2018; Tsutsumi et al. 2018; Zhao et al. 2018), and/or β-sheet arrangements (Hong and Su 2011; Salnikov et al. 2011; Rautenbach et al. 2016a, b; Sychev et al. 2018; Usachev et al. 2017). Thus the insights gained from the studies of cationic amphipathic antimicrobial peptides have stimulated the design of a number of small amphipathic molecules (Arnusch et al. 2012; Ghosh et al. 2014), pseudopeptides (Porter et al. 2002; Patch and Barron 2003; Kuroda and DeGrado 2005; Violette et al. 2006; Makovitzki et al. 2008; Scott et al. 2008; Rotem and Mor 2009; Palermo and Kuroda 2010; Laurencin et al. 2018), and polymers (Rank et al. 2017) with potent antimicrobial properties.

In this review, some of the laboratory-based biophysical techniques that have provided valuable information on the interactions of peptides shall be introduced and the corresponding data presented (Lear et al. 1988; Killian et al. 1998; Harzer and Bechinger 2000; Bechinger 2011) including work on the synergistic activities between PGLa and magainin 2 two members of the magainin family (see Table 4.1 for amino acid sequences). Even after 30 years of biophysical studies on how these AMPs interact with lipid bilayers, new details are revealed, and as a consequence, exiting new research directions open up (Salnikov et al. 2010; Hong and Su 2011). It is now clear that such amphipathic membrane-active peptides are very dynamic in nature and can adopt a large diversity of conformations and topologies whose exchange and interactions are governed by multiple equilibria (Bechinger 2015). These cationic linear peptides are random coil in solution and adopt their three-dimensional amphipathic helical structure only when interacting with membranes. By disrupting the integrity of bacterial and fungal membranes, they inhibit the growth of microorganisms and/or enter into the cell interior (Roversi et al. 2014). This results in complex patterns of metabolic reactions by the bacterial cells (Kozlowska et al. 2014; Cardoso et al. 2017). Because many peptides have been shown to also modulate the immune response of the host organisms, they are also called "host defense peptides" (McCafferty et al. 1999' Holzl et al. 2008; Diamond et al. 2009; Steinstraesser et al. 2010). Such cell biological studies shall not be part of this review.

Table 4.1 Amino acid sequences of selected antimicrobial peptides

Magainin 2	GIGKF LHSAK KFGKA FVGEI MNS
PGLa	GMASK AGAIA GKIAK VALKA L-NH_2
LL37	LLGDF FRKSK EKIGK EFKRI VQRIK DFLRN LVPRT ES
Cecropin P1	SWLSK TAKKL ENSAK KRISE GIAIA IQGGP R
Cecropin A	KWKLF KKIEK VGQNI RDGII KAGPA VAVVG QATQI AK-NH_2
LAH4	KKALL ALALH HLAHL ALHLA LALKK A-NH_2
Melittin	GIGAV LKVLT TGLPA LISWI KRKRQ Q-NH_2
Alamethicin (F50/7)	Ac-Aib-Pro-Aib-Ala-Aib-Aib-Gln-Aib-Val-Aib-Gly-Leu-Aib-Pro-Val-Aib-Aib-Gln-Gln-Phl

The one-letter code is used for peptides made from conventional amino acids. For the alamethicin sequence, the three-letter code with the following nonstandard residues is used: *Aib* α-aminoisobutyric acid, *Phl* L-phenylalaninol, *Ac-* for acetyl- and *-NH_2* for the carboxamide termini, respectively

4.2 Electrophysiological Recordings

Electrophysiological experiments provide an interesting method to observe and characterize membrane interactions of peptides and proteins. One approach is to measure the ionic conductivities across small patches of a lipid bilayer (ranging from several micrometers up to hundreds of micrometers) separating two electrodes connected to a voltage-clamp amplifier (Montal and Mueller 1972). The electrodes can be positioned in different chambers or constitute the inside and outside of dedicated pipettes. When lipids alone are used, a tight electric seal (gigaseal) is established; thus no ions can flow across the lipid bilayer. In the presence of small amounts of pore-forming polypeptides, a decrease in ohmic resistance occurs ("membrane openings"), and ionic current can be recorded, when at the same time the transmembrane voltage is electronically regulated to remain constant ("voltage clamp"). Ideally, single events are observed and characterized in frequency, duration, and conductivity. More recent chip technology working with arrays of freestanding lipid bilayers has simplified the handling and provides an increased throughput for such single-channel measurements (del Rio Martinez et al. 2015).

Notably, discrete single-channel openings/events are rather difficult to observe for magainins and cecropins because most of the time, these peptides lyse the membranes. However, in some electrophysiological experiments, discrete multilevel conductivities have been recorded (Christensen et al. 1988; Duclohier et al. 1989; Cruciani et al. 1991; Watanabe and Kawano 2016), which early on was taken as an indicator for transmembrane helical bundle formation (Tieleman et al. 2002). However, unlike the well-defined ion channels formed by alamethicin, those recorded in the presence of cationic amphipathic peptides are erratic and characterized by large variations (Christensen et al. 1988; Duclohier et al. 1989; Cruciani et al. 1991). Therefore, despite some similarities, electrophysiological recordings between these peptides exhibit rather pronounced differences as do the physicochemical properties of their primary sequences, i.e., number of charges and overall hydrophobicity (Table 4.1).

4.3 Fluorescence Spectroscopy

4.3.1 Fluorophore Release

An alternative method to monitor the formation of membrane openings has been established based on fluorescence spectroscopic approaches. In these experiments vesicles are prepared in the presence of high concentrations of fluorophore such as calcein. At a concentration of some tens of millimolar, the dye molecules are close to each other resulting in the quenching of their radiative emission. Once unilamellar vesicles have been formed, the fluorophore on the outside is exchanged by passage through a gel filtration column. Using this method, LUVs and the encapsulated dye can be easily separated from the outside buffer. Care should be taken that the osmolarities of the buffer solutions at the interior and exterior of the liposomes match each other to assure that there is no pressure gradient across the membrane (Marquette et al. 2008). Under these precautions, vesicles are stable for days or even weeks keeping the dye encapsulated and ready for future experiments.

Upon addition of peptide, the formation of pores can be measured by monitoring the increase in the fluorescence intensity due to the escape of the dye from the vesicles. Fluorescence is continuously measured while irradiating the sample with monochromatic light tuned to the most intense absorption band of the chromophore. Variants of the technique have been developed to measure the kinetics in more detail and to distinguish between all-or-nothing and graded release. For example, in the fluorescence quenching assay, the fluorophore is encapsulated with a quencher. Upon all-or-nothing release, both leak out without changes in fluorescence inside the vesicles. In contrast, a graded release results in a fluorescence increase when the quencher inside the vesicles is diluted (Ladokhin et al. 1995; Clark et al. 2011). Alternatively, due to an about tenfold difference in fluorescence lifetime between entrapped and free dye, their respective proportions can be estimated using biexponential fits. This has been used to develop an approach to distinguish all-or-nothing and graded release (Patel et al. 2014). Yet another method is based on the direct observation of giant unilamellar vesicles under the microscope which can thus be studied one-by-one individually (Tamba and Yamazaki 2005).

Using fluorescence spectroscopy, the formation of magainin pores was investigated, and the kinetics of calcein release from individual giant unilamellar vesicles (GUVs) made from DOPC/DOPG at different molar ratios was monitored

(Islam et al. 2014). Membrane pore formation from DOPC/DOPG GUVs sets in at peptide concentrations of 0.7 mole% (Tamba and Yamazaki 2009). After the addition of peptide, it takes minutes before the release of fluorophores sets in, but then the vesicles, which are several micrometers in diameter, empty within only 30 s. The fast release is suggestive that magainin pore formation in GUVs follows an all-or-nothing mechanism rather than gradual diffusion through the pore. Similar conclusions were drawn from calcein release experiments from suspensions of large unilamellar vesicles made from POPC and POPG (Gregory et al. 2008). However, the topic turns out more complex because the mechanism is lipid dependent and a graded release is observed when the POPG content is reduced from 50 to 20 mole% (Gregory et al. 2008). Once the pore has formed, the subsequent fluorophore release is a two-stage process where an initial fast release is due to an unbalance of the bilayer because as a first step magainin solely associates with the outside monolayer (Tamba et al. 2010). The movement of the peptide from the outer to the inner leaflet results in the transient formation of very large pores concomitant with an equilibration of the peptide density (Tamba et al. 2010). In the following, a slower release of fluorophore sets in, but even these persistent openings are large enough to allow for the passage of molecules with a hydrodynamic radius of 3 nm (Tamba et al. 2010). In follow-up investigations, it was shown that antimicrobial action was associated with a preferential interfacial localization (rather than insertion into the hydrophobic core of the membrane) and correlated to the Gibbs free energy of membrane association rather than membrane insertion (Clark et al. 2011).

4.3.2 Fluorescence: Natural Chromophores and Membrane Partitioning

The intrinsic emission properties of AMPs containing chromophores naturally, by mutagenesis or by chemical attachment, can be exploited to gain insight into peptide-lipid interactions. The amino acid residues phenylalanine, tyrosine, and tryptophan absorb and emit light in the near-UV region. Intensity modulation of emission spectra and/or wavelength shift is generally observed when the amino acids change environments, for example, when they move from aqueous buffers into a lipid membrane. This fluorescence-based approach has the great advantage of being nondestructive for biological samples and require only small amounts of material.

Among the three naturally occurring amino acid chromophores, tryptophan emission has the longest wavelength emission and the highest absorption and emission yields, thereby being most sensitive to measure. Several groups have taken advantage of measuring the blue shift and/or intensity changes in the emission spectrum to quantify an apparent binding constant between peptide and a large variety of model membranes (Matos et al. 2010; Zanin et al. 2013; Michalek et al. 2014). To monitor the membrane-induced changes in fluorescence, a solution of peptide is continuously excited at a fixed wavelength, while the dispersed emission spectra are recorded after each addition of aliquot of vesicle suspension. The wavelength of the maximum emission intensity and/or its amplitude is then plotted against the lipid concentration, and a fitting procedure is applied to extract apparent peptide-to-lipid binding constants. Antimicrobial peptide association to membranes of different composition was used using this method (Vogt and Bechinger 1999), including for magainin 2 and its interactions with PGLa within lipid bilayers (Matsuzaki et al. 1998).

4.3.3 Fluorescence: Depth of Membrane Insertion

Acrylamide and free radicals are known to quench tryptophan fluorescence, and this property has been used to monitor the depth of insertion of AMPs into membranes (Caputo and London 2003). Collisional quenching is enhanced with the contact frequency between the fluorophore and its environment and so with the distance between the fluorescent amino acids and the position of the quencher. Whereas acrylamide is water soluble, the hydrophobic alkyl chain-bound doxyl radicals are positioned within the membrane interior. This method has been calibrated by studying the emission properties of tryptophan residues of different mutants of a membrane-spanning helix (Caputo and London 2003). The quenching efficiency of the emission by acrylamide molecules in solution was found to be decreasing with the distance from the membrane surface; the one by the membrane-inserted radical depends on the penetration depth of the tryptophan. The quenching efficiencies can thus serve as an atomics-scale ruler to localize the position of the tryptophan relative to the membrane surface.

Brome has also been reported as an efficient short-range fluorescence quencher. Brominated phospholipids have been introduced from the beginning of the 1990s to estimate the position of tryptophan residues within the membrane through the use of different labeling positions along the phospholipid fatty acyl chains (Bolen and Holloway 1990). As an example of the method, the membrane penetration depth of tryptophans within the amino-terminal domain of huntingtin was investigated taking advantage of lipids dibrominated at either positions 6 and 7, 9 and 10, or 11 and 12 (Michalek et al. 2014). Fluorescence quenching in the presence of paramagnetic agents allowed the determination of an in-planar alignment of Phe to Trp mutants of magainin 2, where all three tryptophans investigated are localized at about 1 nm from the bilayer center (Matsuzaki et al. 1994).

4.3.4 Fluorescence Self-Quenching

A fluorescence self-quenching effect has been developed to quantitatively evaluate if peptides distribute randomly at the membrane surface or in a more heterogeneous manner (Fig. 4.1). Many models of peptide organization in membranes assume that there are oligomeric structures of peptide monomers. For example, the toroidal pore model assumes the formation of well-defined transmembrane multimers (Ludtke et al. 1996; Matsuzaki 1998). In contrast, the carpet (Shai 1999) or the SMART model (Bechinger 2015) postulate a looser and less well-defined arrangement of the peptides at the membrane sur-

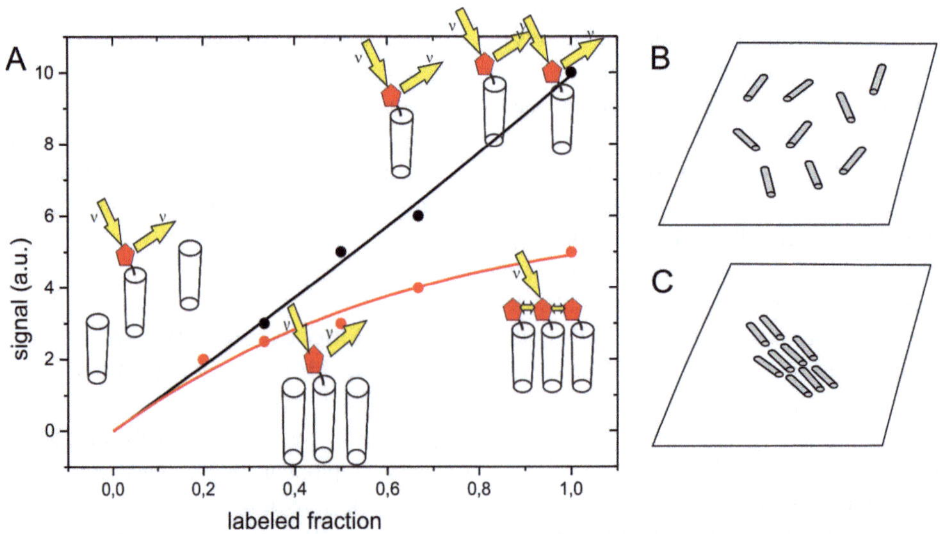

Fig. 4.1 Fluorescence self-quenching to analyze the peptide distribution. The fluorescence signal is measured at a fixed peptide-to-lipid ratio, where the fraction of fluorophore-carrying peptide is increased in a stepwise manner (**a**). When the fluorophore-labeled peptides are distributed randomly (illustrated in panel **b**), a linear increase in fluorescence is observed (black data points and line in **a**). In the presence of peptide aggregates or when peptides are localized in more confined domains (**c**), adding more of the labeled peptide is accompanied in self-quenching of the fluorescence and a nonlinear fluorescence increase (red data points and line in **a**)

face. Nevertheless, these concepts do not exclude that high local order and/or nonrandom peptide assemblies play an important role in enabling and modulating the membrane-disruptive function. Indeed, for the designed antibiotic peptide LAH4, the formation of nematic phases at the membrane surface has been demonstrated which can be modulated by the charge of the lipids and the salt concentration of the surrounding buffer (Fig. 4.1c) (Aisenbrey and Bechinger 2014). Self-quenching of fluorescent molecules acts on a length scale of nanometers and is therefore sensitive to the packing of peptides on the membrane surface at high concentration. Self-quenching is explored by dilution of the label (replacing labeled peptide by unlabeled peptide, Fig. 4.1a), and distances are estimated by fitting the data with a Poisson distribution. Ongoing research indicates an important role of mesophases and clustering for the function of magainin 2 and PGLa (manuscript in preparation).

4.3.5 Förster Resonance Energy Transfer (FRET)

FRET is another fluorescence technique widely used in the field of biophysics that can reveal the proximity between two systems of interest. Most often the proteins and/or peptides have to be labeled with soluble and non-perturbative dyes, so-called acceptor and donor for those who are prone to absorb and emit light, respectively. When they encounter in close vicinity, the excited states of both fluorescent molecules mix resulting in the emission of the acceptor when the donor is excited with light. Because the emission wavelengths of the acceptor are longer than the ones of the donor, this phenomenon is sometimes associated with a considerable shift in the emission spectrum. Detecting the quenching of the emission of the donor and/or the enhancement of the fluorescence of the acceptor reveals the proximity of the dyes. These kinds of experiments can be performed within a conventional spectrofluorometer or under the lens of a microscope which adds spatial resolution to the spectroscopic information. Because of the high absorption coefficients of some well-chosen dyes with almost unity emission yields, single molecules can be detected when highly sensitive instruments are used.

The interaction of PGLa with magainin 2 inside membranes was probed by FRET (Marquette et al. 2015). The peptides were labeled with NBD, and rhodamine fluorophores and LUVs made of POPC/POPS or of POPE/POPG were chosen as model bilayers. At high peptide-to-lipid ratio, FRET is indeed observed because even their random distribution assures that they find neighbors regularly within the Förster radius (≈56 Å for the NBD-rhodamine pair, Medintz and Hildebrandt 2013). However, upon dilution of the peptides in the lipid bilayer, the FRET effect disappears indicating that the peptides do not interact strongly and specifically but distribute in a more stochastic manner along the membrane surface (Marquette et al. 2015).

4.3.6 Fluorescence Imaging

Fluorescence imaging aims to get a deeper understanding in the mechanisms of AMP action by making direct observations on microscopic living or nonliving systems. The main advantage of imaging techniques lies in the fact that temporal and spatial information are made accessible at the same time and sometimes on very dilute systems as low as single fluorescent molecules. When probing bacteria, imaging techniques can provide subcellular spatial resolution and/or discern heterogeneities between cells within a sample population. Over the last decades, time-resolved/high-resolution imaging assays have brought a new dimension to the description and understanding mode of action of AMPs (reviewed in Choi et al. 2016).

To perform time-resolved imaging on single cells with resolution at the millisecond time scale, a time-resolved microscope equipped with high-magnification objectives and phase-contrast detection system has been used. Different schemes of chromophore labeling including visualizing the AMPs themselves have proven their potential in revealing when and where key mechanistic events occur within individual cells (Choi et al. 2016). It is believed that in the near future improvement of time resolution combined with superresolution imaging systems could provide an even more comprehensive picture of how AMPs halt growth and/or kill bacteria.

The binding of antimicrobial peptides to live bacteria was monitored using microscopic imaging revealing a spatiotemporal sequence of events including membrane permeabilization. These images reveal that the human LL37 peptide (Table 4.1) attacks septating *E. coli* cells. The peptide distributes unevenly and preferentially binds to the septum and the curved regions of the outer membrane (Barns and Weisshaar 2013). In non-septating cells, it is found to associate with one of the endcaps. When the AMPs enter the periplasmic space, the cells shrink which is probably due to an osmotic effect. The outer membrane loses the barrier function first, and after a short delay, permeabilization of the cytoplasmatic membrane is observed. The openings of the outer and cytoplasmic membranes monitored in the presence of LL37 are localized and persistent rather than global and transient (Rangarajan et al. 2013). Although many events observed in this manner follow a related schedule, important details vary with the antimicrobial compound when cationic polymers and longer or shorter peptides such as LL37, cecropin A, or melittin, are compared to each other (Yang et al. 2018). From mutagenesis experiments and a comparison of *E. coli* cells grown under aerobic or anaerobic conditions, it has been concluded that LL37 specifically affects the electron transport chain (Choi et al. 2017). On the other hand, human β-defensins tend to concentrate in a few foci that localize at the septum of cell division sites of *Enterococcus faecalis* where they colocalize with PG, CL, Sec A, and sortases, interfering with the activities of the latter proteins (Kandaswamy et al. 2013). In contrast, alamethicin causes a different series of permeabilization events (Barns and Weisshaar 2016). Thus, the detailed cellular response depends on the peptide, the bacterial species, and the growth conditions. Notably, this reflects experiments with membrane model systems where the very details varied with the lipid composition of the bilayers under investigation (Gregory et al. 2008; Cheng et al. 2011).

4.4 Circular Dichroism Spectroscopy

Peptides exhibit multiple chiral centers that are optically active in the mid-ultraviolet spectral range. This can be investigated by circular dichroism where the absorption spectra exhibit pronounced features correlating to the dihedral angles of the polypeptide backbone and thereby the secondary structure (Sreerama and Woody 2000; Miles and Wallace 2006). For example, α-helical folds exhibit two characteristic minima at 208 and 222 nm, while the spectral intensities of random coil structures show a single minimum at 195 nm. Thus, the CD line shape of polypeptides can be deconvoluted into contributions from helices, sheets, turns, and random coil conformations (Sreerama and Woody 2000; Miles and Wallace 2006). When peptides are reconstituted into oriented membranes, different CD line shapes are obtained that provide information on the peptide alignment in membranes (Wu et al. 1990; Perrone et al. 2014). In this context, it should be mentioned that ATR FTIR spectra of oriented membranes were also analyzed to study the bilayer topologies of antimicrobial peptides including magainin 2 (Bechinger et al. 1999).

A refined analysis of CD spectra has even been reported to probe the monomer-dimer equilibrium of a transmembrane helix (Loudet et al. 2005); however, when reproducing such experiments, care should be taken during analysis because related changes can also be due to light scattering.

Because membrane polypeptides exhibit a high local concentration and reside in vesicles that are close to the size of the absorption wavelengths, distortions due to light scattering and absorption flattening artifacts have to be taken into consideration (Wallace and Moa 1984; Miles and Wallace 2016). Recently a new approach has been presented which allows one to quantify and correct for light scattering artifacts (Vermeer et al. 2016). Finally, uncertainties in the determination of peptide concentration due to the presence of counterions and salts or errors in weighing the sample propagate into errors in the secondary structure determination by these algorithms. Therefore, relative intensities have also been used to estimate structural transitions between random coil and helical secondary structure (Bruch et al. 1991).

Because pronounced spectral changes occur when membrane association is accompanied by structural transitions, it is possible to quantitatively determine membrane association constants by titrating vesicles step by step to a peptide solution, also of magainin AMPs (Wieprecht et al. 2000a, b; Voievoda et al. 2015). Furthermore, valuable information on the association kinetics of magainin to whole cells and lipopolysaccharides has been obtained using CD spectroscopy (Avitabile et al. 2014).

As an illustration of the technique, we show in Fig. 4.2 the CD spectra and the corresponding association isotherm of the PGLa peptide to SUVs made of POPE/POPG 3/1 mole/mole at neutral pH. A fitting procedure was applied to each spectrum to estimate the percentage of peptides bound to the membrane as a function of peptide-to-lipid ratio. Based on a simple membrane insertion model (Vogt and Bechinger 1999), an apparent association constant of $K_{ass} \approx 1680\ M^{-1}$ can be calculated, but clearly the experimental data do not correlate well with such an asymptotic binding isotherm. Indeed, it should be noted that this apparent value covers the electrostatic interaction to the membrane surface and the hydrophobic partitioning into the bilayer interface. Therefore, the membrane surface concentration of the positively charged PGLa (nominal charge +4 to +5) is much increased in the proximity of the anionic POPE/POPG bilayer. However, membrane association of the cationic peptide neutralizes the charges of the PG lipids and can even result in repulsive interactions. A more refined analysis separating electrostatic and hydrophobic contributions has been reported in the literature for the binding of PGLa to POPC/POPG 3/1 mole/mole SUVs in the presence of 100 mM NaCl (Wieprecht et al. 2000a, b). Indeed, whereas the apparent membrane association is 50-fold increased for the anionic lipid mixture when compared to pure POPC, a hydrophobic surface partition equilibrium with K_p = 800–1500 M^{-1} was obtained for both membranes.

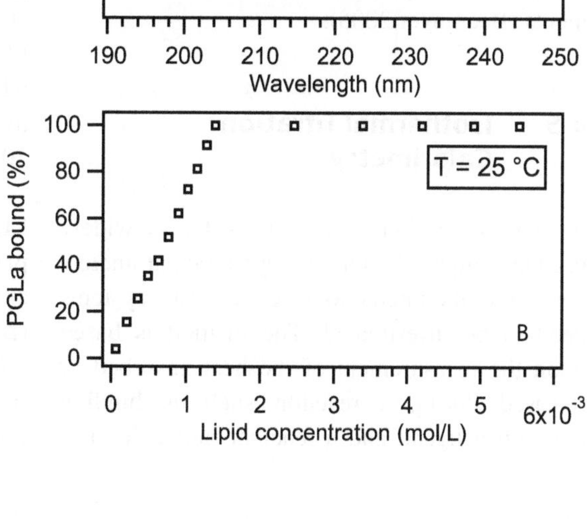

Fig. 4.2 Titration of 50 µM PGLa with increasing amounts of SUVs (POPE/POPG 3/1) in 5 mM Tris-HCl, pH = 7, at 25 °C. The changes in the circular dichroism spectra upon membrane association (**a**) have been used to quantify the percentage of membrane-associated peptide and are reported on the vertical axis of the lower frame panel (**b**)

Therefore, an alternative analysis was performed for the data shown in Fig. 4.2 where focus is on electrostatic interactions. Under these conditions, it was assumed that the amount of peptide that binds to the POPE/POPG 3/1 mole/mole membranes is governed by electrostatic interactions and stops when charge neutrality is reached. Indeed, Fig. 4.2b exhibits a linear increase in bound peptide up to about 1.4 mM total lipid. Under these conditions, the negative charge contribution corresponds to 350 μM POPG; however, almost half of the charges reside in the inner leaflet of the vesicles and during the titration may not or only partially be accessible to the PGLa peptide. At this lipid concentration, 50 μM peptides are used up in the binding reaction, contributing 200–250 μM in positive charges. The P/L ratio at this lipid concentration is about 3.6 mole%.

Such structural investigations show that the random coil structure of magainins in aqueous solution becomes helical once the peptide inserts into membrane environments (Bechinger 1999). This conformational transition has been identified to be a driving force of membrane association (Wieprecht et al. 1999a). Oriented CD spectroscopy confirmed solid-state NMR data obtained at peptide-to-lipid ratios <3 mole% showing that the helix is oriented parallel to the membrane surface (cf. below) and suggested an insertion when the concentration is increased (Ludtke et al. 1994). However, it should be noted that at higher concentrations the membrane supramolecular architecture exhibits pronounced changes probably by a transition to bicellar fragments (Bechinger 2005).

4.5 Isothermal Titration Calorimetry

Isothermal titration calorimetry (ITC) is a widely used technique when thermodynamic parameters describing the interactions between two systems need to be investigated. The method is based upon the measurement of the heat absorbed or released during a reaction such as binding between a ligand and a reactant molecule. The reactant is placed in an isolated reactor, while the ligand is added in a stepwise manner, allowing the determination of enthalpy, binding constants, stoichiometry, and entropy changes, thereby providing a complete thermodynamic picture of the system. The technique is particularly well suited to study water-soluble molecules, but the method was also developed for the determination of the binding parameters between membranes and lipophilic biomolecules such as membrane peptides (Seelig 2004).

The association of magainin 2 and/or PGLa with membranes has been studied by ITC under different conditions and provided valuable insight into their reversible interaction with phospholipid bilayers (Wenk and Seelig 1998; Wieprecht et al. 1999a, b, 2000a, b, 2002). The ITC data reveal apparent partitioning constants for magainin 2 in the 12,000 M^{-1} range for POPC/POPG 3/1 mole/mole. When separated from the electrostatic attractive term, a hydrophobic partitioning of 55 M^{-1} remains (Wenk and Seelig 1998). For pure POPC, values of 2000 M^{-1} were obtained at 30 °C (Wieprecht et al. 1999b). Notably, when 100 nm LUVs and SUVs were compared to each other, the enthalpic and entropic contributions were much different, whereas the Gibbs free energy and therefore the binding constants hardly changed (Wieprecht et al. 2000a, b). By investigating double D-amino acid replacement sequences of magainin 2, the random coil to helix transition at the membrane surface was found to contribute about −0.6 kJ/mole per residue and thereby about 50% of the driving force of the magainin 2 membrane association (Wieprecht et al. 1999a). The hydrophobic contribution of PGLa membrane association is 800–1500 M^{-1} when at the same time the apparent partitioning to POPC/POPG 3/1 mole/mole membranes is 50-fold increased (Wieprecht et al. 2000a, b).

When PGLa or magainin was titrated into LUV suspensions made of POPE/POPG 3/1 at pH 7, the peptide membrane association was characterized by endothermic reaction enthalpies (ΔH) (Marquette and Bechinger 2018). These are relatively small when compared to previous investigations with 30 nm SUVs of different lipid compositions (Wenk and Seelig 1998; Wieprecht et al. 1999a, b, 2000a, 2002). In the presence of both peptides, an additional enthalpy contribution of −8 kJ/mole was observed which correlates with the formation of larger complexes, probably vesicle agglutination (Marquette and Bechinger 2018).

4.6 Solid-State NMR Spectroscopy

4.6.1 Solid-State NMR Investigations of Polypeptides

Nuclear magnetic resonance is a powerful technique to investigate the structure, topology, dynamics, and interactions of biomolecules. Whereas peptides have been investigated by multidimensional ^{1}H-^{1}H solution NMR spectroscopy in membrane-mimetic micelles made from deuterated DPC (Brown 1979; Georgescu et al. 2010), this technique is unsuitable for the investigation of peptides associated with large peptide-bilayer complexes. However, NMR spectroscopy becomes much more powerful when polypeptides can be prepared either by bacterial overexpression and biochemical purification or by chemical synthesis such that stable isotopic labels such as ^{15}N, ^{13}C, and/or ^{2}H were introduced.

For polypeptides associated with large complexes that rotate slowly on the NMR time scales, solid-state NMR spectroscopy has been developed. Because chemical shifts, dipolar and quadrupolar interactions are all dependent on the molecular alignment relative to the magnetic field and the dipolar interactions between nuclei can be strong for immobilized molecules, in a static sample, broad overlapping line shapes are observed that hamper a detailed analysis.

One approach to obtain well-resolved solid-state NMR spectra is to subject the sample to fast rotation around the magic angle. This method results in NMR spectra where only the isotropic chemical shifts remain which resemble those observed in solution, and similar concepts for assignment and structural analysis are used (Das et al. 2015; Eddy et al. 2015; Gopinath and Veglia 2015; Jaipuria et al. 2017; Visscher et al. 2017; Naito et al. 2018).

A second approach relies on exploiting the anisotropies of interactions inherent to solid-state NMR spectra rather than averaging them (Fig. 4.3). When membranes are uniaxially oriented relative to the magnetic field of the NMR spectrometer, a unique molecular alignment is retained and spectral resolution is recovered. The resulting anisotropic chemical shifts, dipolar and quadrupolar interactions provide valuable information about the orientation of bonds, protein domains, and polypeptides as a whole. Thus, the corresponding spectra can be used to analyze the structure, dynamics, and topology of membrane-associated polypeptides (Das et al. 2015; Gopinath et al. 2015; Itkin et al. 2017; Salnikov et al. 2018).

Whereas the ^{15}N chemical shift alone provides an approximate tilt angle of helical domains (Bechinger and Sizun 2003), the combination with ^{2}H solid-state NMR spectra from methyl-deuterated alanines results in accurate tilt and pitch angle information (Fig. 4.3) (Bechinger et al. 2011; Salnikov et al. 2018).

Only few years after the discovery of magainins solid-state NMR experiments on these peptides reconstituted into uniaxially oriented lipid bilayers indicated for the very first time that they adopt stable alignments parallel to the membrane surface (Bechinger et al. 1990, 1991a, b, 1992, 1993). This topology has been confirmed for magainin 2 in all lipid compositions investigated so far (Bechinger 2011), for magainin analogues (Ramamoorthy et al. 2006; Mason et al. 2009), and for several other linear cationic antimicrobial peptides (Resende et al. 2009, 2014; Hayden et al. 2015; Bechinger and Gorr 2017; Sani and Separovic 2018). Furthermore, oriented CD spectra agree with such an alignment of the magainin helix (Ludtke et al. 1994). Fluorescence quenching experiments not only confirm the alignment parallel to the surface but also reveal an interfacial localization of the magainin 2 helix (Matsuzaki et al. 1994). A parallel alignment has also been detected for cecropin P1 (Table 4.1) using ATR FTIR, a topology which was associated with the term “carpet model” (Gazit et al. 1996).

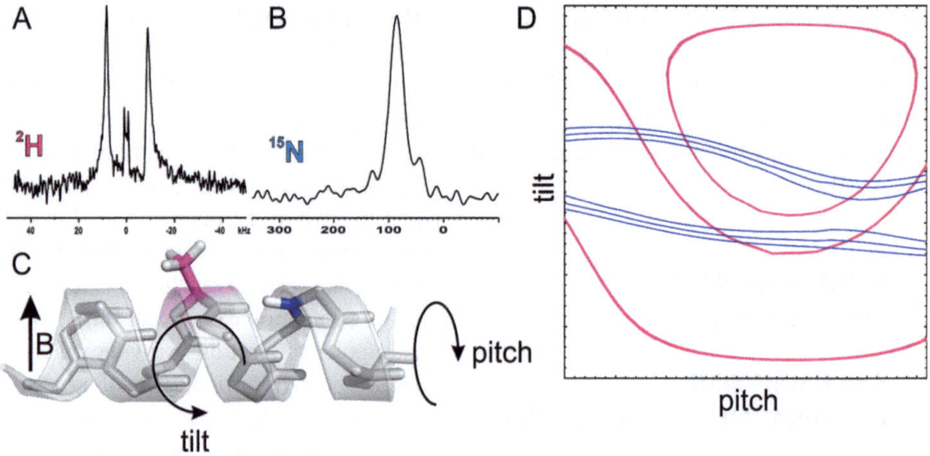

Fig. 4.3 The figure shows the ^{2}H (**a**) and ^{15}N solid-state NMR spectra (**b**) of an amphipathic helical model peptide (Aisenbrey and Bechinger 2004) labeled with $^2H_3$-alanine and a ^{15}N label at a single peptide bond (**c**). The ^{2}H quadrupolar splitting and ^{15}N chemical shift obtained from these solid-state NMR spectra are a function of the alignment of the alanine C_α-C_β bond and of a vector close to parallel to the amide ^{15}N-^{1}H bond, respectively, relative to the magnetic field of the NMR spectrometer (B_0). The bonds are highlighted in red and blue in the helical structure (**c**). This alignment dependence can be used to obtain orientational constraints for the peptide. (**d**) The tilt/pitch angular pairs of the helix that agree with the ^{2}H quadrupolar splitting are shown in red and those that agree with the ^{15}N chemical shift in blue. Both NMR parameters have to agree with the real peptide alignment which leaves only five possible topologies where the two restraints intersect. Data from well-chosen additional positions results in a unique solution for the peptide topology (Bechinger et al. 2011)

This topology assures that the peptide association is reversible (Bechinger 2011) and contrasts findings for alamethicin which is much more hydrophobic and forms stable transmembrane helical bundles when investigated in DMPC or POPC membranes (North et al. 1995; Bak et al. 2001; Milov et al. 2009; Salnikov et al. 2009b, 2016b). Notably, alamethicin also exhibits pronounced differences in electrophysiological recordings where the single-channel events are well defined and reproducible (Sansom 1993; Bechinger 1997). Notably, even for alamethicin, conditions can be found where it adopts in-plane alignments (He et al. 1996; Salnikov et al. 2010) which underlines the dynamic nature of antimicrobial peptide-lipid interactions involving multiple equilibria (Bechinger 2015). Whereas the alignment of magainin 2 has been found parallel to the membrane surface regardless of membrane lipid composition (Matsuzaki et al. 1994; Bechinger 2011), its relative PGLa (Table 4.1) adopts a much wider range of alignments but only when investigated in bilayers composed of fully saturated fatty acyl chains (Tremouilhac et al. 2006a, b; Salnikov and Bechinger 2011). In DMPC the PGLa tilt angle depends on peptide-to-lipid ratio and membrane hydration (Tremouilhac et al. 2006a, b; Salnikov and Bechinger 2011). A continuous range of tilt angles was observed as a function of hydrophobic thickness in fully saturated PC bilayers (Tremouilhac et al. 2006a, b; Salnikov and Bechinger 2011). However, when studied in phospholipid bilayers carrying unsaturations (such as palmitoyl-oleoyl-phospholipids), also this peptide remains stably aligned parallel to the bilayer surface (Bechinger et al. 1991a, 1998; Bechinger 2011; Salnikov and Bechinger 2011; Strandberg et al. 2012).

4.6.2 Solid-State NMR Spectroscopy of Lipids

Notably, solid-state NMR spectroscopy has been used not only to investigate the structure, dynamics, and topology of membrane-associated polypeptides (see ultra) but also provides valuable information on the lipids (Bechinger and Salnikov 2012). Because the lipid packing and phase properties are modulated by interactions with peptides, this information has been particularly valuable when the mechanism of antimicrobial peptides has been investigated. ^{31}P solid-state NMR spectra provide information on the macroscopic phase properties of phospholipid membranes or the orientational order of lipid bilayers. They are particularly straightforward to obatin due to 100% natural abundance of this nucleus (Bechinger and Salnikov 2012). Furthermore, ^{2}H solid-state NMR spectroscopy of deuterated fatty acyl chains has been used to characterize the order parameters (Fig. 4.4), hydrophobic thickness, and packing of lipid bilayers (Harmouche and Bechinger 2018). Finally, conformational changes of the phospholipid head groups have been monitored by ^{2}H and ^{31}P solid-state NMR spectroscopy and thereby allowed to quantitatively follow electrostatic interactions at the membrane interface (Scherer and Seelig 1989).

An amphipathic helix that inserts into the bilayer interface with an alignment parallel to the membrane surface needs to expand the surface at the level of the lipid head groups and in the glycerol region (Matsuzaki et al. 1994). This is paralleled by a loosening of the packing of the hydrophobic region and an increased disorder of the fatty acyl chains (Salnikov et al. 2009a, b; Bortolus et al. 2014). This effect can be even more pronounced when the peptides initially associate with only the outer monolayer, which subsequently allows flip-flop of lipids and pep-

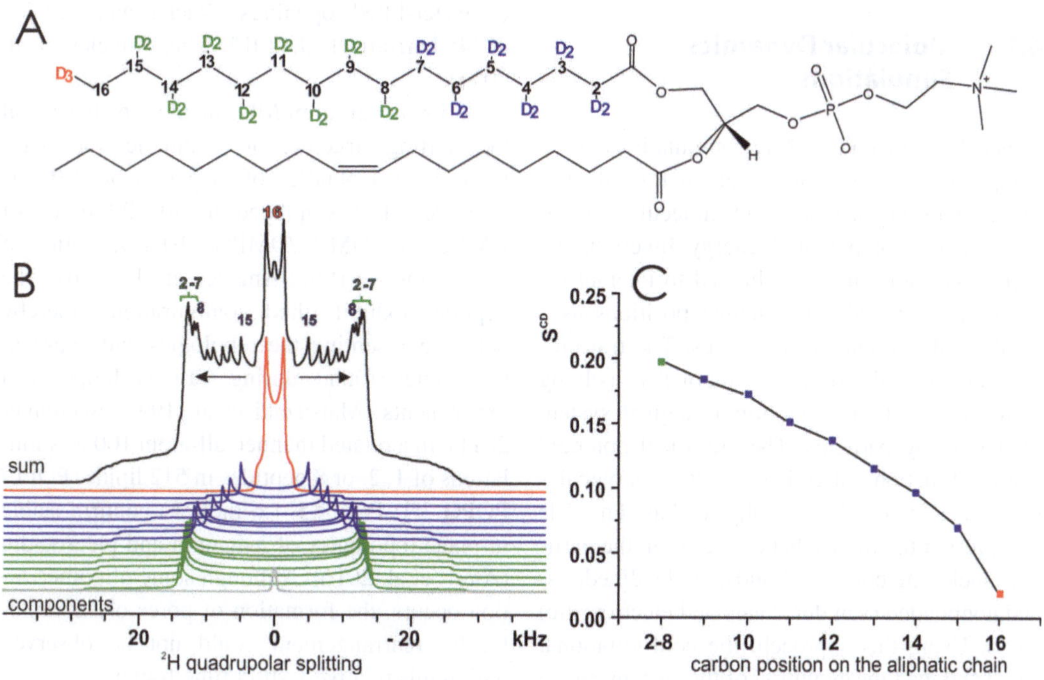

Fig. 4.4 (**a**) Phosphatidylcholine (POPC) where all ^{1}H of the palmitoyl chain have been exchanged with ^{2}H. Upon hydration, the lipid self-assembles into liposomes that tumble slowly in the NMR magnetic field. (**b**) The ^{2}H solid-state NMR spectra are composed of quadrupolar splittings from individual CD_2 functional groups that sum up into a spectrum of superimposed resonances. (**c**) The mobility of the CD_2 segments increases from the bilayer interface to the hydrophobic membrane interior and results in a decrease in quadrupolar splitting and the order parameters S_{CD}

tides across the membrane, thereby relieving the asymmetry-related tension of the outer monolayer (Matsuzaki et al. 1996; Karal et al. 2015; Hasan et al. 2018). The increased disorder at the level of the lipid fatty acyl chains results in the reduction of the membrane thickness (Ludtke et al. 1995; Kim et al. 2009). In other words, while the total volume of the lipid remains constant, in average it expands over an increased area of the bilayer which results in a reduction in hydrophobic thickness.

Pronounced decreases in order parameters especially in the bilayer interior have indeed been observed upon addition of magainin 2, PGLa, and other amphipathic peptides using ^{2}H solid-state NMR of deuterated lipids (Hallock et al. 2002; Salnikov et al. 2009a, b; Grage et al. 2016; Harmouche and Bechinger 2018). Such bilayer disruptive properties have been estimated to extend over 10 nm in diameter (Chen et al. 2003; Mecke et al. 2005).

4.7 Molecular Dynamics Simulations

Molecular dynamics (MD) simulations is a computer-based numerical method for calculating the motions of atoms and molecules. Based on an empirical potential energy function, the atoms and molecules are allowed to interact for short time intervals and change positions as a result of the instantaneous forces. The trajectories are calculated by numerically solving Newton's equations of motion for a given system of interacting particles. The empirical potential energy function, also known as a molecular mechanics force field, uses physics-based models to represent the forces that act between the particles, including bonding (bonds, angle, dihedrals) and nonbonded (van der Waals and electrostatic) terms. Using this approach, the conformational space and the time evolution of the system can be visualized at atomic resolution if a force field representing all atoms is used. An alternative is to use a coarse-grained representation, where atoms are grouped into larger entities. This approach permits long simulations of very large systems (Kmiecik et al. 2016; Harpole and Delemotte 2018).

Molecular dynamics simulations can therefore provide atomistic views of how the molecules change conformation over time. With increasing computer power, lower resolution (coarse grain), and better algorithms for exploiting today's computer architectures, they also reveal how molecules interact and how supramolecular complexes form and evolve. MD simulations have been used, for example, to follow the insertion of hydrophobic peptides into the membrane, where they assemble into transmembrane helical bundles (Tieleman et al. 2002) and how magainin interacts with lipopolysaccharides (Smart et al. 2017). Furthermore, they permitted the visualization of the deformation of the lipid bilayer in the presence of in-plane oriented amphipathic peptides. From the conformational details given by the simulations, it was shown that the positioning of side chains is important for reaching across a bilayer leaflet and to contribute to the formation of water-filled openings (Vacha and Frenkel 2014; Farrotti et al. 2015; Pino-Angeles et al. 2016).

A 5–9 μs all-atom MD calculation shows that the starting structure of tetrameric transmembrane helical bundles of magainin or PGLa is unstable when simulated in 80–120 lipids of DMPC or DMPC/DMPG 3/1(P/L ratio of 3.3–5 mole%) (Pino-Angeles et al. 2016). The peptides exhibit tilted configurations, thereby better representing the topologies and aggregation states found during various biophysical experiments (Matsuzaki et al. 1994; Bechinger 2011). In a related manner, all-atom 100 ns simulations of 1, 2, or 8 peptides in 512 lipids (POPE/POPG 3/1; P/L, 0.5–1.6 mole%) confirm stable in-plane topologies of magainin and pleurocidin (Amos et al. 2016). Although some oligomerization occurs, the formation of pores or supramolecular rearrangement could not be observed within this relatively short time frame.

Finally, the possibility of a double-belt arrangement of peptides oriented parallel to the bilayer plane has been suggested from coarse-grain MD simulations of schematic amphipathic helices (Vacha and Frenkel 2014). Such a model resembles an inversion of the double-belt model that is discussed for apolipoproteins AI particles or related nanoparticles (Gogonea 2015).

From the molecular dynamics point of view, the magainin membrane interactions are characterized by peptides that adopt many different conformations and membrane alignments. Transmembrane alignments have been found unstable, and little direct interactions between the peptides have been observed. Openings can form by stochastic rearrangements of peptides and lipids rather than through well-defined supramolecular assemblies, which could explain the early observations during electrophysiological recordings (Christensen et al. 1988; Duclohier et al. 1989; Cruciani et al. 1991; Watanabe and Kawano 2016). In order to visualize local or global changes in the membrane macroscopic phase or membrane lysis, including the all-or-nothing release observed in dye release experiments (Gregory et al. 2008; Tamba et al. 2010), larger systems would need to be simulated over longer time scales.

Furthermore, a symmetric antiparallel dimer of PGLa has been preassembled and simulated for up to 2 μs using all-atom MD (Ulmschneider et al. 2012). Indeed, dimer formation of membrane-associated antimicrobial peptides is an interesting hypothesis but so far lacks experimental proof, for example, by solid-state NMR distance measurements. Although the GxxxG sequence within the PGLa sequence is suggestive for peptide-peptide interactions, this motif requires a hydrophobic environment to drive dimerization (Russ and Engelman 2000) rather than its experimental orientation along the membrane interface or within a water-filled channel.

4.8 Lipopeptide Biosurfactants with Antimicrobial Properties

Before establishing a more elaborate model for the action of antimicrobial peptides, it is interesting to mention a range of lipopeptides whose amphiphilic character is assured by a long fatty acyl chain attached to a polar peptidic structures (Ines and Dhouha 2015; Otzen 2017; Wu et al. 2017; Zhao et al. 2017). These surface-active compounds are produced by a wide variety of bacteria, fungi, and yeast and exhibit a high structural diversity (Ines and Dhouha 2015). Furthermore, following their natural templates, ultrashort lipopeptides with antimicrobial activities have been engineered (Mangoni and Shai 2011). In particular, the lipopeptides produced by *Bacillus* are small cyclic structures of 7–10 amino acids and a β-hydroxy fatty acid with 13–19 carbon atoms (Zhao et al. 2017). According to the peptidic ring structure, the *Bacillus* lipopeptides are divided into the surfactin, fengycin, and iturin family. They are widely used in agriculture, food, medicine, and feed production due to their antifungal, antibacterial, antitumor, antiviral, and anti-inflammatory activities (Wu et al. 2017; Zhao et al. 2017). Furthermore, they interact with biofilms, have been suggested to be useful in thrombolytic and Alzheimer therapies or to be used to create nanostructures for drug delivery (Wu et al. 2017; Zhao et al. 2017). Many of these activities are thought to be due to the interactions of these biosurfactants with biological membranes (Carrillo et al. 2003; Heerklotz and Seelig 2007). In the context of the molecular shape model, we will focus to present some biophysical work obtained on surfactins and related lipopeptides.

Surfactin is made of seven amino acids (with L- or D-conformation) linked to a fatty acyl chain of C12–C14 that closes the peptide ring by a β-lactone (Wu et al. 2017). Many lipopeptides have been shown to exhibit antimicrobial activities against a range of bacteria (Ines and Dhouha 2015). Importantly, they are some of the most potent and most popular antifungal agents and have been investigated for their anticancer activities (Ines and Dhouha 2015; Wu et al. 2017).

The hydrophobic alkyl chain and the more polar peptidic portion confer an amphiphilic character and thus, in cases where the fatty acyl chain adopts an extended conformation, a pronounced cone shape to the molecule (Otzen 2017). Therefore, these compounds form micelles in aqueous buffer and exhibit surfactant activities such as monolayer formation at the air-water interface (Maget-Dana and Peypoux 1994; Otzen 2017). It should be noted that in contrast to chemical detergents, biosurfactants exhibit a more mosaic-like amphipathic structure (Otzen 2017).

Calcein release from POPC vesicles by surfactin is a cooperative process (index 1.82) (Carrillo et al. 2003). The membrane-perturbing effect of surfactin, measured by calcein release, is attenuated in POPE or cholesterol-containing membranes but accentuated when 25% of DPPC has been mixed into the POPC bilayers (at 25 °C) (Carrillo et al. 2003). Notably as with detergents, the surfactin behavior in aqueous solution has been characterized in terms of CMC (7.5 μM), onset of membrane solubilization (comparatively low when compared to other detergents, Heerklotz and Seelig 2001), and aggregation number (20) (Otzen 2017). The imbalance in the lateral pressure profile at the interface and the hydrophobic portion of the lipid bilayer in the presence of surfactin and chemical detergents has been investigated by ^{2}H solid-state NMR of specifically deuterated POPC membranes (Heerklotz et al. 2004). Whereas $C_{12}EO_6$ and $C_{12}EO_8$ cause the expected disordering of the membrane fatty acyl chains, in the presence of surfactin, the fatty acyl chains tilt, and the head group reorients to accommodate the bulky heptapeptide ring. This difference is probably related to the amphipathic but predominantly hydrophobic character of the peptide moiety which results in a deeper insertion of the peptide ring when compared to the polar groups of the detergents (Heerklotz et al. 2004). To resolve ambiguities about the mechanism how surfactin causes membrane leakage, a series of experiments was performed at non-lytic concentrations correlating data from ITC, ^{31}P solid-state NMR, and leakage assays (Heerklotz and Seelig 2007). This systematic analysis reveals three different mechanisms depending on the surfactin-to-lipid ratio R_b. Leakage starts at $R_b = 0.05$ probably by a mechanism where the surfactin accumulates at the outer membrane leaflet and opens pores transiently to equilibrate with the inner side of the bilayer (bilayer couple mechanism). At $R_b = 0.15$, it is suggested that surfactin-rich clusters form which cause leaks and stabilize the hydrophobic etches of those. Finally, membrane solubilization and micelle formation are observed between $R_b = 0.22$ and 0.42. The same experiments provide a membrane partitioning coefficient of 20,000 M^{-1} (Heerklotz and Seelig 2007). Furthermore, fengycins, lipopeptides sold together with iturins and surfactin for agricultural applications, were investigated in fluorescence lifetime efflux measurements (Patel et al. 2011).

The length of the fatty acyl chain, hydrophobicity, and membrane association of surfactin are directly correlated to its anticancer activities (Wu et al. 2017). Using measurement on lipid monolayers and CD, spectroscopic investigations reveal changes in the membrane insertion process and the proteic structure in the presence of Ca^{2+} (Maget-Dana and Ptak 1995). Furthermore, these investigations demonstrate that electrostatic contributions and the space occupied by the head group play an important role in surfactin membrane insertion (Maget-Dana and Ptak 1995) although the hydrophobic interactions of the fatty acyl chains as well as other factors remain of major importance (Wu et al. 2017). The gel-to-liquid phase transition of DMPG has been shown to be significantly broadened by surfactin an effect that is enhanced in the presence of Ca^{2+} (Grau et al. 1999). Because of this observation, a deeper penetration into the membrane upon complexation of Ca^{2+} with the Glu-1 and Asp-5 residues of the peptide sequence has been suggested (Grau et al. 1999). Surfactin has been incorporated in a wide variety of nanoformulations for materials and biomedical applications (Wu et al. 2017).

Surfactin, iturins, and the closely related bacillomycins and mycosubtilin are of similar built. Iturins are made of circular peptides made of seven L- and D-amino acids connected to a β-amino fatty acid of 14–17 C-atoms. They exhibit strong antifungal but little antibacterial activities. Interestingly, iturin A has been shown to specifically interact with cholesterol and suggested to form specific phospholipid/peptide/cholesterol complexes that are responsible for the measured electrophysiological properties. The latter are strongly dependent on the lipid composition, the physical state of the lipids, and the detailed peptide structure. An interesting observation is that the CMC and the MIC follow the same trend suggesting that large iturin A aggregates are the active component in biological membranes (Maget-Dana and Peypoux 1994). When traces of iturin A are added to black lipid membranes, stepwise conductances, whose characteristics evolve over time, are observed. At P/L ratios above 10^{-7}, the membranes break. Notably iturin A is coproduced with surfactin, and both peptides together exhibit synergistic activities in biological assays (citations in Maget-Dana and Peypoux 1994).

Owing to their membrane activities, several surfactants including iturin A have been shown to exhibit hemolytic and anti-clot-forming activities (Ines and Dhouha 2015). An overview over many more membrane-active surfactants' interesting biomedical and technical activities is presented in Ines and Dhouha (2015).

4.9 The Molecular Shape Concept Explains the Many Different Supramolecular Arrangements of AMPs and Lipids

A number of seemingly contradictory models have been suggested to explain the mechanism of action and the interaction of antimicrobial peptides with membranes. The most cited are the formation of toroidal pores (Ludtke et al. 1996; Matsuzaki 1998), a dense "carpet" of peptides covering the membrane surface that causes lysis (Shai 1999), or small aggregates without specific structure within the membrane (Jenssen et al. 2006). Whereas on the one hand, occasionally channel-like events are recorded in electrophysiological experiment (Christensen et al. 1988; Duclohier et al. 1989; Cruciani et al. 1991), at high peptide concentrations, worm-like structures, disk-shaped particles, or micelles have been observed (Hallock et al. 2002; Bechinger and Lohner 2006; Wolf et al. 2017). This wide variety of observations can be taken into consideration by the differential shape of lipids (Fig. 4.5) and the resulting supramolecular phases which are a function of peptide concentration in the membrane and other environmental parameters (Bechinger 2009). The concept has been developed early on to rationalize the macroscopic phase transitions of lipids (Israelachvili et al. 1980). Because the phosphatidylcholine head groups and two fatty acyl chains expand laterally over an area that is about equivalent at the level of the membrane interface and the hydrophobic interior, the PC molecules when part of a membrane can be described by a cylindrical shape (Fig. 4.5a). In contrast, the head group of phosphatidylethanolamine is much smaller which results in a truncated inverted cone-shaped molecule (Fig. 4.5b), whereas lysolipids with only one fatty acyl chain (or detergents) are best represented by a cone (Fig. 4.5c). For a more quantitative treatment, the critical packing parameter comparing the optimal surface area at the carbon-water interface (a_o), the optimal chain length (l_c), and the hydrocarbon volume (v) is related by the packing parameter $v/l_c a_o$ (Israelachvili et al. 1980). When these geometrical shapes are assembled into supramolecular aggregates, the PC cylinders line up side by side in a phospholipid bilayer (Fig. 4.5f), the cones form micellar assemblies (Fig. 4.5i), and the PE lipids at higher temperatures tend to form hexagonal II phases (Fig. 4.5h). When PE or detergents are forced to be part of a planar lipid bilayer, they are under curvature elastic stress where interactions with the opposite monolayer maintain the bilayer arrangement. However, a spontaneous curvature has been defined for each lipid which takes into account the differences in lateral cross section at the membrane interface when compared to the hydrophobic interior (Kollmitzer et al. 2013). Thus, the intrinsic curvature of POPC is around 0, of POPE it is −0.32, and for lyso-PE, a value of +0.18 has been determined (Kollmitzer et al. 2013; Leber et al. 2018). It should be noted that the shape is not only determined by the van der Waals contacts but can be modulated by other interactions. For example, repulsive electrostatic interactions at the head group level increase its optimal surface area, resulting in a more cone-shaped molecule and a positive curvature (Israelachvili et al. 1980).

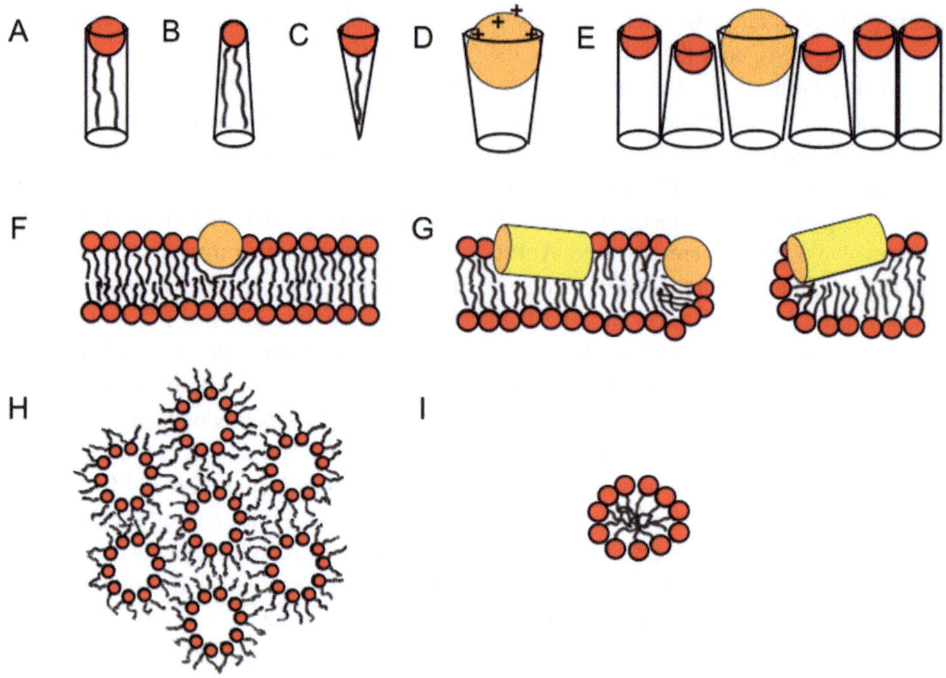

Fig. 4.5 The molecular shape concept and how molecules assemble into supramolecular arrangements. When the lateral cross sections of the lipid head groups and at the level of the fatty acyl chains are compared to each other, phosphatidylcholines and phosphatidylglycerol adopt cylindrical shapes (**a**), phosphatidylethanolamine with its reduced head group size corresponds to a truncated inverted cone (**b**), and lysolipids and detergents resemble a cone (**c**). The corresponding intrinsic curvatures J_o of these lipids are zero, negative, and positive, respectively (Kollmitzer et al. 2013; Leber et al. 2018). (**d**) A highly charged amphipathic peptide partitions into the head group region without filling the hydrophobic region. Thereby it resembles a truncated cone and exerts pronounced positive curvature strain. (**e**) Next to such an amphipathic peptide, the lipids adjust their shape by adding gauche conformations within the alkyl chain, thereby increasing the disorder of their fatty acyl chains. The comparison with unperturbed cylinders illustrates how these conformational changes are accompanied by membrane thinning. (**f**) Cylindrical lipids self-assemble into stable bilayers where by being soft they can adjust to peptides or other external stimuli (Bechinger 2015). (**g**) When the concentration of cone-shaped molecules increases locally, phase transitions and membrane openings occur. (**h**) Inverted cone-shaped molecules self-assemble into hexagonal II phases. (**i**) Cone shapes self-assemble into micelles. In panels D–G, the helices are schematized as yellow/orange cylinders (side view) and circles (front view)

An amphipathic helix that resides at the membrane interface with the long axis approximately parallel to the membrane surface (Bechinger 2009) or surfactin, a cyclic peptide with a long fatty acyl chain (Zhao et al. 2018), occupies a much larger area at the level of the interface without filling the corresponding space at the level of the fatty acyl chain (Fig. 4.5d). This induces considerable positive curvature strain and requires major rearrangement of the lipids in compensation. This shape helps to create the line tension required for membrane pores to form (Hall et al. 2014; Henderson et al. 2016).

Thus, in contrast to the transmembrane alamethicin helix, which is modeled by a cylinder, the shape of these amphipathic cationic peptides has the properties of a large truncated cone (Fig. 4.5d). Notably, the presence of PE, which has intrinsically an inverted cone-shaped and negative curvature (Fig. 4.5b), compensates for the wedge-like properties of the in-plane oriented helices, thereby stabilizing the lipid bilayer arrangement (Batenburg et al. 1988; Hallock et al. 2002). Recently the difference in "molecular shape" of magainin and melittin has been included in a model that explains the distinct activities of these peptides (Paterson et al. 2017).

4.10 The "Molecular Shape" of Membrane Constituents Adapts

When interacting with the membranes, many amphipathic peptides, including magainins, fold into helical conformations and intercalate into the lipid head group region. In this manner, they act as a spacer pushing apart the lipids. This results in positive membrane curvature strain and has been rationalized by a cone molecular shape of the in-plane oriented amphipathic helix (Fig. 4.5d) (Bechinger 2009). However, the peptide does not fill the hydrophobic region completely; thus the lipids respond by increased *trans-gauche* isomerization or chain interdigitation (Bechinger and Lohner 2006). The details of the conformational changes of the lipids are a function of the detailed peptide alignment and penetration depth. The latter depend on the three-dimensional distribution of charges, hydrophobic side chains, i.e., the resulting hydrophobic moment and amino acid distribution. Because both the peptide and the lipids exhibit considerable conformational flexibility, the molecular shape of these membrane constituents is not fully determined, but they adjust to external forces within the supramolecular assembly. Thus, in the neighborhood of the cone of an amphipathic peptide, a PC which is cylindrical in a pure lipid bilayer adopts a more inverted cone-shaped structure, thereby stabilizing the bilayer (Fig. 4.5e). Furthermore, the peptides are not stiff helices and they can adjust their penetration depths. Their alignment relative to the membrane surface can be modulated by a wide range of interaction contributions (Bechinger 1996; Harmouche and Bechinger 2018). All these adjustments correspond to modulations of their "molecular shape" (Fig. 4.5e). The very details of the peptide-lipid supramolecular arrangement thus depend on the peptide sequence, its conformation and resulting hydrophobic moment, the peptide-to-lipid ratio, as well as the detailed lipid composition (Bechinger and Lohner 2006; Bechinger 2009), which are in a delicate balance making predictions on how the macroscopic ensemble behaves rather difficult.

4.11 Soft Membranes Adapt and Respond, Also Transiently

Not only the peptides are highly dynamic with considerable conformational and topological freedom (Cheng et al. 2009, 2011), but also the liquid crystalline lipid bilayer has the capacity to compensate for external influences before their barrier function breaks down. Together they form soft supramolecular assemblies which can change thickness, morphology, and macroscopic phase properties globally or locally. Therefore, in the presence of antimicrobial peptides, Soft Membranes Adapt and Respond, also Transiently, a concept that makes up the SMART model for antimicrobial or other membrane-active peptides (Bechinger 2015). The model thereby extends on the molecular shape concept by taking into account that the lipids have the capacity to respond to the membrane-disruptive properties of the peptides which themselves adjust their properties when associating with lipid membranes. However, once a critical concentration is reached, phase transitions of the membrane are observed locally or globally. Due to lateral diffusion, weak peptide-peptide interactions, transient lipid phase separation, etc., the local peptide concentrations can vary which can explain how stochastic fluctuations of peptide density result in the transient pore events observed in, for example, electrophysiological recordings (Christensen et al. 1988; Duclohier et al. 1989; Cruciani et al. 1991; Watanabe and Kawano 2016). Transient openings also occur during membrane crossing when the peptide density equilibrates between the outer and the inner leaflet of the bilayer (Matsuzaki et al. 1995a, b; Tamba et al. 2010; Wheaten et al. 2013). The different supramolecular morphologies of the SMART model can be nicely represented by phase diagrams where regions corresponding to bilayer, wormholes, tubular structures, bicelle, micelle, or hexagonal phases are represented as a function of the peptide-to-lipid ratio, the detailed membrane composition, temperature, hydration, salt, pH, and other environmental factors (Bechinger and Lohner 2006; Bechinger 2011). Notably, when bacteria are exposed to AMPs, the peptides diffuse to the cell, across the cell wall and other cellular barriers, before they interact with the cellular membranes. Therefore, the peptide local concentrations vary over time, and intermediate states have been observed for magainin 2 where membranes temporarily lyse and recover (Hall et al. 2014).

At low peptide concentrations, the bilayer structure is maintained; however, transient openings may form stochastically due to the lateral diffusion of peptides and lipids concomitant with density alterations. Notably, whereas in most models the membranes mechanically break or water-filled pores form, it seems also possible that the physicochemical properties change in such a manner to allow diffusion of ions through a less densely packed lipid phase or along phase boundaries which form when the peptides are responsible for the lateral phase separation of membrane constituents (Cruzeiro-Hansson and Mouritsen 1988; Jean-Francois et al. 2008; Gallaher et al. 2010; Aisenbrey and Bechinger 2014). Furthermore, it has been pointed out that membrane morphological changes, including the thinning of the lipid bilayer, which has been shown to occur in the presence of magainin 2 (Ludtke et al. 1995), are associated with changes in membrane capacity and concomitantly electrical currents (Heimburg 2012; Laub et al. 2012).

At higher peptide concentrations, the system enters non-bilayer phases (at least locally) which come along with more stable openings (Gregory et al. 2008) and macroscopic phase transitions of the membranes (Fig. 4.5g–i) (Bechinger 2009). The threshold concentrations that have been identified in some biophysical studies thus represent boundaries in phase diagrams (Bechinger and Lohner 2006; Bechinger 2011). The loss of bilayer integrity has been proposed by the "carpet model" (Shai 1999), whereas stochastic and transient openings occur when peptides orient along the membrane surface at low peptide-to-lipid ratios (Christensen et al. 1988; Duclohier et al. 1989; Cruciani et al. 1991).

It should be noted that the peptide concentration at the membrane surface may be different by orders of magnitude from those in bulk solution because the cationic peptides are attracted to the surface of negatively charged membranes (Wenk and Seelig 1998; Wieprecht et al. 1999a, b). Thereby, the apparent partitioning coefficients are also much increased, which provides one explanation why these AMPs kill bacteria which expose a highly negative surface charge. In contrast, the eukaryotic host cells are neutral in charge at their outside monolayer (Matsuzaki et al. 1991; Wenk and Seelig 1998; Bechinger 2004; Klocek and Seelig 2008; Lohner 2009). Electrostatic interactions also play a role when AMPs cause the lateral phase separation of lipids (Mason et al. 2006; Voievoda 2014), when the peptides arrange in mesophase structures along the membrane surface (Aisenbrey and Bechinger 2014), or when peripheral membrane proteins are repelled from the bilayer surface due to a more positive surface charge in the presence of cationic AMPs (Wenzel et al. 2014).

Notably the molecular shape concept and the SMART model have allowed the design of novel peptide and non-peptidic mimetics of AMPs with high antimicrobial efficiency that are cationic or amphipathic and partition into the interface without filling the hydrophobic volume to the same extent such as short peptides (Oyston et al. 2009; Schweizer 2009; Hadley and Hancock 2010; Kindrachuk and Napper 2010; Liu et al. 2010; Mangoni and Shai 2011; Chou et al. 2016; Ahn et al. 2017), foldamers (Porter et al. 2002; Patch and Barron 2003; Kuroda and DeGrado 2005; Violette et al. 2006; Makovitzki et al. 2008; Scott et al. 2008; Rotem and Mor 2009; Palermo and Kuroda 2010), polymers (Rank et al. 2017), and small organic molecules (Ghosh et al. 2014).

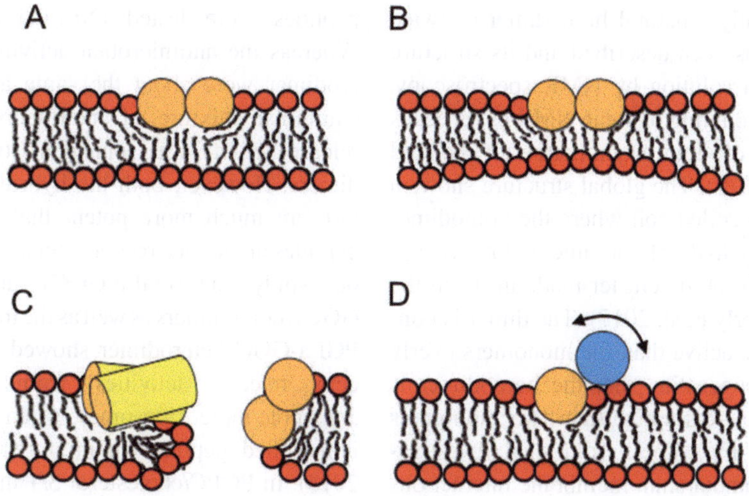

Fig. 4.6 The figure shows how amphipathic helix dimers potentially interact with membranes. (**a**) The disruptive properties of two side-by-side helices should be more pronounced and/or (**b**) result in the approach of the opposite monolayer to interact with the hydrophobic face of the peptides. (**c**) Membrane openings form by bilayer disruptions. (**d**) The two helices of heterodimers can exhibit very different properties and affect the membrane similar to a monomer. Helices are schematized as cylinders and circles, where the combination of different colors is indicative of a heterodimer

4.12 Dimers of Antimicrobial Peptides

In the context of the SMART model, an interesting question arises how the oligomerization of peptides along the membrane surface would affect their activity (Bechinger 1999). Figure 4.6a, b shows possible mechanisms how the membrane can adjust to two tightly packed side-by-side helices. On the one hand, one could imagine a situation where the lipids of the peptide-bearing monolayer compensate for the packing deficiency underneath in-plane oriented helices. Because they are excluded from the peptide-peptide interface, they have to move in exclusively from the opposite helical side affecting the lipids that are neighboring the peptide even more significantly (Fig. 4.6a). On the other hand, the opposite leaflet of the bilayer could be used to cover the hydrophobic face of the peptide dimer (Fig. 4.6b). In yet another situation, a mesophase structure of many helices may form along the bilayer surface (Aisenbrey and Bechinger 2014). In this case the peptides are separated by one or few lipids only (Fig. 4.1c). Notably, previously the membrane-disruptive properties of magainin 2 have been shown to extend over a radius of 5 nm (Chen et al. 2003; Mecke et al. 2005) which involves many lipids of an estimated diameter of about 0.8 nm. Within a mesophase supramolecular arrangement, the close proximity of helices thus results in a concerted destabilization of the membrane (Fig. 4.1c).

Only a few dimers have been identified in nature and have been investigated by biophysical methods. Distinctin is composed of two different polypeptide chains that are linked near their carboxy-terminus by a cystine bond. In solution, the peptide forms a compact four-helix dimer of heterodimers (Raimondo et al. 2005). When interacting with membranes, the solution structure unfolds. Whereas the 25-residue chain 2 partitions into the membrane in a manner similar to magainin with a stable alignment parallel to the membrane surface, chain 1 encompassing 22 residues associates more loosely with the lipid bilayer (Fig. 4.6d) (Resende et al. 2009; Verardi et al. 2011). The slightly increased antimicrobial activity of the dimer when compared to the monomers (Dalla Serra et al. 2008) is thought to be due to the better resistance to proteolysis (Raimondo et al. 2005).

More recently a natural homodimer of twice 24 residues has been described and its structure investigated in solution by NMR spectroscopy. Interestingly doubling of some NMR cross peaks are indicative of slight asymmetries of the fold (Verly et al. 2017). The global structure shows a tightly packed coiled coil where the homodimer is stabilized by hydrophobic interactions encapsulating a hydrophilic cluster made up from the two chains (Verly et al. 2017). The dimer is considerably more active than the monomers (Verly et al. 2017) suggesting that the homodimer is more membrane-disruptive than two monomers (illustrated in Fig. 4.6c). Additional investigations of the homotarsinin membrane interactions are ongoing.

Notably a cystine-linked magainin 2 dimer was also found to exhibit enhanced membrane permeabilization and antimicrobial activities (Dempsey et al. 2003). The higher efficiency was correlated to an enhanced association with negatively charged bilayers and a reduction in the concentration dependence in the dimer when compared to the monomer. On the one hand, the formation of magainin oligomers had been suggested early on; FRET measurements in the presence of membranes did not yield evidence for dimer or higher oligomer formation (Clark et al. 2011). On the other hand, NMR structural investigations of 5 mM magainin in the presence of 0.5 mM DLPC showed a dimer arrangement of the peptide (Wakamatsu et al. 2002). In this densely packed peptide-lipid supramolecular aggregate, aromatic interactions involve mostly the F5Y and F16W sites, which had been introduced artificially for better assignment, and these amino acids are involved in the dimer interface (Wakamatsu et al. 2002). In combination with magainin monomers, the dimer stabilizes the pore formation in egg-PG membranes (Hara et al. 2001). Notably whereas a dimer linked through a carboxy-terminal lysine extension was considerably more active than the monomer, the amino-terminal linkage through glutamic acid has no effect (Lorenzon et al. 2016).

Furthermore, the activities of homodimers of magainin 2 or PGLa carrying GGC extensions and of a heterodimeric hybrid made of the two peptides were tested (Nishida et al. 2007). Whereas the antimicrobial activities of the heterodimer were about the same as those of an equimolar mixture of monomers, the calcein release activity was somewhat stronger for the dimers. However, both the hybrid and the mixture are much more potent that the individual peptides alone. In a recent extension of this previous study, the $(PGLa\text{-}GGC)_2$ and $(magainin\text{-}GGC)_2$ homodimers as well as the magainin-GGC/PGLa-GGC heterodimer showed increased calcein release activities from POPE/POPG 3/1 mole/mole liposomes when compared to unmodified peptides in mixtures (Leber et al. 2018). In POPC/cholesterol 3/1 mole/mole mixtures, only the PGLa-homodimer and the PGLa-magainin heterodimer but not unlinked peptides showed significant release activities at all (Leber et al. 2018). These findings are in line with increased membrane-perturbing properties of larger peptide aggregates (illustrated in Fig. 4.6a–c) (Dempsey et al. 2003; Lorenzon et al. 2016; Verly et al. 2017).

4.13 Synergistic Enhancement of the Activities of Antimicrobial Peptides

Synergistic enhancement of antimicrobial activities has been described (McCafferty et al. 1999; Acar 2000) and includes mixtures of peptides with conventional antibiotics (Chou et al. 2016; Bolosov et al. 2017; Kim et al. 2017; Payne et al. 2017; Rank et al. 2017; Sakoulas et al. 2017). Furthermore, combinations of different peptides from the dermaseptin or the bacteriocin family (Mor et al. 1994; McCafferty et al. 1999), magainin 2 and PGLa (Vaz Gomes et al. 1993), or magainin and the cyclic β-sheet tachyplesin I sequence have been shown to interact in a synergistic manner (Kobayashi et al. 2001). Here we will shortly summarize findings made with magainin 2 and PGLa which have recently been reviewed in more detail (Marquette and Bechinger 2018). This mixture is of particular interest because synergism has been observed in antimicrobial assays but also when model membranes are tested (Westerhoff et al. 1995; Matsuzaki et al. 1998; Leber et al. 2018), suggesting that the mode of action should reveal itself when studying the membrane interactions. In cellular assays, it is also possible that one component helps to path the way of the second component to its active site. In this context it is noteworthy that synergism in calcein release experiments was more pronounced for membranes with high negative intrinsic curvature such as POPE/POPG 3/1 mole/mole (Leber et al. 2018). Furthermore, these peptides naturally occur as a mixture in the skin of *Xenopus laevis* frogs, suggesting that our conventional approach to separate such components to analyze each one individually should be reconsidered.

Early on Matsuzaki et al. suggested that the formation of magainin pores is slow, but once formed, they are more stable than those of PGLa (Matsuzaki et al. 1998). It was also suggested that synergistic vesicle leakage results from optimizing both pore size and distribution among the liposomes present in the suspension (Patel et al. 2014). The openings should be large enough and at the same time abundant in order for all the entrapped dye being released. Such more general concepts are nicely complemented by experiments aiming to reveal the mechanisms of synergism at a molecular, even at an atomistic level.

Solid-state NMR investigations show that in equimolar mixtures and in bilayers whose lipid composition resembles that of membranes that occur in nature, PGLa and magainin exhibit an alignment parallel to the membrane surface (Salnikov and Bechinger 2011; Strandberg et al. 2013; Glattard et al. 2016; Salnikov et al. 2016a). Thus, the helix topology does not change much when compared to investigations of magainin or PGLa individually (Bechinger 2011; Salnikov and Bechinger 2011). It should be mentioned that the PGLa behavior is somewhat different in fully saturated membranes (Salnikov and Bechinger 2011; Strandberg et al. 2013; Harmouche and Bechinger 2018). However, here we focus on studies that were obtained with more biological lipid compositions, carrying an unsaturation. In dye release experiments, mutating the carboxy-terminal residues of magainin showed some effect in reducing the synergistic activity, whereas modifying F5W was neutral (Matsuzaki et al. 1998). Thus, when replacing the negative E19 and the carboxy-terminus of magainin 2, synergism is abolished (Zerweck et al. 2017). Notably, amidation of the magainin 2 carboxyterminus has been shown to increase the activity of the peptide (Cuervo et al. 1988), but within experimental error, the combination with PGLa exhibits a similar degree of synergism (Marquette et al. 2015; Glattard et al. 2016).

When the PGLa sequence was modified, the positively charged K15 and K19 sites had a favorable synergistic effect (Zerweck et al. 2017). Furthermore, the G7, G11, and L18 positions of PGLa are important for the synergistic enhancement of activities between the two peptides (Zerweck et al. 2017). Cross-linking experiments with PGLa and magainin 2 both carrying a GGC extensions indicate that in egg-PC/PG 1/1 mole/mole lipid membranes, parallel dimers preferentially form (Hara et al. 2001).

From fluorescence binding experiments, energies were derived that suggest favorable interactions when magainin and PGLa are added to egg-PG membranes (Matsuzaki et al. 1998). However, FRET experiments do not reveal strong, long-lasting contacts between the peptides when associated with POPE/POPG 3/1 or POPC/POPS 3/1 membranes (Marquette et al. 2015). When investigated by a combination of ITC and dynamic light scattering, an additional exothermic contribution in the presence of both peptides was attributed to the agglutination of the liposomes (Marquette and Bechinger 2018) similar to studies with other amphipathic helical peptides (Marquette et al. 2010; Vermeer et al. 2016). Such processes involving the lipids and/or changes in the supramolecular assembly of peptides and lipids also contribute to the total interaction energies (Bechinger 1996; Harmouche and Bechinger 2018). Considering such supramolecular changes, it should also be mentioned that a reduced bilayer repeat distance has been observed when both peptides are present but not with PGLa alone (Grage et al. 2016).

4.14 Conclusions

During three decades of research, we have moved from a textbook view where AMPs form pores in the shape of transmembrane helical bundles to a more complex vision where lipids play an essential role (Bechinger et al. 1991a, b; Pouny et al. 1992). Thus, in order to understand the mechanism of action not only of AMPs alone but also of their synergism, an in-depth understanding of the interactions within the supramolecular lipid-peptide architectures is required (Bechinger and Lohner 2006; Imura et al. 2008; Kim et al. 2009; Aisenbrey and Bechinger 2014). The molecular shape and SMART models (Bechinger 2009; 2015) provide valuable concepts from where to further explore these activities, where new technological approaches reveal how structure, topology, dynamics, and interactions evolve in a spatiotemporal manner.

Acknowledgments We gratefully acknowledge the discussion with and contributions from many coworkers and colleagues from our own team and from outside. We are grateful to Ekaterina Zaitseva for her feedback on electrophysiology and Roland Stote for much helping to refine the text on molecular dynamics. Over the years, the Agence Nationale de la Recherche (projects TRANSPEP 07-PCV-0018, ProLipIn 10-BLAN-731, membraneDNP 12-BSV5-0012, MemPepSyn 14-CE34-0001-01, InMembrane 15-CE11-0017-01, Biosupramol 17-CE18-0033-3, and the LabEx Chemistry of Complex Systems 10-LABX-0026_CSC), the IRTG Soft Matter Science (Freiburg, Strasbourg), the Marie-Curie Research and Training Network 33439 of the European Commission BIOCONTROL, the University of Strasbourg, the CNRS, the Région Alsace, the RTRA International Center of Frontier Research in Chemistry, and the French Foundation for Medical Research (FRM) have provided financial support. BB thanks the *Institut Universitaire de France* for providing additional time to be dedicated to research.

References

Acar JF (2000) Antibiotic synergy and antagonism. Med Clin N Am 84:1391–1406

Agerberth B, Gunne H, Odeberg J, Kogner P, Boman HG, Gudmundsson GH (1995) FALL-39, a putative human peptide antibiotic, is cysteine-free and expressed in bone marrow and testis. Proc Natl Acad Sci U S A 92:195–199

Ahn M, Gunasekaran P, Rajasekaran G, Kim EY, Lee SJ, Bang G, Cho K, Hyun JK, Lee HJ, Jeon YH, Kim NH, Ryu EK, Shin SY, Bang JK (2017) Pyrazole derived ultra-short antimicrobial peptidomimetics with potent anti-biofilm activity. Eur J Med Chem 125:551–564

Aisenbrey C, Bechinger B (2004) Tilt and rotational pitch angles of membrane-inserted polypeptides from combined 15N and 2H solid-state NMR spectroscopy. Biochemistry 43:10502–10512

Aisenbrey C, Bechinger B (2014) Molecular packing of amphipathic peptides on the surface of lipid membranes. Langmuir 30:10374–10383

Amos ST, Vermeer LS, Ferguson PM, Kozlowska J, Davy M, Bui TT, Drake AF, Lorenz CD, Mason AJ (2016) Antimicrobial peptide potency is facilitated by greater conformational flexibility when binding to gram-negative bacterial inner membranes. Sci Rep 6:37639

Arnusch CJ, Albada HB, van Vaardegem M, Liskamp RMJ, Sahl HG, Shadkchan Y, Osherov N, Shai Y (2012) Trivalent ultrashort lipopeptides are potent pH dependent antifungal agents. J Med Chem 55:1296–1302

Avitabile C, D'Andrea LD, Romanelli A (2014) Circular dichroism studies on the interactions of antimicrobial peptides with bacterial cells. Sci Rep 4:4293

Bak M, Bywater RP, Hohwy M, Thomsen JK, Adelhorst K, Jakobsen HJ, Sorensen OW, Nielsen NC (2001) Conformation of alamethicin in oriented phospholipid bilayers determined by N-15 solid-state nuclear magnetic resonance. Biophys J 81:1684–1698

Barns KJ, Weisshaar JC (2013) Real-time attack of LL-37 on single Bacillus subtilis cells. Biochim Biophys Acta 1828:1511–1520

Barns KJ, Weisshaar JC (2016) Single-cell, time-resolved study of the effects of the antimicrobial peptide alamethicin on Bacillus subtilis. Biochim Biophys Acta 1858:725–732

Batenburg AM, van Esch JH, De Kruijff B (1988) Melittin-induced changes of the macroscopic structure of phosphatidylethanolamines. Biochemistry 27:2324–2331

Bechinger B (1996) Towards membrane protein design: pH-sensitive topology of histidine-containing polypeptides. J Mol Biol 263:768–775

Bechinger B (1997) Structure and functions of channel-forming polypeptides: magainins, cecropins, melittin and alamethicin. J Membr Biol 156:197–211

Bechinger B (1999) The structure, dynamics and orientation of antimicrobial peptides in membranes by multidimensional solid-state NMR spectroscopy. Biochim Biophys Acta 1462:157–183

Bechinger B (2004) Membrane-lytic peptides. Crit Rev Plant Sci 23:271–292

Bechinger B (2005) Detergent-like properties of magainin antibiotic peptides: a 31P solid-state NMR study. Biochim Biophys Acta 1712:101–108

Bechinger B (2009) Rationalizing the membrane interactions of cationic amphipathic antimicrobial peptides by their molecular shape. Cur Opin Colloid Interface Sci Surfactant 14:349–355

Bechinger B (2011) Insights into the mechanisms of action of host defence peptides from biophysical and structural investigations. J Pept Sci 17:306–314

Bechinger B (2015) The SMART model: soft membranes adapt and respond, also transiently, to external stimuli. J Pept Sci 21:346–355

Bechinger B, Gorr SU (2017) Antimicrobial peptides: mechanisms of action and resistance. J Dent Res 96:254–260

Bechinger B, Lohner K (2006) Detergent-like action of linear cationic membrane-active antibiotic peptides. Biochim Biophys Acta 1758:1529–1539

Bechinger B, Salnikov ES (2012) The membrane interactions of antimicrobial peptides revealed by solid-state NMR spectroscopy. Chem Phys Lipids 165:282–301

Bechinger B, Sizun C (2003) Alignment and structural analysis of membrane polypeptides by 15N and 31P solid-state NMR spectroscopy. Concepts Magn Reson 18A:130–145

Bechinger B, Shon K, Eck H, Zasloff M, Opella SJ (1990) NMR studies of magainin peptide antibiotics in membranes. Biol Chem Hoppe Seyler 371:758–758

Bechinger B, Kim Y, Chirlian LE, Gesell J, Neumann JM, Montal M, Tomich J, Zasloff M, Opella SJ (1991a) Orientations of amphipathic helical peptides in membrane bilayers determined by solid- state NMR spectroscopy. J Biomol NMR 1:167–173

Bechinger B, Zasloff M, Opella SJ (1991b) Solid state NMR of magainin and PGLa peptide antibiotics in bilayers. J Cell Biochem Suppl 15G:84

Bechinger B, Zasloff M, Opella SJ (1992) Structure and interactions of magainin antibiotic peptides in lipid bilayers: a solid-state NMR investigation. BiophysJ 62:12–14

Bechinger B, Zasloff M, Opella SJ (1993) Structure and orientation of the antibiotic peptide magainin in membranes by solid-state NMR spectroscopy. Protein Sci 2:2077–2084

Bechinger B, Zasloff M, Opella SJ (1998) Structure and dynamics of the antibiotic peptide PGLa in membranes by multidimensional solution and solid-state NMR spectroscopy. Biophys J 74:981–987

Bechinger B, Ruysschaert JM, Goormaghtigh E (1999) Membrane helix orientation from linear dichroism of infrared attenuated Total reflection spectra. Biophys J 76:552–563

Bechinger B, Resende JM, Aisenbrey C (2011) The structural and topological analysis of membrane-associated polypeptides by oriented solid-state NMR spectroscopy: established concepts and novel developments. Biophys Chem 153:115–125

Bolen EJ, Holloway PW (1990) Quenching of tryptophan fluorescence by brominated phospholipid. Biochemistry 29:9638–9643

Bolosov IA, Kalashnikov AA, Panteleev PV, Ovchinnikova TV (2017) Analysis of synergistic effects of antimicrobial peptide arenicin-1 and conventional antibiotics. Bull Exp Biol Med 162:765–768

Boman HG (1995) Peptide antibiotics and their role in innate immunity. Annu Rev Immunol 13:61–92

Bortolus M, Dalzini A, Toniolo C, Hahm KS, Maniero AL (2014) Interaction of hydrophobic and amphipathic antimicrobial peptides with lipid bicelles. J Pept Sci 20:517–525

Brown LR (1979) Use of fully deuterated micelles for conformational studies of membrane proteins by high resolution 1H nuclear magnetic resonance. Biochim Biophys Acta 557:135–148

Bruch MD, Dhingra MM, Gierasch LM (1991) Side chain-backbone hydrogen bonding contributes to helix stability in peptides derived from an alpha-helical region of carboxypeptidase A. Proteins Struct Funct Genet 10:130–139

Cao P, Yang Y, Uche FI, Hart SR, Li WW, Yuan C (2018) Coupling plant-derived cyclotides to metal surfaces: an antibacterial and antibiofilm study. Int J Mol Sci 19:E793

Caputo GA, London E (2003) Using a novel dual fluorescence quenching assay for measurement of tryptophan depth within lipid bilayers to determine hydrophobic alpha-helix locations within membranes. Biochemistry 42:3265–3274

Cardoso MH, de Almeida KC, Candido ES, Murad AM, Dias SC, Franco OL (2017) Comparative NanoUPLC-MS(E) analysis between magainin I-susceptible and -resistant *Escherichia coli* strains. Sci Rep 7:4197

Carrillo C, Teruel JA, Aranda FJ, Ortiz A (2003) Molecular mechanism of membrane permeabilization by the peptide antibiotic surfactin. Biochim Biophys Acta 1611:91–97

Chang S, Sievert DM, Hageman JC, Boulton ML, Tenover FC, Downes FP, Shah S, Rudrik JT, Pupp GR, Brown WJ, Cardo D, Fridkin SK, Staphylococcus V-R (2003) Infection with vancomycin-resistant *Staphylococcus aureus* containing the vanA resistance gene. N Engl J Med 348:1342–1347

Chen FY, Lee MT, Huang HW (2003) Evidence for membrane thinning effect as the mechanism for peptide-induced pore formation. Biophys J 84:3751–3758

Cheng JTJ, Hale JD, Elliot M, Hancock REW, Straus SK (2009) Effect of membrane composition on antimicrobial peptides aurein 2.2 and 2.3 from Australian Southern Bell Frogs. Biophys J 96:552–565

Cheng JTJ, Hale JD, Elliott M, Hancock REW, Straus SK (2011) The importance of bacterial membrane composition in the structure and function of aurein 2.2 and selected variants. Biochim Biophys Acta 1808:622–633

Choi H, Rangarajan N, Weisshaar JC (2016) Lights, camera, action! Antimicrobial peptide mechanisms imaged in space and time. Trends Microbiol 24:111–122

Choi H, Yang Z, Weisshaar JC (2017) Oxidative stress induced in *E. coli* by the human antimicrobial peptide LL-37. PLoS Pathog 13:e1006481

Chou S, Shao C, Wang J, Shan A, Xu L, Dong N, Li Z (2016) Short, multiple-stranded beta-hairpin peptides have antimicrobial potency with high selectivity and salt resistance. Acta Biomater 30:78–93

Christensen B, Fink J, Merrifield RB, Mauzerall D (1988) Channel-forming properties of cecropins and related model compounds incorporated into planar lipid membranes. Proc Natl Acad Sci USA 85:5072–5076

Clark KS, Svetlovics J, McKeown AN, Huskins L, Almeida PF (2011) What determines the activity of antimicrobial and cytolytic peptides in model membranes. Biochemistry 50:7919–7932

Cruciani RA, Barker JL, Zasloff M, Chen HC, Colamonici O (1991) Antibiotic magainins exert cytolytic activity transformed cell lines through channel formation. Proc Natl Acad Sci U S A 88:3792–3796

Cruzeiro-Hansson L, Mouritsen OG (1988) Passive ion permeability of lipid membranes modelled via lipid-domain interfacial area. Biochim Biophys Acta 944:63–72

Cuervo JH, Rodriguez B, Houghten RA (1988) The magainins: sequence factors relevant to increased antimicrobial activity and decreased hemolytic activity. Pept Res 1:81–86

Dalla Serra M, Cirioni O, Vitale RM, Renzone G, Coraiola M, Giacometti A, Potrich C, Baroni E, Guella G, Sanseverino M, De LS, Scalise G, Amodeo P, Scaloni A (2008) Structural features of distinctin affecting peptide biological and biochemical properties. Biochemistry 47:7888–7899

Das N, Dai J, Hung I, Rajagopalan MR, Zhou HX, Cross TA (2015) Structure of CrgA, a cell division structural and regulatory protein from Mycobacterium tuberculosis, in lipid bilayers. Proc Natl Acad Sci U S A 112:E119–E126

del Rio Martinez JM, Zaitseva E, Petersen S, Baaken G, Behrends JC (2015) Automated formation of lipid membrane microarrays for ionic single-molecule sensing with protein nanopores. Small 11:119–125

Dempsey CE, Ueno S, Avison MB (2003) Enhanced membrane permeabilization and antibacterial activity of a disulfide-dimerized magainin analogue. Biochemistry 42:402–409

Diamond G, Beckloff N, Weinberg A, Kisich KO (2009) The roles of antimicrobial peptides in innate host defense. Curr Pharm Des 15:2377–2392

Duclohier H, Molle G, Spach G (1989) Antimicrobial peptide magainin I from xenopus skin forms anion-permeable channels in planar lipid bilayers. Biophys J 56:1017–1021

Eddy MT, Su Y, Silvers R, Andreas L, Clark L, Wagner G, Pintacuda G, Emsley L, Griffin RG (2015) Lipid bilayer-bound conformation of an integral membrane beta barrel protein by multidimensional MAS NMR. J Biomol NMR 61:299–310

Farrotti A, Bocchinfuso G, Palleschi A, Rosato N, Salnikov ES, Voievoda N, Bechinger B, Stella L (2015) Molecular dynamics methods to predict peptide location in membranes: LAH4 as a stringent test case. Biochim Biophys Acta 1848:581–592

Gallaher J, Wodzinska K, Heimburg T, Bier M (2010) Ion-channel-like behavior in lipid bilayer membranes at the melting transition. Phys Rev E Stat Nonlinear Soft Matter Phys 81:061925

Gazit E, Miller IR, Biggin PC, Sansom MSP, Shai Y (1996) Structure and orientation of the mammalian antibacterial peptide cecropin P1 within phospholipid membranes. J Mol Biol 258:860–870

Georgescu J, Munhoz VHO, Bechinger B (2010) NMR structures of the histidine-rich peptide LAH4 in micellar environments: membrane insertion, pH-dependent mode of antimicrobial action and DNA transfection. Biophys J 99:2507–2515

Ghosh C, Haldar J (2015) Membrane-active small molecules: designs inspired by antimicrobial peptides. ChemMedChem 10:1606–1624

Ghosh C, Manjunath GB, Akkapeddi P, Yarlagadda V, Hoque J, Uppu DSSM, Konai MM, Haldar J (2014) Small molecular antibacterial peptoid mimics: the simpler the better! J Med Chem 57:1428–1436

Ghosh C, Harmouche N, Bechinger B, Haldar J (2018) Aryl-alkyl-lysines interact with anionic lipid components of bacterial cell envelope eliciting anti-inflammatory and anti-biofilm properties. ACS Omega 3:9182–9190

Giovannini MG, Poulter L, Gibson BW, Williams DH (1987) Biosynthesis and degradation of peptides derived from *Xenopus laevis* prohormones. Biochem J 243:113–120

Glattard E, Salnikov ES, Aisenbrey C, Bechinger B (2016) Investigations of the synergistic enhancement of antimicrobial activity in mixtures of magainin 2 and PGLa. Biophys Chem 210:35–44

Gogonea V (2015) Structural insights into high density lipoprotein: old models and new facts. Front Pharmacol 6:318

Gopinath T, Veglia G (2015) Multiple acquisition of magic angle spinning solid-state NMR experiments using one receiver: application to microcrystalline and membrane protein preparations. J Magn Reson 253:143–153

Gopinath T, Mote KR, Veglia G (2015) Simultaneous acquisition of 2D and 3D solid-state NMR experiments for sequential assignment of oriented membrane protein samples. J Biomol NMR 62:53–61

Grage SL, Afonin S, Kara S, Buth G, Ulrich AS (2016) Membrane thinning and thickening induced by membrane-active amphipathic peptides. Front Cell Dev Biol 4:65

Grau A, Gomez Fernandez JC, Peypoux F, Ortiz A (1999) A study on the interactions of surfactin with phospholipid vesicles. Biochim Biophys Acta 1418:307–319

Gregory SM, Cavenaugh A, Journigan V, Pokorny A, Almeida PFF (2008) A quantitative model for the all-or-none permeabilization of phospholipid vesicles by the antimicrobial peptide cecropin A. Biophys J 94:1667–1680

Hadley EB, Hancock RE (2010) Strategies for the discovery and advancement of novel cationic antimicrobial peptides. Curr Top Med Chem 10:1872–1881

Hall K, Lee TH, Mechler AI, Swann MJ, Aguilar MI (2014) Real-time measurement of membrane conformational states induced by antimicrobial peptides: balance between recovery and lysis. Sci Rep 4:5479

Hallock KJ, Lee DK, Omnaas J, Mosberg HI, Ramamoorthy A (2002) Membrane composition determines pardaxin's mechanism of lipid bilayer disruption. Biophys J 83:1004–1013

Hara T, Kodama H, Kondo M, Wakamatsu K, Takeda A, Tachi T, Matsuzaki K (2001) Effects of peptide dimerization on pore formation: antiparallel disulfide-dimerized magainin 2 analogue. Biopolymers 58:437–446

Harmouche N, Bechinger B (2018) Lipid-mediated interactions between the amphipathic antimicrobial peptides magainin 2 and PGLa in phospholipid bilayers. Biophys J 115:1033-1044

Harpole TJ, Delemotte L (2018) Conformational landscapes of membrane proteins delineated by enhanced sampling molecular dynamics simulations. Biochim Biophys Acta 1860:909–926

Harzer U, Bechinger B (2000) The alignment of lysine-anchored membrane peptides under conditions of hydrophobic mismatch: a CD, 15 N and 31 P solid-state NMR spectroscopy investigation. Biochemistry 39:13106–13114

Hasan M, Karal MAS, Levadnyy V, Yamazaki M (2018) Mechanism of initial stage of pore formation induced by antimicrobial peptide magainin 2. Langmuir 34:3349–3362

Hayden RM, Goldberg GK, Ferguson BM, Schoeneck MW, Libardo MD, Mayeux SE, Shrestha A, Bogardus KA, Hammer J, Pryshchep S, Lehman HK, McCormick ML, Blazyk J, Angeles-Boza AM, Fu R, Cotten ML (2015) Complementary effects of host defense peptides piscidin 1 and piscidin 3 on DNA and lipid membranes: biophysical insights into contrasting biological activities. J Phys Chem B 119:15235–15246

He K, Ludtke SJ, Heller WT, Huang HW (1996) Mechanism of alamethicin insertion into lipid bilayers. Biophys J 71:2669–2679

Heerklotz H, Seelig J (2001) Detergent-like action of the antibiotic peptide surfactin on lipid membranes. Biophys J 81:1547–1554

Heerklotz H, Seelig J (2007) Leakage and lysis of lipid membranes induced by the lipopeptide surfactin. Eur Biophys J 36:305–314

Heerklotz H, Wieprecht T, Seelig J (2004) Membrane perturbation by the lipopeptide surfactin and detergents as studied by deuterium. J Phys Chem B 108:4909–4915

Heimburg T (2012) The capacitance and electromechanical coupling of lipid membranes close to transitions: the effect of electrostriction. Biophys J 103:918–929

Henderson JM, Waring AJ, Separovic F, Lee KYC (2016) Antimicrobial peptides share a common interaction driven by membrane line tension reduction. Biophys J 111:2176–2189

Holzl MA, Hofer J, Steinberger P, Pfistershammer K, Zlabinger GJ (2008) Host antimicrobial proteins as endogenous immunomodulators. Immunol Lett 119:4–11

Hong M, Su Y (2011) Structure and dynamics of cationic membrane peptides and proteins: insights from solid-state NMR. Protein Sci 20:641–655

Imura Y, Choda N, Matsuzaki K (2008) Magainin 2 in action: distinct modes of membrane permeabilization in living bacterial and mammalian cells. Biophys J 95:5757–5765

Ines M, Dhouha G (2015) Lipopeptide surfactants: production, recovery and pore forming capacity. Peptides 71:100–112

Islam MZ, Alam JM, Tamba Y, Karal MAS, Yamazaki M (2014) The single GUV method for revealing the functions of antimicrobial, pore-forming toxin, and cell-penetrating peptides or proteins. Phys Chem Chem Phys 16:15752–15767

Israelachvili JN, Marcelja S, Horn RG (1980) Physical principles of membrane organization. Q Rev Biophys 13:121–200

Itkin A, Salnikov ES, Aisenbrey C, Raya J, Raussens V, Ruysschaert JM, Bechinger B (2017) Evidence for heterogeneous conformations of the gamma cleavage site within the amyloid precursor proteins transmembrane domain. ACS Omega 2:6525–6534

Jaipuria G, Leonov A, Giller K, Vasa SK, Jaremko L, Jaremko M, Linser R, Becker S, Zweckstetter M (2017) Cholesterol-mediated allosteric regulation of the mitochondrial translocator protein structure. Nat Commun 8:14893

Jean-Francois F, Castano S, Desbat B, Odaert B, Roux M, Metz-Boutigue MH, Dufourc EJ (2008) Aggregation of cateslytin beta-sheets on negatively charged lipids promotes rigid membrane domains. A new mode of action for antimicrobial peptides? Biochemistry 47:6394–6402

Jenssen H, Hamill P, Hancock RE (2006) Peptide antimicrobial agents. Clin Microbiol Rev 19:491–511

Kandaswamy K, Liew TH, Wang CY, Huston-Warren E, Meyer-Hoffert U, Hultenby K, Schroder JM, Caparon MG, Normark S, Henriques-Normark B, Hultgren SJ, Kline KA (2013) Focal targeting by human beta-defensin 2 disrupts localized virulence factor assembly sites in Enterococcus faecalis. Proc Natl Acad Sci U S A 110:20230–20235

Karal MA, Alam JM, Takahashi T, Levadny V, Yamazaki M (2015) Stretch-activated pore of the antimicrobial peptide, magainin 2. Langmuir 31:3391–3401

Killian JA, de Planque MRR, van der Wel PCA, Salemink I, De Kruijff B, Greathouse DV, Koeppe RE (1998) Modulation of membrane structure and function by hydrophobic mismatch between proteins and lipids. Pure Appl Chem 70:75–82

Kim C, Spano J, Park EK, Wi S (2009) Evidence of pores and thinned lipid bilayers induced in oriented lipid membranes interacting with the antimicrobial peptides, magainin-2 and aurein-3.3. Biochim Biophys Acta 1788:1482–1496

Kim EY, Rajasekaran G, Shin SY (2017) LL-37-derived short antimicrobial peptide KR-12-a5 and its d-amino acid substituted analogs with cell selectivity, anti-biofilm activity, synergistic effect with conventional

antibiotics, and anti-inflammatory activity. Eur J Med Chem 136:428–441
Kindrachuk J, Napper S (2010) Structure-activity relationships of multifunctional host defence peptides. Mini-Rev Med Chem 10:596–614
Kiss G, Michl H (1962) Über das Giftsekret der Gelbbauchunke Bombina variegata L. Toxicon (Oxford) 1:33–39
Klocek G, Seelig J (2008) Melittin interaction with sulfated cell surface sugars. Biochemistry 47:2841–2849
Kmiecik S, Gront D, Kolinski M, Wieteska L, Dawid AE, Kolinski A (2016) Coarse-grained protein models and their applications. Chem Rev 116:7898–7936
Kobayashi S, Hirakura Y, Matsuzaki K (2001) Bacteria-selective synergism between the antimicrobial peptides alpha-helical magainin 2 and cyclic beta-sheet tachyplesin I: toward cocktail therapy. Biochemistry 40:14330–14335
Kollmitzer B, Heftberger P, Rappolt M, Pabst G (2013) Monolayer spontaneous curvature of raft-forming membrane lipids. Soft Matter 9:10877–10884
Kozlowska J, Vermeer LS, Rogers GB, Rehnnuma N, Amos SB, Koller G, McArthur M, Bruce KD, Mason AJ (2014) Combined systems approaches reveal highly plastic responses to antimicrobial peptide challenge in *Escherichia coli*. PLoS Pathog 10:e1004104
Kuroda K, DeGrado WF (2005) Amphiphilic polymethacrylate derivatives as antimicrobial agents. J Am Chem Soc 127:4128–4129
Ladokhin AS, Wimley WC, White SH (1995) Leakage of membrane vesicle contents – determination of mechanisms using fluorescence requenching. Biophys J 69:1964–1971
Laub KR, Witschas K, Blicher A, Madsen SB, Luckhoff A, Heimburg T (2012) Comparing ion conductance recordings of synthetic lipid bilayers with cell membranes containing TRP channels. Biochim Biophys Acta 1818:1123–1134
Laurencin M, Simon M, Fleury Y, Baudy-Floc'h M, Bondon A, Legrand B (2018) Selectivity modulation and structure of alpha/aza-beta(3) cyclic antimicrobial peptides. Chemistry 24:6191–6201
Lear JD, Wasserman ZR, DeGrado WF (1988) Synthetic amphiphilic peptide models for protein ion channels. Science 240:1177–1181
Leber R, Pachler M, Kabelka I, Svoboda I, Enkoller D, Vácha R, Lohner K, Pabst G (2018) Synergism of antimicrobial frog peptides couples to membrane intrinsic curvature strain. Biophys J 114:1945–1954
Leitgeb B, Szekeres A, Manczinger L, Vagvolgyi C, Kredics L (2007) The history of alamethicin: a review of the most extensively studied peptaibol. Chem Biodivers 4:1027–1051
Liu SP, Zhou L, Lakshminarayanan R, Beuerman RW (2010) Multivalent antimicrobial peptides as therapeutics: design principles and structural diversities. Int J Pept Res Ther 16:199–213
Lohner K (2009) New strategies for novel antibiotics: peptides targeting bacterial cell membranes. Gen Physiol Biophys 28:105–116
Lorenzon EN, Santos-Filho NA, Ramos MA, Bauab TM, Camargo IL, Cilli EM (2016) C-terminal lysine-linked magainin 2 with increased activity against multidrug-resistant bacteria. Protein Pept Lett 23:738–747
Loudet C, Khemtemourian L, Aussenac F, Gineste S, Achard MF, Dufourc EJ (2005) Bicelle membranes and their use for hydrophobic peptide studies by circular dichroism and solid state NMR. Biochim Biophys Acta 1724:315–323
Ludtke SJ, He K, Wu Y, Huang HW (1994) Cooperative membrane insertion of magainin correlated with its cytolytic activity. Biochim Biophys Acta 1190:181–184
Ludtke S, He K, Huang H (1995) Membrane thinning caused by magainin 2. Biochemistry 34:16764–16769
Ludtke SJ, He K, Heller WT, Harroun TA, Yang L, Huang HW (1996) Membrane pores induced by magainin. Biochemistry 35:13723–13728
Maget-Dana R, Peypoux F (1994) Iturins, a special class of pore-forming lipopeptides: biological and physicochemical properties. Toxicology 87:151–174
Maget-Dana R, Ptak M (1995) Interactions of surfactin with membrane models. Biophys J 68:1937–1943
Makovitzki A, Baram J, Shai Y (2008) Antimicrobial lipopolypeptides composed of palmitoyl Di- and tricationic peptides: in vitro and in vivo activities, self-assembly to nanostructures, and a plausible mode of action. Biochemistry 47:10630–10636
Mangoni ML, Shai Y (2011) Short native antimicrobial peptides and engineered ultrashort lipopeptides: similarities and differences in cell specificities and modes of action. Cell Mol Life Sci 68:2267–2280
Marquette A, Bechinger B (2018) Biophysical investigations elucidating the mechanisms of action of antimicrobial peptides and their synergism. Biomol Ther 8:E18
Marquette A, Mason AJ, Bechinger B (2008) Aggregation and membrane permeabilizing properties of designed histidine-containing cationic linear peptide antibiotics. J Pept Sci 14:488–495
Marquette A, Lorber B, Bechinger B (2010) Reversible liposome association induced by LAH4: a peptide with potent antimicrobial and nucleic acid transfection activities. Biophys J 98:2544–2553
Marquette A, Salnikov E, Glattard E, Aisenbrey C, Bechinger B (2015) Magainin 2-PGLa interactions in membranes – two peptides that exhibit synergistic enhancement of antimicrobial activity. Curr Top Med Chem 16:65–75
Mason AJ, Martinez A, Glaubitz C, Danos O, Kichler A, Bechinger B (2006) The antibiotic and DNA-transfecting peptide LAH4 selectively associates with, and disorders, anionic lipids in mixed membranes. FASEB J 20:320–322
Mason AJ, Moussaoui W, Abdelrhaman T, Boukhari A, Bertani P, Marquette A, Shooshtarizaheh P, Moulay G,

Boehm N, Guerold B, Sawers RJH, Kichler A, Metz-Boutigue MH, Candolfi E, Prevost G, Bechinger B (2009) Structural determinants of antimicrobial and antiplasmodial activity and selectivity in histidine rich amphipathic cationic peptides. J Biol Chem 284:119–133

Matos PM, Franquelim HG, Castanho MA, Santos NC (2010) Quantitative assessment of peptide-lipid interactions. Ubiquitous fluorescence methodologies. Biochim Biophys Acta 1798:1999–2012

Matsuzaki K (1998) Magainins as paradigm for the mode of action of pore forming polypeptides. Biochim Biophys Acta 1376:391–400

Matsuzaki K, Harada M, Funakoshi S, Fujii N, Miyajima K (1991) Physicochemical determinants for the interactions of magainins 1 and 2 with acidic lipid bilayers. Biochim Biophys Acta 1063:162–170

Matsuzaki K, Murase O, Tokuda H, Funakoshi S, Fujii N, Miyajima K (1994) Orientational and aggregational states of magainin 2 in phospholipid bilayers. Biochemistry 33:3342–3349

Matsuzaki K, Murase O, Fujii N, Miyajima K (1995a) Translocation of a channel-forming antimicrobial peptide, magainin2, across lipid bilayers by forming a pore. Biochemistry 34:6521–6526

Matsuzaki K, Murase O, Miyajima K (1995b) Kinetics of pore formation by an antimicrobial peptide, magainin 2, in phospholipid bilayers. Biochemistry 34:12553–12559

Matsuzaki K, Murase O, Fujii N, Miyajima K (1996) An antimicrobial peptide, magainin 2, induced flip-flop of phospholipid coupled with pore formation and peptide translocation. Biochemistry 35:11361–11368

Matsuzaki K, Mitani Y, Akada K, Murase O, Yoneyama S, Zasloff M, Miyajima K (1998) Mechanism of synergism between antimicrobial peptides magainin 2 and PGLa. Biochemistry 37:15144–15153

McCafferty DG, Cudic P, Yu MK, Behenna DC, Kruger R (1999) Synergy and duality in peptide antibiotic mechanisms. Curr Opin Chem Biol 3:672–680

Mecke A, Lee DK, Ramamoorthy A, Orr BG, Banaszak Holl MM (2005) Membrane thinning due to antimicrobial peptide binding: an atomic force microscopy study of MSI-78 in lipid bilayers. Biophys J 89:4043–4050

Medintz IL, Hildebrandt N (2013) Förster resonance energy transfer: from theory to applications. Wiley-VCH, Weinheim

Michalek M, Aisenbrey C, Bechinger B (2014) Investigation of membrane penetration depth and interactions of the amino-terminal domain of huntingtin: refined analysis by tryptophan fluorescence measurement. Eur Biophys J 43:347–360

Miles AJ, Wallace BA (2006) Synchrotron radiation circular dichroism spectroscopy of proteins and applications in structural and functional genomics. Chem Soc Rev 35:39–51

Miles AJ, Wallace BA (2016) Circular dichroism spectroscopy of membrane proteins. Chem Soc Rev 45:4859–4872

Milov AD, Samoilova RI, Tsvetkov YD, De Zotti M, Formaggio F, Toniolo C, Handgraaf JW, Raap J (2009) Structure of self-aggregated alamethicin in ePC membranes detected by pulsed electron-electron double resonance and electron spin Echo envelope modulation spectroscopies. Biophys J 96:3197–3209

Montal M, Mueller P (1972) Formation of bimolecular membranes from lipid monolayers and a study of their electrical properties. Proc Natl Acad Sci U S A 69:3561–3566

Mor A, Hani K, Nicolas P (1994) The vertebrate peptide antibiotics dermaseptins have overlapping structural features but target specific microorganisms. J Biol Chem 269:31635–31641

Naito A, Matsumori N, Ramamoorthy A (2018) Dynamic membrane interactions of antibacterial and antifungal biomolecules, and amyloid peptides, revealed by solid-state NMR spectroscopy. Biochim Biophys Acta 1862:307–323

Nishida M, Imura Y, Yamamoto M, Kobayashi S, Yano Y, Matsuzaki K (2007) Interaction of a magainin-PGLa hybrid peptide with membranes: insight into the mechanism of synergism. Biochemistry 46:14284–14290

North CL, Barranger-Mathys M, Cafiso DS (1995) Membrane orientation of the N-terminal segment of alamethicin determined by solid-state 15 N NMR. Biophys J 69:2392–2397

Otzen DE (2017) Biosurfactants and surfactants interacting with membranes and proteins: same but different? Biochim Biophys Acta 1859:639–649

Oyston PC, Fox MA, Richards SJ, Clark GC (2009) Novel peptide therapeutics for treatment of infections. J Med Microbiol 58:977–987

Palermo EF, Kuroda K (2010) Structural determinants of antimicrobial activity in polymers which mimic host defense peptides. Appl Microbiol Biotechnol 87:1605–1615

Patch JA, Barron AE (2003) Helical peptoid mimics of magainin-2 amide. J Am Chem Soc 125:12092–12093

Patel H, Tscheka C, Edwards K, Karlsson G, Heerklotz H (2011) All-or-none membrane permeabilization by fengycin-type lipopeptides from Bacillus subtilis QST713. Biochim Biophys Acta 1808:2000–2008

Patel H, Huynh Q, Barlehner D, Heerklotz H (2014) Additive and synergistic membrane permeabilization by antimicrobial (lipo)peptides and detergents. Biophys J 106:2115–2125

Paterson DJ, Tassieri M, Reboud J, Wilson R, Cooper JM (2017) Lipid topology and electrostatic interactions underpin lytic activity of linear cationic antimicrobial peptides in membranes. Proc Natl Acad Sci U S A 114:E8324–E8332

Payne JE, Dubois AV, Ingram RJ, Weldon S, Taggart CC, Elborn JS, Tunney MM (2017) Activity of innate antimicrobial peptides and ivacaftor against clinical cystic fibrosis respiratory pathogens. Int J Antimicrob Agents 50:427–435

Perrone B, Miles AJ, Salnikov ES, Wallace B, Bechinger B (2014) Lipid- interactions of the LAH4, a peptide with antimicrobial and nucleic transfection activities. Eur Biophys J 43:499–507

Pino-Angeles A, Leveritt JM 3rd, Lazaridis T (2016) Pore structure and synergy in antimicrobial peptides of the magainin family. PLoS Comput Biol 12:e1004570

Pirtskhalava M, Gabrielian A, Cruz P, Griggs HL, Squires RB, Hurt DE, Grigolava M, Chubinidze M, Gogoladze G, Vishnepolsky B, Alekseyev V, Rosenthal A, Tartakovsky M (2016) DBAASP v.2: an enhanced database of structure and antimicrobial/cytotoxic activity of natural and synthetic peptides. Nucleic Acids Res 44:D1104–D1112

Porter EA, Weisblum B, Gellman SH (2002) Mimicry of host-defense peptides by unnatural oligomers: antimicrobial beta-peptides. J Am Chem Soc 124:7324–7330

Pouny Y, Rapaport D, Mor A, Nicolas P, Shai Y (1992) Interaction of antimicrobial dermaseptin and its fluorescently labeled analogues with phospholipid membranes. Biochemistry 31:12416–12423

Raimondo D, Andreotti G, Saint N, Amodeo P, Renzone G, Sanseverino M, Zocchi I, Molle G, Motta A, Scaloni A (2005) A folding-dependent mechanism of antimicrobial peptide resistance to degradation unveiled by solution structure of distinctin. Proc Natl Acad Sci U S A 102:6309–6314

Ramamoorthy A, Thennarasu S, Lee DK, Tan A, Maloy L (2006) Solid-state NMR investigation of the membrane-disrupting mechanism of antimicrobial peptides MSI-78 and MSI-594 derived from magainin 2 and melittin. Biophys J 91:206–216

Rangarajan N, Bakshi S, Weisshaar JC (2013) Localized permeabilization of *E. coli* membranes by the antimicrobial peptide cecropin A. Biochemistry 52:6584–6594

Rank LA, Walsh NM, Liu R, Lim FY, Bok JW, Huang M, Keller NP, Gellman SH, Hull CM (2017) A cationic polymer that shows high antifungal activity against diverse human pathogens. Antimicrob Agents Chemother 61:e00204–e00217

Rautenbach M, Troskie AM, Vosloo JA (2016a) Antifungal peptides: to be or not to be membrane active. Biochimie 130:132–145

Rautenbach M, Troskie AM, Vosloo JA, Dathe ME (2016b) Antifungal membranolytic activity of the tyrocidines against filamentous plant fungi. Biochimie 130:122–131

Reijmar K, Edwards K, Andersson K, Agmo Hernandez V (2016) Characterizing and controlling the loading and release of cationic amphiphilic peptides onto and from PEG-stabilized lipodisks. Langmuir 32:12091–12099

Resende JM, Moraes CM, Munhoz VHDO, Aisenbrey C, Verly RM, Bertani P, Cesar A, Pilo-Veloso D, Bechinger B (2009) Membrane structure and conformational changes of the antibiotic heterodimeric peptide distinctin by solid-state NMR spectroscopy. Proc Natl Acad Sci U S A 106:16639–16644

Resende JM, Verly RM, Aisenbrey C, Amary C, Bertani P, Pilo-Veloso D, Bechinger B (2014) Membrane interactions of phylloseptin-1, −2, and −3 peptides by oriented solid-state NMR spectroscopy. Biophys J 107:901–911

Rollins-Smith LA, Doersam JK, Longcore JE, Taylor SK, Shamblin JC, Carey C, Zasloff MA (2002) Antimicrobial peptide defenses against pathogens associated with global amphibian declines. Dev Comp Immunol 26:63–72

Rotem S, Mor A (2009) Antimicrobial peptide mimics for improved therapeutic properties. Biochim Biophys Acta 1788:1582–1592

Roversi D, Luca V, Aureli S, Park Y, Mangoni ML, Stella L (2014) How many AMP molecules kill a bacterium? Spectroscopic determination of PMAP-23 binding to *E. Coli*. ACS Chem Biol 9:2003–2007

Russ WP, Engelman DM (2000) The GxxxG motif: a framework for transmembrane helix-helix association. J Mol Biol 296:911–919

Sakoulas G, Kumaraswamy M, Kousha A, Nizet V (2017) Interaction of antibiotics with innate host defense factors against Salmonella enterica serotype Newport. mSphere 2:e00410–e00417

Salnikov E, Bechinger B (2011) Lipid-controlled peptide topology and interactions in bilayers: structural insights into the synergistic enhancement of the antimicrobial activities of PGLa and magainin 2. Biophys J 100:1473–1480

Salnikov ES, Friedrich H, Li X, Bertani P, Reissmann S, Hertweck C, O'Neil JD, Raap J, Bechinger B (2009a) Structure and alignment of the membrane-associated peptaibols ampullosporin A and alamethicin by oriented 15 N and 31 P solid-state NMR spectroscopy. Biophys J 96:86–100

Salnikov ES, Mason AJ, Bechinger B (2009b) Membrane order perturbation in the presence of antimicrobial peptides by 2H solid-state NMR spectroscopy. Biochimie 91:734–743

Salnikov E, Aisenbrey C, Vidovic V, Bechinger B (2010) Solid-state NMR approaches to measure topological equilibria and dynamics of membrane polypeptides. Biochim Biophys Acta 1798:258–265

Salnikov E, Aisenbrey C, Balandin SV, Zhmak MN, Ovchinnikova AY, Bechinger B (2011) Structure and alignment of the membrane-associated antimicrobial peptide arenicin by oriented solid-state NMR spectroscopy. Biochemistry 50:3784–3795

Salnikov ES, Aisenbrey C, Aussenac F, Ouari O, Sarrouj H, Reiter C, Tordo P, Engelke F, Bechinger B (2016a) Membrane topologies of the PGLa antimicrobial peptide and a transmembrane anchor sequence by Dynamic Nuclear Polarization/solid-state NMR spectroscopy. Sci Rep 6:20895

Salnikov ES, Raya J, De Zotti M, Zaitseva E, Peggion C, Ballano G, Toniolo C, Raap J, Bechinger B (2016b) Alamethicin supramolecular organization in lipid membranes from 19F solid- state NMR. Biophys J 111:2450–2459

Salnikov ES, Anantharamaiah GM, Bechinger B (2018) Supramolecular organization of apolipoprotein A-I – derived peptides within disc-like arrangements. Biophys J 115:467–477 in press

Sani MA, Separovic F (2018) Antimicrobial peptide structures: from model membranes to live cells. Chemistry 24:286–291

Sansom MSP (1991) The biophysics of peptide models of ion channels. Prog Biophysmol Biol 55:139–235

Sansom MS (1993) Alamethicin and related peptaibols – model ion channels. Eur Biophys J 22:105–124

Scherer PG, Seelig J (1989) Electric charge effects on phospholipid headgroups. Phosphatidylcholine in mixtures with cationic and anionic amphiphiles. Biochemistry 28:7720–7727

Schweizer F (2009) Cationic amphiphilic peptides with cancer-selective toxicity. Eur J Pharmacol 625:190–194

Scott RW, DeGrado WF, Tew GN (2008) De novo designed synthetic mimics of antimicrobial peptides. Curr Opin Biotechnol 19:620–627

Seelig J (2004) Thermodynamics of lipid-peptide interactions. Biochim Biophys Acta 1666:40–50

Shai Y (1999) Mechanism of the binding, insertion, and destabilization of phospholipid bilayer membranes by alpha-helical antimicrobial and cell non-selective lytic peptides. Biochim Biophys Acta 1462:55–70

Smart M, Rajagopal A, Liu WK, Ha BY (2017) Opposing effects of cationic antimicrobial peptides and divalent cations on bacterial lipopolysaccharides. Phys Rev E 96:042405

Sreerama N, Woody RW (2000) Estimation of protein secondary structure from circular dichroism spectra: comparison of CONTIN, SELCON, and CDSSTR methods with an expanded reference set. Anal Biochem 287:252–260

Steinstraesser L, Kraneburg U, Jacobsen F, Al-Benna S (2010) Host defense peptides and their antimicrobial-immunomodulatory duality. Immunobiology 216:322–333

Strandberg E, Tiltak D, Ehni S, Wadhwani P, Ulrich AS (2012) Lipid shape is a key factor for membrane interactions of amphipathic helical peptides. Biochim Biophys Acta 1818:1764–1776

Strandberg E, Zerweck J, Wadhwani P, Ulrich AS (2013) Synergistic insertion of antimicrobial magainin-family peptides in membranes depends on the lipid spontaneous curvature. Biophys J 104:L09–L11

Sychev SV, Sukhanov SV, Panteleev PV, Shenkarev ZO, Ovchinnikova TV (2018) Marine antimicrobial peptide arenicin adopts a monomeric twisted beta-hairpin structure and forms low conductivity pores in zwitterionic lipid bilayers. Pept Science 110:e23093

Tamba Y, Yamazaki M (2005) Single giant unilamellar vesicle method reveals effect of antimicrobial peptide magainin 2 on membrane permeability. Biochemistry 44:15823–15833

Tamba Y, Yamazaki M (2009) Magainin 2-induced pore formation in the lipid membranes depends on its concentration in the membrane interface. J Phys Chem B 113:4846–4852

Tamba Y, Ariyama H, Levadny V, Yamazaki M (2010) Kinetic pathway of antimicrobial peptide magainin 2-induced pore formation in lipid membranes. J Phys Chem B 114:12018–12026

Tieleman DP, Hess B, Sansom MS (2002) Analysis and evaluation of channel models: simulations of alamethicin. Biophys J 83:2393–2407

Tremouilhac P, Strandberg E, Wadhwani P, Ulrich AS (2006a) Conditions affecting the re-alignment of the antimicrobial peptide PGLa in membranes as monitored by solid state 2H-NMR. Biochim Biophys Acta 1758:1330–1342

Tremouilhac P, Strandberg E, Wadhwani P, Ulrich AS (2006b) Synergistic transmembrane alignment of the antimicrobial heterodimer PGLa/magainin. J Biol Chem 281:32089–32094

Tsutsumi LS, Elmore JM, Dang UT, Wallace MJ, Marreddy R, Lee RB, Tan GT, Hurdle JG, Lee RE, Sun D (2018) Solid-phase synthesis and antibacterial activity of cyclohexapeptide wollamide B analogs. ACS Comb Sci 20:172–185

Ulmschneider JP, Smith JC, Ulmschneider MB, Ulrich AS, Strandberg E (2012) Reorientation and dimerization of the membrane-bound antimicrobial peptide PGLa from microsecond all-atom MD simulations. Biophys J 103:472–482

Usachev KS, Kolosova OA, Klochkova EA, Yulmetov AR, Aganov AV, Klochkov VV (2017) Oligomerization of the antimicrobial peptide protegrin-5 in a membrane-mimicking environment. Structural studies by high-resolution NMR spectroscopy. Eur Biophys J 46:293–300

Vacha R, Frenkel D (2014) Simulations suggest possible novel membrane pore structure. Langmuir 30:1304–1310

Vaz Gomes A, de Waal A, Berden JA, Westerhoff HV (1993) Electric potentiation, cooperativity, and synergism of magainin peptides in protein-free liposomes. Biochemistry 32:5365–5372

Verardi R, Traaseth NJ, Shi L, Porcelli F, Monfregola L, De Luca S, Amodeo P, Veglia G, Scaloni A (2011) Probing membrane topology of the antimicrobial peptide distinctin by solid-state NMR spectroscopy in zwitterionic and charged lipid bilayers. Biochim Biophys Acta 1808:34–40

Verly RM, Resende JM, Junior EFC, de Magalhães MTC, Guimarães CFCR, Munhoz VHO, Bemquerer MP, Almeida FCL, Santoro MM, Piló-Veloso D, Bechinger B (2017) Structure and membrane interactions of the homodimeric antibiotic peptide homotarsinin. Sci Rep 7:40854

Vermeer LS, Marquette A, Schoup M, Fenard D, Galy A, Bechinger B (2016) Simultaneous analysis of secondary structure and light scattering from circular dichroism titrations: application to vectofusin-1. Sci Rep 6:39450

Violette A, Fournel S, Lamour K, Chaloin O, Frisch B, Briand JP, Monteil H, Guichard G (2006) Mimicking helical antibacterial peptides with nonpeptidic folding oligomers. Chem Biol 13:531–538

Visscher KM, Medeiros-Silva J, Mance D, Rodrigues J, Daniels M, Bonvin A, Baldus M, Weingarth M (2017)

Supramolecular organization and functional implications of K+ channel clusters in membranes. Angew Chem Int Ed Eng 56:13222–13227

Vogt TCB, Bechinger B (1999) The interactions of histidine-containing amphipathic helical peptide antibiotics with lipid bilayers: the effects of charges and pH. J Biol Chem 274:29115–29121

Voievoda N (2014) Biophysical investigations of the membrane and nucleic acids interactions of the transfection peptide LAH4-L1. PhD thesis, University of Strasbourg PhD

Voievoda N, Schulthess T, Bechinger B, Seelig J (2015) Thermodynamic and biophysical analysis of the membrane-association of a histidine-rich peptide with efficient antimicrobial and transfection activities. J Phys Chem B 119:9678–9687

Wakamatsu K, Takeda A, Tachi T, Matsuzaki K (2002) Dimer structure of magainin 2 bound to phospholipid vesicles. Biopolymers 64:314–327

Wallace BA, Moa B (1984) Circular dichroism analyses of membrane proteins: an examination of differential light scattering and absorption flattening effects in large membrane vesicles and membrane sheets. Anal Biochem 142:317–328

Wang G, Li X, Wang Z (2016) APD3: the antimicrobial peptide database as a tool for research and education. Nucleic Acids Res 44:D1087–D1093

Watanabe H, Kawano R (2016) Channel current analysis for pore-forming properties of an antimicrobial peptide, magainin 1, using the droplet contact method. Anal Sci 32:57–60

Wenk M, Seelig J (1998) Magainin 2 amide interaction with lipid membranes: calorimetric detection of peptide binding and pore formation. Biochemistry 37:3909–3916

Wenzel M, Chiriac AI, Otto A, Zweytick D, May C, Schumacher C, Gust R, Albada HB, Penkova M, Kramer U, Erdmann R, Metzler-Nolte N, Straus SK, Bremer E, Becher D, Brotz-Oesterhelt H, Sahl HG, Bandow JE (2014) Small cationic antimicrobial peptides delocalize peripheral membrane proteins. Proc Natl Acad Sci USA 111:E1409–E1418

Westerhoff HV, Zasloff M, Rosner JL, Hendler RW, de Waal A, Vaz G, Jongsma PM, Riethorst A, Juretic D (1995) Functional synergism of the magainins PGLa and magainin-2 in *Escherichia coli*, tumor cells and liposomes. Eur J Biochem 228:257–264

Wheaten SA, Ablan FD, Spaller BL, Trieu JM, Almeida PF (2013) Translocation of cationic amphipathic peptides across the membranes of pure phospholipid giant vesicles. J Am Chem Soc 135:16517–16525

Wieprecht T, Apostolov O, Beyermann M, Seelig J (1999a) Thermodynamics of the alpha-helix-coil transition of amphipathic peptides in a membrane environment: implications for the peptide-membrane binding equilibrium. J Mol Biol 294:785–794

Wieprecht T, Beyermann M, Seelig J (1999b) Binding of antibacterial magainin peptides to electrically neutral membranes: thermodynamics and structure. Biochemistry 38:10377–10378

Wieprecht T, Apostolov O, Beyermann M, Seelig J (2000a) Membrane binding and pore formation of the antibacterial peptide PGLa: thermodynamic and mechanistic aspects. Biochemistry 39:442–452

Wieprecht T, Apostolov O, Seelig J (2000b) Binding of the antibacterial peptide magainin 2 amide to small and large unilamellar vesicles. Biophys Chem 85:187–198

Wieprecht T, Beyermann M, Seelig J (2002) Thermodynamics of the coil-alpha-helix transition of amphipathic peptides in a membrane environment: the role of vesicle curvature. Biophys Chem 96:191–201

Wolf J, Aisenbrey C, Harmouche N, Raya J, Bertani P, Voievoda N, Süss R, Bechinger B (2017) pH-dependent membrane interactions of the histidine-rich cell penetrating peptide LAH4-L1. Biophys J 113:1290–1300

Wu Y, Huang HW, Olah GA (1990) Method of oriented circular dichroism. Biophys J 57:797–806

Wu YS, Ngai SC, Goh BH, Chan KG, Lee LH, Chuah LH (2017) Anticancer activities of surfactin and potential application of nanotechnology assisted surfactin delivery. Front Pharmacol 8:761

Yang D, Zou R, Zhu Y, Liu B, Yao D, Jiang J, Wu J, Tian H (2014) Magainin II modified polydiacetylene micelles for cancer therapy. Nanoscale 6:14772–14783

Yang Z, Choi H, Weisshaar JC (2018) Melittin-induced permeabilization, re-sealing, and re-permeabilization of *E. coli* membranes. Biophys J 114:368–379

Yuksel E, Karakecili A (2014) Antibacterial activity on electrospun poly(lactide-co-glycolide) based membranes via magainin II grafting. Mater Sci Eng C Mater Biol Appl 45:510–518

Zanin LM, Dos Santos Alvares D, Juliano MA, Pazin WM, Ito AS, Ruggiero Neto J (2013) Interaction of a synthetic antimicrobial peptide with model membrane by fluorescence spectroscopy. Eur Biophys J 42:819–831

Zasloff M (1987) Magainins, a class of antimicrobial peptides from Xenopus skin: isolation, characterization of two active forms, and partial cDNA sequence of a precursor. Proc Natl Acad Sci U S A 84:5449–5453

Zasloff M (2002) Antimicrobial peptides of multicellular organisms. Nature 415:389–395

Zerweck J, Strandberg E, Kukharenko O, Reichert J, Burck J, Wadhwani P, Ulrich AS (2017) Molecular mechanism of synergy between the antimicrobial peptides PGLa and magainin 2. Sci Rep 7:13153

Zhao H, Shao D, Jiang C, Shi J, Li Q, Huang Q, Rajoka MSR, Yang H, Jin M (2017) Biological activity of lipopeptides from Bacillus. Appl Microbiol Biotechnol 101:5951–5960

Zhao P, Xue Y, Gao W, Li J, Zu X, Fu D, Bai X, Zuo Y, Hu Z, Zhang F (2018) Bacillaceae-derived peptide antibiotics since 2000. Peptides 101:10–16

Anionic Lipid Clustering Model 5

Richard M. Epand

Abstract

Many molecular features contribute to the antimicrobial activity of peptides. One aspect that contributes to the antimicrobial activity of a peptide, in many cases, results from the fact that many antimicrobial peptides are polycationic and the lipids on the surface of bacteria are often anionic. In certain cases this can result in the clustering of anionic lipids as a result of the binding of the cationic peptide to the surface of the bacterial membrane. This lipid clustering can be detrimental to the viability of the bacteria to which the peptide binds. Several factors, including the charge, size, and conformational flexibility of the peptide, will determine the efficiency of lipid clustering. In addition, the lipid composition of the bacterial membrane is very variable, and it plays a critical role in this mechanism. As a result, one can test the importance of this factor by determining the species specificity of the antimicrobial activity of the peptide. The molecular mechanism by which lipid clustering affects bacterial viability is uncertain in many cases. This phenomenon can be used to increase the antimicrobial potency of peptides in some case and can also predict the bacterial species specificity of some agents.

R. M. Epand (✉)
Department of Biochemistry and Biomedical Sciences, McMaster University, Hamilton, ON, Canada
e-mail: epand@mcmaster.ca

Keywords

Antimicrobial peptides · Anionic lipid clustering · Bacterial membrane lipids · Cardiolipin · Phosphatidylethanolamine · Lipid domain formation

5.1 Description of Model

Gram-negative bacteria have an outer membrane that is negatively charged, but most antimicrobial agents have bactericidal or bacteriostatic action as a result of interactions with the cell membrane, i.e., the inner membrane of Gram-negative bacteria and the cell membrane of Gram-positive bacteria. In both cases these membranes are rich in anionic lipids phosphatidylglycerol and cardiolipin. Antimicrobial peptides have a large range of structures, but the majority of these peptides have several positive charges. This would allow each peptide to bind to several lipids and cluster them in the same location in the membrane. For membranes that also contain zwitterionic or uncharged lipids, this would produce an area in the membrane that is depleted of anionic lipids. A schematic model of this phenomenon is shown in Fig. 5.1.

K. Matsuzaki (ed.), *Antimicrobial Peptides*, Advances in Experimental Medicine and Biology 1117,
https://doi.org/10.1007/978-981-13-3588-4_5

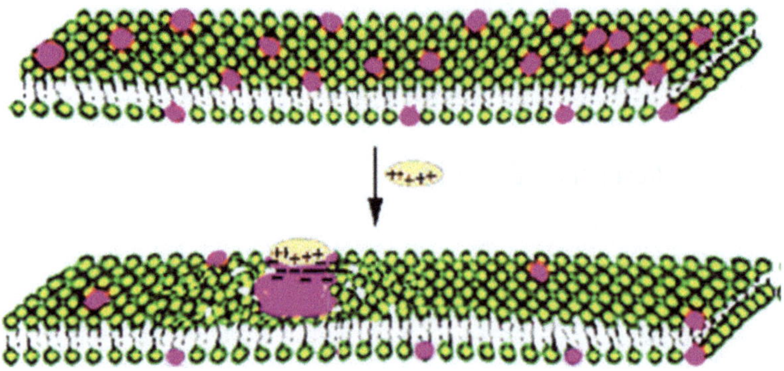

Fig. 5.1 Clustering of anionic lipid in a bilayer composed of anionic (red headgroups) and zwitterionic lipids (green headgroups) in the presence of a cationic agent causes defects in the surrounding area of the bilayer, increasing permeability (Figure taken from Epand et al. 2011)

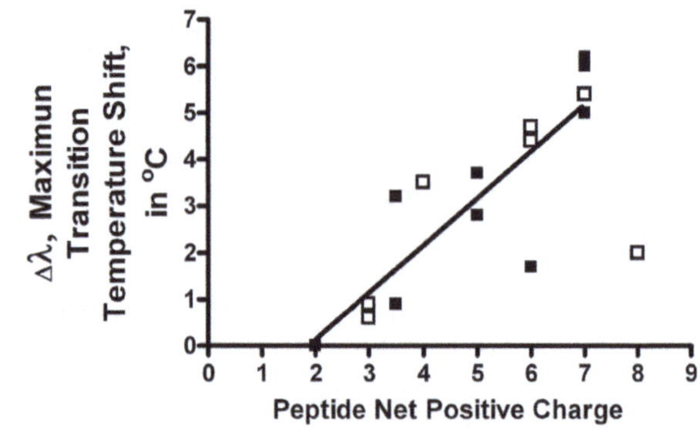

Fig. 5.2 Maximum temperature shift of the main transition obtained by DSC in the presence of peptide, with respect to the lipid mixture in the absence of peptide, $\Delta\lambda$, plotted as a function of the peptide's net positive charge. The short peptides BP100 and HIV-TAT fall out of line. The linear relationship holds for the antimicrobial peptides (*solid squares*) as well as for the cell-penetrating peptides (*open squares*) (Taken from Wadhwani et al. 2012)

5.2 Observations Supporting Model

Several experimental observations support this simple model.

5.2.1 Potency Related to Charge

There are several examples showing a correlation between overall charge and the ability to cluster anionic lipids. I draw attention to a particular study that compared 16 different antimicrobial and cell-penetrating peptides. Anionic lipid clustering was evaluated by differential scanning calorimetry (Wadhwani et al. 2012). As seen in Fig. 5.2, the majority of the peptides followed a simple relationship in which the clustering of anionic lipid was proportional to the number of net positive charges on the peptide.

5.2.2 Bacterial Species Specificity

Our model requires that anionic lipids are clustered away from the remaining lipids in a membrane, i.e., uncharged or zwitterionic lipids. This cannot occur with bacteria that have little uncharged or zwitterionic lipid. Hence certain bacterial species will be more susceptible to anionic lipid clustering. This was first demonstrated in our study of a synthetic oligo-acyl-lysine

(OAK) (Epand et al. 2008b). In this work it was demonstrated that most Gram-negative bacteria had high mole fractions of phosphatidylethanolamine, a zwitterionic lipid, in their membranes. Hence they could cluster anionic lipid in a background of zwitterionic lipid, and therefore they exhibited a high sensitivity (low MIC) to this OAK (Epand et al. 2008b). In contrast, many species of Gram-positive bacteria were largely devoid of phosphatidylethanolamine and required tenfold or higher concentrations of OAK to inhibit cell growth. An exception was bacteria of the genus *Bacillus* that in many cases had high concentrations of phosphatidylethanolamine and exhibited sensitivity to OAKs comparable to that of Gram-negative bacteria (Epand et al. 2008b). The phenomenon of anionic lipid clustering by OAK in lipid mixtures of anionic cardiolipin and zwitterionic phosphatidylethanolamine was demonstrated by differential scanning calorimetry and by ^{31}P-MAS/NMR (Epand et al. 2008b). For another family of synthetic antimicrobial compounds, the ceragenins, it was also found that their toxicity to bacteria was related to the content of phosphatidylethanolamine in the bacteria. In this study, not only were different strains of bacteria utilized, but the loss of susceptibility to this cationic antimicrobial was demonstrated using a mutant form of *E. coli* that could not synthesize phosphatidylethanolamine (Epand et al. 2007). In wild-type strains of *E. coli*, phosphatidylethanolamine comprises 80% of the total lipid. However, phosphatidylethanolamine is not the only bacterial lipid with no overall charge. Some bacteria contain a significant amount of uncharged glycosyl diglycerides. One example is *S. pyogenes*. We have found that *S. pyogenes* has a MIC that is comparable to *Bacilli* that have a high phosphatidylethanolamine mole fraction (Epand et al. 2010).

5.3 Limitation of Model

The proposed charge clustering model is not intended to be a universal model that applies to all antimicrobial agents. Generally antimicrobial agents are toxic to bacteria because of a combination of factors. For a subclass of antimicrobial agents, the clustering of anionic lipids provides an important factor contributing to their potency. Whether a particular antimicrobial agent falls into this subclass can be tested experimentally by determining if the bacterial species that are susceptible to the agent have surface membranes that contain both anionic and either zwitterionic or uncharged lipids, while bacteria that are comprised mainly of anionic lipids are resistant to these agents. Another test can be performed using NMR or DSC to test if the agent induces clustering of anionic lipid, as we have described (Epand and Epand 2009).

It should also be taken into account that there are limitations in the data for the lipid composition of bacterial membranes. The lipid composition will be modified by growth conditions including nutrients, salts and pH, stage of the growth cycle, nature of the acyl chains that are incorporated into lipids, and other factors.

5.3.1 Conformational Flexibility

In addition to charge, conformational flexibility improved antimicrobial potency, possibly allowing the antimicrobial agent to match the distance between positive charge to the distance between negatively charged lipids on the membrane (Epand and Epand 2009). One example of this is a pair of α-/β-peptides having different charge distributions and different conformation flexibilities, but with identical chemical composition. The more flexible peptide had a higher activity against *E. coli* (Schmitt et al. 2004). In addition, a sequence-random cationic polymer that was designed to be unstructured in solution was shown to cluster anionic lipids (Epand et al. 2008a), indicating that a specific conformation of the antimicrobial agent was not required.

5.3.2 Other Factors

5.3.2.1 Hydrophobicity

Since the site of action of antimicrobial agents that cluster anionic lipids is the membrane, the antimicrobial agent has to be sufficiently hydrophobic to partition into a membrane. Partitioning of cationic polymers onto the surface of anionic bacterial membranes is expected. However, the strength of electrostatic binding alone and lack of significant penetration of these peptides into a membrane make such polymers weakly antimicrobial. An example is polycationic amino acids that have antimicrobial activity in the approximate range of 50–250 μg/mL (Venkatesh et al. 2017). Adding acyl groups to a polymer of Lys can increase the membrane partitioning and the antimicrobial potency (Mor 2016).

5.3.2.2 Specific Binding to Component

Of course hydrophobicity is only one factor that will determine the partitioning of a compound to a membrane (Andreev et al. 2018). One additional mechanism is the binding of an antimicrobial molecule to a specific membrane component (Epand et al. 2016; Phoenix et al. 2015; Guder et al. 2000; Schmitt and Rosa 2016). This property is often exhibited by cyclical compounds that have a more restricted conformation.

5.3.2.3 Access to Membrane

The charge cluster model depends on the interaction of the antimicrobial substance with the phospholipids of the cell membrane of the bacteria. For Gram-positive bacteria, the cell membrane is surrounded by a cell wall. This cell wall is generally porous to small molecules, although it does contain some negative charge. There is one example in which the toxicity of an antimicrobial results from it become entrapped in the cell wall and blocking exchange of materials across the cell membrane (Epand et al. 2008a). Generally Gram-negative bacteria are more resistant to antimicrobial agents because they present an additional barrier of the outer membrane to protect the cell membrane from access by antimicrobial agents. A major component of the outer membrane of Gram-negative bacteria is lipopolysaccharide that is anionic and can potentially sequester cationic drugs.

The question of antibiotic resistance is an important separate issue. There are several synthetic agents that have been shown to reverse microbial resistance (Molchanova et al. 2017).

5.3.2.4 Pore Formation

The charge clustering model is one of several models to describe the action of antimicrobial agents. Perhaps the most studied model is that of the membrane pore (Imura et al. 2008) and variations of that model (Shai 2002). There are several observations that support these models which do not assume anything about the rearrangement of membrane components. Hence the lipid composition of the bacteria will not correlate with the potency of the antimicrobial that acts through a pore mechanism. In general, there does not have to be a choice among different proposed mechanisms of antimicrobial action. In most cases the potency of a particular agent will be determined by several properties, some of which will be interrelated and others not. An attractive feature of a pore model is that it directly provides a mechanism for toxicity to bacteria through depolarization of the membrane electrical potential as well as loss of other trans-membrane gradients of ions and small molecules.

5.3.2.5 Phospholipid Acyl Chains

The acyl chains of phospholipids of bacteria differ from those of mammalian systems. In particular there is very little unsaturation in the acyl chains of bacterial lipids and no polyunsaturated chains. In addition, bacterial lipid acyl chains contain single double bonds in the trans configuration, which is virtually nonexistent in higher organisms. Bacterial lipids are also unique in having lipid acyl chains with cyclopropyl groups as well as terminal methyl branching. In methicillin-bacterial lipids are also unique in containing lipid acyl chains with cyclopropyl-containing as well as terminally methyl-branched acyl chains. In methicillin-resistant strains of *E. coli*, the fraction of anteiso-branched acyl chains is elevated relative to susceptible strains (Mitchell et al. 2016). Also, the action of the cytolytic peptide, δ-lysin, strongly depends on the structure of acyl chains in microbial lipids (Pokorny et al. 2008). Model liposome studies have shown that anteiso-branched acyl chains in phospholipids make liposomes more susceptible to lysis by antimicrobial peptides and these lipids have lower gel-to-liquid crystalline phase transition temperatures (Mitchell et al. 2016).

Not only is membrane permeability an important factor for bacterial toxicity, but it is also relevant for the access of antimicrobial agents that have intracellular targets that require that they pass through the cell membrane to access these targets (Chopra 1988). An example of an antimicrobial that has an intracellular ribosomal target is apidaecin 1b (Schmidt et al. 2018). Bacterial lipidomic analysis showed that strains of *E. coli* resistant to apidaecin 1b had lower levels of acyl chains with cyclopropyl groups. Furthermore, knocking out the enzyme responsible for the biosynthesis of the cyclopropyl group also resulted in greater resistance to apidaecin.

The charge clustering model considers only the interaction of the antimicrobial agent with the headgroups of the phospholipids. Clearly there are also other factors affecting sensitivity to antimicrobial agents, and one of these is the nature of the acyl chains, particularly the structures of the acyl chains that are unique to bacteria. One such structure is the trans carbon-carbon double bond. To our knowledge the role of this chemical grouping to bacterial resistance has not been evaluated.

5.3.2.6 Non-membrane Targeting

There is evidence that the bacterial membrane is the site of action of many antimicrobial agents. However, there are also antimicrobials that have intracellular targets. The membrane still plays a role as a barrier of the drug to the cell interior, and the lipid cluster mechanism can play a role in this access. However, the potency of such an antimicrobial will also depend on the affinity and specificity it has for its intracellular target. This will likely have a larger effect on the potency of the antimicrobial and be completely independent of actions at the membrane.

5.4 Molecular Mechanism

We described above the charge cluster model and how certain antimicrobial agents could rearrange bacterial membrane components so as to cluster anionic lipids. However, we did not discuss how this clustering of anionic lipid could result in antibacterial action. This is generally the situation for most antimicrobial agents that are not specific for a particular mechanism and often have contributing factors from several sources. We will consider some of the likely mechanisms by which anionic lipid clustering can be detrimental to bacteria.

5.4.1 Phase Boundary Defect

A possible mechanism by which antimicrobial compounds that function by charge clustering can affect bacterial viability is by forming phase boundary defects between the clustered anionic lipids and the remainder of the membrane (Jean-Francois et al. 2008). However, it is not clear if the formation of these new lipid domains would greatly increase the membrane defects on the bacterial cell surface. There is evidence that bacterial membranes naturally possess domains (Nickels et al. 2017) and hence have phase boundary defects before the addition of any antimicrobial compound. The role of phase boundary defects, in comparison with other membrane defects, as a cause of membrane leakage or instability is difficult to assess quantitatively.

5.4.2 Depletion of Anionic Lipids from Regions of the Membrane

If an antimicrobial agent clusters anionic lipids into a smaller zone in the membrane, of necessity, it will deplete other regions of the membrane from this lipid class. Lipids can modulate the activity of membrane-bound proteins either by modulating the properties of the environment around the protein in a solvent-like effect or by binding to specific sites on the protein (Salas-Estrada et al. 2017; Lee 2003; Yeagle et al. 1988). Both of these mechanisms for changing the behavior of a membrane protein will be affected by the clustering of anionic lipid. This will not happen with many Gram-positive bacteria whose membrane surface is composed largely of anionic lipid; however, with bacteria having in addition to anionic lipid a high fraction of zwitterionic or uncharged lipid, there will be an area depleted of anionic lipids. Anionic lipids often play important roles in membrane function and/or in binding to membrane proteins. These roles cannot be efficiently fulfilled if the anionic lipids are clustered together with an antimicrobial agent. This mechanism appears likely to contribute to the mechanism of action of these anionic lipid clustering agents, but to prove it as a mechanism requires knowledge of the specific roles of anionic lipids in the membrane and how these functions are lost as a consequence of lipid clustering.

5.4.3 Phase Separation Resulting in Redistribution of Membrane Proteins

Recent studies of the mechanism of action of a synthetic cyclical hexapeptide, cWFW, having the structure cyclo(RRRWFW), has bactericidal activity as a result of membrane domain formation (Scheinpflug et al. 2017). Although this peptide has been shown to cluster anionic lipids in model membranes (Arouri et al. 2009; Finger et al. 2015), the mechanism of cell death in vivo is caused by redistribution of proteins in the bacterial membrane caused by domain formation (Scheinpflug et al. 2017). It was found that cWFW does not depolarize the membrane or affect the energy state of the cell, but it does cause a large-scale phase separation of the membrane. It was suggested that the cause of this phase separation was a result of changes in membrane fluidity, rather than clustering of anionic lipids. However, domains in biological membranes are often smaller than they are in model membranes and may therefore be more difficult to detect. However, it was shown that in bacteria the antimicrobial activity was not a consequence of permeation of the membrane but rather a triggering of a stress response resulting in an inhibition of lipid II and cell wall synthesis as well as through increased autolysis triggered by cWFW (Scheinpflug et al. 2017). Future studies will determine how general this mechanism is applicable to other antimicrobial peptides.

5.5 Summary

The charge cluster model is one of a number of mechanisms that can provide antimicrobial activity to a substance. One criterion to test the importance of this factor is to determine the dependence of antimicrobial potency on the nature of the lipid composition of bacterial membranes. Often the only criterion taken into account in evaluating the potency of antimicrobial agents is to compare Gram-positive with Gram-negative bacteria. However, as illustrated in this review and references therein, bacteria vary widely in terms of the nature of their lipid composition. One important example being phosphatidylethanolamine that is the most abundant lipid in the membranes of some bacteria but is present in only trace amounts in other bacteria. Phosphatidylethanolamine and other zwitterionic or uncharged lipids are not clustered by polycationic antimicrobial agents, but they are required to facilitate the clustering of anionic lipids. Thus, discovering agents that function primarily by the charge cluster mechanism will allow the design of antimicrobials that are most suitable for a particular bacterial target.

References

Andreev K, Martynowycz MW, Huang ML, Kuzmenko I, Bu W, Kirshenbaum K, Gidalevitz D (2018) Hydrophobic interactions modulate antimicrobial peptoid selectivity towards anionic lipid membranes. Biochim Biophys Acta Biomembr 1860:1414–1423

Arouri A, Dathe M, Blume A (2009) Peptide induced demixing in PG/PE lipid mixtures: a mechanism for the specificity of antimicrobial peptides towards bacterial membranes? Biochim Biophys Acta 1788:650–659

Chopra I (1988) Molecular mechanisms involved in the transport of antibiotics into bacteria. Parasitology 96:S25–S44

Epand RM, Epand RF (2009) Lipid domains in bacterial membranes and the action of antimicrobial agents. Biochim Biophys Acta Biomembr 1788:289–294

Epand RF, Savage PB, Epand RM (2007) Bacterial lipid composition and the antimicrobial efficacy of cationic steroid compounds (Ceragenins). Biochim Biophys Acta 1768:2500–2509

Epand RF, Mowery BP, Lee SE, Stahl SS, Lehrer RI, Gellman SH, Epand RM (2008a) Dual mechanism of bacterial lethality for a cationic sequence-random copolymer that mimics host-defense antimicrobial peptides. J Mol Biol 379:38–50

Epand RM, Rotem S, Mor A, Berno B, Epand RF (2008b) Bacterial membranes as predictors of antimicrobial potency. J Am Chem Soc 130:14346–14352

Epand RM, Epand RF, Arnusch CJ, Papahadjopoulos-Sternberg B, Wang G, Shai Y (2010) Lipid clustering by three homologous arginine-rich antimicrobial peptides is insensitive to amino acid arrangement and induced secondary structure. Biochim Biophys Acta 1798:1272–1280

Epand RF, Mor A, Epand RM (2011) Lipid complexes with cationic peptides and OAKs; their role in antimicrobial action and in the delivery of antimicrobial agents. Cell Mol Life Sci 68:2177–2188

Epand RM, Walker C, Epand RF, Magarvey NA (2016) Molecular mechanisms of membrane targeting antibiotics. Biochim Biophys Acta Biomembr 1858:980–987

Finger S, Kerth A, Dathe M, Blume A (2015) The efficacy of trivalent cyclic hexapeptides to induce lipid clustering in PG/PE membranes correlates with their antimicrobial activity. Biochim Biophys Acta Biomembr 1848:2998–3006

Guder A, Wiedemann I, Sahl HG (2000) Posttranslationally modified bacteriocins – the lantibiotics. Biopolym Pept Sci Sect 55:62–73

Imura Y, Choda N, Matsuzaki K (2008) Magainin 2 in action: distinct modes of membrane permeabilization in living bacterial and mammalian cells. Biophys J 95:5757–5765

Jean-Francois F, Castano S, Desbat B, Odaert B, Roux M, Metz-Boutigue MH, Dufourc EJ (2008) Aggregation of cateslytin β-sheets on negatively charged lipids promotes rigid membrane domains. A new mode of action for antimicrobial peptides? Biochemistry 47:6394–6402

Lee AG (2003) Lipid-protein interactions in biological membranes: a structural perspective. Biochim Biophys Acta Biomembr 1612:1–40

Mitchell NJ, Seaton P, Pokorny A (2016) Branched phospholipids render lipid vesicles more susceptible to membrane-active peptides. Biochim Biophys Acta Biomembr 1858:988–994

Molchanova N, Hansen PR, Franzyk H (2017) Advances in development of antimicrobial peptidomimetics as potential drugs. Molecules 22(9):E1430

Mor A (2016) Engineered OAKs against antibiotic resistance and for bacterial detection. In: Epand RM (ed) Host defense peptides and their potential as therapeutic agents. Springer, Switzerland, pp 205–226

Nickels JD, Chatterjee S, Stanley CB, Qian S, Cheng X, Myles DAA, Standaert RF, Elkins JG, Katsaras J (2017) The in vivo structure of biological membranes and evidence for lipid domains. PLoS Biol 15:e2002214

Phoenix DA, Harris F, Mura M, Dennison SR (2015) The increasing role of phosphatidylethanolamine as a lipid receptor in the action of host defence peptides. Prog Lipid Res 59:26–37

Pokorny A, Kilelee EM, Wu D, Almeida PFF (2008) The activity of the amphipathic peptide δ-lysin correlates with phospholipid acyl chain structure and bilayer elastic properties. Biophys J 95:4748–4755

Salas-Estrada LA, Leioatts N, Romo TD, Grossfield A (2017) Lipids alter rhodopsin function via ligand-like and solvent-like interactions. Biophys J 114:355–367

Scheinpflug K, Wenzel M, Krylova O, Bandow JE, Dathe M, Strahl H (2017) Antimicrobial peptide cWFW kills by combining lipid phase separation with autolysis. Sci Rep 7:44332

Schmidt R, Yonghong D, Hoffmann R (2018) Phospholipid composition of the outer membrane of *Escherichia coli* influences its susceptibility against antimicrobial peptide apidaecin 1b. Diagn Microbiol Infect Dis 90(4):316–323

Schmitt MA, Weisblum B, Gellman SH (2004) Unexpected relationships between structure and function in α,β-peptides: antimicrobial foldamers with heterogeneous backbones. J Am Chem Soc 126:6848–6849

Schmitt P, Rosa RD, Destoumieux-Garzón D (2016) An intimate link between antimicrobial peptide sequence diversity and binding to essential components of bacterial membranes. Biochim Biophys Acta Biomembr 1858:958–970

Shai Y (2002) Mode of action of membrane active antimicrobial peptides. Biopolym Pept Sci Sect 66:236–248

Venkatesh M, Barathi VA, Goh ETL, Anggara R, Fazil MHUT, Ng AJY, Harini S, Aung TT, Fox SJ, Liu S, Yang L, Barkham TMS, Loh XJ, Verma NK, Beuerman RW, Lakshminarayanan R (2017) Antimicrobial activity and cell selectivity of synthetic and biosynthetic cationic polymers. Antimicrob Agents Chemother 61:e00469–e00417

Wadhwani P, Epand RF, Heidenreich N, Burck J, Ulrich AS, Epand RM (2012) Membrane-active peptides and the clustering of anionic lipids. Biophys J 103:265–274

Yeagle PL, Young J, Rice D (1988) Effects of cholesterol on sodium-potassium ATPase ATP hydrolyzing activity in bovine kidney. Biochemistry 27:6449–6452

Intracellular Antimicrobial Peptides Targeting the Protein Synthesis Machinery

6

Michael Graf and Daniel N. Wilson

Abstract

While antimicrobial peptides (AMPs) are well-known for their disruptive effects on bacterial membranes, the mechanism of many intracellular AMPs is still being elucidated. In the recent years, it has been demonstrated that the subclass of proline-rich AMPs (PrAMPs) can pass through the bacterial membrane and kill bacteria by inhibiting protein synthesis. PrAMPs are a product of the innate immune system and are secreted in response to bacterial infection. So far PrAMPs have been identified in many arthropods, such as beetles, wasps, and flies, as well as some mammals, such as sheep, cows, and goats. PrAMPs show high potency against Gram-negative bacteria, while exhibiting low toxicity in eukaryotes, suggesting that they may represent a promising avenue for the development of future antimicrobial agents to combat the increase of multidrug-resistant bacterial pathogens. Structural and biochemical data have revealed the PrAMP binding sites on the ribosome as well as insight into their mechanisms of action. While the binding site of all so far investigated PrAMPs is situated within nascent polypeptide exit tunnel, the mechanism of action is distinct between class I and II PrAMPs. Specifically, class I PrAMPs, such as Bac7, Onc112, pyrrhocoricin, and metalnikowin, block the delivery of aa-tRNA by EF-Tu to the ribosomal A-site, whereas the class II PrAMPs, such as apidaecin 1b and Api137, act during translation termination and inhibit protein synthesis by trapping of release factors on the 70S ribosome following hydrolysis of the nascent polypeptide chain.

Keywords

Proline-rich antimicrobial peptides · Protein synthesis · Translation · Ribosome

M. Graf · D. N. Wilson (✉)
Institute for Biochemistry and Molecular Biology, University of Hamburg, Hamburg, Germany
e-mail: Daniel.Wilson@chemie.uni-hamburg.de

6.1 Diversity of Antimicrobial Peptide Targets

While AMPs are well-known for their action to kill bacteria via membrane disruption, there is an increasing number of AMPs that have been characterized to pass through the bacterial cell membrane and target intracellular processes (Brogden 2005). Intracellular targets encompass many fundamental processes for the bacterial propagation, such as DNA and RNA replication, mRNA transcription, and protein synthesis (Brogden 2005; Graf et al. 2017). The intrinsic diversity of peptides as well as their ease of synthesis, in combination with the urgent need for novel compounds to fight bacterial infections, makes AMPs

K. Matsuzaki (ed.), *Antimicrobial Peptides*, Advances in Experimental Medicine and Biology 1117,
https://doi.org/10.1007/978-981-13-3588-4_6

a valuable and promising source for future antibiotics to be employed within the clinical setting.

6.1.1 Sources of Proline-Rich Antimicrobial Peptides

The proline-rich AMPs (PrAMPs) are a subclass of cationic peptides, which as their name suggests are characterized by a high content of proline, as well as arginine residues (Graf et al. 2017). The first PrAMP was identified in the late 1980s by HPLC, namely, apidaecin, from the honey bee *Apis mellifera* (Casteels et al. 1989), which was followed by the identification of many other PrAMPs from higher eukaryotes, including some mammals and many arthropods (Fig. 6.1) (Agerberth et al. 1991; Bulet et al. 1993; Casteels et al. 1990; Chernysh et al. 1996; Cociancich et al. 1994; Gennaro et al. 1989; Huttner et al. 1998; Knappe et al. 2010; Mardirossian et al. 2018; Schnapp et al. 1996; Schneider and Dorn 2001; Shamova et al. 1999; Stensvag et al. 2008). Among mammals, PrAMPs have been identified in cows (Gennaro et al. 1989), pigs (Agerberth et al. 1991), sheep (Huttner et al. 1998; Shamova et al. 1999), goats (Shamova et al. 1999), and more recently dolphins (Mardirossian et al. 2018) (Fig. 6.1) but so far not in primates or humans. Arthropod PrAMPs have been found in crabs (Stensvag et al. 2008) and numerous insects, including wasps, bees, flies, and beetles (Fig. 6.1) (Bulet et al. 1993; Casteels et al. 1989, 1990; Chernysh et al. 1996; Cociancich et al. 1994; Knappe et al. 2010; Schneider and Dorn 2001). Analogously to apidaecin (Casteels et al. 1989), the nomenclature of PrAMPs predominantly corresponds to the origin of identification. Accordingly, PrAMPs from *Tursiops truncatus* (bottlenose dolphin) are termed Tur1A and Tur1B (Mardirossian et al. 2018), peptides from *Bos taurus* (cattle) are called bactenecins (Bac) (Gennaro et al. 1989), and *Oncopeltus fasciatus*

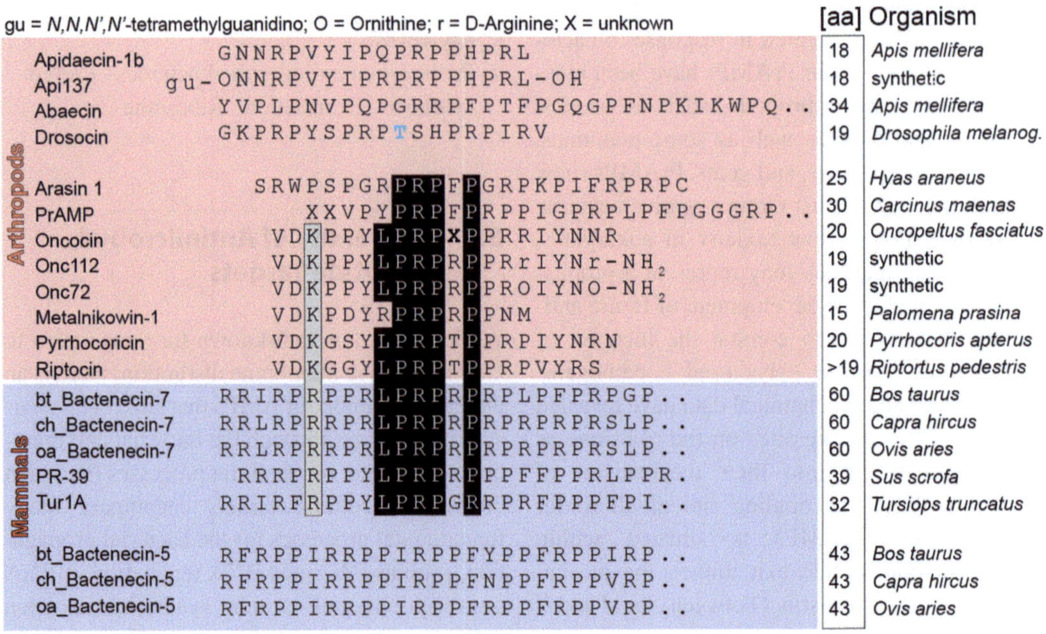

Fig. 6.1 Sequences of natural and synthetic PrAMPs derived from arthropods (insects and crustaceans, salmon) and mammals (purple). The central PrAMPs were aligned based on ribosome-bound structures of Onc112, Pyr, Met, Tur1A, and Bac7 and then on sequence similarity. Similar and identical residues are shown in gray and black, respectively. The O-glycosylation of drosocin is indicated in blue (Thr11). Position 11 of oncocin is unknown and indicated with a bold letter "X." The number of amino acids (aa) representing the mature peptide and the origin (organism) is stated on the right (Figure adapted from Graf et al. 2017)

(milkweed bug)-derived PrAMPs are termed oncocins (Knappe et al. 2010; Schneider and Dorn 2001). Noteworthy, peptides with high sequence similarity are commonly named after the source in which the PrAMP was first discovered. Thus, PrAMPs from *Capra hircus* (goat) and *Ovis aries* (sheep) (Huttner et al. 1998; Shamova et al. 1999), which show similarity to *Bos taurus* peptides, are also referred to as bactenecins.

6.1.2 Synthesis of PrAMPs

PrAMPs are primarily produced upon bacterial infection by cells of the innate immune system, called phagocytes (Zasloff 2002). Phagocyte progenitors, immature myeloid cells in mammals, synthesize PrAMPs as inactive precursors and store these in granules (Fig. 6.2) (Graf et al. 2017; Zanetti et al. 1990, 1991). Storage of PrAMPs as inactive precursors in granules serves different functions. First, it potentially protects the ribosomes of the producing cell from harm by the peptide, and second, it allows for a fast pathogen response by avoiding time-consuming protein synthesis. Activation of the inactive precursors occurs at the target site by proteolytic cleavage (Zanetti et al. 1990, 1991). The respective proteases are stored separately from the inactive precursors in a different set of granules (Fig. 6.2). In mammals, PrAMP-containing granules are called large granules, and the protease-harboring granules are termed azurophil granules (Fig. 6.2). Activation of precursors is induced by simultaneous exocytosis into extracellular space or bacteria-containing phagosomes (Fig. 6.2). However, the overall structure of PrAMP precursors, as well as the resulting path of activation, differs between organisms and PrAMPs (Graf et al. 2017). Although not all mechanisms of peptide activation have been clarified, a single-peptide activation mechanism can be distinguished from a multi-peptide activation mechanism (Graf et al. 2017). All mammalian PrAMPs seem to be activated by the single-peptide mechanism, whereas insect PrAMPs are activated through a multi-peptide activation

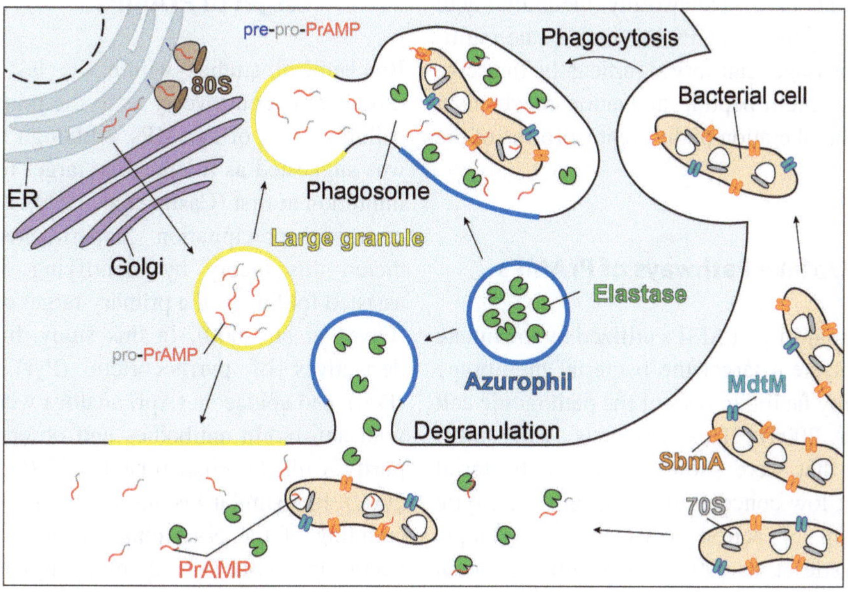

Fig. 6.2 Mammalian PrAMPs are synthesized as pre-/pro-sequences and targeted to large granules. The PrAMPs are activated upon bacterial infection by fusion of pro-PrAMP containing large granules with the elastase-containing azurophil granules and either the plasma membrane or the phagosome. Elastase activates the mature PrAMP by removal of the pro-sequence. The activated PrAMP is transported via SbmA (to less extent by MdtM) into the bacterial cell (Figure adapted from Graf et al. 2017)

mechanism, as described for apidaecin (from *Apis mellifera*) (Casteels-Josson et al. 1993) and riptocin (from *Riptortus pedestris*) (Graf et al. 2017). The single-peptide mechanism involves an mRNA coding for a polypeptide sequence harboring a pre-sequence (~20 to 30 aa), a pro-sequence (~20 to 100 aa), and a Pro-Arg-rich sequence (Bulet et al. 1993; Graf et al. 2017; Storici and Zanetti 1993; Zanetti et al. 1993). The pro-sequence is placed either before (e.g., bactenecin and PR-39) (Storici and Zanetti 1993; Zanetti et al. 1993) or after the PrAMP (e.g., drosocin) (Bulet et al. 1993). The multi-peptide mechanism comprises an mRNA coding for multiple PrAMPs that are interrupted by an inactivating oligopeptide linker, such as "RR-EAEPEAEP" from *Apis mellifera* (Casteels et al. 1989; Casteels-Josson et al. 1993). Proteolytic cleavage sites at the N- and C-terminus of the linker liberate the peptide (Casteels et al. 1989; Casteels-Josson et al. 1993). Similar to single PrAMP-encoding mRNAs, the multi-coding messages harbor a pre- (~15 to 20 aa) and pro-sequence (~13 to 16 aa) at the very beginning of the polypeptide (Casteels et al. 1989; Casteels-Josson et al. 1993). Noteworthy, the amino acid sequences of the PrAMPs encoded by the multi-peptide messages can vary significantly. In accordance, the multi-peptide activation mechanism leads to the liberation of different isoforms of the PrAMP.

6.1.3 Uptake Pathways of PrAMPs

The vast majority of AMPs utilized by the innate immune system target the bacterial membrane and thereby facilitate lysis of the pathogenic cell (Brogden 2005). This contrasts with most PrAMPs that were shown to inhibit bacterial growth at low concentrations using a non-lytic mechanism of action (Casteels and Tempst 1994; Castle et al. 1999; Otvos 2002; Scocchi et al. 2011). The utilization of a non-lytic mechanism initially raised the question as to which uptake pathways are used by PrAMPs. Subsequent mutagenesis experiments led to the identification of two key players responsible for PrAMP uptake, namely, SbmA (Mattiuzzo et al. 2007) and more recently MdtM (Fig. 6.2) (Krizsan et al. 2015). While SbmA seems to be the major transporter for PrAMP uptake, the inner membrane protein MdtM appears to also play an auxiliary role in antimicrobial peptide internalization (Krizsan et al. 2015), especially at higher peptide concentrations. SbmA is a 46.5 kDa protein localized in the inner membrane of Gram-negative bacteria (Mattiuzzo et al. 2007) that internalizes PrAMPs, as well as other antimicrobial agents, using an electrochemical proton gradient (Runti et al. 2013). Besides peptide uptake, a general function for SbmA remains unclear. Among Gram-negative bacteria, homologs of SbmA can be found in *Enterobacteriaceae*, such as *Escherichia* and *Salmonella* species, and *Pseudomonadales*, such as *Acinetobacter baumannii*, as well as in α- and ε-proteobacteria, such as *Neisseria meningitis* and *Campylobacter* species, respectively (Graf et al. 2017).

6.1.4 Discovery of the Intracellular Target of PrAMPs

Biochemical studies within the last 50 years raised some controversy as to the primary intracellular target of PrAMPs. Although translation was suggested as the putative target for PrAMP inhibition at first (Castle et al. 1999), subsequent co-immunoprecipitation experiments contradicted this theory by identifying chaperone-assisted folding as the primary target of PrAMPs (Otvos et al. 2000). In this study, biotinylated derivatives of pyrrhocoricin (Pyr), drosocin (Dro), and apidaecin (Api) analogs were purified with anti-biotin antibodies and observed to co-purify with the chaperone DnaK (Otvos et al. 2000). In a similar fashion to Pyr, Dro, and Api, coupling of the N-terminal 35 amino acids of bactnecin-7 (Bac7) to 2-chlorotrityl resins also resulted in co-purification of DnaK (Scocchi et al. 2009). The direct visualization of PrAMP binding to DnaK was achieved using X-ray crystallography, where PrAMPs were observed to bind to the substrate cleft of DnaK (Knappe et al.

2011; Zahn et al. 2013, 2014). However, soon after structures of PrAMPs in complex with DnaK had been obtained, questions were raised about the validity of the working model since minimal inhibitory concentration (MIC) experiments with *E. coli dnaK* deletion strains yielded susceptibilities comparable to the wild-type *E. coli* strains (Krizsan et al. 2014; Scocchi et al. 2009). Therefore, co-immunoprecipitation experiments with labeled PrAMPs and *E. coli* lysate were again undertaken to search for alternative targets. This led to the identification of ribosomal proteins, suggesting that ribosomes and translation may represent an alternative target of PrAMPs (Krizsan et al. 2014; Mardirossian et al. 2014). Subsequent binding assays, as well as in vitro translation experiments, indicated that protein synthesis is most likely the primary process inhibited by PrAMPs (Krizsan et al. 2014; Mardirossian et al. 2014).

6.1.5 Determination of the Ribosomal Binding Site of PrAMPs

As consequence of the newly discovered ribosome binding and inhibition capacity of PrAMPs, it became interesting to resolve structures of 70S ribosomes in complex with different peptides and to determine the mode of action for each peptide member, for example, to determine whether all PrAMPs bind in the same position on the ribosome and whether they exhibit an identical mechanism of action. The first PrAMP to be resolved on the bacterial ribosome was a derivative of oncocin (Onc) from milkweed bugs (*Oncopeltus fasciatus*), termed Onc112 (Roy et al. 2015; Seefeldt et al. 2015). This was followed by Bac7 from cows (*Bos taurus*), metalnikowin (Met) from the green shield bug (*Palomena prasina*), and Pyr from the firebug (*Pyrrhocoris apterus*) (Gagnon et al. 2016; Seefeldt et al. 2016). The latest resolved structures of PrAMPs are Api137 from the honey bee (*Apis mellifera*) (Florin et al. 2017) and Tur1A from bottlenose dolphins (*Tursiops truncatus*) (Mardirossian et al. 2018). All PrAMPs were found to bind to the polypeptide exit tunnel in an extended conformation, which overlaps with the path of a nascent chain (Florin et al. 2017; Gagnon et al. 2016; Mardirossian et al. 2018; Roy et al. 2015; Seefeldt et al. 2015, 2016). The extended conformation probably results from the high content of interspaced proline residues that prevent helix formation (Graf et al. 2017). As shown by structural studies, binding of PrAMPs involves a series of polar contacts, as well as stacking interactions, between the peptide and the ribosome (Graf et al. 2017). Interactions of Bac7 and Tur1A with the ribosome predominantly involve arginine side chains (Gagnon et al. 2016; Mardirossian et al. 2018; Seefeldt et al. 2016). These structures revealed that two distinct binding modes and mechanisms of action exist for PrAMPs: the type I PrAMPs, including Bac7, Onc, Pyr, Met, and Tur1A, all act as inhibitors of the first elongation step, whereas type II PrAMPs, such as Apidaecin 1b and Api137, act predominantly as translation termination inhibitors.

6.1.6 Interactions of PrAMPs with the Nascent Polypeptide Exit Tunnel

Although class I and class II PrAMPs all bind within the polypeptide tunnel of the ribosome and utilize similar modes of interaction between aromatic and arginine side chains with rRNA nucleotides, the overall orientation of the peptides within the tunnel is different as well as the extent to which they encroach into the ribosomal A-site at the peptidyltransferase center (PTC).

6.1.6.1 Binding of Class I PrAMPs

Class I PrAMPs bind to the ribosome with an inverted orientation compared to a nascent polypeptide, i.e., with the N-terminus located in the A-site and the C-terminus extending down the polypeptide exit tunnel (Fig. 6.3a). The PrAMP binding site on the ribosome can be divided into sections located within the A-site binding pocket, the A-site crevice, and the upper region of the polypeptide exit tunnel. The N-terminal residues reaching into the ribosomal A-site determine the

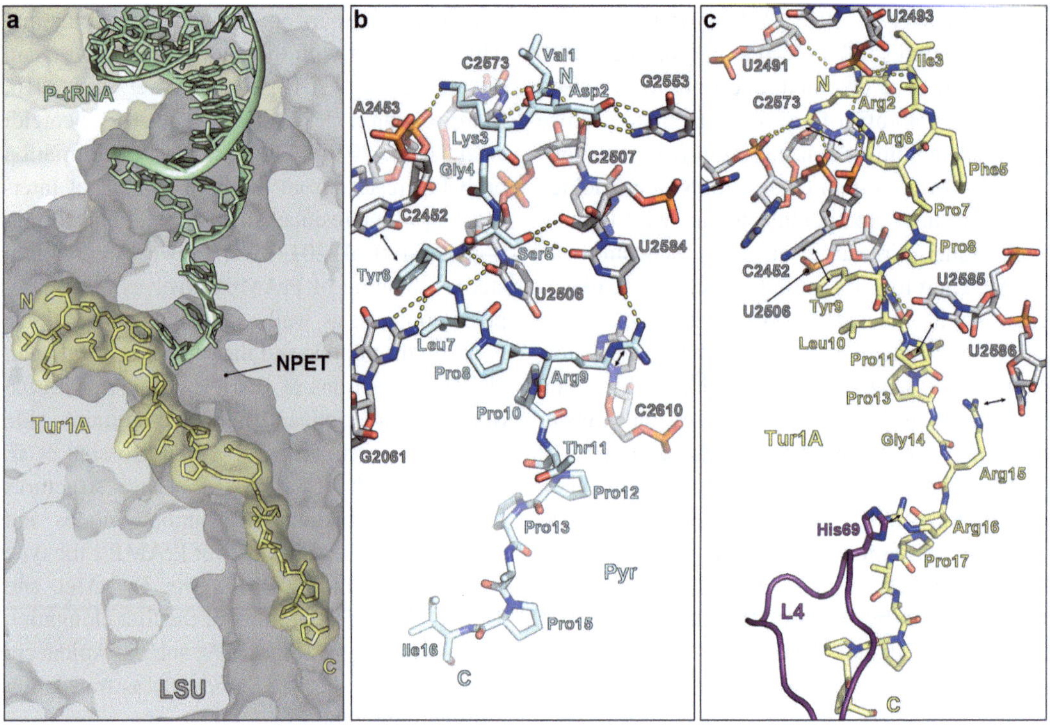

Fig. 6.3 (**a**) Overview showing the binding site of Tur1A within the LSU (gray) polypeptide exit tunnel (NPET) with P-tRNA (green) shown for reference (Mardirossian et al. 2018). The N- and C-termini of Tur1A are indicated by the letters N and C, respectively. (**b**, **c**) Established interactions of (**b**) Pyr (Gagnon et al. 2016; Seefeldt et al. 2016) and (**c**) Tur1A (Mardirossian et al. 2018) within the NPET. Hydrogen bonds are indicated by yellow dashed lines, whereas stacking interactions are indicated by black two-headed arrows. (**b**) Pyr (cyan) establishes interactions solely with 23S rRNA residues (gray) (Gagnon et al. 2016; Seefeldt et al. 2016). (**c**) Tur1A (yellow) establishes interactions with 23S rRNA residues (gray), as well as protein L4 (purple) (Mardirossian et al. 2018)

mechanism of inhibition (Fig. 6.3a). The C-terminal residues reaching into the polypeptide exit tunnel seem to be less crucial for binding and the inhibitory activity of PrAMPs. Noteworthy, in none of the available structures were the residues at the very C-terminus resolved (Gagnon et al. 2016; Mardirossian et al. 2018; Roy et al. 2015; Seefeldt et al. 2015, 2016). In biochemical studies investigating Bac7 fragments, deletion of up to 19 amino acids at the C-terminus did not abolish PrAMP activity (Benincasa et al. 2004). In contrast, deletion of the amino acids valine and aspartate from the N-terminus of oncocins, which are positioned in the A-site binding pocket, leads to strongly reduced activity and underlines the crucial function of the N-terminus for antimicrobial activity (Gagnon et al. 2016). Binding of PrAMPs to the ribosome in general is facilitated through a multitude of polar contacts and stacking interactions with residues of the polypeptide exit tunnel. The most striking similarity between all class I PrAMPs is a conserved core, which harbors a PRP motif. In all resolved class I PrAMP 70S complexes so far, the PRP motif of each PrAMP was located in exactly the same position with exactly the same conformation (Fig. 6.5g) (Gagnon et al. 2016; Mardirossian et al. 2018; Roy et al. 2015; Seefeldt et al. 2015, 2016). However, the amino acid composition, the number of residues, and the number of established contacts within the A-site binding pocket are the most obvious differences among class I PrAMPs (Gagnon et al. 2016; Mardirossian et al. 2018; Roy et al. 2015; Seefeldt et al. 2015, 2016). Consistently, while just four N-terminal amino acids of insect Onc112, Met, and Pyr reach into the A-site binding pocket (Fig. 6.3b) (Gagnon et al. 2016; Roy et al. 2015; Seefeldt et al. 2015, 2016), seven additional amino acids are present in mammalian Bac7 and Tur1A (Fig. 6.3c) (Gagnon et al. 2016; Mardirossian et al. 2018; Seefeldt et al. 2016). The additional residues present in Bac7 and Tur1A form a short loop, which acts like an A-site anchor and seems to be specific for mammalian PrAMPs (Fig. 6.3c) (Gagnon et al. 2016; Mardirossian et al. 2018; Seefeldt et al. 2016).

Binding of Insect PrAMPs Onc112, Metalnikowin, and Pyrrhocoricin The amino acid sequence of the shorter N-terminus of insect PrAMPs is conserved between Onc112, Pyr, and Met (Fig. 6.1) (Graf et al. 2017). Interactions within the A-site binding pocket and A-site crevice involve polar contacts as well as stacking interactions (Fig. 6.3b) (Gagnon et al. 2016; Roy et al. 2015; Seefeldt et al. 2015, 2016). Polar contacts are established by the peptide backbone and side chains of the amino acid residues in the second and third positions (Fig. 6.3b) (Seefeldt et al. 2015, 2016). Consistently, Asp2 of insect PrAMPs contacts the 2′-OH group of C2507 and the base of G2553 (Fig. 6.3b) (Seefeldt et al. 2015, 2016). The side chain of Lys3 establishes a hydrogen bond with the phosphate-oxygen of the 23S rRNA nucleotide A2453 (Fig. 6.3b) (Seefeldt et al. 2015, 2016). Examples for backbone interactions are provided by Val1 (Fig. 6.3b). While the α-carbonyl oxygen of Val1 interacts with the nucleobase of C2573, the α-amine of Val1 contacts the ribose of C2507 (Seefeldt et al. 2015, 2016).

Residues five to seven of insect PrAMPs are located within the A-site crevice (Seefeldt et al. 2015, 2016). Among those residues, Tyr6 is the most conserved and therefore present in all insect PrAMPs (Graf et al. 2017; Seefeldt et al. 2015, 2016). Tyr6 contributes to A-site crevice binding, as well as antimicrobial activity, by establishing a stacking interaction with C2452 (Fig. 6.3b) (Seefeldt et al. 2015, 2016). Replacement of Tyr6 by Ala results in a 32-fold decreased activity (Knappe et al. 2011). This contrasts with positions 5 and 7, which seem to be less critical for binding. While Onc112 harbors a proline residue in position 5, Met and Pyr have in the same position an aspartate and a serine residue, respectively (Fig. 6.1) (Seefeldt et al. 2015, 2016). Furthermore, the leucine residue in position 7 of Onc112 and Pyr is an arginine residue in Met. The alteration of residues 5 and 7 results in different hydrogen bonding patterns within the A-site crevice. Asp5 in Met establishes hydrogen bonds with U2584 and U2585. Arg7 of Met contacts A2503 and the backbone phosphate of G2505. Ser5, which is present in Pyr, interacts with U2584 (Fig. 6.3b). Leu7 does not establish interactions, either in Onc112 or in Pyr. Hydrogen bonds in the A-site crevice, which are common for all insect PrAMPs, involve the peptide backbone. Accordingly, the backbone α-carbonyl oxygen, the α-amine of Tyr6, and the α-amine of Leu7/Arg7 contact the U2506 nucleobase (Fig. 6.3b).

As mentioned before, the number of resolved residues at the C-terminus differs between PrAMPs. With respect to insect PrAMPs, 6 aa of Onc112, 5 aa of Met, and 4 aa of Pyr were not resolved in the reported structures (Seefeldt et al. 2015, 2016). However, the visualized residues at the C-terminus of insect PrAMPs were observed to establish just a single hydrogen bond and a single stacking pair within the upper polypeptide exit tunnel (Fig. 6.3b). Both interactions involve Arg9 of the conserved PRP motif (Fig. 6.3b). The hydrogen bond is established with 23S nucleotide U2584. The stacking interaction is observed with C2610 (Fig. 6.3b). The presence of only two interactions underlines the minor importance of the C-terminus of insect PrAMPs in binding.

Binding of Mammalian PrAMPs Bac7 and Tur1A Binding of mammalian Bac7 and Tur1A to the A-site binding pocket and the A-site crevice of eubacterial 70S ribosomes also involves polar contacts of the peptide backbone and the amino acid side chains (Fig. 6.3c) (Mardirossian et al. 2018; Seefeldt et al. 2016). Four of eight positions in Bac7 and Tur1A harbor arginine residues in the LSU A-site binding pocket which establish an extensive hydrogen bonding network as well as a single stacking interaction.

The stacking interaction is observed between Arg2 and C2573 (Fig. 6.3c). Compared to insect PrAMPs, this stacking interaction replaces the backbone contact of Val1 with the nucleobase of C2573 (Fig. 6.3c) (Mardirossian et al. 2018; Seefeldt et al. 2015, 2016). Hydrogen bonds are evident for all arginine residues in the A-site binding pocket (Fig. 6.3c) (Mardirossian et al. 2018; Seefeldt et al. 2016). The side chain of Arg1 interacts with the nucleobase of U2555. Hydrogen bonds with the 23S rRNA backbone are established by the side chains of Arg2 (G2454 and A2453), Arg4 (C2452 and G2494), and Arg6 (A2453 and U2493). Interactions of the peptide backbone are present for the α-amines of Arg1, Arg2, Ile3, and Arg4 which contact U2491, C2573, U2492, and U2493, respectively.

The overall contacts of Bac7 and Tur1A (Fig. 6.3c) with the A-site crevice are comparable to those of insect PrAMPs (Fig. 6.3b) (Mardirossian et al. 2018; Seefeldt et al. 2015, 2016). Insect PrAMP residues five to seven correspond to mammalian PrAMP residues eight to ten. Backbone interactions involve the α-carbonyl oxygen and α-amine of residue nine as well as the α-amine of residue ten that contacts the U2506 nucleobase (Fig. 6.3c) (Mardirossian et al. 2018; Seefeldt et al. 2016). Similar to Onc112, positions eight and ten harbor proline and leucine residues (Fig. 6.1) (Mardirossian et al. 2018; Seefeldt et al. 2015, 2016). Except for the described backbone interactions, the side chains of both residues do not establish hydrogen bonds with the surrounding tunnel (Fig. 6.3c) (Mardirossian et al. 2018; Seefeldt et al. 2016). Nonetheless, similar to Tyr6 which is present in all insect PrAMPs, residue nine of Bac7 and Tur1A establishes a stacking interaction with C2452 (Fig. 6.3c) (Mardirossian et al. 2018; Seefeldt et al. 2015, 2016). As observed for insect PrAMPs, Tur1A base stacking involves a tyrosine residue as well (Fig. 6.3c). Bac7 instead harbors an arginine residue in this position that stacks upon C2452 (Fig. 6.1) (Seefeldt et al. 2015, 2016).

Due to the enrichment in arginine residues compared to insect PrAMPs, Bac7 and Tur1A establish more interactions within the upper polypeptide exit tunnel (Mardirossian et al. 2018; Seefeldt et al. 2016). The majority of these contacts involve stacking interactions. In the case of Tur1A, two stacking interactions are observed (Fig. 6.3c), such that Arg15 and Arg16 stack upon U2586 and His69 of L4, respectively (Fig. 6.3c) (Mardirossian et al. 2018). The C-terminus of Bac7 establishes three stacking interactions with the surrounding tunnel (Seefeldt et al. 2016). Namely, Arg12, Arg14, and Arg16 form stacking interactions with C2610, U2586, and A2062, respectively. The stacking interactions of Arg12 with C2610 are analogous to the tunnel interaction of Arg9 in insect PrAMPs (Seefeldt et al. 2015, 2016).

6.1.6.2 Binding of Class II PrAMP Api137

Api137, which is derived from wild-type apidaecin 1b from *Apis mellifera* (Fig. 6.1), is a class II PrAMP. In contrast to other class I PrAMPs, Api137 binds to the polypeptide tunnel with a similar orientation to a nascent chain (Fig. 6.4a, b) (Florin et al. 2017). The C-terminal residues Arg17 and Leu18 are located in the A-site crevice but do not reach into the A-site binding pocket of the PTC. Placement of Arg17 and Leu18 in the A-site crevice is crucial for the mechanism of action of Api137. For instance, exchange of Arg17 by Ala leads to decreased activity against *E. coli* ribosomes (Castle et al. 1999). The N-terminal residues pass down the polypeptide tunnel (Fig. 6.4a, c) (Florin et al. 2017). In the available structure of Api137, the last four residues at the N-terminus are not resolved. Nonetheless, compared to class I PrAMPs, binding of Api137 to the polypeptide tunnel is primarily facilitated by stacking interactions (Fig. 6.4c) (Florin et al. 2017; Seefeldt et al. 2015, 2016). Stacking interactions are observed between Tyr7 and A751, Arg12 and C2611, as well as His15 and the nucleobase of G2505 (Fig. 6.4c) (Florin et al. 2017). In addition to stacking interactions, polar contacts are

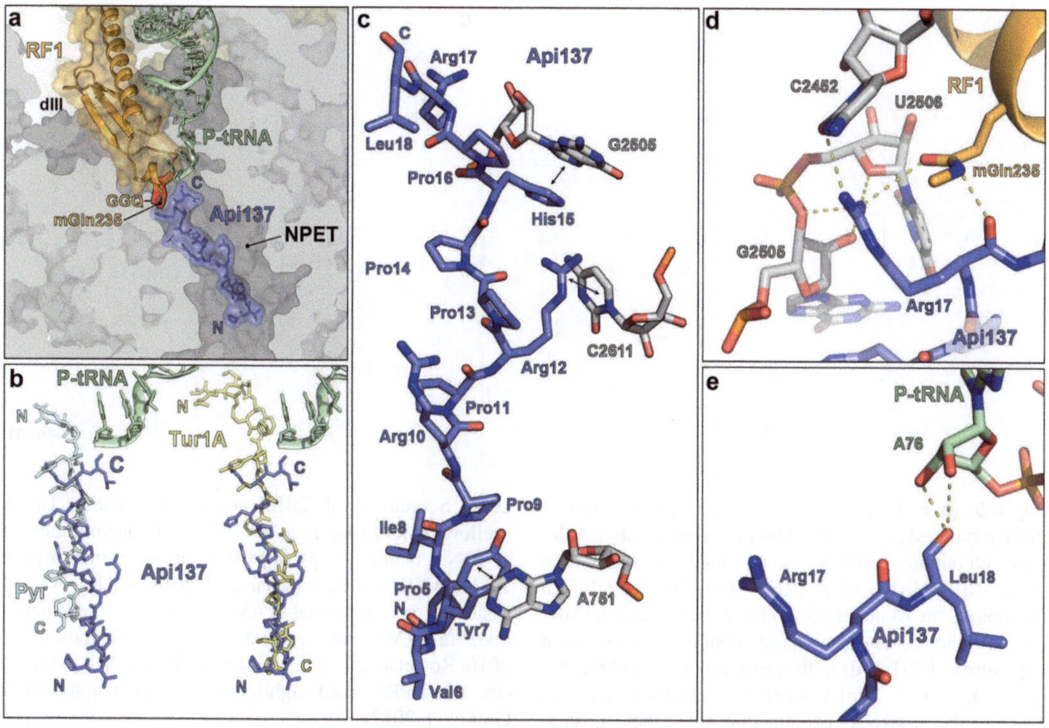

Fig. 6.4 (**a**) Overview showing the binding site of Api137 (slate) within the LSU (gray) polypeptide exit tunnel (NPET) in the presence of P-tRNA (green) and RF1 (orange) (Florin et al. 2017). The GGQ motif is highlighted in red. The N- and C-termini of Api137 are indicated by the letters N and C, respectively. (**b**) Superimposition of Api137 (slate) and Pyr (left; cyan) (Gagnon et al. 2016; Seefeldt et al. 2016) or Tur1A (right; yellow) (Mardirossian et al. 2018). P-tRNA (green) is shown for reference. (**c**) Interactions of Api137 (slate) with 23S rRNA nucleotides (gray) located in the NPET (Florin et al. 2017). Stacking interactions are indicated by black two-headed arrows. (**d, e**) Interactions of Api137 in the presence of (**d**) RF1 (orange) and (**e**) P-tRNA (green) (Florin et al. 2017). (**d**) Arg17 contacts 23S rRNA nucleotides and Gln235 of RF1. (**e**) Leu18 of Api137 contacts A76 of the deacylated tRNA located in the P-site

established in the PTC with RF1 and the deacylated P-site tRNA (Fig. 6.4d, e). The side chain of the penultimate residue Arg17 is coordinated between 23S rRNA residues and Gln235 of the RF1 GGQ motif (Fig. 6.4d). The 23S rRNA contacts comprise hydrogen bonds with the nucleobase of C2452, the 2′- and the 3′-oxygen of G2505, and the ribose of U2506 (Fig. 6.4d). Gln235 interacts with Arg17 through the carbonyl oxygen of the side chain (Fig. 6.4d). Leu18 appears to establish two interactions via the N-terminal OH group (Fig. 6.4e). These interactions include the 2′-OH and the 3′-OH of A76 of deacyl-P-tRNA (Fig. 6.4e). The overall altered interaction network compared to class I PrAMPs forms the basis for the unique mechanism of action of Api137.

6.1.7 Mechanism of Action of Class I and Class II PrAMPs

Under normal conditions ribosomal protein synthesis passes through the phases of translation initiation, translation elongation, and translation termination. Translation initiation involves the placement of an fMet-tRNA^{fMet} over an AUG start codon in the P-site of the 70S ribosome. The CCA-end that carries the fMet moiety is placed in the PTC, and the 3′-end establishes contacts with the P-loop of the LSU. Translation elongation comprises the EF-Tu-mediated delivery and subsequent accommodation of an aa-tRNA to the ribosomal A-site, transpeptidation in the PTC, and translocation catalyzed by EF-G. Accommodation coincides with the estab-

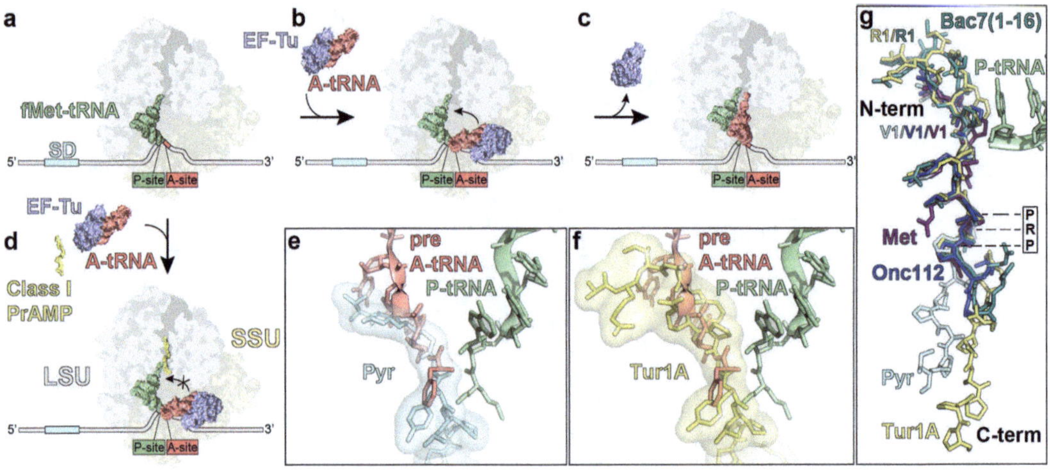

Fig. 6.5 (**a–c**) Canonical translation in the absence of protein synthesis inhibitors, showing (**a**) translation initiation with initiator P-tRNA (green) bound to the ribosomal P-site. (**b**) Delivery of aa-tRNA (light blue) by EF-Tu (salmon) to the A-site, followed by (**c**) tRNA accommodation into the A-site on the large subunit and subsequent departure of EF-Tu. (**d**) In the presence of class I PrAMPs (yellow), such as Tur1A, aa-tRNA delivery can occur; however the aa-tRNA accommodation is blocked. (**e, f**) Superimposition of (**e**) insect Pyr (cyan) (Gagnon et al. 2016; Seefeldt et al. 2016) and (**f**) mammalian Tur1A (yellow) (Mardirossian et al. 2018) with accommodated aa-tRNA (salmon). (**g**) Superimposition of mammalian Tur1A (yellow) (Mardirossian et al. 2018) and Bac7(1-16) (teal) and insect-derived PrAMPs Onc112 (slate), metalnikowin-1 (Met, purple), and Pyr (cyan) (Gagnon et al. 2016; Roy et al. 2015; Seefeldt et al. 2015, 2016), with the conserved PRP motif highlighted. (Figure adapted from Graf et al. 2017)

lishment of interactions between the CCA-end of the A-site tRNA and the A-loop within the 23S rRNA. During translation termination the nascent chain is released from the ribosome with the aid of the peptide chain release factor RF1 or RF2. Both classes of PrAMPs inhibit protein synthesis by binding to the polypeptide tunnel of bacterial 70S ribosomes, yet binding of each class of PrAMPs results in inhibition of different steps of the translation cycle (Florin et al. 2017; Gagnon et al. 2016; Roy et al. 2015; Seefeldt et al. 2015, 2016). In the case of class I PrAMPs, the residues crucial for translation interference are located in the A-site binding pocket (Fig. 6.5) (Seefeldt et al. 2015, 2016). In the case of the class II PrAMP Api137, the important residues are located in the PTC (Figs. 6.4 and 6.6) (Florin et al. 2017).

In the presence of class I PrAMPs, the A-site crevice and the A-site binding pocket are blocked by the N-terminal residues of the PrAMP (Fig. 6.5) (Gagnon et al. 2016; Mardirossian et al. 2018; Roy et al. 2015; Seefeldt et al. 2015, 2016). This allows placement of an fMet-tRNA^{fMet} during translation initiation (Fig. 6.5a) but perturbs delivery of the first aa-tRNA by EF-Tu (Fig. 6.5d) (Mardirossian et al. 2018; Seefeldt et al. 2015, 2016). Although decoding on the SSU is in principle possible, subsequent accommodation following release from EF-Tu is blocked due to the steric hindrance generated by the N-terminal residues of class I PrAMPs (Fig. 6.5d–g). Accordingly, superimposition with pre-attack 70S ribosomes carrying an accommodated aa-tRNA in the A-site shows incompatibility of the A-tRNA CCA-end and the N-terminal residues of PrAMPs (Fig. 6.5e, f) (Graf et al. 2017). The same steric clash also prevents binding of PrAMPs during translation elongation. To avoid steric clashes with a peptidyl-tRNA, binding of class I PrAMPs has to occur between translation termination and initiation (Graf et al. 2017). In general, class I PrAMPs were observed to inhibit the transition from translation initiation to translation elongation as confirmed by biochemical experiments (Mardirossian et al. 2018; Seefeldt et al. 2015, 2016). In accordance, toe-printing assays showed that translation in the presence of class I PrAMPs indeed leads to

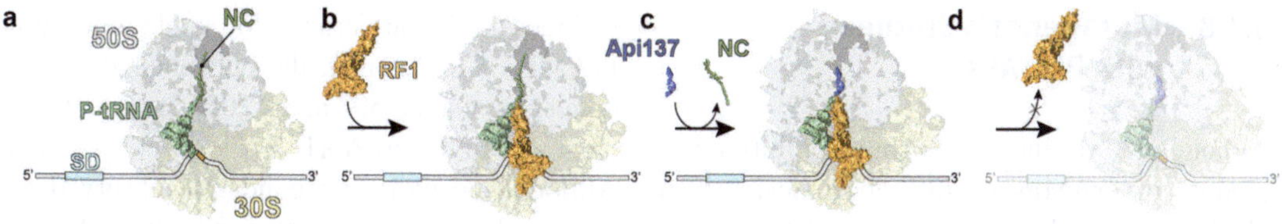

Fig. 6.6 (**a**) Translation terminates when the bacterial 70S ribosome encounters a stop codon. (**b**) RF1 is recruited to the terminating ribosome and mediates the hydrolysis of the nascent chain (NC, green) from P-site tRNA. (**c**) Api137 binds to post-hydrolysis complexes and (**d**) prevents dissociation of release factor from the 70S ribosome

ribosomal stalling at the AUG start codon. Further translation elongation is therefore not possible.

Biochemical toeprinting assays showed that in contrast to class I PrAMPs, class II PrAMPs do not inhibit translation initiation or elongation but rather trap ribosomes with a stop codon in the A-site, suggesting that translation termination is affected (Florin et al. 2017). Subsequent biophysical assays demonstrated that Api137 does not interfere RF binding to the ribosome nor with the RF-mediated release of the nascent chain but instead prevents the dissociation of the RF from the ribosome (Florin et al. 2017). Cryo-EM structures revealed the interactions that are crucial for RF trapping, encompassing the C-terminus of Api137 where Leu18 interacts with the ribose of A76 of the deacyl-tRNA in the P-site and Arg17 establishes a number of hydrogen bonds with the 23S rRNA and Gln235 of the GGQ motif of RF1 (Florin et al. 2017). As consequence, dissociation of RF1 from the post-hydrolysis 70S ribosome is inhibited, even in the presence of RF3, which is known to recycle RF1 and RF2 from the ribosome. Moreover, inhibition of bacterial cell growth by Api137 is not only caused by trapping of RF1 or RF2 on post-hydrolysis ribosomes. Because the number of RFs is 20–100x lower than the number of ribosomes in the cell, Api137 also has an indirect effect to deplete the endogenous pools of free RF1 and RF2, causing ribosomes to become stalled during translation termination due to lack of RFs (Florin et al. 2017). As consequence of this RF depletion, it was shown that Api137 also promotes stop codon read through. Hence, Api137 has a dual mode of inhibition, namely, (i) to directly trap RF1 and RF2 on a small minority of terminating ribosomes and (ii) to trap the majority of ribosomes during termination due to the absence of free RF1 and RF2.

Because of the small time window in which Api137 has to evoke its inhibition, it seems relatively clear how the binding site of Api137 in the polypeptide exit tunnel is accessed. Although the 70S ribosome constitutes a porous complex (Voss et al. 2006), due to the relatively large size of Api137, access to its binding site within the exit tunnel must be gained either from the PTC or via the tunnel exit rather than by diffusion through the ribosomal core. Given that Api137 traps the terminating ribosome with a deacylated tRNA in the P-site and one of the decoding RFs in the A-site (Florin et al. 2017), it seems unlikely that Api137 can gain access to its binding site via the PTC. This suggests that Api137 must access its binding site by entering from the tunnel exit site. Considering the length of ~100 Å and the width of the polypeptide tunnel with a diameter of less than 15 Å at the narrowest section (Nissen et al. 2000), one role of the proline residues may be to maintain an extended conformation of the peptides to facilitate their diffusion up to the ribosomal tunnel. With respect to the timing of binding, Api137 cannot access the binding site at all stages during termination. First, the presence of a nascent chain in the polypeptide exit tunnel during translation elongation represents a steric block that is incompatible with simultaneous binding of Api137. Second, Api137 action requires 70S-bound RF1 and P-tRNA, which interact via Gln235 of the GGQ motif and A76 with Arg17 and Leu18, respectively (Florin et al. 2017). As consequence, binding of Api137 has to occur after peptide chain release and before RF1/2 departure from the post-hydrolysis 70S ribosome.

6.1.8 Mutagenesis Studies with PrAMPs

Although the recent studies show inhibition and binding of PrAMPs to 70S ribosomes, further biochemical evidence was required to confirm that the translation machinery is the primary target of PrAMPs. A tool of choice for this validation is the selection of mutant strains that exhibit resistance to the PrAMP. Bacterial cells exposed to subinhibitory concentrations of a compound in culture acquire rRNA or protein sequence alterations that allow bacteria to escape the inhibitory action. Protein or rRNA alterations in general can affect inhibition either more directly, e.g., by decreasing the affinity to the binding site, or more indirectly, e.g., by preventing compound uptake into the cell. Problematic for the identification of rRNA mutations is the presence of several rRNA operons in *E. coli*. Thus, acquirement of a single mutation within one rRNA operon does not affect the complete pool of rRNAs incorporated into ribosomes. To avoid this issue, the identification of rRNA mutations, which confer resistance to PrAMPs, requires the use of special *E. coli* strains, called Squires (SQ) strains (Asai et al. 1999). The SQ strains exhibit a single rRNA allele that is encoded by an extragenomic plasmid. The genomically encoded operons, coding for 5S, 16S, and 23S rRNA, are absent. By exposing an SQ strain in culture to PrAMPs, it is possible to identify rRNA mutations conferring resistance against antibiotics. This is exemplified macrolide erythromycin resistance mutants, which also exhibit some cross-resistance with the insect PrAMPs Onc112 (Gagnon et al. 2016) and Api137 (Florin et al. 2017), including 23S rRNA nucleotides A2503 and A2059. Single or double mutation of A2503 and/or A2059 was shown to result in a 4- to 16-fold decreased susceptibility of *E. coli* cells against Onc112 and Api137 (Florin et al. 2017; Gagnon et al. 2016). Normally, A2503 and A2059 stack upon each other and stabilize adjacent nucleotides, such as A2062. The decreased susceptibility against both PrAMPs is presumably caused by conformational changes of nucleotides A2503 and A2059 as well as neighboring residues, which are crucial for stable binding of PrAMPs. Consistently, the published structural data show that A2503 and A2059 are situated in close proximity to Api137 (Florin et al. 2017). Although no hydrogen bonds are observed for both nucleotides, the close proximity to Api137 possibly allows van der Waals interactions. Upon mutation these interactions are presumably perturbed and result in a decreased binding affinity of Api137 toward the 70S ribosome. This contrasts the effect of A2503 and A2059 mutation on Onc112 binding (Gagnon et al. 2016). In the presence of Onc112, A2503 hydrogen bonds with A2062 and stabilizes the nucleotide in a 90° rotated conformation, which is crucial for accommodation of Onc112 in its binding site (Roy et al. 2015; Seefeldt et al. 2015). The non-rotated conformation of A2062 represents a steric block that prevents binding of Onc112 in the exit tunnel. However, independent of the effect of the A2503 and A2059 mutation on neighboring rRNA residues, the decreased susceptibility against Onc112 and Api137 is a clear evidence for ribosomes as major target for the inhibition by PrAMPs.

With respect to protein alterations, most of the initially identified mutations, conferring resistance to PrAMPs, were detected in the gene coding for the SbmA transporter. The primary reason for this is possibly the lower cost in fitness for bacteria grown under lab conditions. Accordingly, *sbmA*-deficient *E. coli* cells stay viable and exhibit lower susceptibility against PrAMPs, such as Bac7 (Mattiuzzo et al. 2007). In contrast, sequence alterations within functionally important regions are not well tolerated (Sato et al. 2006). Nonetheless, for Api137 it was possible to identify mutations within proteins by culturing *E. coli* cells that express multiple copies of an *sbmA* coding plasmid (Florin et al. 2017). The presence of multiple copies of the *sbmA* gene makes the acquirement of the same resistance mutation in all plasmids highly unlikely and ensures continuous uptake of PrAMPs into the bacterial cell. Consequently, the bacterial cell is forced to acquire alternative resistance mechanisms, which help to validate the ribosome as primary target for PrAMP action. Overall, it was possible to identify Api137 resistance mutations within ribosomal proteins L3 and L16 as well as in RF1 and RF2. Additionally, mutations within ribosomal proteins L4 and L22 also confer cross-resistance against Api137, specifically, mutation of L4 residue Lys63 to Glu as well as deletion of L22 residues 82–84 (Florin et al. 2017).

Ribosomal proteins L4 and L22 harbor each an extended loop that form together the central constriction within the polypeptide exit tunnel. The observed mutations within L4 and L22 presumably lead to reordering of these loops, which were shown in our structure to directly or indirectly stabilize binding of Api137 in the polypeptide tunnel (Florin et al. 2017). Thus, Arg61 of L4 provides a putative hydrogen bond to binding of Api137. Mutation of Lys63 could reorder the extension, which contains the interacting Arg61, and hence possibly destabilize binding of Api137. In contrast, deletion of L22 residues 82–84 presumably destabilizes Api137 binding more indirectly by contraction of the extended loop of L22. The extended loop of L22 contacts the 23S rRNA backbone and stabilizes H34. H34 harbors nucleotide A751 that stacks upon Api137 residue Tyr6 and is crucial for stable binding. The contraction of the extended loop of L22 most likely causes reordering of H34 and, hence, destabilization of Api137 by the lacking stacking interaction between Tyr6 and 23S rRNA residue A751. Arg81 of L16, which can confer resistance to Api137 upon mutation to Cys, stabilizes P-loop nucleotide G2251 via two hydrogen bonds (Florin et al. 2017). One hydrogen bond is established with the phosphate-oxygen backbone, and another established with the nucleobase of G2251. P-loop nucleotide G2251 Watson-Crick base-pairs with C75 of a P-site tRNA and stabilizes the CCA-end in the PTC for peptide bond formation. In the presence of Api137, the C-terminal residue Leu18 contacts A76 of the P-tRNA. The interaction between Leu18 and the 3′-end of the P-tRNA stabilizes Api137 binding to the ribosome. Upon mutation of Arg81 of L16 to Cys, both polar interactions with nucleotide G2251 are resolved and possibly lead to higher flexibility within the P-loop and the CCA-end of a P-tRNA. As consequence, the binding affinity of Api137 toward the ribosome is reduced.

In addition to ribosomal protein mutations, more resistance mutations against Api137 were identified within RF1 and RF2. In this context, mutation of Arg262 to Cys and Gln280 to Leu in RF2 as well as Asp241 to Gly in RF1 leads to significantly reduced susceptibility of *E. coli* cells against Api137 (Florin et al. 2017). The cause of the decreased susceptibility is most likely a destabilization of the binding of RF1 and RF2 which overcomes the stabilizing interactions of Api137 with the GGQ motif of RF1 and RF2. Accordingly, RF1 binding to the 70S ribosome is stabilized by a single polar contact of 23S nucleotide C2573 with Asp241. Mutation of Asp241 to Gly prevents an interaction with C2573 and hence could promote dissociation of RF1. Similar to RF1, mutation of RF2 residues Arg262 to Cys and Gln280 to Leu resolves interactions with 23S rRNA nucleotides C2556 and U2492, respectively. This could promote dissociation of RF2 since the interactions of Arg17 of Api137 are presumably insufficient to trap RF2 on the ribosome.

It is noteworthy that none of the rRNA or protein mutations conferred complete resistance against Onc112 and Api137. Nonetheless, all mutations taken together are rather strong arguments favoring the ribosome as the major target of these PrAMPs. A major problem for identifying more resistance mutations against PrAMPs is the high conservation of rRNA residues involved in PrAMP binding (RNAcentral 2017). As a consequence, mutation of interacting residues would result in dramatically decreased PTC activity and therefore in reduced viability of bacterial cells (Sato et al. 2006; Thompson et al. 2001). An example is C2452, which is located within the A2451 region (23S nts 2448 to 2554). C2452 establishes a stacking interaction with Tur1A (Mardirossian et al. 2018) and Pyr (Gagnon et al. 2016; Seefeldt et al. 2016) via a Tyr residue and a polar contact with Arg17 of Api137 (Florin et al. 2017). Sato and colleagues (Sato et al. 2006) showed that except for A2448 and A2453, all nucleotides within the A2451 region are essential for ribosomal function. Consistently, it is difficult to select for viable C2452 mutants.

With regard to mammalian PrAMPs, such as Bac7 and Tur1A, no rRNA or ribosomal protein resistance mutations have been identified so far. In general, the identification of mutations conferring resistance to mammalian PrAMPs is more difficult. In contrast to insect PrAMPs, mammalian PrAMP Bac7 and Tur1A are considerably longer (Graf et al. 2017). Mammalian PrAMPs show a full length of 39–60 aa, while insect PrAMPs show a full length between 15 and 34 aa. Although both insect and mammalian PrAMPs target ribosomes and inhibit translation, it was shown for full-length Bac7, which is very similar to Tur1A (Mardirossian et al. 2018), that a lytic mode of action is present (Podda et al. 2006). Only truncated derivatives of Bac7, like Bac7(1-16), solely inhibit protein synthesis without permeabilizing the bacterial membrane (Podda et al. 2006; Seefeldt et al. 2016). Selection for resistance mutations is in principle possible, but it is difficult for bacteria to acquire mutations that affect both modes of action (lytic and nonlytic) of mammalian full-length PrAMPs (Lai and Gallo 2009; Peschel and Sahl 2006). As a consequence, bacteria remain susceptible against full-length AMPs in culture. With respect to the previously described rRNA mutations conferring resistance to Onc112 and Api137 (A2503C and A2059G), no elevated MIC was observed in presence of Tur1A and Bac7 upon mutation of these residues (Gagnon et al. 2016; Mardirossian et al. 2018). Regardless, as described before, the selection for more rRNA resistance mutations is rather difficult, since nucleotide alterations within the exit tunnel that are involved in binding of mammalian PrAMPs could lead to strongly perturbed PTC activity (Sato et al. 2006).

6.1.9 Toxicity of PrAMPs in Eukaryotes

PrAMPs have already been tested using different animal models of infection and shown that they are well tolerated and exert a promising healing activity (Berthold et al. 2013; Knappe et al. 2015; Scocchi et al. 2011). Since PrAMPs target the 70S ribosome and inhibit protein synthesis of eubacteria, it needs clarification whether PrAMPs inhibit human translation as well. To have an inhibitory effect on human cells, it requires, first, the uptake of PrAMPs into the cytoplasm and, second, binding to the eukaryotic ribosome. Because Bac7 and Tur1A are expressed as a pre-pro-peptide, active peptide is not likely to be present in the cytosol of eukaryotic cells. Nevertheless, even if Bac7 and Tur1A gain access to the cytosol, these peptides appear to have a low affinity for the eukaryotic ribosome since the peptides inhibited protein synthesis in an *E. coli* lysate more efficiently than in rabbit reticulocyte system (Mardirossian et al. 2018; Seefeldt et al. 2016). Indeed, the bacteriostatic and even bactericidal concentrations of Bac5 fragments were shown not to be significantly toxic toward eukaryotic cells (Mardirossian et al. 2018).

Similarly, it is necessary to validate in the future whether Api137 can bind and inhibit mammalian 80S ribosomes. The available cryo-EM structure of 70S ribosomes in complex with RF1 and Api137 shows that Api137 binding primarily involves interactions with universally conserved rRNA residues (RNAcentral 2017) and the GGQ motif at the tip of RF1 domain III (Florin et al. 2017). The latter GGQ motif is conserved within bacteria as well as evolutionary unrelated eukaryotic eRF1 (Frolova et al. 1999; Mora et al. 2003; Seit-Nebi et al. 2001; Shaw and Green 2007; Zavialov et al. 2002). Assuming that Api137 is taken up by human cells, the conservation of both rRNA residues and the GGQ motif of the class I release factors implies that Api137 is active against human 80S ribosomes and hence would be unsuitable for use as a therapeutic. Superimposition of a mammalian 80S ribosome in complex with eRF1 (Shao et al. 2016) and the 70S *E. coli* ribosome in complex with RF1 and Api137 (Florin et al. 2017) shows that binding of Api137 is in principle compatible with binding to both eubacterial and eukaryotic ribosomes. All observed contacts between Api137 and the 70S-RF1 complex could be established with an 80S-eRF1 complex as well. Therefore, inhibition of human translation termination by Api137 seems to be likely. Nonetheless, inhibition of human 80S ribosomes would still require uptake of Api137 from the extracellular space. Gram-negative bacteria, like *E. coli*, actively uptake Api137, even at low concentrations, with the SbmA transporter. This transporter is absent from eukaryotic cells. Consistently, uptake of PrAMPs into the cytoplasm of human cells would require diffusion through the plasma membrane, but this is unlikely due to the properties of Api137. Similar to most of the AMPs, Api137 exhibits a positive net charge that has been suggested to interfere with the lytic action or uptake of AMPs into eukaryotic cells (Yeaman and Yount 2003). Bacterial cells exhibit a negatively charged cytoplasmic membrane surface, which promotes adsorption of positively charged AMPs and subsequent insertion into the membrane or uptake into the cell. Eukaryotic membranes, by contrast, exhibit a more neutral surface that does not promote adsorption of AMPs. Future experiments will be required to determine in vivo and in vitro whether Api137 is taken up by human cells and whether it is toxic or even teratogenic.

References

Agerberth B, Lee JY, Bergman T, Carlquist M, Boman HG, Mutt V, Jornvall H (1991) Amino acid sequence of PR-39. Isolation from pig intestine of a new member of the family of proline-arginine-rich antibacterial peptides. Eur J Biochem 202:849–854

Asai T, Zaporojets D, Squires C, Squires CL (1999) An *Escherichia coli* strain with all chromosomal rRNA operons inactivated: complete exchange of rRNA genes between bacteria. Proc Natl Acad Sci U S A 96:1971–1976

Benincasa M, Scocchi M, Podda E, Skerlavaj B, Dolzani L, Gennaro R (2004) Antimicrobial activity of Bac7 fragments against drug-resistant clinical isolates. Peptides 25:2055–2061. https://doi.org/10.1016/j.peptides.2004.08.004

Berthold N et al (2013) Novel apidaecin 1b analogs with superior serum stabilities for treatment of infections by gram-negative pathogens. Antimicrob Agents Chemother 57:402–409. https://doi.org/10.1128/AAC.01923-12

Brogden KA (2005) Antimicrobial peptides: pore formers or metabolic inhibitors in bacteria? Nat Rev Microbiol 3:238–250. https://doi.org/10.1038/nrmicro1098

Bulet P et al (1993) A novel inducible antibacterial peptide of Drosophila carries an O-glycosylated substitution. J Biol Chem 268:14893–14897

Casteels P, Tempst P (1994) Apidaecin-type peptide antibiotics function through a non-poreforming mechanism involving stereospecificity. Biochem Biophys Res Commun 199:339–345. https://doi.org/10.1006/bbrc.1994.1234

Casteels P, Ampe C, Jacobs F, Vaeck M, Tempst P (1989) Apidaecins: antibacterial peptides from honeybees. EMBO J 8:2387–2391

Casteels P et al (1990) Isolation and characterization of abaecin, a major antibacterial response peptide in the honeybee (*Apis mellifera*). Eur J Biochem 187:381–386

Casteels-Josson K, Capaci T, Casteels P, Tempst P (1993) Apidaecin multipeptide precursor structure: a putative mechanism for amplification of the insect antibacterial response. EMBO J 12:1569–1578

Castle M, Nazarian A, Yi SS, Tempst P (1999) Lethal effects of apidaecin on *Escherichia coli* involve sequential molecular interactions with diverse targets. J Biol Chem 274:32555–32564

Chernysh S, Cociancich S, Briand JP, Hetru C, Bulet P (1996) The inducible antibacterial peptides of the Hemipteran insect Palomena prasina: identification of a unique family of prolinerich peptides and of a novel insect defensin. J Insect Physiol 42:81–89. https://doi.org/10.1016/0022-1910(95)00085-2

Cociancich S et al (1994) Novel inducible antibacterial peptides from a hemipteran insect, the sap-sucking bug Pyrrhocoris apterus. Biochem J 300(Pt 2):567–575

Florin T et al (2017) An antimicrobial peptide that inhibits translation by trapping release factors on the ribosome. Nat Struct Mol Biol 24:752–757. https://doi.org/10.1038/nsmb.3439

Frolova L et al (1999) Mutations in the highly conserved GGQ motif of class 1 polypeptide release factors abolish the ability of human eRF1 to trigger peptidyl-tRNA hydrolysis. RNA 5:1014–1020

Gagnon MG, Roy RN, Lomakin IB, Florin T, Mankin AS, Steitz TA (2016) Structures of proline-rich peptides bound to the ribosome reveal a common mechanism of protein synthesis inhibition. Nucleic Acids Res 44:2439–2450. https://doi.org/10.1093/nar/gkw018

Gennaro R, Skerlavaj B, Romeo D (1989) Purification, composition, and activity of two bactenecins, antibacterial peptides of bovine neutrophils. Infect Immun 57:3142–3146

Graf M et al (2017) Proline-rich antimicrobial peptides targeting protein synthesis. Nat Prod Rep 34:702–711. https://doi.org/10.1039/c7np00020k

Huttner KM, Lambeth MR, Burkin HR, Burkin DJ, Broad TE (1998) Localization and genomic organization of sheep antimicrobial peptide genes. Gene 206:85–91

Knappe D et al (2010) Oncocin (VDKPPYLPRPRPPRRIYNR-NH2): a novel antibacterial peptide optimized against gram-negative human pathogens. J Med Chem 53:5240–5247. https://doi.org/10.1021/jm100378b

Knappe D, Zahn M, Sauer U, Schiffer G, Strater N, Hoffmann R (2011) Rational design of oncocin derivatives with superior protease stabilities and antibacterial activities based on the high-resolution structure of the oncocin-DnaK complex. Chembiochem 12:874–876. https://doi.org/10.1002/cbic.201000792

Knappe D, Adermann K, Hoffmann R (2015) Oncocin Onc72 is efficacious against antibiotic-susceptible *Klebsiella pneumoniae* ATCC 43816 in a murine thigh infection model. Biopolymers 104:707–711. https://doi.org/10.1002/bip.22668

Krizsan A, Volke D, Weinert S, Strater N, Knappe D, Hoffmann R (2014) Insect-derived proline-rich antimicrobial peptides kill bacteria by inhibiting bacterial protein translation at the 70S ribosome. Angew Chem Int Ed Eng 53:12236–12239. https://doi.org/10.1002/anie.201407145

Krizsan A, Prahl C, Goldbach T, Knappe D, Hoffmann R (2015) Short proline-rich antimicrobial peptides inhibit either the bacterial 70S ribosome or the assembly of its large 50S subunit. Chembiochem 16:2304–2308. https://doi.org/10.1002/cbic.201500375

Lai Y, Gallo RL (2009) AMPed up immunity: how antimicrobial peptides have multiple roles in immune defense. Trends Immunol 30:131–141. https://doi.org/10.1016/j.it.2008.12.003

Mardirossian M, Grzela R, Giglione C, Meinnel T, Gennaro R, Mergaert P, Scocchi M (2014) The host antimicrobial peptide Bac71-35 binds to bacterial ribosomal proteins and inhibits protein synthesis. Chem Biol 21:1639–1647. https://doi.org/10.1016/j.chembiol.2014.10.009

Mardirossian M et al (2018) The dolphin proline-rich antimicrobial peptide Tur1A inhibits protein synthesis by targeting the bacterial ribosome. Cell Chem Biol 25(5):530–539. https://doi.org/10.1016/j.chembiol.2018.02.004

Mattiuzzo M, Bandiera A, Gennaro R, Benincasa M, Pacor S, Antcheva N, Scocchi M (2007) Role of the *Escherichia coli* SbmA in the antimicrobial activity of proline-rich peptides. Mol Microbiol 66:151–163. https://doi.org/10.1111/j.1365-2958.2007.05903.x

Mora L, Heurgue-Hamard V, Champ S, Ehrenberg M, Kisselev LL, Buckingham RH (2003) The essential role of the invariant GGQ motif in the function and stability in vivo of bacterial release factors RF1 and RF2. Mol Microbiol 47:267–275

Nissen P, Hansen J, Ban N, Moore PB, Steitz TA (2000) The structural basis of ribosome activity in peptide bond synthesis. Science 289:920–930

Otvos L Jr (2002) The short proline-rich antibacterial peptide family. Cell Mol Life Sci 59:1138–1150

Otvos L Jr et al (2000) Interaction between heat shock proteins and antimicrobial peptides. Biochemistry 39:14150–14159

Peschel A, Sahl HG (2006) The co-evolution of host cationic antimicrobial peptides and microbial resistance. Nat Rev Microbiol 4:529–536. https://doi.org/10.1038/nrmicro1441

Podda E, Benincasa M, Pacor S, Micali F, Mattiuzzo M, Gennaro R, Scocchi M (2006) Dual mode of action of Bac7, a proline-rich antibacterial peptide. Biochim Biophys Acta 1760:1732–1740. https://doi.org/10.1016/j.bbagen.2006.09.006

RNAcentral (2017) RNAcentral: a comprehensive database of non-coding RNA sequences. Nucleic Acids Res 45:D128–D134. https://doi.org/10.1093/nar/gkw1008

Roy RN, Lomakin IB, Gagnon MG, Steitz TA (2015) The mechanism of inhibition of protein synthesis by the proline-rich peptide oncocin. Nat Struct Mol Biol 22:466–469. https://doi.org/10.1038/nsmb.3031

Runti G et al (2013) Functional characterization of SbmA, a bacterial inner membrane transporter required for importing the antimicrobial peptide Bac7(1-35). J Bacteriol 195:5343–5351. https://doi.org/10.1128/JB.00818-13

Sato NS, Hirabayashi N, Agmon I, Yonath A, Suzuki T (2006) Comprehensive genetic selection revealed essential bases in the peptidyl-transferase center. Proc Natl Acad Sci U S A 103:15386–15391. https://doi.org/10.1073/pnas.0605970103

Schnapp D, Kemp GD, Smith VJ (1996) Purification and characterization of a proline-rich antibacterial peptide, with sequence similarity to bactenecin-7, from the haemocytes of the shore crab, Carcinus maenas. Eur J Biochem 240:532–539

Schneider M, Dorn A (2001) Differential infectivity of two Pseudomonas species and the immune response in the milkweed bug, *Oncopeltus fasciatus* (Insecta: Hemiptera). J Invertebr Pathol 78:135–140. https://doi.org/10.1006/jipa.2001.5054

Scocchi M, Lüthy C, Decarli P, Mignogna G, Christen P, Gennaro R (2009) The proline-rich antibacterial peptide Bac7 binds to and inhibits in vitro the molecular chaperone DnaK. Int J Pept Res Ther 15:147–155. https://doi.org/10.1007/s10989-009-9182-3

Scocchi M, Tossi A, Gennaro R (2011) Proline-rich antimicrobial peptides: converging to a non-lytic mechanism of action. Cell Mol Life Sci 68:2317–2330. https://doi.org/10.1007/s00018-011-0721-7

Seefeldt AC et al (2015) The proline-rich antimicrobial peptide Onc112 inhibits translation by blocking and destabilizing the initiation complex. Nat Struct Mol Biol 22:470–475. https://doi.org/10.1038/nsmb.3034

Seefeldt AC et al (2016) Structure of the mammalian antimicrobial peptide Bac7(1-16) bound within the exit tunnel of a bacterial ribosome. Nucleic Acids Res 44:2429–2438. https://doi.org/10.1093/nar/gkv1545

Seit-Nebi A, Frolova L, Justesen J, Kisselev L (2001) Class-1 translation termination factors: invariant GGQ minidomain is essential for release activity and ribosome binding but not for stop codon recognition. Nucleic Acids Res 29:3982–3987

Shamova O, Brogden KA, Zhao C, Nguyen T, Kokryakov VN, Lehrer RI (1999) Purification and properties of proline-rich antimicrobial peptides from sheep and goat leukocytes. Infect Immun 67:4106–4111

Shao S, Murray J, Brown A, Taunton J, Ramakrishnan V, Hegde RS (2016) Decoding mammalian ribosome-mRNA states by translational GTPase complexes. Cell 167:1229–1240.e1215. https://doi.org/10.1016/j.cell.2016.10.046

Shaw JJ, Green R (2007) Two distinct components of release factor function uncovered by nucleophile partitioning analysis. Mol Cell 28:458–467. https://doi.org/10.1016/j.molcel.2007.09.007

Stensvag K, Haug T, Sperstad SV, Rekdal O, Indrevoll B, Styrvold OB (2008) Arasin 1, a proline-arginine-rich antimicrobial peptide isolated from the spider crab, *Hyas araneus*. Dev Comp Immunol 32:275–285. https://doi.org/10.1016/j.dci.2007.06.002

Storici P, Zanetti M (1993) A cDNA derived from pig bone marrow cells predicts a sequence identical to the intestinal antibacterial peptide PR-39. Biochem Biophys Res Commun 196:1058–1065. https://doi.org/10.1006/bbrc.1993.2358

Thompson J et al (2001) Analysis of mutations at residues A2451 and G2447 of 23S rRNA in the peptidyltransferase active site of the 50S ribosomal subunit. Proc Natl Acad Sci U S A 98:9002–9007

Voss NR, Gerstein M, Steitz TA, Moore PB (2006) The geometry of the ribosomal polypeptide exit tunnel. J Mol Biol 360:893–906

Yeaman MR, Yount NY (2003) Mechanisms of antimicrobial peptide action and resistance. Pharmacol Rev 55:27–55. https://doi.org/10.1124/pr.55.1.2

Zahn M, Berthold N, Kieslich B, Knappe D, Hoffmann R, Strater N (2013) Structural studies on the forward and reverse binding modes of peptides to the chaperone DnaK. J Mol Biol 425:2463–2479. https://doi.org/10.1016/j.jmb.2013.03.041

Zahn M, Kieslich B, Berthold N, Knappe D, Hoffmann R, Strater N (2014) Structural identification of DnaK binding sites within bovine and sheep bactenecin Bac7. Protein Pept Lett 21:407–412

Zanetti M, Litteri L, Gennaro R, Horstmann H, Romeo D (1990) Bactenecins, defense polypeptides of bovine neutrophils, are generated from precursor molecules stored in the large granules. J Cell Biol 111:1363–1371

Zanetti M, Litteri L, Griffiths G, Gennaro R, Romeo D (1991) Stimulus-induced maturation of probactenecins, precursors of neutrophil antimicrobial polypeptides. J Immunol 146:4295–4300

Zanetti M, Del Sal G, Storici P, Schneider C, Romeo D (1993) The cDNA of the neutrophil antibiotic Bac5 predicts a pro-sequence homologous to a cysteine proteinase inhibitor that is common to other neutrophil antibiotics. J Biol Chem 268:522–526

Zasloff M (2002) Antimicrobial peptides in health and disease. N Engl J Med 347:1199–1200. https://doi.org/10.1056/NEJMe020106

Zavialov AV, Mora L, Buckingham RH, Ehrenberg M (2002) Release of peptide promoted by the GGQ motif of class 1 release factors regulates the GTPase activity of RF3. Mol Cell 10:789–798

Part III

Other Activities of AMPs

Antimicrobial and Cell-Penetrating Peptides: How to Understand Two Distinct Functions Despite Similar Physicochemical Properties

7

Ines Neundorf

Abstract

Antimicrobial and cell-penetrating peptides are both classes of membrane-active peptides sharing similar physicochemical properties. Both kinds of peptides have attracted much attention owing to their specific features. AMPs disrupt cell membranes of bacteria and display urgently needed antibiotic substances with alternative modes of action. Since the multidrug resistance of bacterial pathogens is a more and more raising concern, AMPs have gained much interest during the past years. On the other side, CPPs enter eukaryotic cells without substantially affecting the plasma membrane. They can be used as drug delivery platforms and have proven their usefulness in various applications. However, although both groups of peptides are quite similar, their intrinsic activity is often different, and responsible factors are still in discussion. The aim of this chapter is to summarize and shed light on recent findings and concepts dealing with differences and similarities of AMPs and CPPs and to understand these different functions.

Keywords

Antimicrobial peptides · Cell-penetrating peptides · Plasma membranes · Drug delivery · Lipid-peptide interaction

Abbreviations

AMP	Antimicrobial peptide
CD	Circular dichroism
CPP	Cell-penetrating peptide
CS	Chondroitin sulfate
DSC	Differential scanning calorimetry
EM	Electron microscopy
EPR	Electron paramagnetic resonance
FDA	Food and Drug Administration
FMM	Functional membrane microdomain
GAG	Glycosaminoglycan
GPMV	Giant plasma membrane vesicle
GUV	Giant unilamellar vesicle
HS	Heparan sulfate

I. Neundorf (✉)
Department of Chemistry, Institute for Biochemistry, University of Cologne, Cologne, Germany
e-mail: ines.neundorf@uni-koeln.de

K. Matsuzaki (ed.), *Antimicrobial Peptides*, Advances in Experimental Medicine and Biology 1117, https://doi.org/10.1007/978-981-13-3588-4_7

IR	Infrared
L_d	Liquid disordered
L_o	Liquid ordered
LTA	Lipoteichoic acid
LUV	Large unilamellar vesicle
MALDI	Matrix-assisted laser desorption/ionization
MS	Mass spectrometry
NMR	Nuclear magnetic resonance
OBOC	One bead one compound
PC	Phosphatidylcholine
PE	Phosphatidylethanolamine
PG	Phosphatidylglycerol
PI	Phosphatidylinositol
PS	Phosphatidylserine
QSAR	Quantitative structure-activity relationship
ROS	Reactive oxygen species
STED	Stimulated emission depletion
SUV	Small unilamellar vesicle

7.1 Introduction

Nature has breeded a group of fantastic molecules, namely peptides, which are able to interact with membranes and operate in processes of fundamental importance such as viral fusion, antimicrobial defense mechanisms, membrane poration, delivery across membranes, and hormone-receptor interactions, to name only few. Two groups of peptides fall into focus of this chapter: antimicrobial peptides (AMPs) and cell-penetrating peptides (CPPs). While AMPs are key components of the innate immune system and act against many pathogens, CPPs have been emerged as valuable tools to translocate cargos across biological barriers. Despite their different functions, both groups share a lot of similarities in structure, sequence, and, particularly, membrane activity rising the question how these discrepancies in function can be explained by a common set of physicochemical characteristics. In fact, both groups of peptides are strongly membrane-active and induce membrane fluctuations arguing that the activity of these membrane-active families simply represents different facets of what is a shared energy landscape (Last et al. 2013). Therefore it is not contradictory to notice AMPs that cross plasma membranes and CPPs that offer antimicrobial activity. Moreover, it has been shown in many studies that by careful amino acid substitutions, some CPPs can be turned into AMPs and vice versa, leading to the often posed question: how different are they? The goal of this chapter is to underpin the recently emerged hypothesis that activity of AMPs and CPPs is mainly related to cooperative effects emerging during binding of peptides to the lipid bilayer. Thus, the membrane is not any more seen as a passive layer that is simply affected when peptides approach but actively contributes by its curvature and partitioning into microdomains to the entire peptide-lipid interaction process. In this way it might be possible to explain these different systems by common physicochemical processes. After a short introduction to both groups of peptides, their interplay with biological membranes and resultant biological effects, methods to measure those activities, as well as ways to predict membrane activity and to design novel AMP or CPP sequences will be summarized.

7.1.1 Cell-Penetrating Peptides

Cell-penetrating peptides constitute a family of natural or synthetically generated peptides that are able to translocate across cellular membranes. Usually these peptides are relatively short (smaller than 35 amino acids) and mediate the transport of cargos into cells, which can be either covalently or non-covalently attached to the CPP. A vast sequence variety exists making classification of CPPs difficult. One possibility is to group them depending on their origin in protein-derived, chimeric, or synthetic peptides. Otherwise, they can be ordered based on their physicochemical characteristics into three main classes such as cationic, hydrophobic, and amphipathic. Whereby, this latter classification is more helpful in view of the current understanding of CPPs' cellular entry mechanism. CPPs have emerged as a powerful technique to deliver various types of molecules into cells such as proteins, peptides, siRNA, DNA, liposomes, nanoparticles, or small organic drugs (Kalafatovic and Giralt 2017). Additionally, some CPPs have made it already in clinical applications (Feni and Neundorf 2017).

CPPs enter cells using different modes of action such as energy-dependent (endocytotic) or energy-independent (direct) entry pathways. The latter might be accompanied by toxic membrane activity based on the ability of the CPP to disrupt membranes. Endocytosis on the other side is mainly observed when large cargos are attached to CPPs. Still it is not clear which factors provoke the one or other pathway. Notably, in both cases peptides adsorb at the membrane surface, where they interact with negatively charged surface components, probably glycoconjugates, anionic lipids, or membrane proteins. This adsorption is the result of a structural rearrangement, and sometimes it leads also to an interaction with the interfacial zone of the lipid bilayer. After a critical (threshold) concentration is reached, membrane deformation occurs as a result of elastic stress and mass imbalance (Alvares et al. 2017). Likely, these process parameters are unique for each CPP-cargo complex/conjugate. Of note is that most CPPs are unstructured in aqueous solution but rapidly adopt defined secondary structures when coming in contact with the membrane lipid phase. Frequently, the formation of amphipathic helices is observed in this case (Di Pisa et al. 2015). Although the intracellular uptake of most CPP is not fully understood, it is of wide acceptance that CPP enter cells by using probably both pathways, direct entry and endocytosis, simultaneously and/or depending on physicochemical properties, concentration, charge, and length of the CPP, as well as characteristics of the cargo. Additional influences on the uptake mechanism of CPPs derive from properties, lipid composition, and protein content of the cell membrane. Moreover, CPP concentration is suggested to have an important role, while at low concentrations endocytosis and at high concentrations, direct entry processes are usually observed.

One pitfall when working with CPPs is their mainly endocytotic uptake. Once taken up via endocytosis, the peptide-cargo complex resides in endosomes, from which it has to be released for reaching its target site. Thus, the endocytosis mechanism represents one of the major disadvantages for the further development of CPPs. Problems concerning the efficient escape of CPPs from the endosomes still persist and are in any case present when large molecules are delivered by CPPs. To circumvent these problems, CPPs may be equipped with fusogenic sequences or other endosomolytic molecules (Neundorf et al. 2009). Still many efforts are made to develop more selective and efficient CPP sequences.

7.1.2 Antimicrobial Peptides

Antimicrobial peptides are usually short peptides (<50 amino acids) and present in all forms of life, where they play a major role in the innate immune system and act as the "first line of defense" against invading pathogens. This class of peptides is structurally highly divers and of amphipathic or cationic nature. AMPs share widespread toxicity against bacteria, yeasts, and fungi but are relatively inactive toward host eukaryotic cells at bactericidal concentrations. Membrane permeabilization is their main mechanism of action, but additional mechanisms have been supposed, including membrane destabilization, intracellular translocation, and inhibition of protein or nucleic acid synthesis. Often they display a polycationic nature supporting electrostatic interaction with negatively charged bacterial surface structures such as lipoteichoic acids (LTA). Classical mechanisms of antibiotic action include their penetration of the plasma membrane or cell wall, thus resulting in lysis or disruption of ionic gradients. Specifically, they gain access to the cytoplasmic membrane and interact with lipid bilayers, forming transmembrane pores that disrupt the cell membrane, finally leading to cell death. Several models have been proposed that explain how AMPs induce these membrane-disrupting processes, including the *barrel-stave*, *toroidal pore* and *carpet* model (Sierra et al. 2017).

Although they act preferentially on the membrane level, AMPs may affect multiple biochemical processes in the pathogen. An increasing body of evidence has demonstrated that AMPs have also intracellular targets (Le et al. 2017). For instance, blocking of RNA or protein synthesis and inhibiting enzymes necessary for linking cell wall structural proteins are further mechanisms of AMPs leading to cell death. Some AMPs have the ability to translocate in eukaryotic cells without damage of the plasma membrane. Notably, they target cancer cells and find intracellular targets like mitochondria, where they inhibit, e.g., cellular respiration and induce reactive oxygen species (ROS) formation (da Costa et al. 2015).

Since toxicity of AMPs is in most cases mediated by a non-specific process, bacteria have difficulties to develop resistance against AMPs. However, development of AMP-resistant strains is of course inevitable once they have been put into clinic. Indeed, processes by which microorganisms have produced resistance mechanisms against AMPs have been already reported. Moreover, to survive the bactericidal action of AMPs, bacteria must sense the presence and adapt accordingly by controlling the expression of genes involved in AMP resistance. Generally, bacteria try to change the composition of the outer or inner membrane, or to modify their cell wall composition, thus making principal AMP targets less susceptible. In fact, bacterial defense mechanisms often rely on cell wall modifications, which usually alter the ionic cell wall potential leading to a reduced AMP binding (Maria-Neto et al. 2015). Although clear efforts have been made, more techniques are needed to fully understand bacterial resistance strategies. Nevertheless, owing to their remarkable properties AMPs are one of the most promising drug candidates in a foreseeable future to overcome the alarming rise in microbial drug resistance (da Costa et al. 2015).

7.2 Constitution of Biological Lipid Membranes and Its Relevance to the Activity of AMPs or CPPs

Cellular membranes regulate the in- and outflow of nutrients, give the cell its shape, and are responsible for many other important cellular functions like cell-cell communication and signaling processes. Furthermore, membranes constitute an impermeable barrier for large, charged, or hydrophilic exogenous molecules, like therapeutic oligonucleotides, proteins, or peptides. Still, it is one of the major challenges in pharmaceutical industry to find powerful techniques to overcome this barrier for an efficient drug delivery. The membrane itself is built up of a lipid bilayer that is composed of various lipids, proteins, and sugars. Different phospholipid classes are the most abundant molecules, and besides their structural function, they play important roles in regulating and controlling processes occurring throughout the membrane (Jobin and Alves 2014). Glycosaminoglycans (GAGs) are characteristic for mammalian cells, and especially the presence of heparan sulfate (HS) is thought to be important for CPP-cell interaction. Consistently it has been noticed that AMPs and CPPs act on two main membrane classes: mammalian (eukaryotic) and prokaryotic ones. Moreover, distinct activities for both groups of peptides have been found for tumor cells. All these cells are characterized by certain heterogeneities with differences in lipid bilayer composition, expression of various specific markers, glycosylation profiles at the outer surface layer, or the presence of cell walls in the case of bacteria and will be discussed briefly.

7.2.1 Eukaryotic Cell Membranes

Eukaryotic cell membranes are mainly composed of phospholipids, glycosphingolipids, and cholesterol. The different lipid species are segregated into different domains within the lipid membrane. Moreover, whereas the outer phase of the lipid bilayer is usually characterized by the presence of

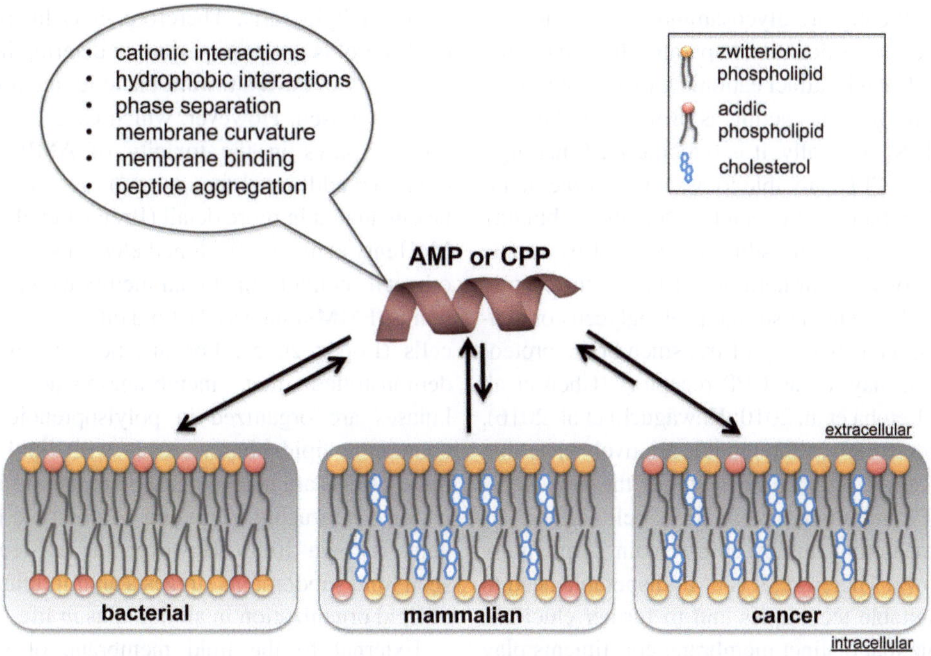

Fig. 7.1 Sketch of different plasma membrane organizations in bacterial, mammalian, and cancer cells. Bacterial outer layers, as well as that of cancer cells, contain more negatively charged phospholipids and attract cationic AMPs or CPPs. Mammalian outer layers present zwitterionic phospholipids and attract cationic or amphipathic peptides by their phase separation and membrane curvature. However, for all membrane systems, the effects can be somehow interchangeable. Peptide orientation and aggregation are further important factors for following peptide association and partitioning within the membrane

zwitterionic phosphatidylcholine (PC), the inner layer contains essentially more negatively charged phosphatidylserine (PS) contributing to a characteristic asymmetry between outer and inner layer (Fig. 7.1). Additionally, cholesterol is a major constituent of eukaryotic plasma membranes and regulates, among other things, plasma membrane fluidity and endocytosis. Together with sphingomyelin, it occurs in so-called liquid-ordered domains, a liquid crystalline phase of a lipid bilayer. Nearly any eukaryotic cell contains a distinct membrane organization in membrane domains, which plays a role in many functional processes related to polarization, signal transduction, and membrane trafficking. Moreover, this temporal and spatial compartmentalization of membrane molecules in assemblies is crucial for membrane function (Simons and Vaz 2004). A number of observations corroborated the idea of the existence of so-called lipid rafts, relatively ordered subdomains, in which lipids and/or proteins are recruited and clustered. Cholesterol and sphingolipids are main components of such domains. With the presence of such lipid raft microdomains, it is hypothesized that lipids do not simply act as passive solvent but play a regulatory role in protein membrane assembly. However, still the raft hypothesis is highly discussed owing to the lack of methods for direct observation of such domains (Sezgin et al. 2017). For many CPPs an involvement of lipid rafts and the relevance of cholesterol for their uptake have been demonstrated (Pae et al. 2014; Watkins et al. 2009). Cholesterol depletion with methyl-β-cyclodextrin significantly affects translocation of many CPPs across the plasma membrane, either by affecting endocytosis pathways such as macropinocytosis or clathrin-/caveolin-mediated uptake or by influencing the overall membrane fluidity and, thus, direct translocation.

Other critical molecules that are exposed at the outer surface of the lipid bilayer and important for

CPP cell entry are glycosaminoglycans and proteoglycans. Particularly, heparan sulfate and other sulfated GAGs attract cationic CPPs by their negative charges, thus acting as primary binding site for CPPs. Generally, it is hypothesized that arginine-rich CPPs are able to associate at the membrane by bidentate-dependent binding at binding sites or partners, possibly represented by sulfate groups of HS. Furthermore, it has been recently shown that heparan sulfate proteoglycans or syndecans, another group of transmembrane proteoglycans, may act as CPP receptors (Chen et al. 2015; Letoha et al. 2010; Kawaguchi et al. 2016).

Interestingly, although it is relatively clear that electrostatic interaction is one of the key events for CPP membrane association, cell surface targets mediating this process remain surprisingly unknown. Therefore, it is of undisputable need to find suitable techniques and to further elucidate the role that distinct membrane constituents play in CPP uptake mechanisms.

7.2.2 Bacterial Cell Membranes

The bacterial cell wall is a complex polymeric structure with essential roles in defense, survival, and pathogenesis. The outer leaflet of the cytoplasmic membrane of both Gram-positive and Gram-negative bacteria is surrounded by a meshlike peptidoglycan sacculus (Caveney et al. 2018). The lipid bilayer of the plasma membrane includes proteins, associated RNA, and the common phospholipids phosphatidylethanolamine (PE), phosphatidylglycerol (PG), and cardiolipin, while PC and phosphatidylinositols (PIs) are less frequent. Cardiolipin and PG are negatively charged and contribute to the negative charge of the membrane. One additional major difference between eukaryotic and prokaryotic cell membranes is the existence of cholesterol in eukaryotic cell membranes and its complete absence in bacterial cell membranes (Fig. 7.1). Although it has been supposed that the presence of cholesterol suppresses the activity of AMPs, recently, this effect was put into perspective when it was shown that AMPs significantly disrupt heterogeneous lipid structures that contain cholesterol-enriched lipid rafts. Therefore, it is likely that cholesterol is not as important in determining the selectivity of AMPs toward bacterial membranes once supposed. However, which exact role cholesterol plays in the toxicity of AMPs or if unknown additional factors are necessary has to be elucidated in more detail (Brender et al. 2012; McHenry et al. 2012). Nonetheless, also prokaryotic cells contain functional membrane microdomains (FMMs) such as the lipid rafts in eukaryotic cells (Lopez 2015). For instance, it has been demonstrated that membrane-bound sensor kinases are organized in polyisoprenoid lipid containing lipid phases. As is with lipid rafts, these FMMs are able to resist detergent disaggregation when using a mixture of nonionic detergents (Brown 2002). However, the existence of such lipid subdomains suggests an important role of lipid organization in all domains in life.

External to the lipid membrane of Gram-positive bacteria exists a thick peptidoglycan layer, whose major constituent is lipoteichoic acid (LTA). The structure of LTA varies between the different species. Owing to its anionic nature caused by the presence of carboxyl and phosphate groups of the LTA as well as carboxyl groups of the muramyl peptides, a first contact with cationic AMPs is feasible. In fact, this electrostatic interaction is thought to be the primary mechanism for antimicrobial activity. Following, AMPs promote membrane damage and cell lysis either by membrane thinning, pore formation, or bilayer disruption (Pushpanathan et al. 2013). Gram-negative bacteria, on the other side, present another outer lipid membrane, which is separated by the periplasm. It consists of phospholipids, lipopolysaccharides, integral membrane proteins, and lipoproteins. In a multistep process, AMPs are first attracted by negatively charged groups of the lipopolysaccharides, following disruption of the outer membrane to gain access to the periplasmic space. As is the case with Gram-positive bacteria, electrostatic attraction by negatively charged groups lead then in a further step to AMP membrane binding. However, owing to the more complex structure, Gram-negative bacteria are usually more difficult to target, and only few exceptional AMPs exist that show activity against Gram-negative bacteria. Additionally, also Gram-negative bacteria have developed various mechanisms to resist AMPs, like proteolytic degradation of AMPs, shielding of the bacterial surface, modification of the bacterial outer membrane, and pumping AMPs in or out of the cell (Gruenheid and Le Moual 2012).

7.2.3 Tumor Cells

Tumor cells are characterized by a phenotypic distinct membrane organization compared to healthy cells. In fact, compared to healthy cells, tumor cells display a higher overall net negative charge on their outer cell surface. This specific phenotype results from a number of key factors making cationic and amphipathic peptides susceptible to tumor cells. One is the presence of an increased amount of anionic phospholipids in the outer layer of the plasma membrane (e.g., PS) (Fig. 7.1) (Ran et al. 2002). Thus, the natural asymmetry between the inner and outer membrane leaflets is lost in tumor cells. Elevated reactive oxygen species (ROS) and hypoxia lead to dysregulation of phospholipid transporters, supporting this imbalance in the maintenance of the plasma asymmetry (Baxter et al. 2017). This involves activation of a putative scramblase and inactivation of a putative ATP-dependent phospholipid translocase. Moreover, several anionic cell surface glycoproteins are highly expressed in cancers and additionally contribute to an increased level of negative surface charge (Utsugi et al. 1991; Ran et al. 2002). Examples include mucins and heparan sulfate proteoglycans, which both promote electrostatic interaction of positively charged peptides at the outer surface of tumor cells. Furthermore, changes in membrane fluidity and pH, increased surface area and transmembrane potential, as well as a higher number of microvilli contribute to this specific characteristic of tumor cells.

7.2.3.1 Anticancer Activity of Membrane-Active Peptides

Cancer is one of the major causes of death worldwide, and although many efforts have been made concerning the development of new anticancer therapeutics, the field is still in problems owing to upcoming resistances and low specificity of currently available drugs. One alternative approach that has come up during the last years is to use anticancer peptides. Many of these sequences derive originally from host-defense peptides (i.e., AMPs), but also for cell-penetrating peptides, an inhibition of cancer cells has been observed. Because of their alternative mode of action including their specificity to cell membranes as their primary target site, resistance and cytotoxicity are less likely to occur.

For AMPs it has become more and more evidenced that they are more than just alternative antibiotic weapons. Indeed, a lot of studies demonstrate a clear anticancer activity, and many anticancer peptides are often based on AMPs (Patel and Akhtar 2017; Deslouches and Di 2017; Roudi et al. 2017). Efforts are being made in order to understand the targeting mechanism of antimicrobial peptides, which would enable an improved design. Again, certainly structure plays a central role in their activity, while AMPs adopt a defined, often alpha-helical, structure when in presence of cancer cell membranes (Felicio et al. 2017). Moreover, the activity of anticancer peptides is in most cases driven by their cationic charge, while the presence of an amphipathic helix, i.e., spatial segregation of cationic and hydrophobic residues, seems also to be essential (Roudi et al. 2017). Notably, AMPs trigger distinct killing mechanisms based on membrane-lytic events or such without membrane-lytic events. In this way, necrosis occurs after cell membrane lysis and apoptosis in the case of mitochondrial membrane lysis. In both cases the presence of anionic lipids such as PS or cardiolipin is indispensable. Binding of AMPs to surface-exposed PS leads probably to membrane depolarization and cell death. Additionally, those anticancer peptides can present other intracellular targets, either targeting essential cell proteins, inhibiting angiogenesis, or recruiting immune cells to attack cancer cells (Wu et al. 2014).

On the other side, CPPs are usually not cell selective and, thus, controlled targeting strategies using internal or external stimuli to selectively increase the activity of CPPs at the target site are necessary. With respect to tumor targeting, several such strategies have been successfully developed making use of activatable CPPs, attachment of ligands to CPPs that act as address labels, or localized hyperthermia, to name only few (Raucher and Ryu 2015; Bergmann et al. 2017; Splith et al. 2012). One other alternative targeting approach for tumor cells is based on the intrinsic properties of cationic CPPs. Their targeting mechanisms rely on the same parameters as those explained for AMP-derived anticancer peptides that act by their positive charge on the negatively charged surface of tumor cells. Several efforts have been made, in which such cationic anticancer peptides were successfully used to target and affect cancerous cells, also in vivo (Szczepanski et al. 2014; Gronewold et al. 2017). Moreover, such anticancer CPPs may also have intracellular targets. Thus, pore formation in the presence of high electrical potential at the mitochondrial membrane might be the basis for their activity (Rodriguez Plaza et al. 2014).

However, for such anticancer peptides, different activity levels might exist, and a careful selection, depending, e.g., on the tumor to be treated, has to be performed. For instance, for many anticancer peptides, it is shown that the interaction with glycosaminoglycans, HS, and chondroitin sulfate (CS), which are present on the outer surface, is one of the key steps during their action. However, it was recently demonstrated that HS at the outer surface of cancer cells sequesters anticancer peptides, in this case bovine lactoferricin, away from the phospholipid bilayer and thereby impede their ability to induce cell lysis (Fadnes et al. 2009). The results let further conclude that poorly differentiated tumors, with low expression of HS, are more susceptible to treatment with anticancer peptides, an interesting hypothesis that should be investigated in in vivo studies. Additionally, by generating modified versions of bovine lactoferricin, the cytotoxic activity against HS- and CS-expressing tumor cells was regained, demonstrating the need of such detailed structure-activity relationship studies (Fadnes et al. 2011).

Other obstacles that have to be faced with for a future application of anticancer peptides are adverse effects such as high toxicity to healthy cells and immune response. For this reason it is still necessary to dissociate the toxicity to mammalian cells from antimicrobial/anticancer activity. Moreover, as is the case with all possible peptide drug candidates, their high susceptibility to proteases is a major challenge that has to be tackled. One solution could be the design of modified peptides and peptide conjugates to increase selectivity and lower proteolytic degradation (Reinhardt and Neundorf 2016; Feni and Neundorf 2017). In this way, some cancer-targeting peptides are now in clinical trials, and the future will show if approvals will arise during the next years.

7.3 Experimental Methods to Classify Membrane-Active Peptides

Studying peptide-membrane interactions constitutes a challenging topic up to now, and besides all processes that have been already uncovered, their detailed understanding is still elusive. This is partially based on the complexity of the membrane composition, and additionally dependent on the favored arrangement of the peptide (also the cargo in the case of CPPs) when in the presence of the lipid phase, and its following membrane insertion at the same time. Regardless, it is of singular importance to reveal the biophysical and biochemical processes behind the function of membrane-active peptides helping to design more potent molecules with tailored functionalities that may be applied as active therapeutics. Several techniques have been developed and applied to biological samples or artificial membrane systems in combination with peptides. They differ in their sensitivity, resolution, and sample preparation and are often combined to gain complementary information. Roughly one can probably divide those methods into the three following categories: (i) quantification methods to unravel, e.g., peptide content in cells, (ii) methods to determine binding constants and affinities when peptides interact with lipids, and (iii) visualization techniques and methods yielding structural information to track membrane-deforming events after peptide binding.

One focus has to be set on the choice of suitable membrane models to examine the specific lipid-peptide interaction. Various different such artificial membrane systems are presented, among those are unilamellar vesicles, mainly giant unilamellar vesicles (GUVs) and large unilamellar vesicles (LUVs), which represent the most relevant model systems to study membrane structures and dynamics. Herein, different lipid compositions are usually tested that mimic either bacterial membranes (containing mainly PG), healthy human cell membranes (containing a mixture out of PC:PE), or cancer cell membranes (containing a mixture out of PC:PE:PG). By adding sphingomyelin or cholesterol, the membrane fluidity can be further modulated, and by including appropriately labeled phospholipids, the vesicle membrane can be stained. As such it is possible to visualize membrane deformation processes or disruption after peptide addition by using fluorescence microscopy. Furthermore, encapsulating dyes within the vesicles allows for performing dye-release assays, which are easily conducted using flow cytometry or fluorescence spectroscopy. In addition to that, many researchers have made use of giant plasma membrane vesicles (GPMVs). GPMVs comprise a biologically more complex model and display a versatile tool to study membrane translocation of CPPs in conditions lacking endocytosis processes (Pae et al. 2014). Since GPMVs are released from cells after chemical induction, their lipid and protein content resembles that of the plasma membrane of living cells. Moreover, by applying low temperature, it is possible to segregate the membrane of GPMVs into different co-existing phases, namely, liquid-ordered (L_o) and liquid-disordered (L_d) membrane microdomains. By cholesterol depletion it is thus possible to determine the influence of these phases (ordered/disordered) to the lipid interaction of membrane-active peptide. For instance, Pae et al. demonstrated that amphiphilic CPP crosses more efficiently membranes that are partially depleted from cholesterol or are less ordered. On the other hand, arginine-rich CPPs translocated dependent on membrane proteinaceous components (Pae et al. 2014).

Table 7.1 Methods to classify membrane-active peptides

Method	Applied in context of…
Determination of binding constants, affinities of peptides to lipids	
Isothermal calorimetry (ITC)	Information about binding constants, affinity, thermodynamic parameters
Differential scanning calorimetry (DSC)	Phase transition measurements to yield thermodynamic parameters
Fluorescence spectroscopy	Peptide insertion into membranes, dye-release assays
Surface plasmon resonance (SPR)	Kinetic profiles of membrane lipid interaction
Quantification methods	
Flow cytometry	Association of peptides to membranes, uptake into cells
Mass spectrometry (MS)	Quantification of peptide cellular uptake or uptake into liposomes
Structural studies and visualization methods	
Infrared (IR) spectroscopy	Secondary structure of peptides in lipid phases
Circular dichroism (CD) spectroscopy	Secondary structure of peptides in lipid phase; possible in the presence of membrane vesicles or bacteria
Nuclear magnetic resonance (NMR) spectroscopy	Topology and three-dimensional structure of peptides in lipid phases; either solid or solution NMR in combination with artificial membranes or membrane lipids
Fluorescence microscopy	Cellular uptake; interaction with membrane vesicles
Electron microscopy (EM)	Membrane organization, peptide distribution at membranes
Atomic force microscopy (AFM)	In-depth membrane structures

There are several biophysical techniques available that help to study membrane-active peptides and their adsorption, location, and orientation relative to a lipid bilayer (Table 7.1).[1]

[1]A few examples are listed in the text and in Table 7.1; however, this list is not exhaustive, and also alternative methods have been used.

Circular dichroism (CD) and infrared (IR) spectroscopy give information about secondary structures of peptides. Nuclear magnetic resonance (NMR) spectroscopy yields three-dimensional structure parameters. Particularly with solid-state NMR, it is possible to gain important insights in peptide depth and positioning within model membranes. Also solution-phase NMR is frequently applied to analyze peptides in the presence of lipid micelles or other membrane mimetics. Herein, small unilamellar vesicles (SUVs) are often used as versatile models, since owing to their small size, they possess a large surface curvature and, thus, differ in their membrane topology from those found in natural membranes. Complementary, electron paramagnetic resonance (EPR) spectroscopy is useful to obtain measures about peptide-lipid interaction on the molecular level (Galdiero et al. 2013; Alves et al. 2010).

Effects on phase transitions, membrane affinities, and binding of peptides can be determined using differential scanning calorimetry (DSC), isothermal calorimetry (ITC), and surface plasmon resonance (SPR). Indeed, SPR has been applied in many studies, in which membrane-like structures are immobilized on sensor chips to measure molecular interaction with peptides. In another method internal tryptophan residues of peptides or fluorophore-labeled peptides or lipids are investigated by fluorescence spectroscopy for determining binding constants and partition coefficients. Moreover, membrane leakage or fusion can be followed by this technique.

Microscopical techniques, such as electron microscopy, atomic force microscopy, or fluorescence microscopy, are useful to track morphological changes of cells or vesicles after binding of peptides. Whereas electron and atomic force microscopies live from their high spatial resolution, classical fluorescence spectroscopy might be limited by the sensitivity of the used dyes and low resolution. However, several recent improvements like stimulated emission depletion (STED) super-resolution microscopy have pushed applications in this latter field forward (Vicidomini et al. 2018).

For quantifying the intracellular peptide content, or the amount of peptides that has crossed the membranes of liposomes, mainly fluorescent methods, such as flow cytometry, or fluorescent microscopy or spectroscopy have been applied. A smart alternative method was offered recently and uses matrix-assisted laser desorption/ionization (MALDI) mass spectrometry (MS) (Walrant et al. 2013). Since MS is not a quantitative method, an internal standard labeled with a stable isotope is added to the sample. Therefore, deuterium-labeled groups are introduced into the peptide sequence.

Another method is to perform electrical measurements on bilayers providing a tool for evaluating the increase in conductance due to the formation of pore structures at low peptide concentrations. By determining the ionic current distribution, it is thus possible to analyze the pore-forming activity according to membrane phospholipid composition, for instance, with respect to cholesterol and sphingomyelin content. In this way, direct evidence for the binding and pore-forming activity of AMPs in either anionic or zwitterionic bilayers can be delivered (dos Santos Cabrera et al. 2012).

7.4 Prediction of AMP or CPP Activity by Bioinformatics and Library Screenings

Recently, several methods have been developed to predict antimicrobial as well as cell-penetrating peptides and to establish a link to their physicochemical properties (Brand et al. 2018; Lee et al. 2018). In silico approaches have further contributed to the design of highly effective engineered peptides with cell-penetrating, antimicrobial, or even anticancer activity (Gautam et al. 2013; Tyagi et al. 2013; Deslouches et al. 2013). For instance, based on the examination of structure-function relationship studies of synthetic peptides, a library of Arg and Val or Trp composed motifs that fold into helical amphipathic structures in the presence of lipid membranes were developed (Deslouches et al. 2005; Deslouches et al. 2013). Additionally, within a recent study, physical properties of around 750 CPPs were analyzed, and it was concluded that they are median 14 residues in length and mainly cationic, the latter promoting their interaction with negatively charged outer plasma membrane constituents.

To facilitate the understanding of membrane interaction and the discovery of cell-penetrating and endosomolytic peptides, Carney et al. reported on a combinatorial library screen with liposomes (Carney et al. 2017). The authors employed an OBOC (one-bead one-compound) approach that was easily adapted to a variety of buffer and liposome compositions mimicking cellular biomembranes. By choosing the right lipid/sterol composition and by measuring the binding to zwitterionic liposomes, it was thus possible to discover several novel peptides and to successfully test them for their siRNA delivery ability. The presented method is easy to expand not only in regard of library size but also with respect to specifying the uptake pathways of the identified peptides.

Machine learning enabled antimicrobial peptide discovery, and design has been summarized recently by Lee et al. (2016). For realization, the authors trained computational models on large and high-quality data sets to perform high-throughput virtual screenings. In principle, such machine learning models fall all under the umbrella of quantitative structure-activity relationship (QSAR) models. Nicely, they help to design novel antimicrobial peptides and to discover membrane activity in diverse peptide families. Introduction of interpretable QSAR models permits also mechanistic understanding of underlying processes and provides predictable data about membrane activity.

On the other hand, several databases have sprouted during the last years that specifically compile sequence, structure, function, activity, etc. of CPPs and AMPs and give more comprehensive information (Table 7.2). Thus, the vast majority of those databases offer additional useful physicochemical parameters, like charge, isoelectric point, hydrophobicity, etc., which might be helpful for own rational peptide design strategies. Some have a particular focus on clinical or patent information or are provided as a tool to design novel AMP or CPP sequences. To date only one extensive database dedicated to CPPs has been developed (Gautam et al. 2013). *CPPsite* has more than 1700 entries, and peptides can be searched by their sequence, name, and source or defined by delivered cargo, modification, or used cell lines. Furthermore, also cyclized peptides are included in this collection. It has to be pointed out that increasing interest lies in the development of cell-permeable, small, often cyclized peptides, which, after cellular entry, interact selectively with proteins. Actually, inhibitors of protein-protein interactions in cells indeed count to a rapid developing field in pharmaceutical research (Kauffman et al. 2015). Thus, researchers have to find and determine the physical properties for a peptide that provides it with adequate membrane activity promoting binding at the lipid interface and favorable perturbance of the membrane structure. Such parameters may be found within this database and used for further peptide design. Notably, the same authors also offer another database including FDA-approved proteins and peptides (*THPdb*) (Usmani et al. 2017). Herein, some membrane-active peptides can be recovered, as, for instance, from the gramicidin family, and thus, this database might be a worthwhile extension. The number of available AMP databases is relatively large, and only some of them are listed in Table 7.2. They differ mainly in their collection of AMPs concerning source (APD offers mainly natural AMP sequences) or additional including information about, e.g., pharmacokinetic parameters and therapeutic index. Actually, it seems that these databases belonging to AMPs do not act as a network but rather in competition. It is a pity, since by merging and by updating all these data, differences and redundancies in entries would be diminished. As a consequence, researchers would greatly benefit by a more comprehensive picture about structure and function of given peptides or peptide families and their cellular mechanisms.

Table 7.2 Useful databases for membrane-active peptides

Database	Description	Link	Reference
Cell-penetrating peptides			
CPPsite	Contains around 1700 CPPs along with their structures, delivered cargos, and used cell lines	http://crdd.osdd.net/raghava/cppsite/	Gautam et al. (2013)
CPC Scientific	Commercial supplier that provides tool to assist in custom CPP design. Adapted to THPdb and CPPsite	https://www.cpcscientific.com/resources/cell-penetrating-peptide-database/	Gautam et al. (2013) and Usmani et al. (2017)
Antimicrobial peptides			
The Antimicrobial Peptide Database (APD)	Contains 2987 peptide entries from six kingdoms, mainly natural AMPs	http://aps.unmc.edu/AP/main.php https://omictools.com/apd-tool	Wang et al. (2016)
Collection of Antimicrobial Peptides (CAMP)	Design of new AMPs, links databases for sequence alignment	http://www.camp.bicnirrh.res.in/	Waghu et al. (2016)
Database of Antimicrobial Activity and Structure of Peptides (DBAASP)	Manually curated database providing information and analytical resources to develop antimicrobial compounds with high therapeutic index	https://dbaasp.org/	Pirtskhalava et al. (2016)
Data repository of antimicrobial peptides (DRAMP)	Collection of AMPs with special focus on patent and clinical information containing 17,608 entries (thereof 4833 general AMPs, 12,704 patents)	http://dramp.cpu-bioinfor.org/	Liu et al. (2017)
Other useful databases			
THPdb	Collection of FDA-approved therapeutic peptides and proteins providing information on sequence, indication, mechanism of action, pharmacodynamics, toxicity, metabolism, absorption, half-life, etc.	http://crdd.osdd.net/raghava/thpdb/index.html	Usmani et al. (2017)

To mention is further that the accuracy of these in silico tools highly depends on the applied template structure, quality of sequence alignment, and prediction method. Hence, careful analysis of the data is necessary to allow a better understanding of physicochemical and functional properties. Nonetheless, these tools are highly valuable to expand the knowledge about AMPs and CPPs and to provide new leads for the translational design of new-generation antibiotics or drug transporters. Particularly in view of production costs, concerning future clinical applications, such in silico tools offer valuable alternatives to modulate and improve natural peptide sequences.

7.5 Functional and Mechanistic Redundancy of CPPs and AMPs

Although CPPs and AMPs share physicochemical properties, like their often alpha-helical structure, cationic charge, and amphipathicity, it is still not clear why these peptides display different activities. CPPs have only limited toxicity to eukaryotic cells and the ability to cross cellular plasma membranes in both energy-independent and endocytosis processes. Although many of these share also amphipathic characteristics, the overall interfacial hydrophobicity of most CPPs is not favorable for spontaneous partitioning into membranes. Concluding that the observed activities are far more dependent on the chemical structure of a CPP, and that membrane lipid composition, applied concentration of CPPs but also lipid concentration and other bilayer physical properties play important factors for peptide-membrane interaction. Indeed, certain threshold values are indispensable for cellular uptake, even for direct translocation or for endocytosis. On the other hand, AMPs target and effectively kill bacterial cells. Arguing that the presence of bacterial cells might switch the activity of a CPP, several CPPs have indeed been tested and demonstrated to have antibacterial activity, too (Splith and Neundorf 2011). Interestingly, the same mechanistic models leading to pore formation, which have been proposed for AMPs, are postulated for CPPs when acting on bacterial cells. In many cases, helical structure or hydrophobic content is tuned by small sequence modifications leading to the one or other activity. Hereby, substitutions with positively charged arginine or aromatic residues play an important role (Piotrowska et al. 2017). However, considering their therapeutic potential, both groups share some disadvantages that often come along when using peptides such as toxicity, stability, and cost issues for their clinical and commercial development.

In spite of this observation, one might ask if the distinct membranes (prokaryotic versus eukaryotic) modulate their function. In fact, studies using artificial membrane vesicles illustrate the need of certain phospholipids present at the water-lipid interface directing the activity of the peptides in the one or other direction. Typically, CPPs and AMPs have only weak interaction with zwitterionic, synthetic membranes and strongly interact with anionic membranes. In addition, dependent on the CPP sequence, different effects after membrane binding can occur, such as vesicle aggregation and fusion, curvature formation, lipid flip-flop, membrane disruption, and many others. Indeed, some AMPs and CPPs share lipid-mixing abilities, concluding that they might be capable of triggering membrane fusion in vitro (Wadhwani et al. 2012). However, within this study of Wadhwani et al., fusion activity could not be correlated with the obtained secondary structure of membrane-bound peptides. Thus, it was hypothesized that not the particular type of secondary structure might be crucial but rather the extent of the conformational change upon membrane binding that correlates with the fusion activity. It was concluded that the driving force for fusion is the energy released when a formerly disordered, soluble peptide binds to a vesicle and acquires a secondary structure by H-bond formation (Wadhwani et al. 2012). Although pre-folded peptides are suggested not to promote much vesicle fusion, this well-defined fold might be nonetheless supportive for other functions of CPPs, like lipid-phase interaction and cargo delivery of cyclic peptides (Horn et al. 2016; Lattig-Tunnemann et al. 2011). Also charge distribution patterns were found to play a significant role for peptide-membrane interaction, particularly for membrane insertion of lytic and cell-penetrating peptides (Chen et al. 2017). In fact, charge distribution alongside the amphipathic helix might be the key factor affecting peptide insertion ability and thus bilayer disruption. Moreover, surface pressure is likely increased by both groups of peptides, CPPs and lytic peptides, whereas for the latter the pressure declines owing to molecule rearrangements after peptide insertion into the membrane.

On the other side, many AMPs have mainly deforming effects on the lipid phase, leading often to total membrane rupture. The response of AMPs to synthetic lipid compositions can be easily modulated by the presence of distinct molecules like cholesterol, sphingomyelin, or cardiolipin within the lipid phase. By the presence of cholesterol, eukaryotic membranes can be mimicked, and usually the activity of AMPs is decreased (Reinhardt et al. 2014). More recent findings have already demonstrated that for some AMPs their lack in cell selectivity might be based on aggregation behavior in solution and propensity to proteolytic degradation. For instance, LL-37 exists in equilibrium between monomers and oligomers in solution and is highly resistant to proteolytic degradation when bound to both zwitterionic and negatively charged membranes. The oligomerized state might support a detergent-like effect also in the presence of zwitterionic membranes (Oren et al. 1999).

One more parameter that is strongly linked to membrane insertion activity is the spontaneous lipid curvature triggering response of membrane-active peptides. Recently, Koller and Lohner perfectly discussed how interfacially active peptides induce membrane curvature and which factors facilitate its formation (Koller and Lohner 2014). Membrane curvature is a result of a complex interplay between membrane proteins, lipids, and physical forces that are applied to the membrane surface. Nevertheless, membrane phospholipids itself have an intrinsic property to adopt planar or curved lipid molecular shapes leading to the formation of positive or negative curvature. Since the bilayer contains an asymmetrical lipid distribution, the different physical parameters of, e.g., PE at the inner monolayer and PC at the outer monolayer, induce curvature to a different extent. Membrane-active peptides induce curvature after incorporation in a bilayer; however, curvature induction by a peptide may differ for different lipid systems. Several parameters play a role, like H-bonding, electrostatic repulsion, monolayer surface area, and lateral pressure. Otherwise membrane curvature triggers peptide response, what becomes noticeable in peptide secondary structure formation. Thus, not only is membrane curvature spontaneously changed upon incorporation of a guest peptide, but lipid curvature can also influence the preference of a given peptide to insert into the host lipid matrix. For instance, the orientation of a peptide in the membrane is a key parameter determining the subsequent mechanism of action (*toroidal pore*, *barrel stave*, etc.). Hence, different modes of peptide-lipid interaction can be expected depending on the different cell types, which differ in terms of lipid composition and as a result in the amount of lamellar and non-lamellar phase-forming lipids present in the target membrane. PE exhibits a negative spontaneous curvature and is more prone to membrane disruption by interfacial active peptides. Given that PE is more abundant in the cytoplasmic membrane of Gram-negative bacteria and less present at the outer monolayer of eukaryotic cells, the selectivity of such peptides targeting membrane curvature can be nicely explained.

7.6 Conclusions

Membrane-active peptides are a class of peptides with undisputable relevant functions. Antimicrobial peptides are negotiated as highly promising new weapons against bacterial infections, while cell-penetrating peptides act as versatile delivery tools. Although both groups of peptides share physicochemical properties, their intrinsic function and mechanism of action can only hardly be predicted. It is relatively certain that in both cases lipid composition and constitution play major roles in peptide interaction and that the nature of the lipid phase somehow defines the orientation of the peptides. So, size of the phospholipid headgroups and bilayer elastic properties and dynamics are important determinants. On the other side, it is clearly demonstrated that charge is a further important key to attract cationic peptides to membrane surfaces. One way to achieve peptide selectivity would thus be to equip it with positive charge, making binding to anionic bacterial membranes (or to that of tumoral cells) more probable. Additionally, structural rearrangements during peptide-lipid interplay that help to bind and to position the peptide within the lipid bilayer are certainly essential. As already mentioned, folding into amphipathic helices might stabilize the membrane-bound state and increase the probability of the peptide to partition into the bilayer. Following, the interaction of AMPs or CPPs with lipid bilayers causes either local rupture (AMP) or transient permeation (CPP). How one can predict and exactly determine these two functions has to be one of the main foci within this research field in the future.

However, several tools that link those different functions are nowadays available, like databases, molecular dynamic studies, and other experimental classification systems. In this regard, it might be possible to use AMPs as a constructive template to design CPPs and, of course, vice versa. Therefore, it is hoped that in the near future, a more obvious picture can be drawn to unravel the differences of AMPs and CPPs and that this knowledge will help in the engineering and discovery of more efficient and safer peptide sequences.

References

Alvares DS, Viegas TG, Ruggiero Neto J (2017) Lipid-packing perturbation of model membranes by pH-responsive antimicrobial peptides. Biophys Rev 9(5):669–682. https://doi.org/10.1007/s12551-017-0296-0

Alves ID, Jiao CY, Aubry S, Aussedat B, Burlina F, Chassaing G, Sagan S (2010) Cell biology meets biophysics to unveil the different mechanisms of penetration internalization in cells. Biochim Biophys Acta 1798(12):2231–2239. https://doi.org/10.1016/j.bbamem.2010.02.009

Baxter AA, Lay FT, Poon IKH, Kvansakul M, Hulett MD (2017) Tumor cell membrane-targeting cationic antimicrobial peptides: novel insights into mechanisms of action and therapeutic prospects. Cell Mol Life Sci 74(20):3809–3825. https://doi.org/10.1007/s00018-017-2604-z

Bergmann R, Splith K, Pietzsch J, Bachmann M, Neundorf I (2017) Biological characterization of novel nitroimidazole-peptide conjugates in vitro and in vivo. J Pep Sci Off Publ Eur Pep Soc 23(7–8):597–609. https://doi.org/10.1002/psc.2995

Brand GD, Ramada MHS, Genaro-Mattos TC, Bloch C Jr (2018) Towards an experimental classification system for membrane active peptides. Sci Rep 8(1):1194. https://doi.org/10.1038/s41598-018-19566-w

Brender JR, McHenry AJ, Ramamoorthy A (2012) Does cholesterol play a role in the bacterial selectivity of antimicrobial peptides? Front Immunol 3:195. https://doi.org/10.3389/fimmu.2012.00195

Brown DA (2002) Isolation and use of rafts. Curr Protoc Immunol Chapter 11:Unit 11:10. https://doi.org/10.1002/0471142735.im1110s51

Carney RP, Thillier Y, Kiss Z, Sahabi A, Heleno Campos JC, Knudson A, Liu R, Olivos D, Saunders M, Tian L, Lam KS (2017) Combinatorial library screening with liposomes for discovery of membrane active peptides. ACS Comb Sci 19(5):299–307. https://doi.org/10.1021/acscombsci.6b00182

Caveney NA, Li FK, Strynadka NC (2018) Enzyme structures of the bacterial peptidoglycan and wall teichoic acid biogenesis pathways. Curr Opin Struct Biol 53:45–58. https://doi.org/10.1016/j.sbi.2018.05.002

Chen CJ, Tsai KC, Kuo PH, Chang PL, Wang WC, Chuang YJ, Chang MDT (2015) A heparan sulfate-binding cell penetrating peptide for tumor targeting and migration inhibition. Biomed Res Int:237969. https://doi.org/10.1155/2015/237969

Chen L, Zhang Q, Yuan X, Cao Y, Yuan Y, Yin H, Ding X, Zhu Z, Luo SZ (2017) How charge distribution influences the function of membrane-active peptides: lytic or cell-penetrating? Int J Biochem Cell Biol 83:71–75. https://doi.org/10.1016/j.biocel.2016.12.011

da Costa JP, Cova M, Ferreira R, Vitorino R (2015) Antimicrobial peptides: an alternative for innovative medicines? Appl Microbiol Biotechnol 99(5):2023–2040. https://doi.org/10.1007/s00253-015-6375-x

Deslouches B, Di YP (2017) Antimicrobial peptides with selective antitumor mechanisms: prospect for anticancer applications. Oncotarget 8(28):46635–46651. https://doi.org/10.18632/oncotarget.16743

Deslouches B, Phadke SM, Lazarevic V, Cascio M, Islam K, Montelaro RC, Mietzner TA (2005) De novo generation of cationic antimicrobial peptides: influence of length and tryptophan substitution on antimicrobial activity. Antimicrob Agents Chemother 49(1):316–322. https://doi.org/10.1128/AAC.49.1.316-322.2005

Deslouches B, Steckbeck JD, Craigo JK, Doi Y, Mietzner TA, Montelaro RC (2013) Rational design of engineered cationic antimicrobial peptides consisting exclusively of arginine and tryptophan, and their activity against multidrug-resistant pathogens. Antimicrob Agents Chemother 57(6):2511–2521. https://doi.org/10.1128/AAC.02218-12

Di Pisa M, Chassaing G, Swiecicki JM (2015) Translocation mechanism(s) of cell-penetrating peptides: biophysical studies using artificial membrane bilayers. Biochemistry 54(2):194–207. https://doi.org/10.1021/bi501392n

dos Santos Cabrera MP, Arcisio-Miranda M, Gorjao R, Leite NB, de Souza BM, Curi R, Procopio J, Ruggiero Neto J, Palma MS (2012) Influence of the bilayer composition on the binding and membrane disrupting effect of Polybia-MP1, an antimicrobial mastoparan peptide with leukemic T-lymphocyte cell selectivity. Biochemistry 51(24):4898–4908. https://doi.org/10.1021/bi201608d

Fadnes B, Rekdal O, Uhlin-Hansen L (2009) The anticancer activity of lytic peptides is inhibited by heparan sulfate on the surface of the tumor cells. BMC Cancer 9:183. https://doi.org/10.1186/1471-2407-9-183

Fadnes B, Uhlin-Hansen L, Lindin I, Rekdal O (2011) Small lytic peptides escape the inhibitory effect of heparan sulfate on the surface of cancer cells. BMC Cancer 11:116. https://doi.org/10.1186/1471-2407-11-116

Felicio MR, Silva ON, Goncalves S, Santos NC, Franco OL (2017) Peptides with dual antimicrobial and anticancer activities. Front Chem 5:5. https://doi.org/10.3389/fchem.2017.00005

Feni L, Neundorf I (2017) The current role of cell-penetrating peptides in cancer therapy. Adv Exp Med Biol 1030:279–295. https://doi.org/10.1007/978-3-319-66095-0_13

Galdiero S, Falanga A, Cantisani M, Vitiello M, Morelli G, Galdiero M (2013) Peptide-lipid interactions: experiments and applications. Int J Mol Sci 14(9):18758–18789. https://doi.org/10.3390/ijms140918758

Gautam A, Chaudhary K, Kumar R, Sharma A, Kapoor P, Tyagi A, Open Source Drug Discovery Consortium, Raghava GP (2013) In silico approaches for designing highly effective cell penetrating peptides. J Transl Med 11:74. https://doi.org/10.1186/1479-5876-11-74

Gronewold A, Horn M, Randelovic I, Tovari J, Munoz Vazquez S, Schomacker K, Neundorf I (2017) Characterization of a cell-penetrating peptide with potential anticancer activity. ChemMedChem 12(1):42–49. https://doi.org/10.1002/cmdc.201600498

Gruenheid S, Le Moual H (2012) Resistance to antimicrobial peptides in Gram-negative bacteria. FEMS Microbiol Lett 330(2):81–89. https://doi.org/10.1111/j.1574-6968.2012.02528.x

Horn M, Reichart F, Natividad-Tietz S, Diaz D, Neundorf I (2016) Tuning the properties of a novel short cell-penetrating peptide by intramolecular cyclization with a triazole bridge. Chem Commun (Camb) 52(11):2261–2264. https://doi.org/10.1039/c5cc08938g

Jobin ML, Alves ID (2014) On the importance of electrostatic interactions between cell penetrating peptides and membranes: a pathway toward tumor cell selectivity? Biochimie 107(Pt A):154–159. https://doi.org/10.1016/j.biochi.2014.07.022

Kalafatovic D, Giralt E (2017) Cell-penetrating peptides: design strategies beyond primary structure and amphipathicity. Molecules 22(11):1929. https://doi.org/10.3390/molecules22111929

Kauffman WB, Fuselier T, He J, Wimley WC (2015) Mechanism matters: a taxonomy of cell penetrating peptides. Trends Biochem Sci 40(12):749–764. https://doi.org/10.1016/j.tibs.2015.10.004

Kawaguchi Y, Takeuchi T, Kuwata K, Chiba J, Hatanaka Y, Nakase I, Futaki S (2016) Syndecan-4 is a receptor for Clathrin-mediated endocytosis of arginine-rich cell-penetrating peptides. Bioconjug Chem 27(4):1119–1130. https://doi.org/10.1021/acs.bioconjchem.6b00082

Koller D, Lohner K (2014) The role of spontaneous lipid curvature in the interaction of interfacially active peptides with membranes. Biochim Biophys Acta 1838(9):2250–2259. https://doi.org/10.1016/j.bbamem.2014.05.013

Last NB, Schlamadinger DE, Miranker AD (2013) A common landscape for membrane-active peptides. Protein Sci 22(7):870–882. https://doi.org/10.1002/pro.2274

Lattig-Tunnemann G, Prinz M, Hoffmann D, Behlke J, Palm-Apergi C, Morano I, Herce HD, Cardoso MC (2011) Backbone rigidity and static presentation of guanidinium groups increases cellular uptake of arginine-rich cell-penetrating peptides. Nat Commun 2:453. https://doi.org/10.1038/ncomms1459

Le CF, Fang CM, Sekaran SD (2017) Intracellular targeting mechanisms by antimicrobial peptides. Antimicrob Agents Chemother 61(4). https://doi.org/10.1128/AAC.02340-16

Lee EY, Fulan BM, Wong GC, Ferguson AL (2016) Mapping membrane activity in undiscovered peptide sequence space using machine learning. Proc Natl Acad Sci U S A 113(48):13588–13593. https://doi.org/10.1073/pnas.1609893113

Lee EY, Wong GCL, Ferguson AL (2018) Machine learning-enabled discovery and design of membrane-active peptides. Bioorg Med Chem 26(10):2708–2718. https://doi.org/10.1016/j.bmc.2017.07.012

Letoha T, Keller-Pinter A, Kusz E, Kolozsi C, Bozso Z, Toth G, Vizler C, Olah Z, Szilak L (2010) Cell-penetrating peptide exploited syndecans. BBA-Biomembranes 1798(12):2258–2265. https://doi.org/10.1016/j.bbamem.2010.01.022

Liu S, Fan L, Sun J, Lao X, Zheng H (2017) Computational resources and tools for antimicrobial peptides. J Pep Sci Off Publ Eur Pep Soc 23(1):4–12. https://doi.org/10.1002/psc.2947

Lopez D (2015) Molecular composition of functional microdomains in bacterial membranes. Chem Phys Lipids 192:3–11. https://doi.org/10.1016/j.chemphyslip.2015.08.015

Maria-Neto S, de Almeida KC, Macedo ML, Franco OL (2015) Understanding bacterial resistance to antimicrobial peptides: from the surface to deep inside. Biochim Biophys Acta 1848(11 Pt B):3078–3088. https://doi.org/10.1016/j.bbamem.2015.02.017

McHenry AJ, Sciacca MF, Brender JR, Ramamoorthy A (2012) Does cholesterol suppress the antimicrobial peptide induced disruption of lipid raft containing membranes? Biochim Biophys Acta 1818(12):3019–3024. https://doi.org/10.1016/j.bbamem.2012.07.021

Neundorf I, Rennert R, Hoyer J, Schramm F, Lobner K, Kitanovic I, Wolfl S (2009) Fusion of a short HA2-derived peptide sequence to cell-penetrating peptides improves cytosolic uptake, but enhances cytotoxic activity. Pharmaceuticals (Basel) 2(2):49–65. https://doi.org/10.3390/ph2020049

Oren Z, Lerman JC, Gudmundsson GH, Agerberth B, Shai Y (1999) Structure and organization of the human antimicrobial peptide LL-37 in phospholipid membranes: relevance to the molecular basis for its non-cell-selective activity. Biochem J 341(Pt 3):501–513

Pae J, Saalik P, Liivamagi L, Lubenets D, Arukuusk P, Langel U, Pooga M (2014) Translocation of cell-penetrating peptides across the plasma membrane is controlled by cholesterol and microenvironment created by membranous proteins. J Control Release 192:103–113. https://doi.org/10.1016/j.jconrel.2014.07.002

Patel S, Akhtar N (2017) Antimicrobial peptides (AMPs): the quintessential 'offense and defense' molecules are more than antimicrobials. Biomed Pharmacother 95:1276–1283. https://doi.org/10.1016/j.biopha.2017.09.042

Piotrowska U, Sobczak M, Oledzka E (2017) Current state of a dual behaviour of antimicrobial peptides-therapeutic agents and promising delivery vectors. Chem Biol Drug Des 90(6):1079–1093. https://doi.org/10.1111/cbdd.13031

Pirtskhalava M, Gabrielian A, Cruz P, Griggs HL, Squires RB, Hurt DE, Grigolava M, Chubinidze M, Gogoladze G, Vishnepolsky B, Alekseyev V, Rosenthal A, Tartakovsky M (2016) DBAASP v.2: an enhanced database of structure and antimicrobial/cytotoxic activity of natural and synthetic peptides. Nucleic

Acids Res 44(13):6503. https://doi.org/10.1093/nar/gkw243

Pushpanathan M, Gunasekaran P, Rajendhran J (2013) Antimicrobial peptides: versatile biological properties. Int J Pept 2013:675391. https://doi.org/10.1155/2013/675391

Ran S, Downes A, Thorpe PE (2002) Increased exposure of anionic phospholipids on the surface of tumor blood vessels. Cancer Res 62(21):6132–6140

Raucher D, Ryu JS (2015) Cell-penetrating peptides: strategies for anticancer treatment. Trends Mol Med. https://doi.org/10.1016/j.molmed.2015.06.005

Reinhardt A, Neundorf I (2016) Design and application of antimicrobial peptide conjugates. Int J Mol Sci 17(5):701. https://doi.org/10.3390/ijms17050701

Reinhardt A, Horn M, Schmauck JP, Brohl A, Giernoth R, Oelkrug C, Schubert A, Neundorf I (2014) Novel imidazolium salt – peptide conjugates and their antimicrobial activity. Bioconjug Chem 25(12):2166–2174. https://doi.org/10.1021/bc500510c

Rodriguez Plaza JG, Morales-Nava R, Diener C, Schreiber G, Gonzalez ZD, Lara Ortiz MT, Ortega Blake I, Pantoja O, Volkmer R, Klipp E, Herrmann A, Del Rio G (2014) Cell penetrating peptides and cationic antibacterial peptides: two sides of the same coin. J Biol Chem 289(21):14448–14457. https://doi.org/10.1074/jbc.M113.515023

Roudi R, Syn NL, Roudbary M (2017) Antimicrobial peptides as biologic and immunotherapeutic agents against cancer: a comprehensive overview. Front Immunol 8:1320. https://doi.org/10.3389/fimmu.2017.01320

Sezgin E, Levental I, Mayor S, Eggeling C (2017) The mystery of membrane organization: composition, regulation and roles of lipid rafts. Nat Rev Mol Cell Biol 18(6):361–374. https://doi.org/10.1038/nrm.2017.16

Sierra JM, Fuste E, Rabanal F, Vinuesa T, Vinas M (2017) An overview of antimicrobial peptides and the latest advances in their development. Expert Opin Biol Ther 17(6):663–676. https://doi.org/10.1080/14712598.2017.1315402

Simons K, Vaz WL (2004) Model systems, lipid rafts, and cell membranes. Annu Rev Biophys Biomol Struct 33:269–295. https://doi.org/10.1146/annurev.biophys.32.110601.141803

Splith K, Neundorf I (2011) Antimicrobial peptides with cell-penetrating peptide properties and vice versa. Eur Biophys J EBJ 40(4):387–397. https://doi.org/10.1007/s00249-011-0682-7

Splith K, Bergmann R, Pietzsch J, Neundorf I (2012) Specific targeting of hypoxic tumor tissue with nitroimidazole-peptide conjugates. ChemMedChem 7(1):57–61. https://doi.org/10.1002/cmdc.201100401

Szczepanski C, Tenstad O, Baumann A, Martinez A, Myklebust R, Bjerkvig R, Prestegarden L (2014) Identification of a novel lytic peptide for the treatment of solid tumours. Genes Cancer 5(5–6):186–200

Tyagi A, Kapoor P, Kumar R, Chaudhary K, Gautam A, Raghava GP (2013) In silico models for designing and discovering novel anticancer peptides. Sci Rep 3:2984. https://doi.org/10.1038/srep02984

Usmani SS, Bedi G, Samuel JS, Singh S, Kalra S, Kumar P, Ahuja AA, Sharma M, Gautam A, Raghava GPS (2017) THPdb: database of FDA-approved peptide and protein therapeutics. PLoS One 12(7):e0181748. https://doi.org/10.1371/journal.pone.0181748

Utsugi T, Schroit AJ, Connor J, Bucana CD, Fidler IJ (1991) Elevated expression of phosphatidylserine in the outer membrane leaflet of human tumor cells and recognition by activated human blood monocytes. Cancer Res 51(11):3062–3066

Vicidomini G, Bianchini P, Diaspro A (2018) STED super-resolved microscopy. Nat Methods 15(3):173–182. https://doi.org/10.1038/nmeth.4593

Wadhwani P, Reichert J, Burck J, Ulrich AS (2012) Antimicrobial and cell-penetrating peptides induce lipid vesicle fusion by folding and aggregation. Eur Biophys J EBJ 41(2):177–187. https://doi.org/10.1007/s00249-011-0771-7

Waghu FH, Barai RS, Gurung P, Idicula-Thomas S (2016) CAMPR3: a database on sequences, structures and signatures of antimicrobial peptides. Nucleic Acids Res 44(D1):D1094–D1097. https://doi.org/10.1093/nar/gkv1051

Walrant A, Matheron L, Cribier S, Chaignepain S, Jobin ML, Sagan S, Alves ID (2013) Direct translocation of cell-penetrating peptides in liposomes: a combined mass spectrometry quantification and fluorescence detection study. Anal Biochem 438(1):1–10. https://doi.org/10.1016/j.ab.2013.03.009

Wang G, Li X, Wang Z (2016) APD3: the antimicrobial peptide database as a tool for research and education. Nucleic Acids Res 44(D1):D1087–D1093. https://doi.org/10.1093/nar/gkv1278

Watkins CL, Schmaljohann D, Futaki S, Jones AT (2009) Low concentration thresholds of plasma membranes for rapid energy-independent translocation of a cell-penetrating peptide. Biochem J 420(2):179–189. https://doi.org/10.1042/BJ20090042

Wu D, Gao Y, Qi Y, Chen L, Ma Y, Li Y (2014) Peptide-based cancer therapy: opportunity and challenge. Cancer Lett 351(1):13–22. https://doi.org/10.1016/j.canlet.2014.05.002

Synthetic Anti-lipopolysaccharide Peptides (SALPs) as Effective Inhibitors of Pathogen-Associated Molecular Patterns (PAMPs)

8

Wilmar Correa, Lena Heinbockel, Guillermo Martinez-de-Tejada, Susana Sánchez, Patrick Garidel, Tobias Schürholz, Walter Mier, Aline Dupont, Mathias Hornef, Thomas Gutsmann, Karl Mauss, Günther Weindl, and Klaus Brandenburg

Abstract

Antimicrobial peptides (AMPs) are in the focus of scientific research since the 1990s. In most cases, the main aim was laid on the design of AMP to kill bacteria effectively, with particular emphasis on broadband action and independency on antibiotic resistance. However, so far no approved drug on the basis of AMP has entered the market.

Our approach of constructing AMP, called synthetic anti-lipopolysaccharide peptides (SALPs), on the basis of inhibiting the inflammatory action of lipopolysaccharide (LPS, endotoxin) from Gram-negative bacteria was focused on the neutralization of the decisive toxins. These are, beside LPS from Gram-negative bacteria, the lipoproteins (LP) from Gram-positive origin. Although some of the SALPs have an antibacterial action, the most

W. Correa · T. Gutsmann
LG Biophysik, Forschungszentrum Borstel, Leibniz Lungenzentrum, Borstel, Germany

L. Heinbockel
Forschungszentrum Borstel, Leibniz Lungenzentrum, Klinische und Experimentelle Pathologie, Borstel, Germany

G. Martinez-de-Tejada · S. Sánchez
Dep. de Microbiologia, Universidad de Navarra, Pamplona, Spain

P. Garidel
Universität Halle-Wittenberg, Halle/Saale, Germany

T. Schürholz
Klinik&Poliklinik für Anästhesiologie und Intensivtherapie, Universitätsmedizin Rostock, Rostock, Germany

W. Mier
Radiopharmazeutische Chemie, Universitätsklinikum Heidelberg, Heidelberg, Germany

A. Dupont · M. Hornef
Universitätsklinik Aachen, Institute of Medical Microbiology, Aachen, Germany

K. Mauss
Asklepios-Klinik Hamburg-Altona, Hamburg, Germany

G. Weindl
Freie Universität Berlin, Institut für Pharmazie, Berlin, Germany

K. Brandenburg (✉)
Brandenburg Antiinfektiva GmbH, c/o Forschungszentrum Borstel, Borstel, Germany
e-mail: kbrandenburg@fz-borstel.de

K. Matsuzaki (ed.), *Antimicrobial Peptides*, Advances in Experimental Medicine and Biology 1117,
https://doi.org/10.1007/978-981-13-3588-4_8

important property is the high-affinity binding to LPS and LP, whether as constituent of the bacteria or in free form which prevents the damaging inflammation, that could otherwise lead to life-threatening septic shock. Most importantly, the SALP may inhibit inflammation independently of the resistance status of the bacteria, and so far the repeated use of the peptides apparently does not cause resistance of the attacking pathogens.

In this chapter, an overview is given over the variety of possible applications in the field of fighting against severe bacterial infections, from the use in systemic infection/inflammation up to various topical applications such as anti-biofilm action and severe skin and soft tissue infections.

8.1 General Remarks

According to the WHO, about 700,000 people die every year due to an infection induced by resistant bacteria. These bacteria have developed resistance against most commonly used antibiotics (Molchanova et al. 2017). A number of bacteria have developed a strong resistance; the WHO has recently established a priority list, classifying bacteria against which antibiotics have to be developed (Global priority list of antibiotic-resistant bacteria to guide research, discovery, and development of new antibiotics) (Tacconelli et al. 2018). Table 8.1 lists the most critical bacteria and their drug resistance.

Since millions of years, host defense peptides, known as antimicrobial peptides (AMPs), are present in the animal, plant, and bacterial world, to protect the organism. The strengths of these naturally occurring AMP are (i) the broad spectrum of biological activity, (ii) the quasi-absence of antimicrobial resistance development, (iii) the possibility of a "cocktail" of AMP with improved activity, and (iv) an additional LPS-neutralizing activity of some AMP (Boto et al. 2018; Brandenburg et al. 2011; Brandenburg et al. 2016). Although it is described that bacterial resistance against AMP is low, different studies have already shown the opposite, namely, that bacteria have developed resistance against antimicrobial peptides, such as the resistance observed for *S. aureus* against pexiganan (Fox 2013; Pfalzgraff et al. 2018b and references cited therein). The potential mechanisms that might lead to bacterial resistance to AMP are listed in Table 8.2.

These bacterial resistance mechanisms include:

- Surface charge modification
- Cell surface structure modification
- Active efflux
- Protease release
- AMP sequestering
- Host AMP downregulation

Although a few commercialized AMP or peptide analogous are already available, the progresses made in the last decades have been modest (Table 8.3) (Mishra et al. 2017). In fact, the first commercialized AMP was put on the market more than half a century ago.

One of the most critical clinical pathologies is sepsis, triggered by microbial infection and leads to malfunctions of the lungs, kidneys, or other vital organs. It is a quite common, potentially fatal systemic illness with high mortalities (close to 50%) (Andra et al. 2006).

The development of therapeutic agents for the treatment of sepsis has been fostered for the last four decades with very limited therapeutic success so far. With few exceptions, the results of these clinical trials have been disappointing, and no specific therapeutic agent is currently approved for the treatment of sepsis. Strategies to improve drug development for sepsis were reviewed recently by Fink and Warren (2014). Their conclusions to improve upon this dismal record were focused on (i) the need for investigators to identify more suitable therapeutic targets, (ii) improvement of the strategies and approaches for selecting appropriate candidate compounds for clinical development, and (iii) adoption and adjustment of the clinical trial design and treatments.

Therefore, it was necessary to characterize the target structures (Brandenburg et al. 2003) and

Table 8.1 WHO priority pathogens for research and development of new antibiotics (Tacconelli et al. 2018)

Priority classification	Pathogen	Resistance
1. Critical[a]	*Acinetobacter baumannii*	Carbapenem-resistant
	Pseudomonas aeruginosa	Carbapenem-resistant
	Enterobacteriaceae[b]	Carbapenem-resistant
		Third-generation cephalosporin-resistant
2. High	*Enterococcus faecium*	Vancomycin-resistant
	Staphylococcus aureus	Methicillin-resistant
		Vancomycin intermediate and resistant
	Helicobacter pylori	Clarithromycin-resistant
	Campylobacter	Fluoroquinolone-resistant
	Salmonella spp.	Fluoroquinolone-resistant
	Neisseria gonorrhoeae	Third-generation cephalosporin-resistant, fluoroquinolone-resistant
3. Medium	*Streptococcus pneumoniae*	Penicillin-non-susceptible
	Haemophilus influenzae	Ampicillin-resistant
	Shigella spp.	Fluoroquinolone-resistant

[a]Mycobacteria (including *Mycobacterium tuberculosis*, the cause of human tuberculosis) were not subjected to review for inclusion in this prioritization exercise as they are already a globally established priority for which innovative new treatments are urgently needed
[b]*Enterobacteriaceae* include *Klebsiella pneumoniae*, *Escherichia coli*, *Enterobacter* spp., *Serratia* spp., *Proteus* spp., *Providencia* spp., and *Morganella* spp.

Table 8.2 Summary of potential bacterial resistance mechanisms to reduce the therapeutic effect of AMP (Sierra et al. 2017)

Bacterial resistance mechanisms against antimicrobial peptides	
Possible resistance mechanisms	Consequence
Net surface charge modification with an increase of the positive charge potential	Reduction AMP attraction, especially relevant for electrostatic interactions between the host and the peptide
Gram-negative bacteria: main modification of LPS headgroup charge	
Gram-positive bacteria: modification of lipoteichoic acid	
Protease: release of proteases, such as metalloproteases or serine endopeptidases and/or increase of protease concentration	Increased proteolytic degradation of the peptides
	Inactivation of AMP by bacterial proteases strongly depends on the peptide structure, given a higher susceptibility to degradation of linear peptides compared to cyclic peptides containing disulfide bonds
Modification of cell surface structure: changes the phospholipid composition and thus alters the cytoplasmic membrane structure and packing characteristics by chemical modification of specific lipids	Masking the anionic phosphate group with cationic primary amines and decreasing AMP attraction and membrane insertion
For example, amino acylation of phosphatidylglycerol headgroups	
General resistance to AMP	Can further reduce susceptibility to antibiotics
Active efflux modulation: efflux pumps belonging to the RND (resistance-nodulation cell division) family transporters in Gram-negative bacteria	Increase the AMP transport out of the cell and reduce the AMP concentration inside
Downregulation of host AMP production: by interference with or suppression of AMP production	Reduction of host AMP concentration
AMP sequestration: e.g., of HNP-1 and HNP-2 sequestration by *S. aureus* staphylokinase	Reduce the active AMP concentration

AMP antimicrobial peptide, *LPS* lipopolysaccharide

Table 8.3 Selected examples of commercialized AMP

INN and molar mass/gmol^{-1}	Brand names	Admin.	Therapeutic indication and relevant information
Anidulafungin	Eraxis	i.v.	Semisynthetic echinocandin used as an antifungal drug
MM: 1140	Ecalta		Treatment of invasive *Aspergillus* infections
	Pfizer		FDA approved in 2006
Bacitracin	Baciim	top.	Mixture of related cyclic peptides disrupting Gram-positive bacteria by interfering with the call wall and peptidoglycan synthesis
MM: 1423	Neosporin	i.m.	Used against skin infections
			Available as bacitracin zinc salt and used in "triple" ointment composed of additionally PMB and neomycin
			Used in humans and chickens/turkey
			FDA approved: 1948
Colistin (polymyxin E)	Colomycin	top.	Acute/chronic infections due to sensitive strains of specific gram-negative microorganism
MM: 1155	Coly-Mycin	oral	Last-in-line treatment
	Promixin	i.v.	Highly toxic
	ColistiFlex	inhal.	Available since 1959
Dalbavancin	Dalvance (US)	i.v.	Second-generation lipoglycopeptide antibiotic; semisynthetic, belongs to the same class as vancomycin
MM: 1817	Xydalba (EU)		One of the few treatments available to patients infected with MRSA
			FDA approved in 2014 for the treatment of ABSSIs, including MRSA and *Streptococcus pyogenes* infections
			EMA approved in 2006
Daptomycin	Cubicin		First marketed cyclic lipopeptide
MM: 1621			Solely active against Gram-positive bacteria
			Complex skin and skin structure infections caused by susceptible strains of Gram-positive bacteria
Enfuvirtide	Fuzeon	s.c.	Antiretroviral drugs for the treatment of AIDS/HIV-1 infection
MM: 4492	Roche		FDA approved in 2003 as the first HIV fusion inhibitor
Oritavancin	Orbactiv	i.v.	Semisynthetic glycopeptide for the treatment of serious Gram-positive bacterial infections
MM: 1793	Eli Lilly		FDA approved in 2014 for treatments of skin infections
			EMA approved in 2015 for the treatment of acute bacterial skin and skin structure infections in adults
Teicoplanin	Targocid (Sanofi)	i.v.	A semisynthetic glycopeptide antibiotic used in the prophylaxis and treatment of serious infections caused by Gram-positive bacteria, incl. MRSA and *Enterococcus*
MM: 1564–1908	Ticocin (Cipla)	i.m.	
Telavancin	Vibativ	i.v	Bactericidal lipoglycolipid for use in MRSA or other Gram-positive infections
MM: 1756			Telavancin: semisynthetic derivative of vancomycin
			FDA approved in 2009 for cSSI
Vancomycin	Vancocin (as Hydro chlorid, Eli Lilly)	i.v.	Glycopeptide antibiotic
MM: 1449		top.	Treatment for complicated skin, bloodstream, endocarditis bone and joint infections, and meningitis caused by susceptible strains of methicillin-resistant (beta-lactam-resistant) *Staphylococci*
			First sold in 1954

ABSSSIs acute bacterial skin and skin structure infections, *Admin.* administration, *cSSI* complicated skin and skin structure infections, *EMA* European Medicines Agency, *FDA* Food and Drug Administration, *i.m.* intramuscular, *inhal.* inhalation, *i.v.* intravenous, *MM* molar mass expressed in g/mol, *MRSA* Methicillin-resistant *Staphylococcus aureus*, *PMB* polymyxin B, *s.c.* subcutaneous, *top.* topical

Table 8.4 Pharmaceutical challenges for AMP

Pharmaceutical challenges
Physical and chemical degradation
Biological and enzymatic degradation
Colloidal instabilities and the formation of associates and aggregates
Solubility
Ability to generate conformers
Limited half-life
Fast elimination
Reduced membrane permeability
Limited or complex drug delivery and overall drug product stability
Synthesis/expression, purification and manufacturing (CMC issues)
Costs

understand the biophysics of the target-AMP interaction (Andra et al. 2005; Howe et al. 2007; Brandenburg et al. 2005, 2010, b; Kaconis et al. 2011). A deeper understanding of the mechanism of action of current peptide and new candidates is still necessary (Andra et al. 2007; Garidel and Brandenburg 2009; Mirski et al. 2017). Optimization of naturally occurring host defense AMP is performed via, e.g., structure-activity relationship studies (Mishra et al. 2017; Sierra et al. 2017).

Peptide molecules are in general quite labile structures. In order to develop a therapeutic drug, a number of pharmaceutical challenges, such as stability, formulation, and drug delivery, have to be addressed (Katarzyna and Małgorzata 2017; Nordström and Malmsten 2017; Kuhlmann et al. 2018). Table 8.4 summarizes the most prominent pharmaceutical development challenges that have to be undertaken in order to develop a successful therapeutic drug.

In order to reduce resistance, synthetic antimicrobial peptides are under investigations as anti-lipopolysaccharide peptides (Habersetzer et al. 2013; Garidel et al. 2007).

8.2 Development of SALP

The development of synthetic anti-lipopolysaccharide peptides (SALPs) was based on a systematic sequence variation of the LPS-binding structure of the *Limulus* anti-LPS factor (LALF), also called endotoxin-neutralizing protein (ENP) in recombinant form. The LPS-binding polypeptide of LALF22 ($LALF_{31-52}$), a polypeptide with the sequence HYRIKPTFRRLKWKYKGKFWCG, was taken as starting template. Multiple structural variants, all in a cyclated form bridged by N- and C-terminal cysteine linkages, have been analyzed with respect to their antimicrobial and anti-inflammatory activity, respectively, all in a cyclized form bridged by the N- and C-terminal cysteine linkages (Andra et al. 2004, 2007). Some of these peptides were able to significantly inhibit the LPS-induced cytokine production in human mononuclear cells. The most active of these compounds, however, needed a 100- to 1000-fold excess molar ratio [Peptide]:[LPS] to act with high affinity and thus did not fulfill the claim of extreme selectivity.

These findings were the starting point of systematic analysis of the structure-activity relationships with sequence variations of the N- and C-terminal ends of the peptides, in a way that the N-terminal end was assigned to have a polar/positive charge and the C-terminal end a hydrophobic character. In fact, it was assumed that binding of the peptide's N-terminal end to the negatively charged backbone of the lipid should be the most adequate way to counteract the lipid A portion of LPS (Gutsmann et al. 2010). For an optimal binding and inhibition of the LPS-induced cytokine production, the exact number and position of each amino acid (aa) within the polypeptide plays a decisive role for optimizing the neutralizing activity. In Fig. 8.1, the SALP Pep19-2 (GCKKYRRFRWKFKGKFWFWCG), Pep19-8 (GRRYKKFRWK FKGRWFWFG), Pep19-2.5KO (KFGKWRFGKYRFCWKFRGWK), and Pep19-2.5 (GCKKYRRFRWKFKGKFWFWG) were compared in their ability to inhibit the inflammation signal. Cleary, the compound Pep19-2.5, called Aspidasept®, has by far the strongest anti-LPS activity. It is remarkable that the omittance of only the cysteine at the C-terminal end of Pep19-2, leading to Pep19-2.5, was responsible for the dramatic increase in neutralizing action. In the same way, it becomes clear that the exact sequence is

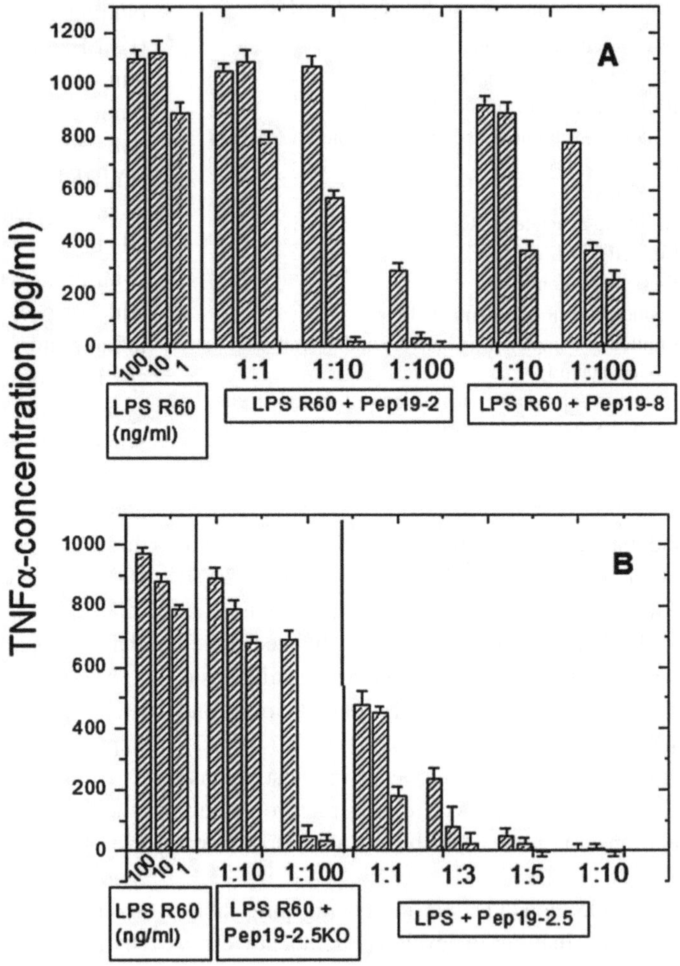

Fig. 8.1 LPS-induced secretion of tumor necrosis factor α at three different concentrations 100, 10, and 1 ng/ml in the presence of the four peptides Pep19-2, Pep19-8, Pep19-2.5KO, and Pep19-2.5 at different weight ratios. (From Gutsmann et al. 2010, reproduced with permission)

important, since the scrambled version of Pep19-2.5, Pep19-2.5KO, with the same aa but in an arbitrary sequence order, shows only activity at high excess weight ratio. Similarly, Pep19-8, which has a similar physicochemical characteristic as Pep19-2.5, loses dramatically the ability to reduce the inflammation signal in comparison to the latter. Therefore, from these and other results, it was concluded that Pep19-2.5 represents the optimum sequence for LPS neutralization and its binding constant to LPS was determined by isothermal titration calorimetry (ITC) to 2.8×10^8 M^{-1}, (Fig. 8.2). Further sequence variations did not lead to further improvements (Kaconis et al. 2011).

Besides ITC, further techniques were applied to get a complete characterization of the LPS-peptide binding. These investigations comprise the analysis of the LPS aggregate structure by small-angle X-ray scattering (SAXS) with synchrotron radiation as well as with freeze-fracture transmission electron microscopy (FFTEM), the use of a Zetasizer for determination of the surface charges (Zeta potential) of LPS layers, the analysis of the gel to liquid crystalline phase transition of the methylene groups of the lipid A portion of LPS by Fourier-transform infrared spectroscopy (FTIR), and the secondary structure of the peptide using also FTIR. It was found that LPS in the presence of the peptide is converted from a non-lamellar bilayered aggregate structure into a multilamellar form, as evidenced by SAXS and FFTEM, connected with an exothermic reaction between LPS and peptide with saturation characteristic (Fig. 8.2) and increase of the Zeta potential of LPS from

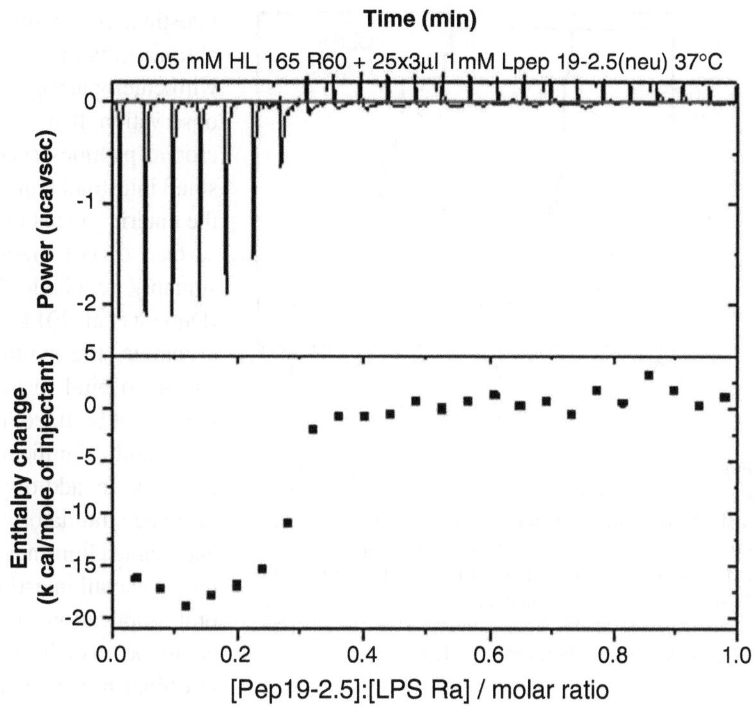

Fig. 8.2 Isothermal calorimetric titration of LPS from *Salmonella minnesota* strain R60 with Pep19-2.5 (Aspidasept) indicating an exothermic reaction of the two compounds with saturation, which takes place at a molar ratio [Pep19-2.5]:[LPS] 0.3, i.e., three peptide molecules bind and neutralize ten LPS molecules. (Adapted from Kaconis et al. 2011)

−45 mV to +10 mV in saturation. The presence of the peptide leads to a fluidization of the lipid A acyl chains as seen by FTIR. In contrast, the ß-like structure of the peptide's secondary structure is not changed during the binding process.

Interestingly, much less active compounds such as Pep19-8 neutralize the surface charges in the same way or even better than Pep19-2.5, and this is similarly true for the fluidization of the acyl chains of lipid A. From these data, it can be concluded that as physical determinants of anti-LPS activity, a high-affinity exothermic reaction associated with a change of the supramolecular aggregate structures of LPS into a multilamellar organization is necessary and sufficient for its neutralization.

8.3 SALPs Inhibit Sepsis-Induced Cardiac Dysfunction

An impairment of cardiac function is a key feature of the cardiovascular failure associated with sepsis. A study was designed to evaluate the effects of Pep19-2.5 in (i) a murine model of polymicrobial sepsis using cecal ligation and puncture (CLP) to induce septic cardiomyopathy and (ii) an in vitro model of cardiomyocytes exposed to human sepsis serum for a translational approach. After discovering a Pep19-2.5-induced weakening of cardiomyopathy caused by sepsis, the effects of the peptide on SERCA2 expression were then analyzed (Martin et al. 2015, 2016). It was assumed that the administration of Pep19-2.5 may attenuate the cardiac dysfunction in murine polymicrobial sepsis through regulating SERCA2 expression. Actually, the infusion of Pep19-2.5 reduced the impaired systolic and diastolic contractility and improved the survival time in polymicrobial sepsis (Fig. 8.3).

It is important to note that the survival benefit was obtained without any antibiotics by continuous intravenous administration of Pep19-2.5. Furthermore, preservation of cardiac function in sepsis by Pep19-2.5 was associated with inhibition of the activation of NF-κB and of Akt/eNOS survival pathways. Most interestingly, the peptide prevented the downregulation of SERCA2 expression in (i) murine heart samples from mice with sepsis and (ii) in cardiomyocytes exposed to

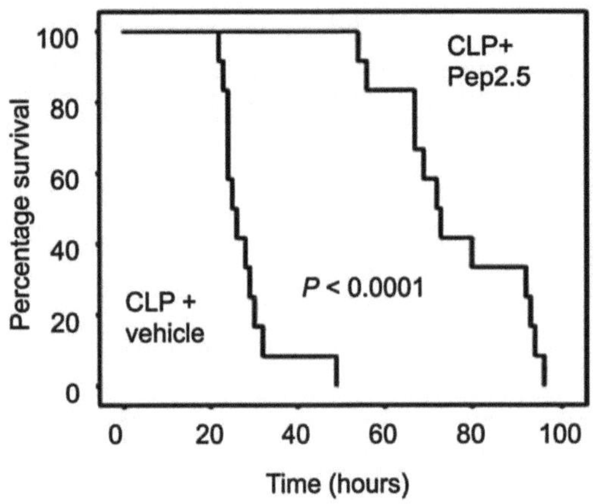

Fig. 8.3 Effect of cecal ligation and puncture (CLP) and treatment with Pep19-2.5 on survival rate of 2-month-old male NMRI mice (2.0 μg/h in saline 0.9%) or vehicle (100 μl/h saline 0.9%). The following groups were studied: CLP + vehicle (n = 12); CLP + Pep19-2.5 (n = 12). Results were globally analyzed by means of Kaplan-Meier survival analysis for n number of observations. (Reproduced from Martin et al. 2015)

serum from septic shock patients. From these data, it can be anticipated that Pep19-2.5 is able to prevent downregulation of cardiac SERCA2 expression in patients with sepsis. This can in turn improve cardiac function and outcome in these patients.

8.4 SALPs Inhibit Inflammation in Intestinal Cells

In addition to the systemic route, also local application of synthetic anti-lipopolysaccharide peptides to outer or inner body surfaces might provide significant health benefits. Indeed, the local application of SALP derivatives to the skin was already successfully applied (see later paragraph). Likewise, Pep19-2.5 was shown to inhibit lipoprotein- and lipopolysaccharide-induced cytokine secretion by keratinocytes, fibroblasts, and dendritic cells and to promote wound healing (Pfalzgraff et al. 2016). Further analysis showed that Pep19-2.5 retained its anti-lipopolysaccharide activity in a cream formulation (Kuhlmann et al. 2018).

In principle, the potent anti-inflammatory and antibacterial activity of SALP might also reduce intestinal tissue inflammation in the context of chronic inflammatory bowel diseases or infection with enteropathogens. This was suggested by the observation that cryptdins, endogenous antimicrobial peptides secreted into the gut lumen by small intestinal Paneth cells, were concentrated in the enteric mucus layer, where they bound to the surface of colonizing commensal bacteria presumably reducing their inflammatory potential (Dupont et al. 2014, 2015). Similar to endogenous cryptdins, the synthetic peptide Pep19-2.5 was shown to bind to the bacterial cell surface of a number of both, commensal and pathogenic, bacteria and significantly reduce their viability in vitro. In addition, it efficiently reduced the immune stimulatory potential of both heat-killed and viable Gram-negative and Gram-positive bacteria on small intestinal epithelial cells. Following oral application, rhodamine-labeled Pep19-2.5 could be visualized within the proximal small intestinal mucus layer where it colocalized with LPS aggregates. Using small-angle X-ray scattering (SAXS) analysis, we noted that Pep19-2.5 bound to and converted LPS to a multilamellar structure. This phenomenon was also observed in a mucus matrix suggesting a similar effect also in vivo. We therefore hypothesized that Pep19-2.5 would reduce the concentration of active LPS molecules present within the intestinal mucus layer, in close proximity of the intestinal epithelial cell surface, and reduce the overall inflammatory potential (Dupont et al. 2015). Therefore, we next tested the influence of oral Pep19-2.5 administration on mice kept under homeostatic conditions. Whereas the total bacterial load was unaffected in both the small and the large intestine following daily oral administration of 100 μg of Pep19-2.5 for 4 days, a significant alteration of the colonic microbiota composition was noted. The oral administration of Pep19-2.5 to mice infected with *Salmonella enterica sv.* Typhimurium (*S. typhimurium*) had no effect on bacterial organ counts or the total mouse body weight. However, consistent with its potent anti-inflammatory activity, oral Pep19-2.5 administration reduced the local mucosal pro-inflammatory response, as witnessed by a lower, albeit not significant, epithelial *Cxcl2* mRNA expression in the small intestine. Peptide degradation and binding to bacterial surfaces within the gut lumen might have reduced the fraction of bioactive peptide at the mucosal surface. Future work should aim at improving peptide stability in the intestine and its targeted release at a close proximity to the mucosal surface. Oral Pep19-2.5 administration might then be able to greatly influence mucosal inflammation, wound healing, and microbiota composition and thereby exert a beneficial effect on the infected host.

8.5 SALPs Efficiently Neutralize LPS In Vivo

The first evidence that SALPs could neutralize LPS in vivo was reported by Gutsmann and collaborators (Gutsmann et al. 2010) using a mouse model of endotoxemia induced by intraperitoneal (i.p.) co-inoculation of LPS and galactosamine. Immediately after this endotoxin challenge, a single dose of the treatment (i.e., the peptide) was administered as bolus by the same route. Although this model seems rather unsophisticated, it allowed the authors to rank SALPs according to their endotoxin-neutralizing ability. Thus, whereas Pep 19-2, Pep 19-2.5, Pep 19-12, and Pep 19-2.2 were as potent as polymyxin B (PMB), Pep 19-5 and Pep 19-8 granted no detectable protection and a third group of peptides, exemplified by Pep19-4, and conferred an intermediate level of protection against endotoxemia.

Using the same mouse model, Martinez de Tejada and collaborators confirmed the high antiendotoxic potency of Pep 19-2, Pep 19-12, and Pep 19-2.2, while showing that shorter derivatives of this peptide (Pep 17-1 and Pep 17-2) granted almost no protection (Martínez de Tejada et al. 2012). Since Pep 17-1 and Pep 17-2 lack the N-terminal polar residue characteristic of Pep 19 series, this observation suggests the importance of this structural feature for anti-LPS activity. An alternative (although compatible) explanation is that peptides must have a minimum length to guarantee appropriate binding to LPS.

To better characterize the therapeutic potential of SALPs, Heinbockel and collaborators tested if peptides could protect mice when their administration was delayed 30 min or 60 min with respect to the LPS challenge (Heinbockel et al. 2013). These experiments were performed with Pep 19-2.5, the peptide showing the highest LPS-binding activity in vitro. Interestingly, this compound conferred a very significant level of protection when given 30 min after LPS challenge, whereas it exhibited no therapeutic activity when administered 60 min post-challenge. In the same report, Pep 19-2.5 appeared to possess a low half-life in vivo, since a single dose of the compound administered 30 min prior to the endotoxin challenge gave no protection to mice. However, keeping in mind that antisepsis treatments are not given as bolus but as intravenous infusion, this observation does not detract from the clinical application of SALPs. Actually, Schuerholz and collaborators demonstrated that Pep 19-2.5 efficiently protects against sepsis when given as infusion using a mouse model of cecal ligation and puncture (Schuerholz et al. 2013).

8.6 SALPs Neutralize PAMPs from Gram-Positive Bacteria In Vivo

Since only approximately 50% of sepsis cases result from Gram-negative origin, we extended the investigations also to the interaction of Pep19-2.5 with various germs of Gram-positive origin (Tabah et al. 2012). Surprisingly, Pep19-2.5 exhibited a better antimicrobial activity against a variety of Gram-positive source (e.g., *Bacillus megaterium*, *Streptococcus agalactiae*, *Clostridium perfringens*) as compared to Gram-negative source, such as *E. coli* (EPEC), *Salmonella enterica*, and *Stenotrophomonas maltophilia* (Dupont et al. 2015). These data suggest a hint that Pep19-2.5 could also be efficient against the Gram-positive inflammation-inducing toxins, which were described to be peptidoglycans (PG), lipoteichoic acids (LTA), and lipoproteins/peptides (LP). In a systematic study, the influence of Pep19-2.5 on the stimulatory capacity of these compounds was analyzed. Out of all these compounds, only LP induced a strong cytokine activity in human mononuclear cells, and the peptide inhibited the cytokine secretion considerably. Then, using the above-mentioned animal model, Martinez de Tejada and collaborators showed that Pep19-2.5 efficiently neutralizes FSL-1 (fibroblast-stimulating lipopeptide) in vivo (Fig. 8.4) (Martinez de Tejada et al. 2015). This synthetic compound is homolog to the N-terminal part of a lipoprotein from *Mycoplasma salivarium*. Notably, the level of protection against lethal toxemia (80%) was achieved with a single bolus of Pep 19-2.5 and lasted for the entire time of monitoring (at least 4 days). Researchers had previously shown that the inoculation of FSL-1 into animals caused the characteristic symptoms of septic shock in a dose-dependent manner. In the same report, this group demonstrated that lethality due to FSL-1 was associated with high serum levels of TNF-alpha and IL-6. Importantly, administration of Pep 19-2.5 to animals challenged with the lipopeptide completely inhibited secretion of TNF-alpha and caused a slight reduction of IL-6. It is worth noting that animals only received the lipopeptide (i.e., without galactosamine) in this set of experiments.

8.7 SALPs Block Cell-Bound Toxins

Heinbockel et al. (2013) extended the preclinical investigations of compound Pep19-2.5 to the direct interaction of this peptide with Gram-negative (*Salmonella*) and Gram-positive (methicillin-resistant *Staphylococcus aureus*) bacteria (Heinbockel et al. 2013). They found similar exothermic reactions of Pep19-2.5 with both bacterial species including saturation character, which was indicative that a binding of Pep19-2.5 took place also with bound bacterial toxins. The bacterial binding was correlated with the suppression of an immune reaction, i.e., a decrease of cytokine secretion. Also, the addition of Pep19-2.5 to surgically remove human lung tissue stimulated with LPS and with MRSA was again able to completely inhibit this immune reaction. Finally, in the study the therapeutic efficacy of combinations of various antibiotics, i.e., amoxicillin, erythromycin, ciprofloxacin, imipenem, ceftriaxone, and tetracycline, with Pep19-2.5 was checked in the mouse model for bacteremia. For all combinations the animals had a high level of TNFα in the absence of the peptide, and this level decreased considerably when Pep19-2.5 was added. This means that for anti-sepsis therapies, antibiotics alone are not sufficient to control the inflammation. This may be the reason that the treatment of sepsis patients with antibiotics alone frequently fails due to the inability to reduce inflammation.

8.8 Mode of Action of SALP Against PAMPs

Summarizing these data, the mode of interaction of the peptide with LPS can be described as a Coulomb interaction of the positively charged aa of the positively charged residues (R and K) of the N-terminal region of the peptide with the negatively charged phosphate and carboxylates within the LPS head group in a first step, followed by a folding of the nonpolar hydrophobic interaction of the C-terminal region of the peptide into the lipid A acyl chain moiety. It should be mentioned that the latter step is absolutely necessary, since a shortened variant of the peptide lacking the last 5 aa of the C-terminal end was unable to bind and neutralize LPS effectively. The final result of this two-step mechanism is a kind of sequestering of LPS, similarly what has been observed for lipopolyamines (Sil et al. 2013). This leads to a blocking of the binding of LPS to their cell receptors as outlined in Fig. 8.5. As illustrated, SALPs are able to block the bacterial toxins as constituents of the bacterial cells or in free form and thus inhibit the overbordering immune reaction.

8.9 SALP Synergizes with Antibiotics In Vivo

Candidate therapies against sepsis are not expected to replace antibiotics but to work in tandem with these latter drugs. In this context, any

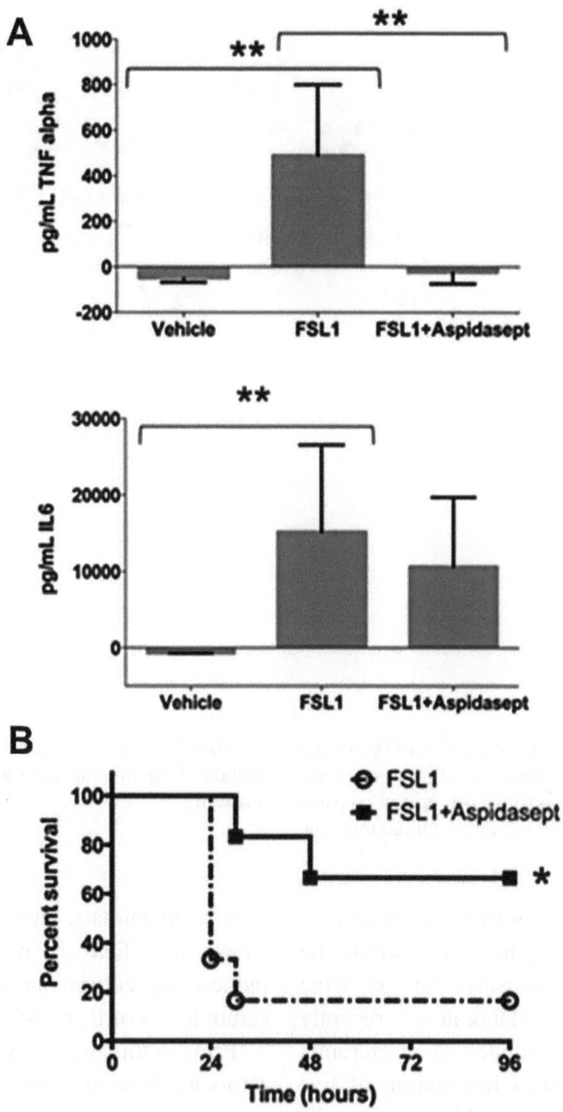

Fig. 8.4 (**a**) Serum levels of tumor necrosis factor α (upper panel) and interleukin-6 (lower panel) in a group of BALB/c mice ($n = 4$) intraperitoneally (i.p.) inoculated with 40 μg of FSL-1 (central bars) compared to another group receiving i.p. 400 μg of Pep19-2.5 (Aspidasept®) immediately after an identical FSL-1 challenge (right bars; $n = 5$) and a third group administered only with pyrogen-free saline i.p. (vehicle; left bars; $n = 5$). TNFα (top) and IL-6 (bottom) levels were measured at 1.5 h or 4 h after challenge, respectively. Double asterisks denote significant statistical differences between the two groups indicated by the bracket ($p < 0.01$; Mann-Whitney U test). (**b**) Survival rate of mice intraperitoneally inoculated with FSL-1 (open circles; $n = 6$) or receiving FSL-1 and then treated with Pep19-2.5 (solid squares; $n = 6$). All the animals received FSL-1 (4 μg/mouse) co-administered with D-galactosamine (18 mg/mouse) intraperitoneally. A group of animals was treated with Pep19-2.5 (200 μg/mouse) immediately after FSL-1 challenge at a different site of the peritoneum, and animal mortality was monitored every 4 h for 4 days. Results were globally analyzed by means of a Kaplan-Meier survival analysis. Asterisk denotes significant statistical differences ($p < 0.05$)

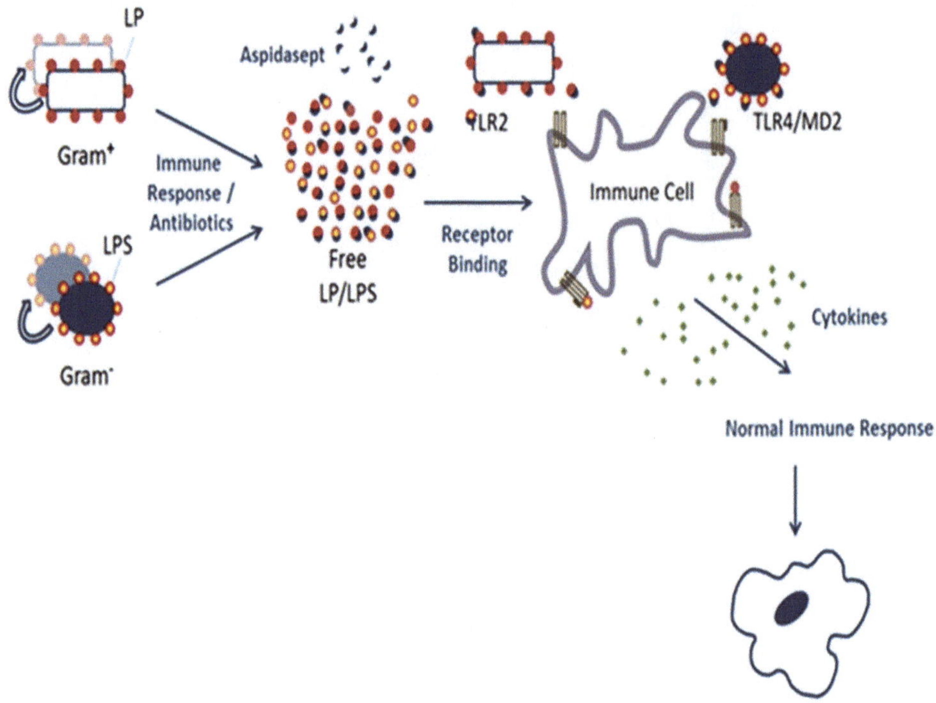

Fig. 8.5 Summary of the therapeutic action of Pep19-2.5 by blocking the PAMPs of bacteria, LPS from Gram-negative and LP from Gram-positive origin, and converting the excessive immune response into a normal reaction. The SALPs have no direct influence on the cytokine production of the immune cells and thus do not impair their function

mutual enhancement of activity between antimicrobials and antisepsis compounds would be highly advantageous. Precisely, this is what Barcena-Varela and collaborators recently showed using a rabbit model of bacteremia induced by the intravenous inoculation of live *Salmonella enterica* serovar Minnesota cells (Barcena-Varela et al. 2017). This rabbit model reproduces symptoms associated with human sepsis including leukopenia, hyperlactatemia, hyperglycemia, hypothermia, splenomegaly, and hyperproduction of TNF-alpha and IL-6.

Barcena-Varela and collaborators demonstrated that a combination of Pep19-2.5 and ceftriaxone administered intravenously to the rabbits killed bacteria and eliminated bacteremia more rapidly than any of the components of the combination when given alone. In addition, the combined treatment was the only one capable of reverting hypothermia and giving rise to temperature values indistinguishable from levels in uninfected animals. Interestingly, the peptide was shown to efficiently neutralize the endotoxin induced by ceftriaxone in vivo and to reduce serum levels of IL-6 and TNF-alpha.

Prior to this report, Heinbockel and collaborators made similar observations using a mouse model of peritonitis induced by i.p. injection of live *Salmonella enterica* serovar Minnesota cells. Specifically, these researchers showed that administration of Pep19-2.5 combined with antibiotics efficiently protected mice against sepsis (Heinbockel et al. 2013). Notably, neither the antibiotic nor the peptide by themselves had any therapeutic efficacy. Their experimental setting allowed the authors to conclude that the components responsible for the bactericidal and PAMP-neutralizing activities were the antibiotic and the peptide, respectively. However, synergism between the SALP and the antibiotics was not observed, contrary to the previous case.

8.10 SALPs Synergize with Anti-inflammatory Agents In Vivo

Although the use of nonsteroidal anti-inflammatory drugs (NSAIDs) as part of sepsis treatment is under discussion, a survival benefit after NSAD therapy has been reported in several animal models of endotoxemia. These anti-inflammatory drugs inhibit the activity of prostaglandin-synthesizing enzyme cyclooxygenase-2 (COX-2) and most of them also COX-1. On the other hand, due to its ability to neutralize PAMPs, Pep19-2.5 is known to prevent Toll-like receptor TLR-4- and TLR-2-dependent activation (Martinez de Tejada et al. 2015).

To test whether these two pro-inflammatory pathways could be simultaneously inhibited, Heinbockel and collaborators evaluated the anti-sepsis potential of a combined SALP-NSAID treatment (Heinbockel et al. 2015). For this purpose, they used a mouse model of endotoxemia induced by the inoculation of a high amount of LPS without galactosamine. Their experiments demonstrated that ibuprofen and Pep 19.2-5 acted in synergy to protect mice against lethal toxemia. Specifically, they showed that neither ibuprofen nor Pep19.2-5 protected mice when administered alone 1 h after the LPS challenge. In contrast, the combination of both compounds afforded a high level of protection that lasted for the entire experimental duration (at least 4 days). Interestingly, authors showed that animals receiving the combined treatment had reduced concentrations of both TNF-α and PGE2, compared to mice treated with either compound.

To better characterize this cooperativity at the molecular level, the same research group performed a transcriptome analysis of human monocytes exposed to the combined treatment and confirmed downregulation of TNF-α and PGE2 coding transcripts. Finally, this experiment showed that other transcripts involved in inflammatory cascades were affected by the combination of ibuprofen and Pep 19-2.5, such as those related with innate immune response, pattern recognition receptor networks, and TLR signaling pathways. These observations highlight the potential of the triple combination of Pep 19.2-5, antibiotics, and NSAIDs in human sepsis therapy.

8.11 SALPs Inhibit Intracellular LPS Responses

The LPS-induced sepsis via TLR4 was thought to be the critical signaling pathway to drive sepsis. However, a novel intracellular LPS sensing mechanism, which proved to be highly relevant for LPS-triggered sepsis independently of TLR4, has been identified, providing a new paradigm for sepsis development. Inflammatory caspases, such as caspase-11 in mice and caspase-4/caspase-5 in humans, recognize LPS in the cytosol (Shi et al. 2014; Hagar et al. 2013). Binding of LPS to inflammatory caspases induces their oligomerization, followed by pyroptosis, an inflammatory form of cell death, and NLRP3 inflammasome- and caspase-1-dependent IL-1β and IL-18 release. LPS-triggered pyroptosis was found to be the main driver for sepsis development and could explain, at least partially, why TLR4 inhibitors failed in clinical trials as anti-sepsis drugs (Opal et al. 2013; Rice et al. 2010). In line with this, the TLR4 inhibitor TAK-242 reduced intracellular LPS-mediated IL-1β release but failed to reduce pyroptosis (Pfalzgraff et al. 2017). In contrast, Pep19-2.5 inhibited intracellular LPS-induced caspase-1 activation, IL-1β production, as well as pyroptosis as demonstrated by reduced HMGB (high mobility group box) 1 secretion and LDH (lactate dehydrogenase) release. Since Pep19-2.5 neutralizes also LP from Gram-positive bacteria, a similar mode of action can be expected for LP such as fibroblast-stimulating lipid 1 (FSL-1) which activates the NLRP7 inflammasome in the cytosol (Khare et al. 2012). Indeed, transfection of THP-1 monocytes with FSL-1 leads to IL-1β release which is inhibited by Pep19-2.5 (manuscript in revision).

While the intracellular detection of LPS by the noncanonical inflammasome pathway is now well established, it has only recently been discovered how LPS gains access to inflammatory caspases in the cytosol since most Gram-negative bacteria are not cytosolic. LPS is delivered into the cytosol via outer membrane vesicles (OMVs) (Vanaja et al. 2016) that are secreted by all Gram-negative bacteria during cell growth or as an adaptive response to stress such as antibiotic treatment or when exposed to the host. OMVs contain several PAMPs found within the parent bacterium including LPS and lipoproteins indicating that OMVs potentially mediate pro-inflammatory effects and may contribute to sepsis (Park et al. 2010). SALPs are also able to suppress OMV-induced responses. In macrophages, Pep19-2.5 as well as polymyxin B reduced IL-1β and TNF release as well as pyroptosis induced by OMVs, while TAK-242 suppressed OMV-induced TNF and IL-1β secretion, but not pyroptosis (manuscript in revision). The exact underlying mechanisms are not fully understood; however, Pep19.2-5 may act extracellularly by directly binding to OMVs or intracellularly by neutralizing LPS after internalization of OMVs and Pep19-2.5. Thus, given the neutralizing mode of action of SALPs, the peptides inhibit not only extracellular TLR2/TLR4 signaling but also intracellular signaling cascades mediated by inflammasomes (Fig. 8.6). This holds even true for more physiological conditions as demonstrated by inhibition of OMV-induced activation of the inflammasome/IL-1 axis by Pep19-2.5.

8.12 SALP for Topical Treatment of Wounds and Skin Infections

Endogenous AMPs play an important role during the different stages of wound healing which can be divided into an inflammatory, a proliferative and remodeling phase (Werner and Grose 2003; Mangoni et al. 2016). Inadequate wound healing due to infection or bacterial colonization can lead to the formation of chronic wounds which are challenging to heal and often remain in the inflammatory stage (Frykberg and Banks 2015). Chronic wounds are mainly polymicrobial and most frequently caused by *S. aureus* and *Pseudomonas aeruginosa* which represent a therapeutic challenge given the increasing resistance against common antibiotics including topical agents such as mupirocin and fusidic acid (Serra et al. 2015). Consequently, new treatment options are urgently needed not only for systemic but also for non-systemic infections. AMPs show important wound healing-promoting activities which render them promising therapeutic options for the treatment of wounds and skin and soft tissue infections (SSTIs) (Pfalzgraff et al. 2018b). Yet, for topical treatment of noninfected and infected wounds, AMPs should have low cytotoxicity and high stability in the host environment with high salt concentrations and proteases present at the wound site.

Besides a broad spectrum of antimicrobial activity, AMPs may also show immunomodulatory properties such as anti-inflammatory actions which are often physiologically more relevant (Hilchie et al. 2013). As mentioned earlier, neutralization of TLR2- and TLR4-mediated responses induced by cell wall-derived inflammatory toxins of Gram-positive and Gram-negative bacteria has also been confirmed for Pep19-2.5 and Pep19-4LF in skin cells such as keratinocytes, dermal fibroblasts, and dendritic cells (Pfalzgraff et al. 2016).

In addition, natural and synthetic AMPs have been demonstrated to induce pivotal processes of wound healing such as cell migration and proliferation as well as angiogenesis. Various peptides promote cell migration, an important mechanism during reepithelialization through transactivation of the epidermal growth factor receptor (EGFR). Pep19-2.5 and Pep19-4LF potently accelerated artificial wound closure in keratinocytes via purinergic P2X7 receptors (P2X7R) leading to calcium influx and mitochondrial reactive oxygen species (ROS) release followed by metalloprotease-dependent transactivation of the EGFR and downstream activation of ERK1/ERK2 (Pfalzgraff et al. 2016, 2018a) (Fig. 8.7). Pep19-2.5-mediated keratinocyte migration was reduced in the presence of the ATPase hexokinase; however, Pep19-2.5 failed to increase extracellular levels of the P2X7R ligand, adenosine triphosphate (ATP). This indicates that Pep19-2.5 either indirectly activates the P2X7R or increases the sensitivity of the P2X7 receptor to its ligand

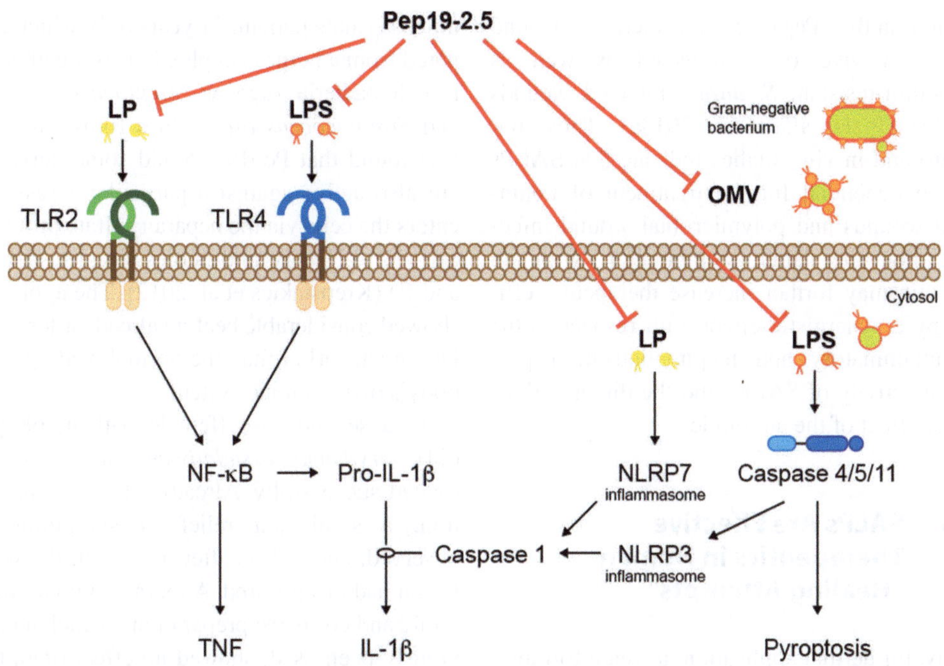

Fig. 8.6 Pep19-2.5 inhibits signaling pathways induced by LP and LPS via transmembrane and cytosolic PRRs. The signaling cascades lead to inflammation and pyroptosis. *LP* lipopeptides, *LPS* lipopolysaccharides, *OMV* outer membrane vesicle. (Adapted from Pfalzgraff et al. (2018a, b). Use permitted under the Creative Commons Attribution License CC BY 4.0)

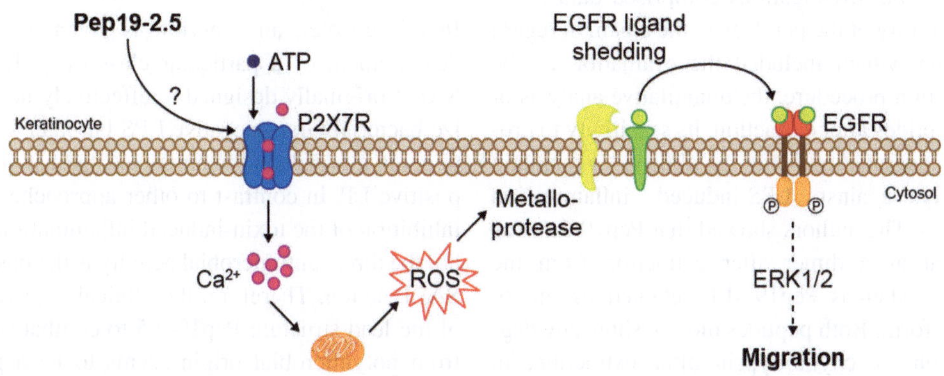

Fig. 8.7 Potential mode of action of Pep19-2.5-induced keratinocyte migration. Pep19-2.5 may activate the P2X7R indirectly to increase sensitivity of the extracellular ligand ATP. Activation of P2X7R leads to calcium influx and subsequently mitochondrial ROS is released which triggers metalloprotease-mediated EGFR transactivation and ERK1/ERK2 activation. (From Pfalzgraff et al. (2018a). Reproduced with permission)

ATP at least at low ATP concentrations. Whether Pep19-2.5 acts as an allosteric modulator of the P2X7R similar to polymyxin B is currently not known (Ferrari et al. 2004). In contrast to other AMPs such as human beta defensins and melittin which stimulate cell migration and proliferation via EGFR transactivation, SALPs promote keratinocyte migration but not proliferation (Pfalzgraff et al. 2016). It remains to be determined whether Pep19.2-5 modulates other wound healing-promoting activities such as induction of angiogenesis.

Importantly, Pep19-2.5 accelerated wound closure in vivo of noninfected as well as methicillin-resistant *S. aureus*-infected wounds in mice (Pfalzgraff et al. 2018a). Thus, our in vitro and in vivo studies indicate that SALPs may be beneficial for the treatment of noninfected wounds and polymicrobial wound infections. For the latter, combination with an antibiotic may further increase therapeutic efficacy by a synergistic effect with respect to the anti-inflammatory and reepithelialization promoting activity of SALPs and the direct antimicrobial effect of the antibiotic.

8.13 SALPs Are Effective Therapeutics in Human "Healing Attempts"

Finally, for dermal application, a cream formulation of Pep19-2.5 and the second lead structure Pep19-4LF were manufactured by dispersing the lyophilized peptide into a DAC base cream (Deutscher Apotheker Codex) (Kuhlmann et al. 2018). The investigations comprised data about the stability of the peptides in the cream in regard to time, which included the evaluation of the extraction procedure, the quantitative analysis of the peptides after extraction, its sensitivity to protease degradation, and its ability to maintain activity against LPS-induced inflammation in vitro. The authors showed that Pep19-2.5 was present as a dimer after extraction from the cream, whereas Pep19-4LF retained its monomeric form. Both peptides did not show any degradation by chymotrypsin after extraction, in contrast to the peptides dissolved in buffer. The formation of Pep19-2.5 into a dimer structure, however, did not decline its ability to inhibit the LPS-induced inflammation reaction in human mononuclear cells.

The formulation in DAC base cream was applied in "healing attempts" (Heilversuch according to German Arzneimittelgesetz §4b). As example for this, the base cream formulation of Pep19-2.5 was applied in a 0.01% crème formulation against a coinfection of the mouth mucosa (male patient, 31 years old), which originated from a herpes simplex I virus together with mouth bacteria such as *Streptococcus sanguis* and *Streptococcus oralis*. In a former report, it was found that Pep19-2.5 and some derivatives are also active against a particular virus which enters the cells via the heparan sulfate proteoglycan (HIV, Hep B, and herpes simplex viruses I and II) (Krepstakies et al. 2012). The application showed considerable healing already after 2 days, i.e., much earlier than the normal healing due to body's own immune system.

In a second case (female patient, 68 years old), *erythema exsudativum multiforme* was treated successfully. Already 24 h after application, a significant relief of symptoms was observed, and 7 days after treatment, the wound lesion had disappeared. A therapy with antihistaminic and cortisone preparations, which had previously been used, showed no effect (data to be published elsewhere).

8.14 Conclusions

In this chapter, an overview is given over the development of a particular class of AMP, the SALP, originally designed to effectively neutralize bacterial Gram-negative LPS but later shown to act in a broadband manner also against Gram-positive LP. In contrast to other approaches, the inhibition of the toxin-induced inflammation and not the direct antimicrobial activity is the basis of SALP action. Therefore, the clinical application of the lead structure Pep19-2.5 to combat sepsis from polymicrobial origin seems to be a great chance for the development of an effective medicament. This is even strengthened by the results from the intracellular action of the SALP indicating also an inhibition of the TLR-4 independent toxin-induced cellular stimulation, with which a gap is closed which may be responsible for former failed antisepsis approaches using pure TLR-4 antagonists. Furthermore, the suitability of the SALP for fighting against chronical wounds including severe skin and tissue infections (SSTIs) seems to represent a big chance for their topical administration to resolve an unmet medical need worldwide, also considering that the development of resistance seems to be improbable.

References

Andrä J, Lamata M, Martinez de Tejada G, Bartels R, Koch MHJ, Brandenburg K (2004) Cyclic antimicrobial peptides based on *Limulus* anti lipopolysaccharide factor for neutralization of lipopolysaccharide. Biochem Pharmacol 68(7):1297–1307. https://doi.org/10.1016/j.bcp.2004.05.054

Andrä J, Lohner K, Blondelle SE, Jerala R, Moriyon I, Koch MH, Garidel P, Brandenburg K (2005) Enhancement of endotoxin neutralization by coupling of a C12-alkyl chain to a lactoferricin-derived peptide. Biochem J 385(Pt 1):135–143. https://doi.org/10.1042/BJ20041270

Andrä J, Gutsmann T, Garidel P, Brandenburg K (2006) Mechanisms of endotoxin neutralization by synthetic cationic compounds. J Endotoxin Res 12(5):261–277. https://doi.org/10.1179/096805106X118852

Andrä J, Howe J, Garidel P, Rössle M, Richter W, Leiva-Leon J, Moriyon I, Bartels R, Gutsmann T, Brandenburg K (2007) Mechanism of interaction of optimized limulus derived cyclic peptides with endotoxins: thermodynamic, biophysical and microbiological analysis. Biochem J 406(2):297–307. https://doi.org/10.1042/BJ20070279

Barcena-Varela S, Martinez de Tejada G, Martin L, Schuerholz T, Gil-Royo AG, Fukuoka S, Goldmann T, Droemann D, Correa W, Gutsmann T, Brandenburg K, Heinbockel L (2017) Coupling killing to neutralization: combined therapy with ceftriaxone/Pep19-2.5 counteracts sepsis in rabbits. Exp Mol Med 49(6):e345. https://doi.org/10.1038/emm.2017.75

Boto A, Perez de la Lastra JM, Gonzalez CC (2018) The road from host-defense peptides to a new generation of antimicrobial drugs. Molecules 23(2). https://doi.org/10.3390/molecules23020311

Brandenburg K, Andrä J, Müller M, Koch MHJ, Garidel P (2003) Physicochemical properties of bacterial glycopolymers in relation to bioactivity. Carbohydr Res 338(23):2477–2489. https://doi.org/10.1016/j.carres.2003.08.008

Brandenburg K, David A, Howe J, Koch MHJ, Andrä J, Garidel P (2005) Temperature dependence of the binding of endotoxins to the polycationic peptides polymyxin B and its nonapeptide. Biophys J 88(3):1845–1858. https://doi.org/10.1529/biophysj.104.047944

Brandenburg K, Garidel P, Fukuoka S, Howe J, Koch MHJ, Gutsmann T, Andrä J (2010) Molecular basis for endotoxin neutralization by amphipathic peptides derived from the alpha helical cationic core region of NK lysin. Biophys Chem 150(1–3):80–87. https://doi.org/10.1016/j.bpc.2010.01.009

Brandenburg K, Howe J, Sanchez-Gomez S, Garidel P, Roessle M, Andrä J, Jerala R, Zweytick D, Lohner K, Rappolt M, Blondelle S, Moriyon I, Martinez de Tejada GM (2010b) Effective antimicrobial and antiendotoxin activity of cationic peptides based on lactoferricin: a biophysical and microbiological study. Anti-Infect Agents Med Chem 9(1):9–22. https://doi.org/10.2174/187152110790886736

Brandenburg K, Andrä J, Garidel P, Gutsmann T (2011) Peptide- based treatment of sepsis. Appl Microbiol Biotechol 90(3):799–808. https://doi.org/10.1007/s00253-011-3185-7

Brandenburg K, Heinbockel L, Correa W, Lohner K (2016) Peptides with dual mode of action: killing bacteria and preventing endotoxin-induced sepsis. Biochim Biophys Acta 1858(5):971–979. https://doi.org/10.1016/j.bbamem.2016.01.011

Dupont A, Heinbockel L, Brandenburg K, Hornef MW (2014) Antimicrobial peptides and the enteric mucus layer act in concert to protect the intestinal mucosa. Gut Microbes 5(6):761–765. https://doi.org/10.4161/19490976.2014.972238

Dupont A, Kaconis Y, Yang I, Albers T, Woltemate S, Heinbockel L, Andersson M, Suerbaum S, Brandenburg K, Hornef MW (2015) Intestinal mucus affinity and biological activity of an orally administered antibacterial and anti-inflammatory peptide. Gut 64(2):222–232. https://doi.org/10.1136/gutjnl-2014-307150

Ferrari D, Pizzirani C, Adinolfi E, Forchap S, Sitta B, Turchet L, Falzoni S, Minelli M, Baricordi R, Di Virgilio F (2004) The antibiotic polymyxin B modulates P2X7 receptor function. J Immunol 173(7):4652–4660. https://doi.org/10.4049/jimmunol.173.7.4652

Fink MP, Warren HS (2014) Strategies to improve drug development for sepsis. Nat Rev Drug Discov 13:741. https://doi.org/10.1038/nrd4368

Fox JL (2013) Antimicrobial peptides stage a comeback. Nat Biotechnol 31:379. https://doi.org/10.1038/nbt.2572

Frykberg RG, Banks J (2015) Challenges in the treatment of chronic wounds. Adv Wound Care (New Rochelle) 4(9):560–582. https://doi.org/10.1089/wound.2015.0635

Garidel P, Brandenburg K (2009) Current understanding of polymyxin B applications in bacteraemia/sepsis therapy prevention: clinical, pharmaceutical, structural and mechanistic aspects. Antinfect Agents Med Chem 8(4):367–385

Garidel P, Andrä J, Howe J, Gutsmann T, Brandenburg K (2007) Novel antiinflammatory and antiinfective agents. Anti-infect Agents Med Chem 6(3):185–200. https://doi.org/10.2174/187152107781023647

Gutsmann T, Razquin-Olazaran I, Kowalski I, Kaconis Y, Howe J, Bartels R, Hornef M, Schürholz T, Rössle

M, Sanchez-Gomez S, Moriyon I, Martinez de Tejada G, Brandenburg K (2010) New antiseptic peptides to protect against endotoxin-mediated shock. Antimicrob Agents Chemother 54(9):3817–3824. https://doi.org/10.1128/AAC.00534-10

Habersetzer F, Leboeuf C, Doffoël M, Zeisel MB, Baumert TF (2013) Synthetic anti-lipopolysaccharide peptides and hepatitis C virus infection. Exp Opin Invest Drugs 22(7):853–862. https://doi.org/10.1517/13543784.2013.794218

Hagar JA, Powell DA, Aachoui Y, Ernst RK, Miao EA (2013) Cytoplasmic LPS activates caspase-11: implications in TLR4-independent endotoxic shock. Science 341(6151):1250–1253. https://doi.org/10.1126/science.1240988

Heinbockel L, Sanchez-Gomez S, Martinez de Tejada G, Dömming S, Brandenburg J, Kaconis Y, Hornef M, Dupont A, Marwitz S, Goldmann T, Ernst M, Gutsmann T, Schürholz T, Brandenburg K (2013) Preclinical investigations reveal the broad-spectrum neutralizing activity of peptide Pep19-2.5 on bacterial pathogenicity factors. Antimicrob Agents Chemother 57(3):1480–1487. https://doi.org/10.1128/AAC.02066-12

Heinbockel L, Marwitz S, Barcena Varela S, Ferrer-Espada R, Reiling N, Goldmann T, Gutsmann T, Mier W, Schürholz T, Drömann D, Brandenburg K, Martinez de Tejada G (2015) Therapeutical administration of peptide Pep19-2.5 and ibuprofen reduces inflammation and prevents lethal sepsis. PLoS One 10(7):e0133291. https://doi.org/10.1371/journal.pone.0133291

Hilchie AL, Wuerth K, Hancock RE (2013) Immune modulation by multifaceted cationic host defense (antimicrobial) peptides. Nat Chem Biol 9(12):761–768. https://doi.org/10.1038/nchembio.1393

Howe J, Andrä J, Conde R, Iriarte M, Garidel P, Koch MHJ, Gutsmann T, Moriyon I, Brandenburg K (2007) Thermodynamic analysis of the lipopolysaccharide dependent resistance of gram-negative bacteria against polymyxin B. Biophys J 92(8):2796–2805. https://doi.org/10.1529/biophysj.106.095711

Kaconis Y, Kowalski I, Howe J, Brauser A, Richter W, Razquin-Olazaran I, Inigo-Pestana M, Garidel P, Rössle M, Martinez de Tejada G, Gutsmann T, Brandenburg K (2011) Biophysical mechanisms of endotoxin neutralization by cationic amphiphilic peptides. Biophys J 100(11):2652–2661. https://doi.org/10.1016/j.bpj.2011.04.041

Katarzyna EG, Małgorzata D (2017) Antimicrobial peptides under clinical trials. Curr Top Med Chem 17(5):620–628. https://doi.org/10.2174/156802661666 66160713143331

Khare S, Dorfleutner A, Bryan NB, Yun C, Radian AD, de Almeida L, Rojanasakul Y, Stehlik C (2012) An NLRP7-containing inflammasome mediates recognition of microbial lipopeptides in human macrophages. Immunity 36(3):464–476. https://doi.org/10.1016/j.immuni.2012.02.001

Krepstakies M, Lucifora J, Nagel CH, Zeisel MB, Holstermann B, Hohenberg H, Kowalski I, Gutsmann T, Baumert TF, Brandenburg K, Hauber J, Protzer U (2012) A new class of synthetic peptide inhibitors blocks attachment and entry of human pathogenic viruses. J Infect Dis 205(11):1654–1664. https://doi.org/10.1093/infdis/jis273

Kuhlmann N, Heinbockel L, Correa W, Gutsmann T, Goldmann T, Englisch U, Brandenburg K (2018) Peptide drug stability: the anti-inflammatory drugs Pep19-2.5 and Pep19-4LF in cream formulation. Eur J Pharm Sci 115:240–247. https://doi.org/10.1016/j.ejps.2018.01.022

Mangoni ML, McDermott AM, Zasloff M (2016) Antimicrobial peptides and wound healing: biological and therapeutic considerations. Exp Dermatol 25(3):167–173. https://doi.org/10.1111/exd.12929

Martin L, De Santis R, Koczera P, Simons N, Haase H, Heinbockel L, Brandenburg K, Marx G, Schuerholz T (2015) The synthetic antimicrobial peptide 19-2.5 interacts with heparanase and heparan sulfate in murine and human sepsis. PLoS One 10(11):e0143583. https://doi.org/10.1371/journal.pone.0143583

Martin L, Horst K, Chiazza F, Oggero S, Collino M, Brandenburg K, Hildebrand F, Marx G, Thiemermann C, Schuerholz T (2016) The synthetic antimicrobial peptide 19-2.5 attenuates septic cardiomyopathy and prevents down-regulation of SERCA2 in polymicrobial sepsis. Sci Rep 6:37277. https://doi.org/10.1038/srep37277

Martínez de Tejada GM, Sanchez-Gomez S, Razquin-Olazaran I, Kowalski I, Kaconis Y, Heinbockel L, Andrä J, Schürholz T, Hornef M, Dupont A, Garidel P, Lohner K, Gutsmann T, David SA, Brandenburg K (2012) Bacterial cell wall compounds as promising targets of antimicrobial agents I. Antimicrobial Peptides and Lipopolyamines. Curr Drug Targets 13(9):1121–1130

Martinez de Tejada G, Heinbockel L, Ferrer-Espada R, Heine H, Alexander C, Barcena-Varela S, Goldmann T, Correa W, Wiesmüller KH, Gisch N, Sanchez-Gomez S, Fukuoka S, Schürholz T, Gutsmann T, Brandenburg K (2015) Lipoproteins/peptides are sepsis-inducing toxins from bacteria that can be neutralized by synthetic anti-endotoxin peptides. Sci Rep 5:14292. https://doi.org/10.1038/srep14292

Mirski T, Niemcewicz M, Bartoszcze M, Gryko R, Michalski A (2017) Utilisation of peptides against microbial infections – a review. Ann Agric Environ Med 25(2):205–210. https://doi.org/10.26444/aaem/74471

Mishra B, Reiling S, Zarena D, Wang G (2017) Host defense antimicrobial peptides as antibiotics: design and application strategies. Curr Opin Chem Biol 38:87–96. https://doi.org/10.1016/j.cbpa.2017.03.014

Molchanova N, Hansen PR, Franzyk H (2017) Advances in development of antimicrobial peptidomimetics as potential drugs. Molecules 22(9). https://doi.org/10.3390/molecules22091430

Nordström R, Malmsten M (2017) Delivery systems for antimicrobial peptides. Adv Colloid Interf Sci 242:17–34. https://doi.org/10.1016/j.cis.2017.01.005

Opal SM, Laterre P, Francois B et al (2013) Effect of eritoran, an antagonist of md2-tlr4, on mortality in patients with severe sepsis: the access randomized trial. JAMA 309(11):1154–1162. https://doi.org/10.1001/jama.2013.2194

Park KS, Choi KH, Kim YS, Hong BS, Kim OY, Kim JH, Yoon CM, Koh GY, Kim YK, Gho YS (2010) Outer membrane vesicles derived from *Escherichia coli* induce systemic inflammatory response syndrome. PLoS One 5(6):e11334. https://doi.org/10.1371/journal.pone.0011334

Pfalzgraff A, Heinbockel L, Su Q, Gutsmann T, Brandenburg K, Weindl G (2016) Synthetic antimicrobial and LPS-neutralising peptides suppress inflammatory and immune responses in skin cells and promote keratinocyte migration. Sci Rep 6:31577. https://doi.org/10.1038/srep31577

Pfalzgraff A, Heinbockel L, Su Q, Brandenburg K, Weindl G (2017) Synthetic anti-endotoxin peptides inhibit cytoplasmic LPS-mediated responses. Biochem Pharmacol 140:64–72. https://doi.org/10.1016/j.bcp.2017.05.015

Pfalzgraff A, Barcena-Varela S, Heinbockel L, Gutsmann T, Brandenburg K, Martinez de Tejada G, Weindl G (2018a) Antimicrobial endotoxin-neutralizing peptides promote keratinocyte migration via P2X7 receptor activation and accelerate wound healing in vivo. Br J Pharmacol. https://doi.org/10.1111/bph.14425

Pfalzgraff A, Brandenburg K, Weindl G (2018b) Antimicrobial peptides and their therapeutic potential for bacterial skin infections and wounds. Front Pharmacol 9:281. https://doi.org/10.3389/fphar.2018.00281

Rice TW, Wheeler AP, Bernard GR, Vincent JL, Angus DC, Aikawa N, Demeyer I, Sainati S, Amlot N, Cao C, Ii M, Matsuda H, Mouri K, Cohen J (2010) A randomized, double-blind, placebo-controlled trial of TAK-242 for the treatment of severe sepsis. Crit Care Med 38(8):1685–1694. https://doi.org/10.1097/CCM.0b013e3181e7c5c9

Schuerholz T, Doemming S, Hornef M, Martin L, Simon TP, Heinbockel L, Brandenburg K, Marx G (2013) The anti inflammatory effect of the synthetic antimicrobial peptide 19-2.5 in a murine sepsis model: a prospective randomized study. Crit Care 17(1):R3. https://doi.org/10.1186/cc11920

Serra R, Grande R, Butrico L, Rossi A, Settimio UF, Caroleo B, Amato B, Gallelli L, de Franciscis S (2015) Chronic wound infections: the role of *Pseudomonas aeruginosa* and *Staphylococcus aureus*. Expert Rev Anti-Infect Ther 13(5):605–613. https://doi.org/10.1586/14787210.2015.1023291

Shi J, Zhao Y, Wang Y, Gao W, Ding J, Li P, Hu L, Shao F (2014) Inflammatory caspases are innate immune receptors for intracellular LPS. Nature 514(7521):187–192. https://doi.org/10.1038/nature13683

Sierra JM, Fusté E, Rabanal F, Vinuesa T, Viñas M (2017) An overview of antimicrobial peptides and the latest advances in their development. Expert Opin Biol Ther 17(6):663–676. https://doi.org/10.1080/14712598.2017.1315402

Sil D, Heinbockel L, Kaconis Y, Rössle M, Garidel P, Gutsmann T, David SA, Brandenburg K (2013) Biophysical mechanisms of the neutralization of endotoxins by lipopolyamines. Open Biochem J 7:82–93. https://doi.org/10.2174/1874091X01307010082

Tabah A, Koulenti D, Laupland K, Misset B, Valles J, Bruzzi de Carvalho F, Paiva JA, Cakar N, Ma X, Eggimann P, Antonelli M, Bonten MJ, Csomos A, Krueger WA, Mikstacki A, Lipman J, Depuydt P, Vesin A, Garrouste-Orgeas M, Zahar JR, Blot S, Carlet J, Brun-Buisson C, Martin C, Rello J, Dimopoulos G, Timsit JF (2012) Characteristics and determinants of outcome of hospital-acquired bloodstream infections in intensive care units: the EUROBACT international cohort study. Intensive Care Med 38(12):1930–1945. https://doi.org/10.1007/s00134-012-2695-9

Tacconelli E, Carrara E, Savoldi A, Harbarth S, Mendelson M, Monnet DL, Pulcini C, Kahlmeter G, Kluytmans J, Carmeli Y, Ouellette M, Outterson K, Patel J, Cavaleri M, Cox EM, Houchens CR, Grayson ML, Hansen P, Singh N, Theuretzbacher U, Magrini N, Aboderin AO, Al-Abri SS, Awang Jalil N, Benzonana N, Bhattacharya S, Brink AJ, Burkert FR, Cars O, Cornaglia G, Dyar OJ, Friedrich AW, Gales AC, Gandra S, Giske CG, Goff DA, Goossens H, Gottlieb T, Guzman Blanco M, Hryniewicz W, Kattula D, Jinks T, Kanj SS, Kerr L, Kieny M-P, Kim YS, Kozlov RS, Labarca J, Laxminarayan R, Leder K, Leibovici L, Levy-Hara G, Littman J, Malhotra-Kumar S, Manchanda V, Moja L, Ndoye B, Pan A, Paterson DL, Paul M, Qiu H, Ramon-Pardo P, Rodríguez-Baño J, Sanguinetti M, Sengupta S, Sharland M, Si-Mehand M, Silver LL, Song W, Steinbakk M, Thomsen J, Thwaites GE, van der Meer JWM, Van Kinh N, Vega S, Villegas MV, Wechsler-Fördös A, Wertheim HFL, Wesangula E, Woodford N, Yilmaz FO, Zorzet A (2018) Discovery, research, and development of new antibiotics: the WHO priority list of antibiotic-resistant bacteria and tuberculosis. Lancet Infect Dis 18(3):318–327. https://doi.org/10.1016/s1473-3099(17)30753-3

Vanaja SK, Russo AJ, Behl B, Banerjee I, Yankova M, Deshmukh SD, Rathinam VA (2016) Bacterial outer membrane vesicles mediate cytosolic localization of LPS and caspase-11 activation. Cell 165(5):1106–1119. https://doi.org/10.1016/j.cell.2016.04.015

Werner S, Grose R (2003) Regulation of wound healing by growth factors and cytokines. Physiol Rev 83(3):835–870. https://doi.org/10.1152/physrev.2003.83.3.835

Anticancer Activities of Natural and Synthetic Peptides

9

A. L. Hilchie, D. W. Hoskin, and M. R. Power Coombs

Abstract

Anticancer peptides (ACPs) are cationic amphipathic peptides that bind to and kill cancer cells either by a direct- or indirect-acting mechanism. ACPs provide a novel treatment strategy, and selected ACPs are currently in phase I clinical trials to examine their safety and overall benefit in cancer patients. Increasing the selectivity of ACPs is important so that these peptides kill cancer cells without harming normal cells. Peptide sequence modifications may help to improve ACP selectivity. ACPs also have immune-modulatory effects, including the release of danger signals from dying cancer cells, induction of chemokine genes, increasing T-cell immune responses, and inhibiting T regulatory cells. These effects ultimately increase the potential for an effective anticancer immune response that may contribute to long-term benefits and increased patient survival. Packaging ACPs in nanoparticles or fusogenic liposomes may be beneficial for increasing ACP half-life and enhancing the delivery of ACPs to tumor target cells. Additionally, engineering ACP-producing oncolytic viruses may be an effective future treatment strategy. Overall research in this area has been slow to progress, but with ongoing ACP-based clinical trials, the potential for ACPs in cancer treatments is closer to being realized. The integration of basic research with computer modeling of ACPs is predicted to substantially advance this field of research.

Keywords

Anticancer peptides · Cytotoxicity · Immune modulation · Nanoparticles · Selectivity · Therapeutic

A. L. Hilchie
Department of Microbiology and Immunology, Dalhousie University, Halifax, Nova Scotia, Canada

D. W. Hoskin (✉)
Department of Microbiology and Immunology, Dalhousie University, Halifax, Nova Scotia, Canada

Department of Pathology, Dalhousie University, Halifax, Nova Scotia, Canada
e-mail: d.w.hoskin@dal.ca

M. R. Power Coombs
Department of Biology, Acadia University, Wolfville, Nova Scotia, Canada

9.1 Introduction

9.1.1 Anticancer Therapies and the Need for Alternative Treatment Strategies

Despite decades of research and progress in the field of cancer therapy, conventional chemotherapy remains the most commonly used treatment modality for most cancers. Chemotherapy functions by indiscriminately killing rapidly dividing

K. Matsuzaki (ed.), *Antimicrobial Peptides*, Advances in Experimental Medicine and Biology 1117, https://doi.org/10.1007/978-981-13-3588-4_9

cells. As a consequence of this mechanism of action, chemotherapy cannot discriminate normal proliferating cells from cancer cells, and as a result, it is unable to target indolent or dormant cancers (Donnelly 2004; Naumov et al. 2003). Furthermore, the acquisition by cancer cells of a chemo-resistant phenotype further reduces the therapeutic value of chemotherapeutic compounds (Bush and Li 2002). Importantly, certain chemotherapeutic compounds (e.g., cyclophosphamide) are associated with the development of secondary malignancies (Choi et al. 2014). This issue is particularly problematic in pediatric cancers in which secondary malignancies, as well as lifelong consequences of toxicities, represent the most severe long-term complications of chemotherapy (Kebudi and Ozdemir 2017). For example, alkylating agents (e.g., cyclophosphamide and ifosfamide) are commonly used to treat pediatric hematologic malignancies and solid tumors and as preconditioning treatment regimens for hematopoietic stem cell transplantation (Choi et al. 2014). However, these same drugs are known to cause therapy-related acute myelogenous leukemia (Thirman and Larson 1996). To address the many limitations of chemotherapy, significant research efforts over the last decade led to the identification of "targeted therapies" (e.g., trastuzumab) that function by selectively targeting and killing cancer cells while sparing normal healthy cells, regardless of their rate of growth. Unfortunately, cancer cell resistance to these targeted therapies was reported shortly after their introduction to the clinic (Nagy et al. 2005).

Researchers and clinicians alike are now recognizing that novel treatment strategies harnessing the power of the immune system may lead to improved clinical outcomes. Indeed, the use of neutralizing antibodies targeting the immune checkpoints cytotoxic T-lymphocyte-associated protein 4 (CTLA-4; e.g., ipilimumab) and programmed cell death protein 1 (PD-1; e.g., pembrolizumab and nivolumab) has enjoyed considerable clinical success (Seidel et al. 2018; Jean et al. 2017; Furue et al. 2018). These therapies are now used as first- and second-line therapies for the treatment of inoperable advanced melanoma and non-small cell lung cancer, respectively. However, these therapies are not without side effects (e.g., severe diarrhea, colitis, inflammation pneumonitis), and patients with advanced disease often do not respond to treatment or relapse thereafter (Seidel et al. 2018; Jean et al. 2017; Furue et al. 2018; Pillai et al. 2018). Collectively, these issues highlight the ongoing need for novel, broad-spectrum anticancer compounds capable of selectively killing cancer cells. Ideally, these new therapies would also harness the power of the immune system by initiating protective antitumor immune responses in patients.

9.1.2 Anticancer Peptides

Cationic anticancer peptides (ACPs) represent a promising alternative to conventional chemotherapy. ACPs are small peptides that contain several cationic and hydrophobic amino acids, giving them an overall positive charge and amphipathic structure (Hoskin and Ramamoorthy 2008). Most ACPs are inherently antimicrobial in nature. In fact, cationic peptides isolated from various organisms were historically assessed for antimicrobial activities and were studied as such prior to their first being described as potent anticancer agents in 1985 (Sheu et al. 1985). In addition to their antimicrobial and anticancer activities, these so-called host defense peptides (HDPs) exhibit many other biological properties, including antiviral (Wang et al. 2008; Bergman et al. 2007), anti-biofilm (Overhage et al. 2008; de la Fuente-Núñez et al. 2012), wound healing (Steinstraesser et al. 2012), anti-parasitic (Couto et al. 2018), adjuvant (Kindrachuk et al. 2009), and immune-modulatory activities (Madera and Hancock 2012; Nijnik et al. 2010) (Fig. 9.1). Peptides are ideal drug candidates due to their low cost of production, the ease with which they can be modified, and relatively high tissue penetration (e.g., compared to antibody-based therapies) (Soman et al. 2009; Richardson et al. 2009; Hilchie et al. 2012, 2013a, 2015).

ACPs are often classified based on the structure that they adopt upon contact with a biological membrane. Three main classes exist, namely,

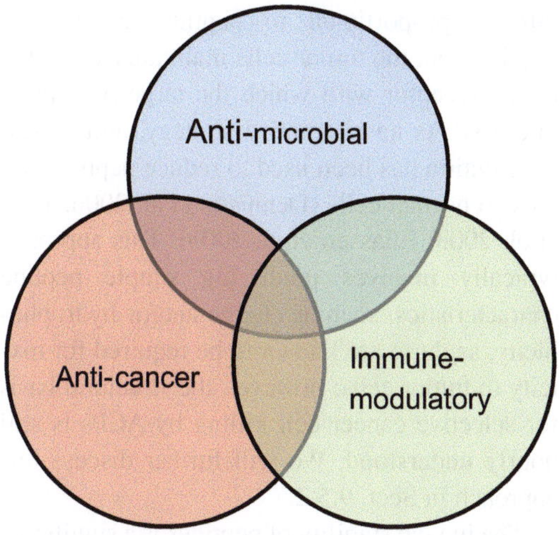

Fig. 9.1 Biological activity of cationic amphipathic peptides. Cationic amphipathic peptides may exhibit any combination of anti-microbial, anti-cancer, or immune-modulatory properties. While many still refer to anti-cancer peptides as cationic antimicrobial peptides, or host defense peptides (i.e., immune-modulatory peptides), it is important to appreciate that these biological activities may be completely independent of each other, and thus should be examined on an individual basis

α-helical (e.g., magainin - Baker et al. 1993; Nguyen et al. 2011), β-sheet (e.g., lactoferricin - Nguyen et al. 2011; Mader et al. 2005), and extended (e.g., LfcinB6 - Richardson et al. 2009; Nguyen et al. 2011). These structures, which are all amphipathic in nature, typically consist of a predominantly cationic face and a hydrophobic face. This is necessary to facilitate peptide interaction with the target cell. ACPs can also be classified on the basis of their mechanism of action, of which two classes exist: direct-acting (i.e., lytic) or indirect-acting (i.e., apoptosis-inducing) (Hilchie and Hoskin 2010), both of which will be discussed in further detail in Sect. 9.2.

9.1.3 Advantages of ACPs Over Conventional Chemotherapy

Due to their unique mechanism of action, ACPs, and particularly direct-acting ACPs (DAAs), have many advantages over conventional chemotherapy. Unlike conventional chemotherapy, many ACPs kill slow-growing as well as multidrug-resistant (MDR) cancer cells (Hilchie et al. 2011; Kim et al. 2003; Johnstone et al. 2000). Several different peptides, including the pleurocidin NRC-03, act as chemosensitizing agents by reducing the EC_{50} of several different chemotherapeutic drugs (Hilchie et al. 2011; Kim et al. 2003; Johnstone et al. 2000; Hui et al. 2002). These chemosensitizing activities suggest that ACPs may work in a synergistic fashion with conventional anticancer drugs. Indeed, we recently showed that the wasp venom peptide mastoparan synergizes with chemotherapeutic compounds both in vitro and in vivo (Hilchie et al. 2016). Many ACPs, including DAAs, destroy primary tumors and their metastases without causing undue harm to normal tissues (Hansel et al. 2007). Moreover, preclinical studies show that DAAs exert antitumor effects when delivered by intratumoral, intraperitoneal, or intravenous injection (Hilchie et al. 2016; Berge et al. 2010; Camilio et al. 2014a). Importantly, the work of others shows that, in addition to their ability to destroy the primary tumor, certain DAAs initiate an antitumor immune response that protects the mouse from tumor rechallenge (Berge et al. 2010; Camilio et al. 2014a). These activities will be discussed in more detail in Sect. 9.3. Furthermore, tumor resistance to DAAs is predicted to be difficult to achieve because DAAs do not rely on unique receptors or a specific signal transduction pathway for their action. Indeed, we investigated cancer cell resistance to DAAs and found that continuous exposure (i.e., more than 1 year) to increasing concentrations of ACPs only generated cancer cells with low-level resistance to lytic peptides (manuscript in preparation). Importantly, these peptide-resistant cancer cells maintained susceptibility to chemotherapeutic drugs and, to our surprise, were unable to establish tumors in immune-deficient mice.

9.1.4 Limitations to the Clinical Use of ACPs

Until recently, the clinical use of ACPs was limited by their high cost of production. However, since their discovery as novel anticancer agents (Sheu et al. 1985), the cost of producing peptides at high purity (i.e., >95%) by high-performance liquid chromatography (HPLC) has undergone a substantial decline. Moreover, the cost of synthesizing large amounts of good manufacturing practice (GMP)-grade peptide is declining as more and more peptide synthesis companies enter the marketplace. The use of recombinant technology, which is a useful method for synthesizing large amounts of peptides, has to date been very difficult because most ACPs exhibit antimicrobial activities (Greenshields et al. 2008). To address this issue, Ishida et al. recently developed a unique method whereby calmodulin is used as a carrier protein to express several different antimicrobial peptides (Ishida et al. 2016). In this approach, the toxic (i.e., antimicrobial) activities are masked, and the peptide is protected from degradation during peptide expression and purification. Others have taken an alternate approach of identifying truncated forms of the parent peptide that maintain their biological activities (Richardson et al. 2009; Mader et al. 2005). Moreover, identifying combinations of ACPs and chemotherapeutic compounds that synergize in vivo is expected to reduce the dose of each compound that is required for a biological effect, thereby reducing any treatment-related toxicities and overall treatment cost. Collectively, these research endeavors, as well as a competitive marketplace, significantly reduce the financial burden of novel peptide-based therapies.

One of the most significant shortcomings of ACP-based therapies is their toxicity to normal cells at *high* peptide concentrations. Many research groups have attempted to reduce off-target toxicity by adding a targeting sequence to their peptide of choice (Liu et al. 2011; Zitzmann et al. 2002; Leuschner and Hansel 2004). To this end, small targeting moieties that interact with specific cell surface molecules overexpressed on cancer cells are added to the peptide of interest, typically using a glycine-glycine linker. To date, this strategy, which will be discussed in more detail in Sect. 9.5.1, has shown mixed results. It is important to note that this strategy increases the cost of production, as synthesis costs are directly proportional to peptide length, and requires that the tumor cells maintain expression of the receptor with which the targeted peptide interacts. As an alternative strategy, amino acid substitution has been used to reduce peptide toxicity to normal cells (Dennison et al. 2006; Yang et al. 2003; Eliassen et al. 2003). This approach typically involves modifying simple peptide characteristics, such as charge and/or hydrophobicity, as these are known to be required for toxicity to tumor cells; however, the structural basis for selective cancer cell killing by ACPs is still poorly understood. We will further discuss this approach in Sect. 9.5.2.

The in vivo stability of peptides is a significant shortcoming of many ACP-based therapeutics. Unpublished work by the Hancock group suggests that small cationic peptides rapidly distribute to all tissues in the body and possess a half-life of approximately 2 min in the blood (discussed in Hilchie et al. 2013a). While many see this as an issue, others argue that this problem is negated by the speed with which many different ACPs exert their toxic effects to cancer cells and that, by limiting peptide half-life, the likelihood of off-target toxicities is also reduced. Nevertheless, there are several reports that peptidomimetics show improved stability in vitro. Moreover, various nanoparticle-based delivery strategies show considerable promise. These strategies will be discussed in further detail in Sects. 9.4.1 and 9.4.2.

In our opinion, the biggest issue facing ACP-based therapies is the loss of momentum that this field of research is experiencing. Time and time again, researchers identify new ACPs and describe their mechanism of action and perhaps their spectrum of activity (i.e., which cancer cell types are susceptible to peptide-mediated killing); however, there is little follow-up work. Thus, with few exceptions, little has been done to thoughtfully address the shortcomings of ACP-based therapies. Here, our aim is to describe the anticancer potential of these molecules and their mechanism(s) of action and discuss ways in which momentum can be regained in an otherwise promising field of research.

9.2 Direct-Acting Versus Indirect-Acting ACPs

ACPs are classified as direct- or indirect-acting based on their mechanism of action (Hilchie and Hoskin 2010). DAAs bind to and kill cancer cells by causing irreparable membrane damage followed by cell lysis (Fig. 9.2). In contrast, exposure to indirect-acting ACPs results in cell death by apoptosis, which occurs in the absence of extensive membrane damage.

DAAs do not require access to the cytosol in order to kill the cell. As a consequence of this, DAAs have many advantages over conventional chemotherapy (see Sect. 9.1.3). DAAs tend to be highly potent and maintain a relatively broad spectrum of activity (i.e., kill a wide variety of cancer cells) in comparison to indirect-acting ACPs. Indirect-acting ACPs tend to be less potent than DAAs, and they often target mitochondria, thereby killing cancer cells by initiating mitochondrial-dependent (i.e., intrinsic) apoptosis (Mader et al. 2005; de Azevedo et al. 2015). While these two mechanisms vary considerably, both are initiated by the selective binding of ACPs to cancer cell membranes.

9.2.1 Factors That Contribute to Selective Peptide Binding to Cancer Cell Membranes

ACPs are thought to selectively bind to cancer cell membranes because of differences in membrane composition (i.e., charge), surface area, transmembrane potential, and membrane fluidity (reviewed in (Hoskin and Ramamoorthy 2008; Hilchie and Hoskin 2010; Mader and Hoskin 2006; Yeaman and Yount 2003; Bhutia and Maiti 2008; Giuliani et al. 2007)). To our knowledge, no study has definitively elucidated the mechanism by which ACPs selectively bind to cancer cell membranes. However, experts agree that membrane composition (i.e., charge) appears to be the most significant factor in this process. Thus, we will limit our discussion to the importance of membrane composition, as the other factors have been reviewed elsewhere (Hoskin and

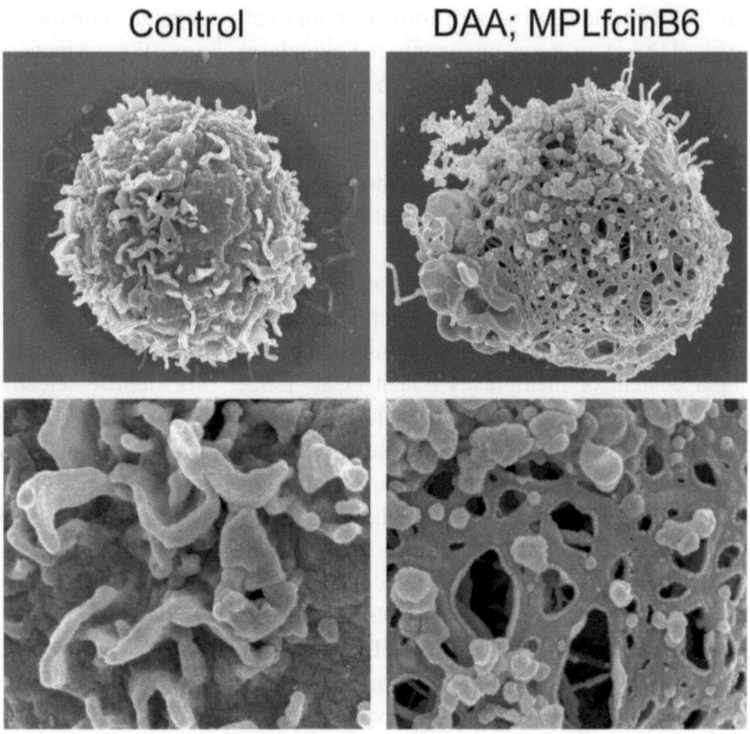

Fig. 9.2 Direct-acting ACPs rapidly lyse human multiple myeloma cells. MPLfcinB6 (50 μM) or its vehicle control were added to U226 human multiple myeloma cells for 2 h. The cells were subsequently fixed, processed, and visualized by scanning electron microscopy. The top and bottom images were captured under 7000 and 40,000× magnification, respectively

Ramamoorthy 2008; Hilchie and Hoskin 2010; Mader and Hoskin 2006; Yeaman and Yount 2003; Bhutia and Maiti 2008; Giuliani et al. 2007).

Owing to the presence of zwitterionic phosphatidylcholine, phosphatidylethanolamine, and sphingomyelin, normal cell membranes are neutral in charge (Zachowski 1993). In contrast, the outer membrane leaflet of cancer cells carries a net negative charge due to increased levels of anionic phosphatidylserine, O-glycosylated mucins, heparan and chondroitin sulfate proteoglycans, and sialylated glycoproteins (Utsugi et al. 1991; Bafna et al. 2010; Koo et al. 2008; van Beek et al. 1973; Iida et al. 1996). Collectively, these differences are thought to contribute to the selective attraction of ACPs to cancer cell membranes. Following the initial stages of peptide binding, ACPs are thought to anchor to the membrane via insertion of hydrophobic residues into the hydrophobic core of the plasma membrane (Hoskin and Ramamoorthy 2008; Hilchie and Hoskin 2010; Mader and Hoskin 2006; Yeaman and Yount 2003; Bhutia and Maiti 2008; Giuliani et al. 2007). Once the peptide is securely bound to the membrane, it either causes membrane instability followed by pore formation and cell lysis (DAAs), or it penetrates into the cytoplasm without substantially damaging the membrane, wherein the peptide initiates apoptosis (indirect-acting ACPs). There are several different models to describe how ACPs cause membrane instability. These models are thoughtfully described elsewhere (Nguyen et al. 2011).

Many studies have used artificial membranes as model systems to show that peptide binding and membrane perturbation are influenced by the lipid content of membranes (Gazit et al. 1995; Matsuzaki et al. 1989). However, it is considerably more difficult to determine the factors that are involved in ACP binding to eukaryotic membranes due to the complexity of the membrane. We demonstrated that the DAAs NRC-03 and NRC-07 exhibit 100- and 50-fold, respectively, greater binding to breast cancer cells than to normal untransformed fibroblasts (Hilchie et al. 2011). In this case, peptide binding was influenced by, but not dependent on, several different anionic surface molecules. Our own work revealed that hundreds of genes are differentially expressed in cancer cells that are refractory to these DAAs (manuscript in preparation). Importantly, these factors appear to influence the toxicity of several DAAs, suggesting a common mechanism of membrane perturbation. Further to this, decreased susceptibility to these DAAs impacted the tumorigenicity of the malignant cells. This work also suggests that dozens of components of the extracellular matrix are likely involved in peptide binding to, and disruption of, the target cell membrane. It is clear that we are only beginning to comprehend the complexity of this process.

9.2.2 Factors That Influence the Mechanism of ACP-Mediated Anticancer Activity

To our knowledge, there is no evidence to suggest that specific structural determinants are responsible for rendering an ACP direct-acting or indirect-acting. Interestingly, in select cases, the mechanism of peptide-mediated cytotoxicity is dependent on the cancer cell line under investigation. For instance, bovine lactoferricin induces apoptosis in human leukemia, lymphoma, and breast cancer cells (Mader et al. 2005; Furlong et al. 2010; Furlong et al. 2006), whereas the same ACP is lytic to fibrosarcoma, melanoma, colorectal cancer, and neuroblastoma cells (Eliassen et al. 2002; Eliassen et al. 2006). In other cases, ACPs may be selectively toxic for one cancer cell type but devoid of effects on another cancer cell type. For example, the ACP MPLfcinB6 selectively lyses leukemia and lymphoma cells (Hilchie et al. 2013b) but is not cytotoxic for breast cancer cells (unpublished). This may be the result of many fundamental differences in the complexity of the membranes of these different types of cancer cells, as we consistently note that cancer cells in suspension (e.g., Jurkat T leukemia cells) are much more susceptible to killing by ACPs than are cancer cells grown as monolayers (e.g., MDA-MB-231 breast carcinoma cells). For instance, the pleurocidins

NRC-03 and NRC-07, as well as the wasp venom peptide mastoparan, are roughly two- to fourfold more toxic to leukemia and myeloma cells than they are to breast carcinoma cells (Hilchie et al. 2011, 2013c, 2016). These findings are further supported by our ongoing quantitative structure/activity relationship studies, which are discussed briefly in Sect. 9.5.2.

As ACPs are small amphipathic molecules with defined secondary structures, it stands to reason that alterations in the amino acid composition of ACPs may affect their potency and mechanism of action. While this has not been studied extensively, we recently noted a striking difference in the mechanism of action of mastoparan by simple C-terminal amidation. Others have shown that unamidated mastoparan kills cancer cells by induction of apoptosis; in contrast, we showed that mastoparan that incorporates a C-terminal amide is much more potent and kills cancer cells by inducing cell lysis (Hilchie et al. 2016; de Azevedo et al. 2015). This finding not only demonstrates the importance of the primary amino acid sequence in determining the mechanism of action of a given ACP but also provides hope that detailed structure/activity relationship studies may reveal next-generation peptides with improved selectivity for cancer cells, thereby addressing one of the most significant limitations to peptide-based therapeutics.

9.3 ACP-Mediated Immune Activation

9.3.1 Cationic Amphipathic Peptides as Modulators of Immune Function

Many cationic peptides were initially characterized for their antimicrobial activities. More recent research shows that many of these same peptides exhibit immune-modulatory activities, as previously reviewed (Hilchie et al. 2013a). Importantly, the antimicrobial activities of many of these peptides are lost in the presence of serum, whereas the immune-modulatory functions of these peptides are maintained under physiologically relevant conditions (Hilchie et al. 2013a).

Some synthetic peptides that have the ability to modulate the immune system are known as innate defense regulators (IDR). Although some of the antibacterial properties of these peptides are lost under physiological conditions, these peptides are still bioactive through immune-modulating effects such as increasing chemokine production (Hilchie et al. 2013a). Monocyte migration in response to chemokines shows a further increase in the presence of the peptide IDR-1002 via a mechanism involving integrins and AKT signaling (Madera and Hancock 2012). IDR-1002 also activates the immune response by increasing chemokine production and by recruiting leukocytes to the site of infection (Nijnik et al. 2010).

Some peptides have the ability to promote antibody production. Immunizing mice with the peptide HH2 along with pertussis toxoid and CpG 10101 significantly increases the titer of toxoid-specific antibodies, indicating that this peptide enhances antibody production in this mouse vaccination model (Kindrachuk et al. 2009).

Despite the many actions of ACPs that promote immune responses, the bioactive peptide lactoferricin B decreases superantigen-mediated interleukin-2 production by mouse splenocytes (Hayworth et al. 2009). This finding indicates that certain ACPs modulate immune function; however, the nature of that modulation may be dependent on the variables present in a given situation.

Mast cells are prominent within tissues and exert multiple effects on the vasculature as a result of their degranulation. Pleurocidins, IDR-1018, and other HDPs induce mast cell degranulation, intracellular calcium mobilization, and the release of prostaglandins (Pundir et al. 2014; Yanashima et al. 2017; Gupta et al. 2016). These peptides may therefore act on mast cells to promote vascular permeability and vasodilation, subsequently shaping the developing immune response.

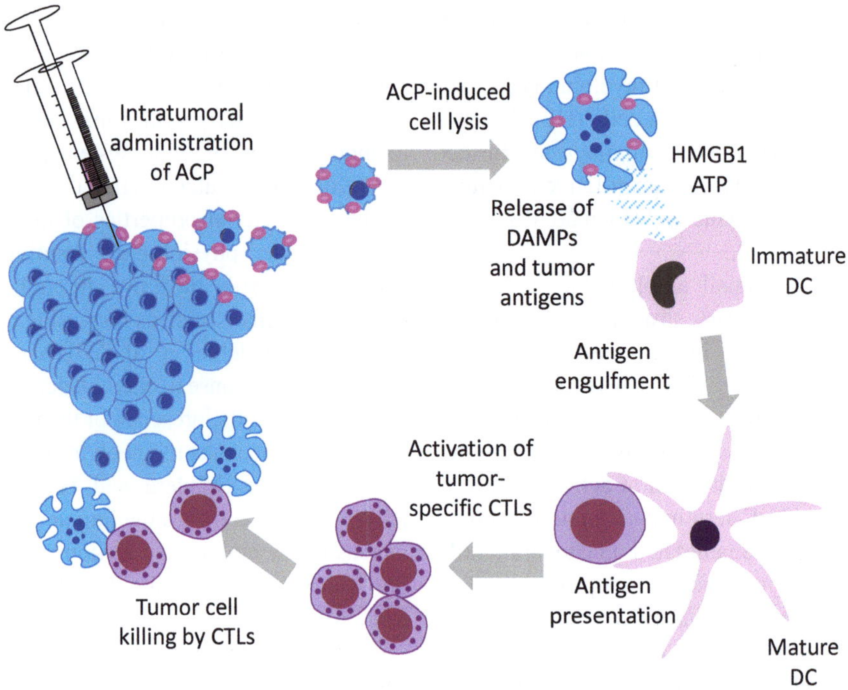

Fig. 9.3 Activation of a protective anti-tumor cytotoxic T lymphocyte (CTL) response by ACPs. In vivo data suggests that a protective immune response develops after intratumoral administration of an ACP. The ACP kills tumor cells, resulting in the release of danger-associated molecular pattern molecules (DAMPs) such as ATP and high mobility group box protein 1 (HMGB1) that promote tumor antigen uptake by dendritic cells (DCs), which then mature and present antigen to T cells. Tumor-specific cytotoxic T lymphocytes (CTLs) are generated that kill tumor cells

9.3.2 Induction of Antitumor Immune Responses by Immunogenic Cell Death

Some ACPs release danger signals from the cell that are thought to be immunogenic. The release of danger-associated molecular patterns (DAMPs) like calreticulin, ATP, and high mobility group box protein 1 (HMGB1) from dying cancer cells results in the induction of an immune response to tumor antigens (Fig. 9.3) (Camilio et al. 2014b).

Intratumoral administration of the bovine lactoferricin-derived ACP LTX-302 to A20 lymphoma-bearing immune-competent mice results in tumor necrosis and inflammatory cell infiltration, followed by complete tumor regression as well as tumor-specific protection against tumor rechallenge. In vitro treatment of these lymphoma cells with LTX-302 results in an increase in HMGB1 release from these cells (Berge et al. 2010). Taken together, these findings suggest that ACP-mediated lysis of malignant cells induces anticancer immunity.

The study of B16 melanoma-bearing mice showed that intratumoral treatment with DAA LTX-315 results in tumor regression and significantly increased survival following tumor rechallenge (Camilio et al. 2014a). In these animals, T cells are recruited to LTX-315-treated tumors, and inflammatory cytokine gene expression is elevated following LTX-315 treatment. Mice that were previously cured of palpable melanoma with LTX-315 treatment are protected from rechallenge with B16 melanoma cells (Camilio et al. 2014a). In vitro LTX-315 treatment of melanoma cells releases DAMPs that include ATP, cytochrome C, reactive oxygen species (ROS), and HMGB1 (Camilio et al. 2014a; Eike et al. 2015).

ACPs that induce a local immune response in tumors may also trigger a systemic immune response that removes all neoplastic cells, including those that have spread to other parts of the body. This immune response is activated by the ACP-induced release of DAMPs. Inhibition of local regulatory T cells (Tregs) at the tumor site is another aspect to consider since inhibiting these cellular regulators of the immune response is known to promote anticancer immune responses. In tumor beds, LTX-315 increased the number of $CD4^+$ (Th1 and Th17) and $CD8^+$ T cells while decreasing Treg numbers (Yamazaki et al. 2016). The cationic peptide LL-37, which has both pro- and anticancer effects depending on the cancer (Chen et al. 2018), also inhibits $CD25^+CD4^+FOXP3^+$ T regulatory cells and so may be helpful in promoting an anticancer immune response (Mader et al. 2011).

Administration of LTX-315 increased CTLA-4 expression on $CD8^+$ T cells while decreasing PD-1 expression, suggesting that using this ACP in combination with an inhibitor of CTLA-4 (ipilimumab) may improve treatment outcome (Yamazaki et al. 2016). Initial experiments with immune checkpoint inhibitors such as ipilimumab suggest that timing of the treatments may be critical as administration of the CTLA-4 neutralizing agent prior to the treatment with the ACP, LTX-315, may be needed to achieve a therapeutic benefit. The need for treatment with ipilimumab in advance of ACP administration is explained by the fact that CTLA-4 is involved in down-regulating T-cell activation.

9.3.3 Comparison of ACPs to Oncolytic Virus Therapy

Oncolytic viruses are another class of novel therapeutics being investigated for the management of various cancers. Oncolytic viruses may fail to kill tumor cells in an individual if the virus is quickly eliminated as the result of triggering an innate immune response (Chiocca and Rabkin 2014). The first oncolytic virus derived from a genetically modified type 1 herpes simplex virus (HSV) has been approved for use and shows therapeutic benefit to melanoma patients (Andtbacka et al. 2015). However, only a moderate increase in survival is reported with this oncolytic virus therapy, indicating other treatments are needed. Another oncolytic virus, vaccinia JX-594, has been used to treat liver cancer patients, in which the virus was shown to be oncolytic and increase patient survival with some evidence of an immune-activating mechanism (Heo et al. 2013). Therefore, evidence exists that treatment with certain oncolytic viruses is able to increase the survival of cancer patients.

Oncolytic peptides, as a result of their short half-life, may provide a safer alternative for patients in comparison to oncolytic viruses. There are safety concerns when patients are administered a virus that may persist long term and has the potential to mutate into a harmful variant. Some ACPs are active against drug-resistant cancer cells and are not lytic for red blood cells, making them potential candidates for development as future treatments for cancer. Since some ACPs kill cancer cells and induce an anticancer immune response (Haug et al. 2016), injection of LTX-315 into transdermally accessible tumors is currently in phase I clinical trials to assess safety, dosing, pharmacokinetics, and immune response development (ClinicalTrials.gov (NCT01058616) 2010). LTX-315 is also now being assessed in multiple cancers as a monotherapy or in combination with ipilimumab or pembrolizumab (ClinicalTrials.gov (NCT01986426) 2013). Discovery research has identified these oncolytic peptides and has revealed their in vitro and in vivo activities. These ACPs are now being examined for clinical efficacy. Even in phase I clinical trials, there is assessment of ACP antitumor activity as indicated by complete and partial response rates, overall response rate, and progression-free survival. The results of these trials will begin to answer questions regarding the effectiveness of ACPs and the potential benefit of enhanced delivery of these oncolytic peptides.

9.4 Strategies to Enhance ACP Delivery

Since peptides typically undergo rapid degradation in the body, a delivery platform may be needed to ensure that ACPs get to their desired target. This may not be necessary for fast-acting peptides; nevertheless, methods to package ACPs so that they reach the tumor microenvironment include the use of nanoparticles and fusogenic liposomes. Peptide modification strategies can also be used to promote tumor cell targeting; however, this approach will be discussed in Sect. 9.5.

9.4.1 Nanoparticles

Nanoparticles provide a mechanism for drug delivery to the correct location in patients, including those with drug-resistant cancers. Many different nanoparticle formulations have been considered. The nanoparticles themselves need to be stable and nontoxic, and they must be targetable in order to deliver the drug of interest to the correct cell/tissue.

Perfluorocarbon nanoparticles have been of particular interest for drug delivery because these nanoparticles are biologically inert, stable, nontoxic, and can be monitored using different imaging platforms. Perfluorocarbon nanoparticles can carry large quantities of drugs, and their delivery to target sites can be observed in vivo (Winter 2014; Chen et al. 2013). Since ACPs are small, it is feasible to put ACPs on/in these perfluorocarbon nanoparticles to enhance their delivery to a primary tumor and metastatic lesions. Studies that used perfluorocarbon nanoparticles loaded with melittin, a cytolytic peptide from bee venom, have revealed that the combination of an ACP and nanoparticle delivery system is able to significantly decrease B16 melanoma tumor volumes in vivo (Soman et al. 2009; Pan et al. 2011).

Since the microenvironment of many solid tumors is acidic (Tannock and Rotin 1989), some nanoparticle delivery approaches have been engineered to function best at low pH. For example, a CPMSN nanocarrier bearing the arginine-glycine-aspartic acid (RGD) peptide on its surface is taken up by breast cancer cells via integrin receptor-mediated endocytosis and has been engineered to subsequently degrade in the acidic endosomal compartment, resulting in the intracellular release of cytotoxic anticancer drugs (Murugan et al. 2016). It should be possible to use a similar approach to deliver ACPs directly into the acidic tumor microenvironment.

Some nanocarriers are toxic on their own, especially those made with polyacrylic acids, indicating the need for nontoxic nanocarriers. In this regard, doxorubicin has been encapsulated within a nanocarrier made of 30% oxidized starch and decorated with an integrin-targeting peptide attached with a polyethylene glycol (PEG) linker in order to selectively target integrin-overexpressing cancer cells (Jiang et al. 2018). Such a starch-based nanocarrier is likely to be less toxic than other nanoparticle formulations.

Another strategy for delivering ACPs to cancer cells is through the use of fusogenic liposomes, which are able to deliver hydrophobic or hydrophilic drugs directly into a target cell without risking degradation by the endocytic pathway (Kube et al. 2017). Fusogenic liposomes that effect membrane fusion have been used to deliver LfcinB6 into the cytoplasmic compartment of both leukemia cells and breast cancer cells, resulting in rapid cytotoxicity (Richardson et al. 2009). In addition to the potential for tumor cell targeting, fusogenic liposomes are expected to protect ACPs from proteolysis long enough for them to reach effective concentrations in the tumor site.

Clearly, there are multiple strategies that can be employed to effectively deliver ACPs to cancer cells. The next critical step will be to evaluate the safest approach in phase I clinical trials. Indeed, multiple clinical trials in which nanoparticles are being used to deliver different chemotherapeutic agents (e.g., paclitaxel) are underway. Since there are already ongoing clinical trials with ACPs, such as LTX-315, future use of nanoparticles as delivery vehicles for ACPs may be an effective strategy to increase the half-life of oncolytic peptides.

9.4.2 Peptides with Altered Stereochemistry

One potential problem with ACP-based treatments is that these peptides can be easily degraded by proteolytic enzymes present in the digestive system and blood plasma (Vlieghe et al. 2010). Susceptibility to degradation is dependent on the peptide sequence (e.g., trypsin cleaves arginine and lysine); however, altering the stereochemistry of an ACP may render it unrecognizable by proteolytic enzymes. In this regard, since amino acids occur naturally as an "L" stereoisomer, D-isomers are not susceptible to proteolytic degradation (Hilchie et al. 2015). For example, an all-D-amino acid variant of pleurocidin that is based on the L form of the cationic antimicrobial peptide from winter flounder resists degradation by trypsin, plasmin, and carboxypeptidase (Jung et al. 2007). Findings such as this indicate that the stereochemistry of a peptide is relevant with respect to its susceptibility to degradation.

9.4.3 Potential for ACP-Expressing Oncolytic Virus Therapy

If clinical trials continue to show that oncolytic viruses are safe and effective anticancer agents, it may be advantageous to engineer an oncolytic virus that also expresses a direct- or indirect-acting ACP. Since the mechanism of oncolysis is different between oncolytic viruses and ACPs, the potential exists for an enhanced cytotoxic effect by an ACP-expressing oncolytic virus. Administration of an oncolytic virus that also codes for an oncolytic peptide is predicted to increase the likelihood of killing all cancer cells in a given tissue, including cancer stem cells, and activating a long-lasting anticancer immune response that will protect against cancer recurrence.

9.5 Strategies to Enhance ACP Selectivity for Cancer Cells

ACPs, particularly those that are direct-acting, have many advantages over conventional chemotherapeutic agents; however, ACPs continue to be limited by their toxicity to normal human cells at *high* peptide concentrations. Several strategies have been used to improve ACP selectivity for cancer cells. Many of these strategies involve optimizing the delivery of peptide to tumor cells through the use of nanoparticle-based delivery systems, as discussed in Sect. 9.4.1. Here, we will briefly review how alterations in the primary amino acid sequence influence the selectivity of ACPs for cancer cells.

9.5.1 Generating Tumor-Specific ACPs Through the Addition of Peptide-Targeting Motifs

As noted in Sect. 9.1.4, ACP selectivity for cancer cells can be enhanced through the addition of so-called targeting sequences. This strategy involves the use of a glycine-glycine linker to conjugate the ACP with a peptide sequence that recognizes specific molecules that are overexpressed by cancer cells. The targeting motif then promotes ACP binding to the tumor cell, after which the cytotoxic portion of the peptide triggers cell death. There are dozens of examples of targeting sequences and many instances in which this strategy has been used to improve ACP selectivity – some have been successful, whereas others have not enjoyed success. Here, we will provide an example of each strategy for the purpose of illustration.

Bombesin is a 14-residue tumor-homing peptide that binds several receptors that are overexpressed by many cancer cell types (Anastasi et al. 1971; Reubi et al. 2002; Cornelio et al. 2007). Significant improvements in tumor cell killing were noted when magainin 2 was conjugated to bombesin (Liu et al. 2011). In comparison to the parental peptide (magainin 2), the IC_{50} of the hybrid peptide for cancer cells was at least tenfold lower, which was substantially lower than the IC_{50} for normal cells. This finding suggests that the increase in potency of the hybrid peptide was not at the expense of cancer cell selectivity.

Phage display libraries can be used to identify novel targeting sequences for ACPs. For example, a screen of phage display libraries was used to identify the sequence LTVSPWY, which has been successfully used to deliver oligonucleotides to SKBR3 breast cancer cells (Shadidi and Sioud 2003). However, this sequence did not improve the cytotoxicity of LfcinB6 for a different breast cancer cell line (unpublished), indicating the need to screen for broad applicability of targeting sequences in a particular type of cancer.

9.5.2 Enhancing ACP Selectivity Through Amino Acid Substitution/Modification

It is no secret that slight alterations to the primary amino acid sequence can drastically affect the potency of an ACP. In some cases, a minor alteration may even change the mechanism of action of the peptide. For example, the addition of a C-terminal amide causes the wasp venom peptide mastoparan to become lytic, whereas in its unamidated form, mastoparan induces mitochondria-dependent apoptosis (Hilchie et al. 2016; de Azevedo et al. 2015).

Many groups have attempted to improve peptide selectivity through amino acid substitution. The vast majority of these studies use hypothesis-driven, small-scale approaches, whereby charge and/or hydrophobicity of the parent ACP is modified, based on the knowledge that these features are required for cancer cell killing (Hoskin and Ramamoorthy 2008). Such studies generate a very small peptide library that is subsequently screened for cytotoxic activity against cancer cells and normal cells. This approach has been used to identify novel peptides with slightly increased selectivity for cancer cells (Dennison et al. 2006; Yang et al. 2003; Eliassen et al. 2003; Arias et al. 2017). Often, many incremental improvements are needed before one obtains an ACP with significantly improved selectivity relative to the parent peptide, likely because we still do not understand how the overall structure of the ACP affects its selectivity for cancer cells. To our knowledge, only one study has examined the effect of altered charge and hydrophobicity on cancer cell selectivity in the context of the overall structure of the ACP (Yang et al. 2002). In this study, helical wheel diagrams of the parent peptide were used to show that positively charged amino acids cluster into two spatially separated regions, termed the major and minor sector, that contain four and two cationic amino acids, respectively. Moving the two cationic amino acids from the minor sector to the major sector increased cancer cell killing at the expense of cancer cell selectivity, suggesting that the presence of a minor sector may reduce ACP toxicity to normal cells. The authors also noted that increasing the overall charge of the ACP by the addition of two additional cationic amino acids to the major sector resulted in reduced potency; however, selectivity for malignant cells was maintained, most likely because the addition of these two amino acids occurred at the expense of two hydrophobic amino acids.

In spite of numerous efforts to generate next-generation ACPs with improved selectivity for cancer cells, we still do not really understand the structural basis for cancer cell selectivity. It is our opinion that this is due to the lack of available datasets that are sufficiently large to conduct thorough structure/activity relationship (SAR) studies. To this end, we have used SPOT array technologies to create a massive peptide library (n = 210), which we then screened for cytotoxic activity against cancer cells and normal cells (manuscript in preparation). We found that single amino acid substitutions may eliminate cytotoxicity for both cancer cells and normal cells, eliminate selectivity for cancer cells, and/or improve selectivity for leukemia and/or breast cancer cells. Our goal is to use an artificial intelligence approach to predict highly selective next-generation ACPs through computer modeling of quantitative structure/activity relationships (QSAR), which yielded hundreds of peptides predicted to be more selective for cancer cells than the parent ACP. Efforts to screen this new peptide library are underway. While this study is in its infancy, we are confident that highly selective ACPs will be identified as this approach has successfully delivered novel peptides with improved antimicrobial and anti-biofilm activities (Hilpert et al. 2005; Haney et al. 2018).

9.5.3 Improving Tumor Selectivity Through Histidine Substitution

It is well established that the microenvironment of solid tumors is acidic in comparison to most normal tissues due to lactic acid buildup coupled with inadequate washout of acidic products as a result of inadequate vascularization (Tannock and Rotin 1989; Newell et al. 1993; Vaupel et al. 1989). Yechiel Shai's group has used these differences in tumor microenvironment to optimize ACP selectivity through the use of histidine substitutions (Makovitzki et al. 2009). In this innovative approach, three or six lysine residues in the ACP [D]-K_6L_9 (pKa ~10.5) were replaced with histidine residues (pKa ~6.1), generating [D]-$K_3H_3L_9$ and [D]-H_6L_9, respectively. Unlike [D]-K_6L_9, neither [D]-$K_3H_3L_9$ nor [D]-H_6L_9 had adverse toxic side effects when delivered to mice via intravenous injection, and both ACPs caused a reduction in the growth of prostate tumor xenografts in mice. These results provide the intriguing possibility of customizing peptides for selective targeting of the solid tumor microenvironment, thereby sparing healthy tissues from potential adverse side effects. Despite these exciting results, to our knowledge, this proof of concept work has not been replicated with other ACPs. However, it is worth noting that our QSAR analysis predicts that peptide selectivity for breast carcinoma cells often involves histidine substitutions (manuscript in preparation).

9.6 The Future of ACP Research

Cancer cells are increasingly resistant to conventional treatment modalities. As patient survival increases, so does the risk of recurrent disease in a form that is resistant to previously used drugs. Peptide-based therapies have the potential to treat many different cancers, including those that are multidrug resistant or slow growing and therefore not susceptible to conventional chemotherapy. Unfortunately, research in this area has been slow to progress. The anticancer potential of magainins has been appreciated since at least the 1980s without any significant progress to clinical trials. Areas in which further study is essential include ascertaining the immuno-modulatory properties of ACPs and improving their selectivity for cancer cells under physiologically relevant conditions. Indeed, ACP-mediated induction of danger signals and the subsequent development of anticancer immune responses may be essential for long-term benefit and increased patient survival. Moreover, improved selectivity is essential for future clinical trials to ensure the safety and efficacy of ACP administration to cancer patients. ACPs must be able to kill cancer cells without adverse toxicities. Although it is possible that a single ACP may be effective against all cancer types, it is more likely that different ACPs will be needed to treat different cancers. Computer modeling may help to advance this area of research so that future treatments can be identified and assessed at a more rapid pace. In addition, computer models may predict highly selective ACPs that also activate antitumor immune responses.

Treating patients with combinational therapies including novel drugs like ACPs is going to be essential to combat multidrug-resistant cancers. Generation of ACP-producing oncolytic viruses may be an additional area for future study to combat recurrence and multidrug-resistant cancers. Finally, the use of tumoricidal ACPs in combination with conventional cytotoxic drugs is likely to improve patient survival by more effective lysis of tumor cells with reduced treatment-related toxicities and a reduction in the risk of tumor recurrence as the result of the generation of long-lasting tumor-specific immunity.

References

Anastasi A, Erspamer V, Bucci M (1971) Isolation and structure of bombesin and alytesin, 2 analogous active peptides from the skin of the European amphibians Bombina and Alytes. Experientia 27:166–167

Andtbacka RHI et al (2015) Talimogene Laherparepvec improves durable response rate in patients with advanced melanoma. J Clin Oncol 33:2780–2788

Arias M et al (2017) Anticancer activities of bovine and human lactoferricin-derived peptides. Biochem Cell Biol 95:91–98

Bafna S, Kaur S, Batra SK (2010) Membrane-bound mucins: the mechanistic basis for alterations in the growth and survival of cancer cells. Oncogene 29:2893–2904

Baker MA, Maloy WL, Zasloff M, Jacob LS (1993) Anticancer efficacy of magainin2 and analogue peptides. Cancer Res 53:3052–3057

Berge G et al (2010) Therapeutic vaccination against a murine lymphoma by intratumoral injection of a cationic anticancer peptide. Cancer Immunol Immunother 59:1285–1294

Bergman P, Walter-Jallow L, Broliden K, Agerberth B, Söderlund J (2007) The antimicrobial peptide LL-37 inhibits HIV-1 replication. Curr HIV Res 5:410–415

Bhutia SK, Maiti TK (2008) Targeting tumors with peptides from natural sources. Trends Biotechnol 26:210–217

Bush JA, Li G (2002) Cancer chemoresistance: the relationship between p53 and multidrug transporters. Int J Cancer 98:323–330

Camilio KA, Berge G, Ravuri CS, Rekdal O, Sveinbjørnsson B (2014a) Complete regression and systemic protective immune responses obtained in B16 melanomas after treatment with LTX-315. Cancer Immunol Immunother 63:601–613

Camilio KA, Rekdal O, Sveinbjörnsson B (2014b) LTX-315 (Oncopore™): a short synthetic anticancer peptide and novel immunotherapeutic agent. Oncoimmunology 3:e29181

Chen J, Pan H, Lanza GM, Wickline SA (2013) Perfluorocarbon nanoparticles for physiological and molecular imaging and therapy. Adv Chronic Kidney Dis 20:466–478

Chen X et al (2018) Roles and mechanisms of human cathelicidin LL-37 in cancer. Cell Physiol Biochem 47:1060–1073

Chiocca EA, Rabkin SD (2014) Oncolytic viruses and their application to cancer immunotherapy. Cancer Immunol Res 2:295–300

Choi DK, Helenowski I, Hijiya N (2014) Secondary malignancies in pediatric cancer survivors: perspectives and review of the literature. Int J Cancer 135:1764–1773

ClinicalTrials.gov (NCT01058616) (2010)

ClinicalTrials.gov (NCT01986426) (2013)

Cornelio DB, Roesler R, Schwartsmann G (2007) Gastrin-releasing peptide receptor as a molecular target in experimental anticancer therapy. Ann Oncol 18:1457–1466

Couto J et al (2018) Antiplasmodial activity of tick defensins in a mouse model of malaria. Ticks Tick-Borne Dis 9:844–849

de Azevedo RA et al (2015) Mastoparan induces apoptosis in B16F10-Nex2 melanoma cells via the intrinsic mitochondrial pathway and displays antitumor activity in vivo. Peptides 68:113–119

de la Fuente-Núñez C et al (2012) Inhibition of bacterial biofilm formation and swarming motility by a small synthetic cationic peptide. Antimicrob Agents Chemother 56:2696–2704

Dennison SR, Whittaker M, Harris F, Phoenix DA (2006) Anticancer alpha-helical peptides and structure/function relationships underpinning their interactions with tumour cell membranes. Curr Protein Pept Sci 7:487–499

Donnelly JG (2004) Pharmacogenetics in cancer chemotherapy: balancing toxicity and response. Ther Drug Monit 26:231–235

Eike L-M, Yang N, Rekdal Ø, Sveinbjørnsson B (2015) The oncolytic peptide LTX-315 induces cell death and DAMP release by mitochondria distortion in human melanoma cells. Oncotarget 6:34910–34923

Eliassen LT et al (2002) Evidence for a direct antitumor mechanism of action of bovine lactoferricin. Anticancer Res 22:2703–2710

Eliassen LT, Haug BE, Berge G, Rekdal O (2003) Enhanced antitumour activity of 15-residue bovine lactoferricin derivatives containing bulky aromatic amino acids and lipophilic N-terminal modifications. J Pept Sci 9:510–517

Eliassen LT et al (2006) The antimicrobial peptide, lactoferricin B, is cytotoxic to neuroblastoma cells in vitro and inhibits xenograft growth in vivo. Int J Cancer 119:493–500

Furlong SJ, Mader JS, Hoskin DW (2006) Lactoferricin-induced apoptosis in estrogen-nonresponsive MDA-MB-435 breast cancer cells is enhanced by C6 ceramide or tamoxifen. Oncol Rep 15:1385–1390

Furlong SJ, Mader JS, Hoskin DW (2010) Bovine lactoferricin induces caspase-independent apoptosis in human B-lymphoma cells and extends the survival of immune-deficient mice bearing B-lymphoma xenografts. Exp Mol Pathol 88:371–375

Furue M et al (2018) Melanoma and immune checkpoint inhibitors. Curr Oncol Rep 20:29

Gazit E, Boman A, Boman HG, Shai Y (1995) Interaction of the mammalian antibacterial peptide cecropin P1 with phospholipid vesicles. Biochemistry (Mosc) 34:11479–11488

Giuliani A, Pirri G, Nicoletto S (2007) Antimicrobial peptides: an overview of a promising class of therapeutics. Cent Eur J Biol 2:1–33

Greenshields AL, Knickle LC, Syvitski R, Douglas SE (2008) Strategies for recombinant expression of small, highly disulphide-bonded, cationic antimicrobial peptides. Protein Pept Lett 15:985–994

Gupta K, Subramanian H, Ali H (2016) Modulation of host defense peptide-mediated human mast cell activation by LPS. Innate Immun 22:21–30

Haney EF et al (2018) Computer-aided discovery of peptides that specifically attack bacterial biofilms. Sci Rep 8:1871

Hansel W, Enright F, Leuschner C (2007) Destruction of breast cancers and their metastases by lytic peptide conjugates in vitro and in vivo. Mol Cell Endocrinol 260–262:183–189

Haug BE et al (2016) Discovery of a 9-mer cationic peptide (LTX-315) as a potential first in class oncolytic peptide. J Med Chem 59:2918–2927

Hayworth JL et al (2009) Attenuation of massive cytokine response to the staphylococcal enterotoxin B superantigen by the innate immunomodulatory protein lactoferrin. Clin Exp Immunol 157:60–70

Heo J et al (2013) Randomized dose-finding clinical trial of oncolytic immunotherapeutic vaccinia JX-594 in liver cancer. Nat Med 19:329–336

Hilchie AL, Hoskin DW (2010) Application of cationic antimicrobial peptides in cancer treatment: laboratory investigations and clinical potential. In: Emerging cancer therapy: microbial approaches and biotechnological tools. Wiley, Hoboken, pp 309–332

Hilchie AL et al (2011) Pleurocidin-family cationic antimicrobial peptides are cytolytic for breast carcinoma cells and prevent growth of tumor xenografts. Breast Cancer Res 13:R102

Hilchie AL, Power Coombs MR, Hoskin DW (2012) Obstacles and solutions to the use of cationic antimicrobial peptides in the treatment of cancer. In: Small wonders: peptides for disease control. American Chemical Society/Oxford University Press, Washington, DC/Oxford, pp 61–78

Hilchie AL, Wuerth K, Hancock REW (2013a) Immune modulation by multifaceted cationic host defense (antimicrobial) peptides. Nat Chem Biol 9:761–768

Hilchie AL, Vale R, Zemlak TS, Hoskin DW (2013b) Generation of a hematologic malignancy-selective membranolytic peptide from the antimicrobial core (RRWQWR) of bovine lactoferricin. Exp Mol Pathol 95:192–198

Hilchie AL et al (2013c) Pleurocidin-family cationic antimicrobial peptides mediate lysis of multiple myeloma cells and impair the growth of multiple myeloma xenografts. Leuk Lymphoma 54:2255–2262

Hilchie AL, Haney EF, Pinto DM, Hancock REW, Hoskin DW (2015) Enhanced killing of breast cancer cells by a d-amino acid analog of the winter flounder-derived pleurocidin NRC-03. Exp Mol Pathol 99:426–434

Hilchie AL et al (2016) Mastoparan is a membranolytic anti-cancer peptide that works synergistically with gemcitabine in a mouse model of mammary carcinoma. Biochim Biophys Acta 1858:3195–3204

Hilpert K, Volkmer-Engert R, Walter T, Hancock REW (2005) High-throughput generation of small antibacterial peptides with improved activity. Nat Biotechnol 23:1008–1012

Hoskin DW, Ramamoorthy A (2008) Studies on anticancer activities of antimicrobial peptides. Biochim Biophys Acta 1778:357–375

Hui L, Leung K, Chen HM (2002) The combined effects of antibacterial peptide cecropin A and anti-cancer agents on leukemia cells. Anticancer Res 22:2811–2816

Iida J, Meijne AM, Knutson JR, Furcht LT, McCarthy JB (1996) Cell surface chondroitin sulfate proteoglycans in tumor cell adhesion, motility and invasion. Semin Cancer Biol 7:155–162

Ishida H, Nguyen LT, Gopal R, Aizawa T, Vogel HJ (2016) Overexpression of antimicrobial, anticancer, and transmembrane peptides in *Escherichia coli* through a calmodulin-peptide fusion system. J Am Chem Soc 138:11318–11326

Jean F, Tomasini P, Barlesi F (2017) Atezolizumab: feasible second-line therapy for patients with non-small cell lung cancer? A review of efficacy, safety and place in therapy. Ther Adv Med Oncol 9:769–779

Jiang Y et al (2018) TEMPO-oxidized starch nanoassemblies of negligible toxicity compared with polyacrylic acids for high performance anti-cancer therapy. Int J Pharm 547:520–529

Johnstone SA, Gelmon K, Mayer LD, Hancock RE, Bally MB (2000) In vitro characterization of the anticancer activity of membrane-active cationic peptides. I. Peptide-mediated cytotoxicity and peptide-enhanced cytotoxic activity of doxorubicin against wild-type and p-glycoprotein over-expressing tumor cell lines. Anticancer Drug Des 15:151–160

Jung HJ et al (2007) Fungicidal effect of pleurocidin by membrane-active mechanism and design of enantiomeric analogue for proteolytic resistance. Biochim Biophys Acta 1768:1400–1405

Kebudi R, Ozdemir GN (2017) Secondary neoplasms in children treated for cancer. Curr Pediatr Rev 13:34–41

Kim S, Kim SS, Bang YJ, Kim SJ, Lee BJ (2003) In vitro activities of native and designed peptide antibiotics against drug sensitive and resistant tumor cell lines. Peptides 24:945–953

Kindrachuk J et al (2009) A novel vaccine adjuvant comprised of a synthetic innate defence regulator peptide and CpG oligonucleotide links innate and adaptive immunity. Vaccine 27:4662–4671

Koo C-Y, Sen Y-P, Bay B-H, Yip GW (2008) Targeting heparan sulfate proteoglycans in breast cancer treatment. Recent Pat Anticancer Drug Discov 3:151–158

Kube S et al (2017) Fusogenic liposomes as nanocarriers for the delivery of intracellular proteins. Langmuir 33:1051–1059

Leuschner C, Hansel W (2004) Membrane disrupting lytic peptides for cancer treatments. Curr Pharm Des 10:2299–2310

Liu S et al (2011) Enhancement of cytotoxicity of antimicrobial peptide magainin II in tumor cells by bombesin-targeted delivery. Acta Pharmacol Sin 32:79–88

Mader JS, Hoskin DW (2006) Cationic antimicrobial peptides as novel cytotoxic agents for cancer treatment. Expert Opin Investig Drugs 15:933–946

Mader JS, Salsman J, Conrad DM, Hoskin DW (2005) Bovine lactoferricin selectively induces apoptosis in human leukemia and carcinoma cell lines. Mol Cancer Ther 4:612–624

Mader JS, Ewen C, Hancock REW, Bleackley RC (2011) The human cathelicidin, LL-37, induces granzyme-mediated apoptosis in regulatory T cells. J Immunother 34:229–235

Madera L, Hancock REW (2012) Synthetic immunomodulatory peptide IDR-1002 enhances monocyte migration and adhesion on fibronectin. J Innate Immun 4:553–568

Makovitzki A, Fink A, Shai Y (2009) Suppression of human solid tumor growth in mice by intratumor and systemic inoculation of histidine-rich and pH-dependent host defense-like lytic peptides. Cancer Res 69:3458–3463

Matsuzaki K et al (1989) Magainin 1-induced leakage of entrapped calcein out of negatively-charged lipid vesicles. Biochim Biophys Acta 981:130–134

Murugan C et al (2016) Combinatorial nanocarrier based drug delivery approach for amalgamation of anti-tumor agents in breast cancer cells: an improved nanomedicine strategies. Sci Rep 6:34053

Nagy P et al (2005) Decreased accessibility and lack of activation of ErbB2 in JIMT-1, a herceptin-resistant, MUC4-expressing breast cancer cell line. Cancer Res 65:473–482

Naumov GN et al (2003) Ineffectiveness of doxorubicin treatment on solitary dormant mammary carcinoma cells or late-developing metastases. Breast Cancer Res Treat 82:199–206

Newell K, Franchi A, Pouysségur J, Tannock I (1993) Studies with glycolysis-deficient cells suggest that production of lactic acid is not the only cause of tumor acidity. Proc Natl Acad Sci U S A 90:1127–1131

Nguyen LT, Haney EF, Vogel HJ (2011) The expanding scope of antimicrobial peptide structures and their modes of action. Trends Biotechnol 29:464–472

Nijnik A et al (2010) Synthetic cationic peptide IDR-1002 provides protection against bacterial infections through chemokine induction and enhanced leukocyte recruitment. J Immunol 184:2539–2550

Overhage J et al (2008) Human host defense peptide LL-37 prevents bacterial biofilm formation. Infect Immun 76:4176–4182

Pan H, Soman NR, Schlesinger PH, Lanza GM, Wickline SA (2011) Cytolytic peptide nanoparticles ('NanoBees') for cancer therapy. Wiley Interdiscip Rev Nanomed Nanobiotechnol 3:318–327

Pillai RN et al (2018) Comparison of the toxicity profile of PD-1 versus PD-L1 inhibitors in non-small cell lung cancer: a systematic analysis of the literature. Cancer 124:271–277

Pundir P, Catalli A, Leggiadro C, Douglas SE, Kulka M (2014) Pleurocidin, a novel antimicrobial peptide, induces human mast cell activation through the FPRL1 receptor. Mucosal Immunol 7:177–187

Reubi JC, Wenger S, Schmuckli-Maurer J, Schaer J-C, Gugger M (2002) Bombesin receptor subtypes in human cancers: detection with the universal radioligand ^{125}I-[D-TYR(6), β-ALA(11), PHE(13), NLE(14)] bombesin(6-14). Clin Cancer Res 8:1139–1146

Richardson A, de Antueno R, Duncan R, Hoskin DW (2009) Intracellular delivery of bovine lactoferricin's antimicrobial core (RRWQWR) kills T-leukemia cells. Biochem Biophys Res Commun 388:736–741

Seidel JA, Otsuka A, Kabashima K (2018) Anti-PD-1 and anti-CTLA-4 therapies in cancer: mechanisms of action, efficacy, and limitations. Front Oncol 8:86

Shadidi M, Sioud M (2003) Identification of novel carrier peptides for the specific delivery of therapeutics into cancer cells. FASEB J 17:256–258

Sheu MJ, Baldwin WW, Brunson KW (1985) Cytotoxicity of rabbit macrophage peptides MCP-1 and MCP-2 for mouse tumor cells. Antimicrob Agents Chemother 28:626–629

Soman NR et al (2009) Molecularly targeted nanocarriers deliver the cytolytic peptide melittin specifically to tumor cells in mice, reducing tumor growth. J Clin Invest 119:2830–2842

Steinstraesser L et al (2012) Innate defense regulator peptide 1018 in wound healing and wound infection. PLoS One 7:e39373

Tannock IF, Rotin D (1989) Acid pH in tumors and its potential for therapeutic exploitation. Cancer Res 49:4373–4384

Thirman MJ, Larson RA (1996) Therapy-related myeloid leukemia. Hematol Oncol Clin North Am 10:293–320

Utsugi T, Schroit AJ, Connor J, Bucana CD, Fidler IJ (1991) Elevated expression of phosphatidylserine in the outer membrane leaflet of human tumor cells and recognition by activated human blood monocytes. Cancer Res 51:3062–3066

van Beek WP, Smets LA, Emmelot P (1973) Increased sialic acid density in surface glycoprotein of transformed and malignant cells – a general phenomenon? Cancer Res 33:2913–2922

Vaupel P, Kallinowski F, Okunieff P (1989) Blood flow, oxygen and nutrient supply, and metabolic microenvironment of human tumors: a review. Cancer Res 49:6449–6465

Vlieghe P, Lisowski V, Martinez J, Khrestchatisky M (2010) Synthetic therapeutic peptides: science and market. Drug Discov Today 15:40–56

Wang G, Watson KM, Buckheit RW (2008) Anti-human immunodeficiency virus type 1 activities of antimicrobial peptides derived from human and bovine cathelicidins. Antimicrob Agents Chemother 52:3438–3440

Winter PM (2014) Perfluorocarbon nanoparticles: evolution of a multimodality and multifunctional imaging agent. Scientifica 2014:746574

Yamazaki T et al (2016) The oncolytic peptide LTX-315 overcomes resistance of cancers to immunotherapy with CTLA4 checkpoint blockade. Cell Death Differ 23:1004–1015

Yanashima K et al (2017) Innate defense regulator IDR-1018 activates human mast cells through G protein-, phospholipase C-, MAPK- and NF-κB-sensitive pathways. Immunol Res 65:920–931

Yang N, Stensen W, Svendsen JS, Rekdal Ø (2002) Enhanced antitumor activity and selectivity of lactoferrin-derived peptides. J Pept Res 60:187–197

Yang N, Lejon T, Rekdal O (2003) Antitumour activity and specificity as a function of substitutions in the lipophilic sector of helical lactoferrin-derived peptide. J Pept Sci 9:300–311

Yeaman MR, Yount NY (2003) Mechanisms of antimicrobial peptide action and resistance. Pharmacol Rev 55:27–55

Zachowski A (1993) Phospholipids in animal eukaryotic membranes: transverse asymmetry and movement. Biochem J 294(Pt 1):1–14

Zitzmann S, Ehemann V, Schwab M (2002) Arginine-glycine-aspartic acid (RGD)-peptide binds to both tumor and tumor-endothelial cells in vivo. Cancer Res 62:5139–5143

10 Antimicrobial Host Defence Peptides: Immunomodulatory Functions and Translational Prospects

Anne M. van der Does, Pieter S. Hiemstra, and Neeloffer Mookherjee

Abstract

Cationic host defence peptides (CHDPs), also known as antimicrobial peptides, exhibit a wide range of activities contributing to immune responses and resolution of infections. CHDPs are expressed across diverse species, are generally amphipathic with less than 50 amino acids in length, and differ significantly in sequence and structure. This chapter focuses on the role of these peptides in immunity. CHDPs are known to function in both innate and adaptive immune responses. These peptides exert both pro- and anti-inflammatory properties, which are likely context dependent based on cell and tissue type, concentration of the peptides, and its interaction with other factors in the microenvironment. Furthermore, the crosstalk between CHDPs and the microbiome and how this may influence mucosal immunity is a rapidly emerging field of research. Overall, the immunomodulatory functions of CHDPs play an important role in the control of infections, regulation of inflammation, and maintaining immune homeostasis. It is thus not surprising that dysregulation of expression of CHDPs is implicated in the susceptibility, pathology, and progression of various diseases. In this chapter, we summarize the immunomodulatory functions of CHDPs, its clinical relevance, and the translational opportunities that these peptides provide for the development of new therapies.

Keywords

Inflammation · Immunomodulation · Cathelicidin · Defensin · LL-37 · IDR peptides

A. M. van der Does · P. S. Hiemstra
Department of Pulmonology, Leiden University Medical Center, Leiden, The Netherlands
e-mail: a.van_der_does@lumc.nl; p.s.hiemstra@lumc.nl

N. Mookherjee (✉)
Manitoba Centre for Proteomics and Systems Biology, Department of Internal Medicine, University of Manitoba, Winnipeg, Canada
e-mail: neeloffer.mookherjee@umanitoba.ca

10.1 Introduction

Cationic host defence peptides (CHDPs), also known as antimicrobial peptides (AMPs), are immune effector molecules with antimicrobial functions. These natural peptides are critical for resolution of infections and exhibit a wide range of immune functions. CHDPs are typically amphipathic, less than 50 amino acids in length with a net positive charge of +2 to +9 at physiological pH 7.4 (Powers and Hancock 2003; Epand and Vogel 1999; Choi and Mookherjee 2012). CHDPs are expressed across diverse

K. Matsuzaki (ed.), *Antimicrobial Peptides*, Advances in Experimental Medicine and Biology 1117,
https://doi.org/10.1007/978-981-13-3588-4_10

species including microbes, plants, insects, crustaceans, amphibians, reptiles, and mammals (Ganz 2003a, b; Nakamura et al. 1988; Cowland et al. 1995; Fernandez de Caleya et al. 1972; Simmaco et al. 1993, 1996; Hultmark et al. 1982). The first report on AMPs was by Kiss and Michl in the 1960s with the description of the peptide bombinin from speckled frog *Bombina variegata* (Simmaco et al. 2009). In the 1980s, cecropin peptide was isolated from moths by Boman's group (Steiner et al. 1981) followed by the characterization of defensins from human neutrophils by Ganz and Lehrer (Ganz et al. 1985) and magainins from amphibians by Zasloff et al. (Zasloff 1987). Over the last four decades, various studies have defined broad-spectrum antimicrobial activity of natural cationic peptides against bacteria, viruses, fungi, and parasites (Powers and Hancock 2003; Epand and Vogel 1999; Ganz 2003a, b; Hancock and Lehrer 1998; Barlow et al. 2011; Larrick et al. 1995). With the rise of microbial resistance to antibiotics and simultaneous lack of development of new antibiotics, there has been an intense focus on the development of new antimicrobials using cationic antimicrobial peptides. Several mechanisms underlying the direct antimicrobial functions of AMPs have been proposed over the years, including direct electrostatic interaction of the peptide with microbial membranes leading to membrane permeabilization and destabilization, intracellular translocation of the peptide and inhibition of microbial DNA/RNA and protein synthesis, and triggering autolysis (Seil et al. 2010; Wang et al. 2014; Sierra et al. 2017). However, studies in the last two decades have demonstrated that the direct microbicidal functions of certain cationic peptides, e.g., the human cathelicidin peptide LL-37 and β-defensin, are compromised in the presence of host factors such as physiological salt concentrations (divalent cations Mg^{2+}, Ca^{2+}) and anionic polysaccharides (Scott et al. 2002; Bowdish et al. 2005). Despite the antagonizing properties of physiological host factors, these peptides are essential for resolution of infections. Concomitant with this, studies have demonstrated effects of endogenous cationic peptides on the host immune system and wound repair (discussed below). Therefore, the term CHDPs is now widely used to encompass both the direct antimicrobial functions and immunomodulatory properties exhibited by these natural cationic peptides. The fundamental functions and emerging themes of clinical applications of CHDPs, with focus on mammalian cathelicidins and defensins, are reviewed in this chapter.

10.1.1 Expression and Processing of CHDPs

CHDPs differ significantly in sequence and structures and can be broadly classified into four groups based on conformational structures: those with linear α-helical structure (e.g., cathelicidin), β-sheet structure with disulfide bridges (e.g., α-defensin), cyclic structures (e.g., catestatin), and extended and flexible loop structures (e.g., indolicidin) (Epand and Vogel 1999; Ganz 2003b; Hancock and Lehrer 1998; Hancock et al. 2016; Wang et al. 2016). The two most well-characterized families of CHDPs in mammals are the cathelicidins and defensins. Cathelicidins are characterized by a conserved cathelin-like N-terminal domain which forms a random-coil conformation in hydrophilic environment, whereas the C-terminal domain corresponding to the mature peptide exhibits diversity in structures which is predominantly α-helical or can adopt elongated or beta-hairpin conformations (De Smet and Contreras 2005; Tomasinsig and Zanetti 2005). Defensins are peptides with six cysteine residues that form intramolecular disulfide bridges and β-sheet conformation (De Smet and Contreras 2005). CHDPs can be expressed constitutively or induced depending on the cell type, tissue, and extracellular stimulus. These peptides are expressed throughout the body such as in different cell types of hematopoietic origin (e.g., lymphocytes, macrophages, neutrophils) and in structural cells (e.g., airway epithelial cells and keratinocytes) and can be found in various body fluids such as sweat, breast milk, plasma, and saliva (Doss et al. 2010).

Humans express only one cathelicidin, hCAP-18/LL-37, which is encoded by the *camp* gene, transcription of which results in the generation of a pre-propeptide and an inactive precursor hCAP-

18 (Cowland et al. 1995; Agerberth et al. 1995). hCAP-18 undergoes proteolytic cleavage by proteases such as proteinase 3 in neutrophils (Sorensen et al. 2001) or kallikrein 5 and 7 in skin (Yamasaki et al. 2006) which leads to the loss of the N-terminal cathelin domain and the release of the active, mature 37 amino acid peptide LL-37 from the C-terminus. LL-37 can be further cleaved by proteases and unknown molecules (also derived from the microbiome (Sieprawska-Lupa et al. 2004)) to smaller fragments, some of which exhibit increased antimicrobial activity but reduced immunomodulatory properties (Murakami et al. 2004). hCAP-18/LL-37 levels are highest in neutrophils where the inactive pro-form is stored in secondary or specific granules (Cowland et al. 1995; Gudmundsson et al. 1996). LL-37 is also expressed by other immune cells including monocytes, NK cells, and lymphocytes and by several types of epithelial cells (Agerberth et al. 2000; Frohm et al. 1997; Chromek et al. 2006; Bals et al. 1998). *Camp* expression in these cell types is usually relatively low during homeostasis, but transcription can be strongly increased upon activation by pathogens such as *Salmonella enterica* serovar Dublin or *E. coli* (Hase et al. 2002) or by pathogen-derived molecules like LPS or LTA (Nell et al. 2004). Furthermore, signals not associated with "danger" can also regulate *camp* expression, including vitamin D_3 (Yim et al. 2007), curcumin (Guo et al. 2013), and several histone deacetylase inhibitors (HDACi) (Schauber et al. 2004). In addition, specific levels of endoplasmic reticulum stress can promote transcription of LL-37 (Park et al. 2011). *Camp* gene expression is not always increased upon activation of the cell with pathogens or associated molecules, for example, infections with Shigella (Islam et al. 2001) or Neisseria (Bergman et al. 2005) result in a decreased expression in infected cells.

Defensins can be classified, based on the location of their disulfide bonds, into α-, β-, and θ-defensins, of which only α- and β-defensins are expressed by humans (Ganz 2003a) and θ-defensins are only expressed in nonhuman primates. Six α-defensins (HNP1–4 and HD5–6) and 11 β-defensins (hBD1–6 and hBD25–29) have so far been identified in humans. α- and β-Defensins are encoded by distinct genes and similar to cathelicidins are also produced as pre-propeptides, which require one or more proteolytic step to generate the mature peptides. Various proteases have been reported to be involved in the processing of the pre-propeptide for the generation of the mature defensins such as neutrophil-derived proteinase 3 and elastase process HNP1 (Valore and Ganz 1992; Tongaonkar et al. 2012), trypsin is involved in the processing of HD5 (Ghosh et al. 2002), whereas in mice α-defensins are processed by metalloproteinase 7 or matrilysin (Wilson et al. 1999). α-Defensin expression seems to be cell-type specific; HNP1–4 are primarily expressed by human neutrophils, where they localize in primary or azurophilic granules (Ganz 2003a). Other immune cell types, for example, NK cells and T-cell subsets, also express HNP1–4 (Obata-Onai et al. 2002). HD5 and HD6 are expressed by Paneth cells of the small intestine (Jones and Bevins 1992; Jones and Bevins 1993), and HD5 is also expressed in the endometrium and fallopian tubes (Quayle et al. 1998). β-Defensins are expressed more ubiquitously relative to α-defensin, although mostly by epithelial cells, hBD3 and hBD4 are also expressed in the endometrium (King et al. 2003), hBD5 and 6 can be found in the epididymis (Yamaguchi et al. 2002), and hBD25–29 can be found primarily in the male genital tract (Rodriguez-Jimenez et al. 2003). Expression levels of defensins are differently regulated; for example, hBD1 is expressed in both a constitutive and inducible manner, whereas hBD2 and 3 are usually expressed in low levels during homeostasis but transcription increases in response to, e.g., microbial exposure or inflammation (Harder et al. 2004; Kuwano et al. 2006; Harder et al. 2001). A recent study has shown that levels of α-defensin can be increased in the lungs in response to inhaled air pollution (Piyadasa et al. 2018b). In contrast, virulence factors of pulmonary pathogens and air pollution-related particulate matter can decrease the expression of defensin genes (Laube et al. 2006).

Physiological concentrations of CHDPs vary within the body, with higher concentrations (in the range of mg/mL) found in the granules of leukocytes and at the bottom of intestinal crypts where Paneth cells are located and lower concentrations (in the range of ng/mL to µg/mL) found in the mucosa and in circulation (Murakami et al. 2002, 2004; Frohm et al. 1997; Gordon et al. 2005; Agerberth et al. 1999; Woo et al. 2003). Likely these concentrations are higher in the close vicinity of the cells that produce these peptides, such as in the pericellular layer overlying the airway epithelial cells. As mentioned above, concentrations of these peptides are typically increased in response to infectious challenge and inflammatory signals. Epithelial wound repair has also been shown to increase the expression of CHDPs (Sorensen et al. 2003). A wide range of immunomodulatory effects of CHDPs contribute to regulation of inflammation and immunity and thus resolution of infections. Immunity-related functions of these peptides include induction of cytokine and chemokine production, promotion of leukocyte recruitment to sites of infection and inflammation, and modulation of dendritic cell and lymphocyte proliferation and differentiation (Choi and Mookherjee 2012; Hancock et al. 2016; Bucki et al. 2010; Hemshekhar et al. 2016; Hiemstra 2015). These immunomodulatory activities of CHDPs are discussed in the next section.

10.1.2 CHDPs in Inflammation and Immunity

Most of the initial CHDPs identified were isolated using activity-guided purification procedures and assessment of antimicrobial activity to identify active components. Later on in silico approaches were also used to identify novel CHDPs. This explains the initial focus in studies on the biology of CHDPs on their antimicrobial activity against a wide range of microorganisms and parasites. However, relatively early on other activities were also discovered, including their ability to modulate inflammation and immunity (Chertov et al. 1996; Van Wetering et al. 1997). These and subsequent studies demonstrated that CHDPs may regulate inflammation and immunity through their direct action on inflammatory and immune cells but also indirectly via induction of mediators in other cell types (Hancock et al. 2016; Hiemstra 2015; Choi et al. 2012). The findings on the impact of CHPDs on the immune system have raised discussions related to the primary function of these peptides and the relative contribution of these functions versus direct antimicrobial activity of the peptides in host defense against infections. Furthermore, the cellular receptor systems employed by CHDPs to mediate the diverse range of effects on the host is also an area of intense investigation in recent years. Nevertheless, the pleiotropic activity of these peptides results in both pro- and anti-inflammatory roles and enhances immune activation as well as contributes to immune regulation, thus complicating the interpretation of the role of CHDPs. It is likely that the exact contribution of CHDPs to the regulation of inflammation and immunity is context dependent.

10.1.2.1 Pro- and Anti-inflammatory Role of CHDPs in the Regulation of Innate Immunity and Inflammation

Neutrophils are a rich source of various CHDPs, including the human cathelicidin hCAP-18/LL-37 and α-defensins. Degranulation and other forms of release from neutrophils may therefore generate high local concentrations of CHDPs that can either be released into the environment or become immobilized in neutrophil extracellular traps (NETs). In addition, CHDPs, such as LL-37, can also facilitate the formation of NETs (Neumann et al. 2014). CHDPs from other cellular sources can accumulate locally at high concentrations, especially if these are stored in granules such as in the intestinal Paneth cells or mast cells. At high concentrations such peptides are found to display cytotoxic activity toward cultured cells (Okrent et al. 1990). Subsequent in vitro and in vivo studies showed that this may also contribute to accumulation of inflammatory cells by direct and/or indirect chemoattraction (Rehaume and Hancock 2008; Hosoda et al. 2017). Furthermore, it was shown that CHDPs display a range of pro-inflammatory and anti-inflammatory activities toward both recruited leukocytes and tissue resident cells (Hancock et al. 2016; Suarez-Carmona et al. 2015; Agier et al. 2015). Following the characterization of human neutrophil α-defensins as T-cell chemoattractants (Chertov et al. 1996), other studies reported chemotactic activity toward additional cell types including macrophages and mast cells (Suarez-Carmona et al. 2015). These defensins were also found to indirectly induce inflammatory cell accumulation, e.g., by increasing expression of the neutrophil chemoattractant IL-8/CXCL8 in epithelial cells (Van Wetering et al. 1997). The relevance of these activities was demonstrated in various reports of in vivo animal models, including one demonstrating that neutrophil α-defensins may aggravate experimental colitis in a mouse model (Hashimoto et al. 2012). Following the pioneering work of Yang et al. demonstrating that β-defensins also display direct chemotactic activities (Yang et al. 1999), a variety of studies showed that β-defensins display similar activities as outlined above for α-defensins (Hancock et al. 2016; Suarez-Carmona et al. 2015).

The activities of the human cathelicidin LL-37 in inflammation and innate immunity display a remarkable overlap with those outlined for α- and β-defensins, although the anti-inflammatory activities of LL-37 have been better defined (Hancock et al. 2016; Agier et al. 2015). LL-37 is chemotactic for neutrophils, eosinophils, monocytes, T cells, and mast cells (Yang et al. 2000; Tjabringa et al. 2006), which in part is mediated through formyl peptide receptors to modulate the activity of these cells (reviewed in Hancock et al. 2016; Agier et al. 2015). However, at the site of inflammation, LL-37 may dampen further chemotaxis by internalization of CXCR2 on neutrophils and monocytes (Zhang et al. 2009), regulate tissue half-life and clearance of neutrophils by suppressing apoptosis (Nagaoka et al. 2006), and induce non-inflammatory secondary necrosis of apoptotic neutrophils (Li et al. 2009). The ability of LL-37 to regulate neutrophil recruitment through direct neutrophil chemotactic activity and induction of neutrophil attracting chemokines was found to contribute to host defense against respiratory tract infection with *Pseudomonas aeruginosa* (Beaumont et al. 2014). Interestingly, LL-37 was also found to redirect macrophage polarization toward macrophages with a pro-inflammatory M1 phenotype, which preferentially expresses this CHDP when compared to the anti-inflammatory pro-repair M2 cells (van der Does et al. 2010). In addition, LL-37 displays a range of anti-inflammatory activities; it neutralizes the pro-inflammatory activity of selected TLR-ligands such as lipopolysaccharide (LPS) which was initially thought to be mediated in part by direct LPS binding (Scott et al. 2000). The direct neutralizing interaction of CHDPs such as LL-37 with pro-inflammatory microbial components, most notably LPS, has prompted the search for CHDPs or their mimetics as anti-endotoxin compounds. However, subsequent studies have shown that LPS neutralization by direct interaction between LL-37 and LPS alone does not fully explain the inhibitory effect of LL-37 on LPS-induced cell activation, and that LL-37-mediated modulation of intracellular pathways such as the TLR-to-NFκB pathway likely contributes to the anti-endotoxin effects of the peptide (Mookherjee et al. 2006). At high concentrations (~20 μg/ml), LL-37 can contribute to the local regulation of inflammation by inducing the expression of the anti-inflammatory cytokine IL-10 in monocytes/macrophages, dendritic cells, and B and T cells (Mookherjee et al. 2009). In line with these data, LL-37 has also been shown to increase the production of IL-1 receptor antagonist (IL-1RA) in monocytes and neutrophils (Choi et al. 2014; Zhang et al. 2008). These studies demonstrate that CHDPs may affect a range of cell types that play crucial roles in inflammation and innate immunity. As CHDPs also influence related systems involved in these processes including the complement system (Hiemstra 2015) and antigen-presenting cells (APCs), this broadens the scope of immunological effects of CHDPs beyond innate immunity.

10.1.2.2 Role of CHPDs in Shaping Adaptive Immunity

CHDPs play an important role in the link between innate and adaptive immunity. This function is primarily mediated by the ability of CHDPs to recruit APCs such as monocyte/macrophages and dendritic cells (DCs) to the site of infection and/or inflammation. As mentioned above, Yang et al.

in 1999 demonstrated that human β-defensin hBD2 is directly chemotactic to immature dendritic cells (iDCs) and T cells, thus establishing the concept that CHDPs function in the interplay between the innate and adaptive arms of the immune system (Yang et al. 1999). Subsequent studies have established that CHDPs such as defensins and cathelicidins can recruit APCs (Suarez-Carmona et al. 2015; Kim et al. 2015). The direct chemotactic activity of CHDPs is mediated by several immune receptors such as chemokine receptors (e.g., CCR6 and CCR2), pattern recognition receptors (TLR4 and TLR1/2), and G-protein-coupled receptors (Suarez-Carmona et al. 2015; De et al. 2000; Kurosaka et al. 2005). Apart from promoting migration and recruitment of APCs to local sites of infection and/or inflammation, CHDPs also influence adaptive immune response by activating APCs and modulating the differentiation of lymphocytes. Human defensin hBD3 induces the expression of co-stimulatory molecules CD80, CD86, and CD40 on monocytes and myeloid DCs via interaction with TLRs, thereby enhancing adaptive immune response (Funderburg et al. 2007). Similarly, hBD2 and hBD3 can promote the production of IFN-α from plasmacytoid DCs in a TLR-9-dependent manner and the consequent initiation of a T-cell immune response (Tewary et al. 2013). Likewise, LL-37 promotes the maturation of iDCs, enhances their expression of co-stimulatory molecules and phagocytic activity, and influences polarization of T -cells toward a Th1-skewed response (Davidson et al. 2004). A recent study has also shown that LL-37 can activate follicular DCs of Peyer's patches primarily by engaging formyl peptide receptors and can enhance the proliferation and activation of B cells (Kim et al. 2017). The ability of CHDPs to enhance adaptive immunity indicates the potential to use these peptides as adjuvants. Indeed, CHDPs have been shown to function as adjuvants exerting antigen-specific immune responses. For example, hBD3 and murine cathelicidin CRAMP exhibited adjuvant effects in ovalbumin-challenge models (Kurosaka et al. 2005; Tewary et al. 2013). Similarly, LL-37 was shown to mediate adjuvant effect in an oral vaccine formulation inducing both mucosal and systemic antigen-specific responses (Kim et al. 2017).

Overall, CHDPs are involved in the interplay between many cell types of the human body, and this is crucial for the ability of CHDPs to influence and regulate immunity. Additional discoveries that these peptides are expressed by intestinal Paneth cells which play a role in maintaining host-microbiome homeostasis have raised interest in exploring the crosstalk between CHDPs and the microbiome, as discussed in the next section.

10.1.3 CHDPs and the Microbiome

The role of CHDPs in microbiome composition is a relatively new emerging field of research. Several studies have demonstrated that CHDPs are important contributors to the host-microbiome interactions. For example, enteric α-defensins expressed by Paneth cells may be essential in shaping the murine gut microbiome (Salzman et al. 2010), as transgenic mice expressing a human defensin, HD5, exhibit reduced *Firmicutes* and increased *Bacteroidetes* gut microbiome levels compared to wild-type controls, whereas deficiency of matrilysin which is essential for murine α-defensin processing results in the opposite effect on murine gut microbiome composition (Salzman et al. 2010). Further indications that CHDPs play an important role in shaping the microbiome is derived from a study demonstrating that differences in copy number of a specific β-defensin gene cluster correlate with alterations in the microbiota of the nasopharynx associated with otitis media infection-related pathogens (Jones et al. 2014). Cullen and co-workers showed that dominant members of the gut microbiota develop a selective resistance against specific CHDPs that are enhanced during inflammation, suggesting that co-evolvement of this microbial resistance mechanism with the host may have allowed stabilization of the gut microbiome during inflammation (Cullen et al. 2015). Apart from CHDPs influencing the composition of the microbiome, recent studies have also demonstrated that the host microbiome can in turn impact CHDP expression. Direct effects of microbial signaling through pattern recognition receptors have been shown to enhance expression of CHDPs such as defensins in the gut (Ostaff et al. 2013). In addition, bacterial metabolites also contribute to the regulation of expression of CHDPs. For example, several short-chain fatty acids such as butyrate produced by gut bacteria upon fermentation of dietary fibers can regulate the expression of specific CHDPs like cathelicidins (Schauber et al. 2004; Zhao et al. 2018), and commensal bacteria can promote CHDP expression in the skin (Meisel et al. 2018). Furthermore, host-derived CHDPs such as cathelicidins can exhibit synergistic functions with the microbiome. For example, antimicrobials produced by the skin microbiome such as *Staphylococcus hominis Sh*-lantibiotic-α and *Sh*-lantibiotic-β can synergistically enhance antimicrobial activity against *Staphylococcus aureus* with LL-37 (Nakatsuji et al. 2017). Overall, recent reports clearly demonstrate the complex interplay between CHDPs and the microbiome in maintaining immune homeostasis and the control of pathogens (Fig. 10.1). Therefore, it is not surprising that disturbance of the balance between microbiome and CHDPs can have severe consequences for the host (Sun et al. 2015). The recognition of the importance of the microbiome in human health has opened up a new avenue of research to examine the therapeutic possibility of targeting the microbiome with synthetic derivatives or mimetics of CHDPs (Stone and Xu 2017; Kaplan et al. 2011).

10.1.4 Clinical Relevance and Therapeutic Potential of CHDPs

As discussed above, CHDPs elicit antimicrobial functions and exhibit a wide range of roles in immunity (summarized in Fig. 10.2). Thus, there

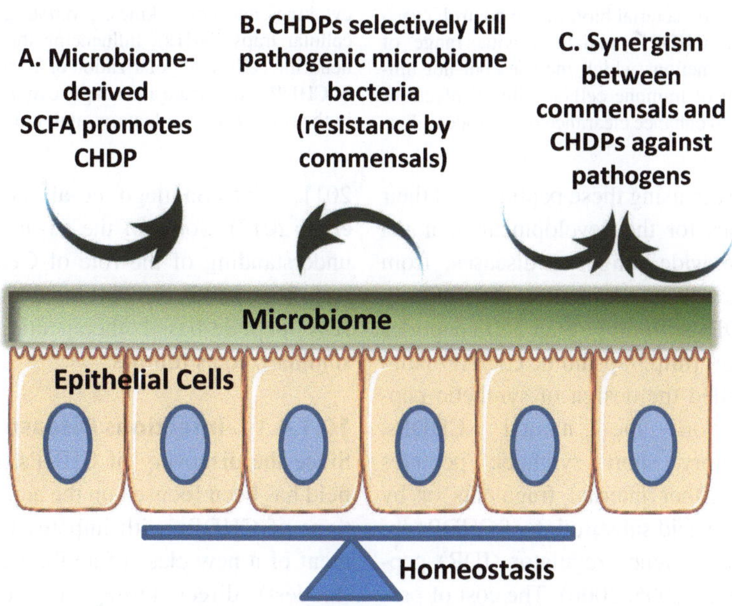

Fig. 10.1 CHDP-microbiome interaction potentially contributes to homeostasis. Cationic host defence peptides (CHDPs) may contribute to host-microbiome homeostasis. (**A**) Short-chain fatty acids (SCFA) produced by microbiome bacteria upon fermentation of dietary fibers were demonstrated to promote HDP expression by epithelium, as were commensal bacteria and microbial products. These effects on CHDP expression could aid in controlling pathogenic bacterial strains. (**B**) Resistance was demonstrated by commensal bacterial strains against CHDPs that are specifically expressed during inflammation, which could potentially stabilize the microbiome during inflammation. (**C**) CHDPs exhibit synergistic antimicrobial activities with skin microbiome-derived antimicrobials against pathogenic bacterial strains

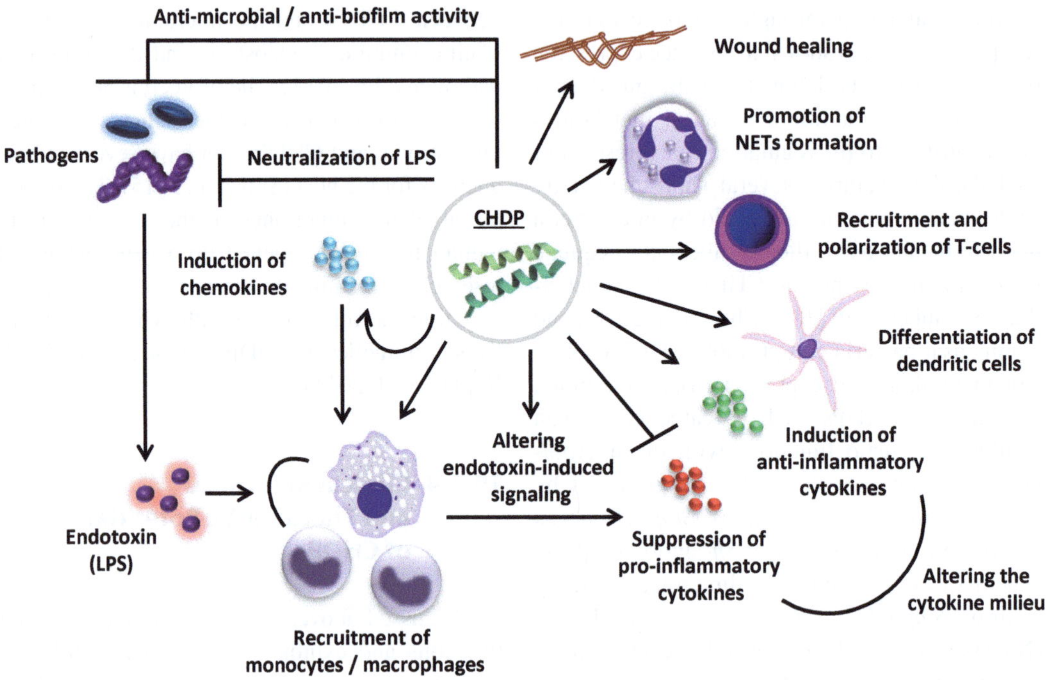

Fig. 10.2 CHDP-mediated activity controls infections and inflammation. CHDPs exhibit antimicrobial activities, including toward bacterial biofilms, to control a variety of infections. CHDPs mediate a wide range of immunity-related functions which includes but not limited to recruitment of immune cells to site of infections which contributes to enhance clearance of microbes, altering endotoxin-mediated signaling, suppression of pro-inflammatory cytokines, induction of anti-inflammatory cytokines and chemokines, promoting neutrophil extracellular traps (NETs), influencing the differentiation of dendritic cells, and polarization of T -cells. The functions of CHDPs link innate and adaptive immunity to contribute in the resolution of infections and immune regulation

is a keen interest in using these peptides and their synthetic mimics for the development of novel therapies for a wide range of diseases, from infectious disease to autoimmunity. The diversity of natural CHDPs with more than 2600 peptides defined to date (http://aps.unmc.edu/AP/main.php) has propelled the design of synthetic peptides based on the natural CHDPs. Immunomodulatory short synthetic peptides designed from either internal fragments or by systematic amino acid substitution of CHDPs are known as innate defence regulator (IDR) peptides (Hilpert et al. 2005, 2006). The cost of production of shorter IDR peptides is less compared to CHDPs (Scott et al. 2007; Nijnik et al. 2010; Cherkasov et al. 2009). Moreover, IDR peptides have negligible toxicity and are not immunogenic, thus making these valuable therapeutic candidates (Scott et al. 2007; Nijnik et al. 2010; Achtman et al. 2012; Molhoek et al. 2009; Steinstraesser et al. 2012; Turner-Brannen et al. 2011; Rivas-Santiago et al. 2013; Niyonsaba et al. 2013). Some of the advances made in the understanding of the role of CHDPs in disease pathology and in the development of CHDPs and IDR peptide-based therapeutic strategies are summarized as follows.

10.1.4.1 Infectious Disease

Since the discovery of CHDPs, research in the field has been focused on the antimicrobial functions of CHDPs with impetus for the development of a new class of antibiotics. Whereas the (modest) direct killing activities of natural CHDPs show some possibilities for this development, the expanding insight into the immunomodulatory activities of CHDPs gives rise to many more therapeutic opportunities. Current paradigm suggests that CHDPs modulate the host immune response to facilitate clearance of infections under physiological conditions. However, it should be noted that the immunomodulatory activities of CHDPs that aid in the resolution of infections are extensive and depend on peptide concentrations, source, and target cell. Several studies have demonstrated the potential of CHDPs in controlling infections in various models; α-defensins were shown to promote recruitment of macrophages and lymphocytes to the peritoneal cavity upon *Klebsiella pneumoniae* infection in mice (Welling et al. 1998). LL-37 was demonstrated to augment phagocytosis by human macrophages (Wan et al. 2014) and facilitate airway epithelial barrier function to prevent *Pseudomonas aeruginosa* invasion (Byfield et al. 2011). Davidson et al. showed that administration of LL-37 promoted clearance of *P. aeruginosa* from infected mice by enhancing the early neutrophil response (Beaumont et al. 2014). Similar protective effects were obtained by overexpression of endogenous cathelicidin in the lungs of mice (Bals et al. 1999). LL-37 was also shown to diminish *Bacillus anthracis* spore-induced death of mice by promoting the recruitment of neutrophils to the site of infection (Lisanby et al. 2008). Administration of the neutrophil α-defensin HNP1 was demonstrated to protect against *M. tuberculosis* in a murine mouse model, with in vitro mechanistic studies clearly demonstrating beneficial effects to support the possible use as anti-infective in tuberculosis (Sharma et al. 2001). Similarly, beneficial effects of exogenous CHDPs administration have also been achieved against viral pathogens. For example, administration of LL-37 was demonstrated to be effective against influenza A virus and respiratory syncytial virus in a murine model (Barlow et al. 2011; Currie et al. 2016).

Another promising strategy being explored for anti-infective therapy is by promoting the expression of endogenous CHDP expression using inducers such as vitamin D_3 (Gombart et al. 2005), sodium (phenyl)butyrate (Steinmann et al. 2009; Schauber et al. 2003), and other recently discovered compounds (Miraglia et al. 2016; Fischer et al. 2016). Inducers of CHDPs are being explored to restore pathogen-induced downregulation of CHDPs, which is a virulence strategy of pathogens such as *Shigella flexneri (*Islam et al. 2001*)* and *Neisseria gonorrhoeae* (Bergman et al. 2005). In line with this, adjunct therapy with (phenyl)butyrate has shown beneficial effects in humans infected with tuberculosis (Mily et al. 2015) and Shigella (Raqib et al. 2012) and in rabbits infected with *E. coli* (Al-Mamun et al. 2013). However, in these studies it remains unclear if the pathogens are cleared as a consequence of the microbicidal activity of the peptides or is it due to CHDP-induced immune responses in the host. Furthermore, it should be noted that inducers of CHDPs such as phenylbutyrate and vitamin D may also control infections independent of CHDPs by influencing the immune system based on in vitro studies demonstrating that these compounds can promote autophagy and intracellular killing of bacteria (Rekha et al. 2015).

Exogenous administration of synthetic IDR peptides has also been shown to control infections in various models, including antibiotic-resistant MRSA (Scott et al. 2007; Nijnik et al. 2010; Rivas-Santiago et al. 2013; Hirsch et al. 2008; Mookherjee and Hancock 2007; Hou et al. 2013). Recent studies have demonstrated the use of IDR peptides on the control of biofilm infections that are typically recalcitrant to antibiotics (Haney et al. 2015). The recent advances in the development of IDR peptides for treatment of infectious diseases are encouraging as some of these peptide therapies are currently in phase II/III clinical trials (Cherkasov et al. 2009; Yeung et al. 2011; Hancock et al. 2012). A distinct advantage of developing CHDP-based antimicrobial therapeutics is thought to be the potential to avoid emergence of microbial resistance. This is speculated primarily due to two reasons: (i) the selective pressure on pathogens is reduced as CHDPs may resolve infections by modulating host immunity rather than directly targeting the microbe, and (ii) when applicable the direct microbicidal activity of these peptides is directed at microbial targets that are indispensable for microbial survival. However, recent studies have indicated several bacterial resistance mechanisms to antimicrobial peptides (reviewed in Joo et al. 2016), and so it remains unclear if CHDP-based therapy can indeed avoid the emergence of microbial resistance.

10.1.4.2 Chronic Inflammatory Lung Disease

The lung is exposed daily to large numbers of inhaled pathogens, and CHDPs play a central role in the protection of the host from such infections. CHDPs are produced by a range of cells in the lungs, including the airway epithelial cells and neutrophils. As discussed above, these peptides exert prominent effects on host immune responses, and therefore it is not surprising that specific CHDPs are implicated in chronic inflammatory lung diseases. In this context, it should be noted that a relative deficiency of CHDPs resulting from decreased expression, degradation, or impairment of function, as well as their excessive activity, both have been demonstrated to contribute to disease pathology in the lung.

Deficient expression or activity of CHDP has been linked to the frequent respiratory infections observed in smokers (Herr et al. 2009), patients with chronic obstructive pulmonary disease (COPD) (Pace et al. 2012; Amatngalim et al. 2017), and those with cystic fibrosis (CF) (Chen et al. 2004; Pezzulo et al. 2012). Previous studies have used airway epithelial cell cultures to show that smoking, the main risk factor for COPD in industrialized societies, reduces the expression of specific CHDP (Herr et al. 2009; Amatngalim et al. 2017). Furthermore, the expression of some of these peptides is also decreased in airway epithelial cell cultures from COPD patients (Amatngalim et al. 2017). In addition, expression of specific CHDPs is also decreased in allergic airway inflammation which may be linked to decreased pulmonary host defenses (Beisswenger et al. 2006). Furthermore, a recent study demonstrates that specific CHDPs are decreased in the lungs in response to inhaled air pollutants known to exacerbate asthma (Piyadasa et al. 2018b). In contrast, it is also evident that neutrophil-derived CHDPs such as the neutrophil α-defensins and hCAP-18/LL-37 may be increased in neutrophil-dominated airway inflammation in COPD and CF and during rhinovirus infection in asthma (Chen et al. 2004; Paone et al. 2011; Merkel et al. 2005; Rohde et al. 2014). It was demonstrated that COPD exacerbations are associated with changes in the pattern of CHDP in sputum, with increased levels of hCAP-18/LL-37 and decreased levels of the antimicrobial proteinase inhibitor secretory leukocyte proteinase inhibitor (SLPI), and that higher hCAP-18/LL-37 levels in stable COPD disease are associated with an increased risk of exacerbations of the disease (Persson et al. 2017). It is thus tempting to speculate that specific CHDPs may contribute to dysregulated innate and adaptive immune responses in inflammatory lung disease due to their wide range of effects on immunity.

Enhancing CHDP levels in the lungs may be an attractive strategy to counteract deficient endogenous expression and combat the respiratory infections that play a detrimental role in patients with CF, COPD, and asthma (Hiemstra et al. 2016). However, control of inflammation and restoration of dysregulated immune responses is also a major aim of treatment. The dual face of CHDP in providing direct antimicrobial activity and regulating inflammation and immunity has to be taken into account when considering the use of CHDP as therapeutics in the treatment of chronic inflammatory lung diseases. Therefore, vitamin D treatment aimed to increase endogenous expression of CHDPs such as hCAP-18/LL-37 while at the same time dampening unwanted immune responses is an attractive strategy. Indeed, results of two recent trials in COPD have shown that vitamin D treatment may serve to decrease exacerbations in patients with a vitamin D deficiency (Lehouck et al. 2012; Martineau et al. 2015). Other therapeutic strategies may include direct administration of synthetic IDR peptides with an optimal balance of activities, to facilitate both antimicrobial activities and immunomodulatory functions in the lungs. For example, administration of a bovine cathelicidin-based IDR peptide may alleviate airway inflammation and pathophysiology of asthma (Piyadasa et al. 2018a).

10.1.4.3 Skin Disease

The skin provides a protective interface between internal organs and environmental factors including pathogens. Apart from being a physical barrier, it is also an active immune organ. Thus, substantial research has focused on trying to understand the role of CHDPs in antimicrobial defenses of the skin and in unraveling the role of these peptides in chronic skin diseases such as psoriasis and atopic dermatitis. The variety of constitutively expressed and inducible CHDPs produced by keratinocytes and other cells in the skin provides an important layer of defense (Niyonsaba et al. 2017). Whereas most studies have focused on the role of β-defensins and hCAP-18/LL-37, other CHDPs such as S100A7/psoriasin, the RNase family member RNase 7, and dermcidin are likely also important for local defense. Various experimental models have been used to uncover the role of CHDPs in host defense, including knockout models showing increased susceptibility to necrotizing skin infections with group A streptococci of mice deficient in CRAMP, the mouse homologue of the gene encoding the human cathelicidin hCAP-18/LL-37 (Nizet et al. 2001). Also, treatment with a synthetic CHDP-inspired peptide (SAAP-148) was found to eradicate acute and established biofilm infections with antibiotic-resistant *Staphylococcus aureus* and *Acinetobacter baumannii* in wounded human skin ex vivo and mouse skin in vivo (de Breij et al. 2018). Moreover, patient studies in atopic dermatitis and psoriasis, two frequent chronic inflammatory skin diseases, indicate that the role of CHDP in diseases may be more complex.

Initial studies showed an association of deficient expression of selected CHDPs in lesional skin of patients with atopic dermatitis who suffer from frequent bacterial skin infections (Ong et al. 2002). The ability of Th2 cytokines, key in allergic inflammation in diseases such as atopic dermatitis, to suppress expression of CHDPs offered an attractive explanation for the frequently observed skin infections with *Staphylococcus aureus* in patients with atopic dermatitis. Furthermore, these studies suggested that the relative deficiency of CHDPs in atopic dermatitis and high expression in psoriasis might help to explain the differential occurrence of skin infections in these patients (Ong et al. 2002). However, whether decreased levels of CHDPs are a common characteristic and main factor in determining susceptibility of the skin in atopic dermatitis is a matter of debate, as CHDPs also play an important pro-inflammatory and immunoregulatory role in atopic dermatitis (Kopfnagel et al. 2013). Furthermore, studies in various models and in atopic dermatitis and psoriasis have demonstrated a role for CHDPs in wound repair (Mangoni et al. 2016). This resulted in a clinical trial exploring the topical use of LL-37 to treat venous leg ulcers. Topical application of LL-37 to these chronic wounds was found to be safe and well tolerated, and initial results suggested a beneficial effect on wound healing (Gronberg et al. 2014). The observation that high β-defensin gene copy numbers are associated with an increased risk of developing psoriasis also indicates that CHDPs such as β-defensins are not merely protective against skin infections in inflammatory skin diseases (Hollox et al. 2008). Aligned with this, it has been shown that genetic knockout of β-defensins in a bioengineered skin-humanized mouse model for psoriasis has beneficial effects (Bracke et al. 2014). Other CHDPs, including hCAP18/LL-37, have been shown to contribute to skin inflammation in psoriasis, and in contrast protective roles for CHDPs in skin barrier function and wound healing and in reducing inflammation have also been reported (Niyonsaba et al. 2017). Collectively these studies suggest a complex role of CHDPs in skin disease, with both protective antimicrobial, immunomodulatory, and wound repair enhancing functions and disease-promoting inflammatory activities with dysregulation of innate and adaptive immune response. Therefore, a better insight into the role of CHDPs is essential for the development of novel CHDP-based therapies for skin diseases.

10.1.4.4 Autoimmune Disease

Previous studies provide contradictory evidences demonstrating CHDPs as both effectors and regulators of autoimmune diseases such as in systemic lupus erythematosus (SLE) and rheumatoid arthritis (RA). Serum levels of CHDPs, e.g., defensins (hBD-1, hBD-2, HNP1–3) and psoriasin, are increased, with the circulating levels of these defensins associated with the disease activity in SLE (Kahlenberg and Kaplan 2013; Kreuter et al. 2011; Sthoeger et al. 2009; Vordenbaumen et al. 2010). Release of nuclear autoantigens within neutrophil extracellular traps (NETs) is a characteristic feature of SLE (Radic 2014). As LL-37 can complex with nuclear autoantigens within NETs, it has been suggested that LL-37 may facilitate subsequent pro-inflammatory responses (Lande et al. 2011; Kahlenberg et al. 2013; Favilli et al. 2009), as well as prevent the degradation of NETs, and facilitate the formation of NETs (Neumann et al. 2014), thus collectively promoting autoantibody formation and generation of immune complexes in SLE (Lande et al. 2011). Similarly, levels of other CHDPs such as defensins and S100 peptides are increased in the synovial fluid of patients with RA (Baillet et al. 2010; Bokarewa et al. 2003). In vitro studies using cells such as synoviocytes, osteoclasts, and granulocytes isolated from RA patients also show an increased expression of specific CHDPs such as hBD-2, HNP-1, and LL-37 (Hoffmann et al. 2013; Varoga et al. 2006, 2009; Ahn et al. 2013; Kienhofer et al. 2014). This is corroborated by animal model studies demonstrating a correlation with increased cathelicidin expression and disease pathology in arthritis (Hoffmann et al. 2013). These studies indicate that specific CHDPs may contribute to the disease pathology and progression in autoimmune diseases such as SLE and RA. In contrast, recent studies suggest that CHDPs may function in the control of inflammation and tissue damage in autoimmune disease; Miles et al. showed anti-inflammatory mechanisms mediated by the release of α-defensins HNP1–3 from neutrophils (Miles et al. 2009), thus suggesting that the release of specific CHDPs from apoptotic or necrotic neutrophils in SLE may contribute to the control of inflammation. Similarly, LL-37 and its synthetic derivative peptide can inhibit pro-inflammatory responses and signaling mechanisms related to RA (Choi et al. 2014; Xu et al. 2013). Moreover, LL-37 was shown to prevent osteoclastogenesis (Supanchart et al. 2012) thus indicating a role of LL-37 in preventing bone loss in RA. Aligned with this, a LL-37-derived synthetic peptide was shown to prevent disease progression, inflammation, and articular tissue damage in a murine model of RA (Chow et al. 2014). There are limited studies examining the role of CHDPs in other autoimmune conditions, which also provide contradictory evidence. Increased circulating levels of α-defensins have been demonstrated in type I diabetes, suggesting that these peptides may contribute to the pathogenesis of this disease (Saraheimo et al. 2008; Joseph et al. 2008; Nemeth et al. 2014). In contrast, levels of β-defensins hBD-1 and hBD-2 were shown to be decreased in salivary glands of patients with Sjogren's syndrome (Kaneda et al. 2009), and CHDPs can limit inflammation and prevent infections in this disease (Luciano et al. 2015). Similarly, LL-37 was shown to be beneficial in the control of keratinopathy and polymicrobial infected corneal wounds, common in Sjogren's syndrome (Yin and Yu 2010; Duplantier and van Hoek 2013). Similar to what was discussed above for other diseases, there are contradictory evidences in various studies examining the role of CHDPs in autoimmunity. It is plausible that the function of these peptides may be disease specific and/or depend on the disease stage. Nevertheless, it is important to explore the role of CHDPs in autoimmunity, which may contribute to the development of new therapies for the spectrum of diseases in this category.

10.1.4.5 Cancer

The involvement of CHDPs in cancer development and progression is increasingly being examined (Deslouches and Di 2017; Roudi et al. 2017). Reports demonstrate that CHDPs such as defensins and cathelicidins exert both anticancer and pro-tumorigenic activity. For example, LL-37 was demonstrated to be overexpressed in melanoma (Kim et al. 2010), breast cancer (Heilborn et al. 2005), prostate cancer (Hensel et al. 2011), and some lung cancers, but not in small cell lung cancers and carcinoid tumors (von Haussen et al. 2008). It was also shown in in vitro studies that LL-37 can promote proliferation and/or migration of malignant cell lines such as melanoma (Jia et al. 2017), keratinocytes (Weber et al. 2009), kidney cells (Weber et al. 2009), prostate cancer cells (Hensel et al. 2011), and several lung cancer cells (von Haussen et al. 2008), via different signaling pathways such as MAPK, EGFR, Y-box binding protein 1, and NF-κB. Indirect effects have been demonstrated in the case of ovarian cancer, where LL-37 mediated the recruitment of mesenchymal stem cells to promote tumor development (Coffelt et al. 2009). Conversely, anticancer effects have also been suggested for CHDPs in colorectal and gastric cancer; loss of LL-37 expression was associated with tumor growth, thus suggesting a protective function of LL-37 in these cancers (Cheng et al. 2015; Ren et al. 2012). Wu et al. further demonstrated that LL-37 can inhibit the proliferation of gastric cell lines in vitro and gastric cancer xenografts in vivo (Wu et al. 2010), and pro-apoptotic effects of this peptide were shown in Jurkat T-leukemia cells (Mader et al. 2009). Similarly, several studies demonstrated loss of β-defensins at tumor sites including in salivary gland tumor (Kesting et al. 2012), renal and prostatic cancer (Donald et al. 2003), and oral squamous cell carcinoma (Joly et al. 2009), suggesting a protective role for these defensins in cancer. This was further highlighted by a study examining defensin expression in 29 different tumor-derived cell lines, which showed a low or no transcript expression of β-defensins, while α-defensin expression was maintained in these cell lines (Winter et al. 2016). Aligned with this, several studies have demonstrated direct protective effects for β-defensins. For example, hBD-1 was shown to inhibit cell migration and invasion in oral squamous cell carcinoma (Han et al. 2014); the sequence of hBD-3 was found to contain an oncolytic binding motif promoting cytolysis of tumor cells (Phan et al. 2016); hBD-3 was found to be cytotoxic to human alveolar basal epithelial cell line A549; the murine orthologue of hBD-3, Defb14, was cytotoxic to Lewis lung carcinoma cell line-derived tumors in mice (Hanaoka et al. 2016); and hBD-3 was demonstrated to suppress the migration of head and neck cancer cells (Wang et al. 2012) and several colon cancer cell lines (Uraki et al. 2015). However, similar to LL-37, contradictory results have also been shown for defensins, for example, high expression of hBD-3 was associated with oral squamous cell carcinoma (Kesting et al. 2009). Interestingly, it has been suggested that CHDPs may also influence cancer development indirectly by interacting with bacteria or viruses. Bacteria associated with the development of oral cancers actively dysregulate transcript levels of α- and β-defensins in oral tumor cells, which in its turn has been suggested to promote the proliferation of these cells (Hoppe et al. 2016). In contrast, a possible protective role for α-defensins in cancer was suggested by the antiviral activities of HNP1–3 and HD-5 against human papillomavirus (Buck et al. 2006). CHDPs are also being examined for a possible role as biomarkers due to their dysregulated expression in cancers, often as a result of a significant loss of expression in tumors (Hong et al. 2017), except for α-defensins which are often highly expressed by tumors, e.g., in colorectal cancer (Melle et al. 2005), renal cell carcinomas (Muller et al. 2002), and bladder cancer (Holterman et al. 2006). Overall, it remains unclear whether CHDPs are protective or contribute to the pathology of cancer. It is likely that this is dependent on the type of cancer, the specific CHDP, and concentration of the specific peptide. Nevertheless, the therapeutic potential of HDPs for cancer is being examined (Roudi et al. 2017). An hBD-2-based gene therapy was suggested based on a study demonstrating that hBD-2 can promote recruitment of DCs to the tumor site resulting in enhanced antitumor immunity and a reduction in tumor size (Malinovschi et al. 2014). A combination of LL-37 and CpG-oligodeoxynucleotides was shown to be successful in a murine model of ovarian cancer (Chuang et al. 2009). Vragniau and co-authors have showed high expression of HD5 in ovary, endometrium, and lung cancer and suggested a role of HD5 in preventing oncolytic adenovirus replication and spread (Vragniau et al. 2017). Even though the diverse activities of CHDPs provide potential for their use in cancer therapy, contradictory effects of these peptides complicate the development of CHDP-based therapies for cancers. It is possible that future studies using synthetic IDR peptides designed with defined activity could be explored to target specific tumors.

10.1.5 Challenges in the Development of CHDP-Based Therapy

Despite the impressive and wide repertoire of activities, most natural CHDP may not be ideally suited as drugs that can be used for direct application. Direct antimicrobial activity requires high local concentrations, and many CHDPs display pro-inflammatory and cytotoxic activities at such concentrations. This needs to be taken into account when developing therapies using CHDPs to target infections requiring local concentrations high enough to display direct antimicrobial activity. The route of administration is also an important consideration, and so far most studies using CHDP-based therapies have employed topical application, e.g., on skin, or local administration in the respiratory tract. Another challenge is the local activity of CHDPs, since especially at sites of intense inflammation and infection the local microenvironment may not allow full activity of the peptides. For example, local mucosal pH (Pezzulo et al. 2012), degradation by microbial and host proteases (Mallia et al. 2012), and impairment of CHDP-mediated antimicrobial activity by salt, F-actin, DNA, mucus, and microbial saccharides (Bucki et al. 2007) are among the factors that can alter/impair CHDP activity. Thus stability and bioavailability are among the limiting factors in the use of CHDPs as therapeutics. However, this can potentially be mitigated by exploring delivery systems with formulations using liposomes, polymeric nanoparticles, carbon nanotubes, and other materials. Another challenge in the development of CHDP therapy is the cost of production. Based on these challenges, alternative strategies are being explored to develop CHDP-based treatments. These include (i) stimulation of endogenous CHDP production by inducers such as vitamin D or butyrate, (ii) improving the local conditions that may otherwise impair the activity of CHDPs, and (iii) the development of novel synthetic peptides inspired by the structure and/or sequence of natural CHDPs, such as IDR peptides.

10.2 Summary

The wide repertoire of functions of CHDPs from its role in the control of infections to diverse functions in immunity underscores the importance of these peptides in health and disease. It is now well appreciated that CHDPs exhibit both innate and adaptive immune functions, and influence regulation of inflammation by direct effects on immune cells and via induction of immune mediators by other cell types. Impact of CHDPs on the immune system has significantly broadened the scope of research in this field beyond antimicrobial activity. However, the duality of immunomodulatory functions exhibited by these peptides, by both pro- and anti-inflammatory functions contributing to immune activation and regulation, and the interplay of CHDPs and microbiome in mucosal immunity and immune homeostasis, highlights the complexity of interpreting the exact role of these peptides in the context of immune response. It is thus also not surprising that the dysregulation of expression of these peptides is associated with disease pathology from infectious disease to autoimmunity and various cancers. Diversity of natural CHDPs and its repertoire of functions have propelled interest in the use of these peptides and their synthetic mimics in the development of new therapies for various diseases. Despite challenges associated with the development of CHDP-based therapy, the applicability of CHDP-based immunomodulatory therapies offers an entirely new therapeutic approach for the resolution of infections by enhancing host immune response and on the other hand the control of inflammatory diseases without compromising the patients' ability to resolve infections. In this context, current research indicates that synthetic variants of CHDPs may be promising for future clinical use.

Acknowledgment AD is supported by an EU Marie Curie Global Fellowship (#748569). Studies in the laboratory of PSH on CHDPs are supported by grants from the Lung Foundation Netherlands, the Eurostars program, The Netherlands Organisation for Health Research and Development (ZonMw), and Galapagos NV. NM is supported by Canadian Institutes of Health Research (CIHR) and Natural Sciences and Engineering Research Council of Canada (NSERC) for peptide research.

References

Achtman AH, Pilat S, Law CW, Lynn DJ, Janot L, Mayer ML et al (2012) Effective adjunctive therapy by an innate defense regulatory Peptide in a pre-clinical model of severe malaria. Sci Transl Med 4(135):135ra64

Agerberth B, Gunne H, Odeberg J, Kogner P, Boman HG, Gudmundsson GH (1995) FALL-39, a putative human peptide antibiotic, is cysteine-free and expressed in bone marrow and testis. Proc Natl Acad Sci U S A 92(1):195–199

Agerberth B, Grunewald J, Castanos-Velez E, Olsson B, Jornvall H, Wigzell H et al (1999) Antibacterial components in bronchoalveolar lavage fluid from healthy individuals and sarcoidosis patients. Am J Respir Crit Care Med 160(1):283–290

Agerberth B, Charo J, Werr J, Olsson B, Idali F, Lindbom L et al (2000) The human antimicrobial and chemotactic peptides LL-37 and alpha-defensins are expressed by specific lymphocyte and monocyte populations. Blood 96(9):3086–3093

Agier J, Efenberger M, Brzezinska-Blaszczyk E (2015) Cathelicidin impact on inflammatory cells. Cent Eur J Immunol 40(2):225–235

Ahn JK, Huang B, Bae EK, Park EJ, Hwang JW, Lee J et al (2013) The role of alpha-defensin-1 and related signal transduction mechanisms in the production of IL-6, IL-8 and MMPs in rheumatoid fibroblast-like synoviocytes. Rheumatology (Oxford) 52(8):1368–1376

Al-Mamun A, Mily A, Sarker P, Tiash S, Navarro A, Akter M et al (2013) Treatment with phenylbutyrate in a pre-clinical trial reduces diarrhea due to enteropathogenic *Escherichia coli*: link to cathelicidin induction. Microbes Infect 15(13):939–950

Amatngalim GD, Schrumpf JA, Henic A, Dronkers E, Verhoosel RM, Ordonez SR et al (2017) Antibacterial defense of human airway epithelial cells from chronic obstructive pulmonary disease patients induced by acute exposure to nontypeable haemophilus influenzae: modulation by cigarette smoke. J Innate Immun 9(4):359

Baillet A, Trocme C, Berthier S, Arlotto M, Grange L, Chenau J et al (2010) Synovial fluid proteomic fingerprint: S100A8, S100A9 and S100A12 proteins discriminate rheumatoid arthritis from other inflammatory joint diseases. Rheumatology (Oxford) 49(4):671–682

Bals R, Wang X, Zasloff M, Wilson JM (1998) The peptide antibiotic LL-37/hCAP-18 is expressed in epithelia of the human lung where it has broad antimicrobial activity at the airway surface. Proc Natl Acad Sci U S A 95(16):9541–9546

Bals R, Weiner DJ, Moscioni AD, Meegalla RL, Wilson JM (1999) Augmentation of innate host defense by expression of a cathelicidin antimicrobial peptide. Infect Immun 67(11):6084–6089

Barlow PG, Svoboda P, Mackellar A, Nash AA, York IA, Pohl J et al (2011) Antiviral activity and increased host defense against influenza infection elicited by the human cathelicidin LL-37. PLoS One 6(10):e25333

Beaumont PE, McHugh B, Gwyer Findlay E, Mackellar A, Mackenzie KJ, Gallo RL et al (2014) Cathelicidin host defence peptide augments clearance of pulmonary Pseudomonas aeruginosa infection by its influence on neutrophil function in vivo. PLoS One 9(6):e99029

Beisswenger C, Kandler K, Hess C, Garn H, Felgentreff K, Wegmann M et al (2006) Allergic airway inflammation inhibits pulmonary antibacterial host defense. J Immunol 177(3):1833–1837

Bergman P, Johansson L, Asp V, Plant L, Gudmundsson GH, Jonsson AB et al (2005) Neisseria gonorrhoeae downregulates expression of the human antimicrobial peptide LL-37. Cell Microbiol 7(7):1009–1017

Bokarewa MI, Jin T, Tarkowski A (2003) Intraarticular release and accumulation of defensins and bactericidal/permeability-increasing protein in patients with rheumatoid arthritis. J Rheumatol 30(8):1719–1724

Bowdish DM, Davidson DJ, Scott MG, Hancock RE (2005) Immunomodulatory activities of small host defense peptides. Antimicrob Agents Chemother 49(5):1727–1732

Bracke S, Carretero M, Guerrero-Aspizua S, Desmet E, Illera N, Navarro M et al (2014) Targeted silencing of DEFB4 in a bioengineered skin-humanized mouse model for psoriasis: development of siRNA SECosome-based novel therapies. Exp Dermatol 23(3):199–201

Buck CB, Day PM, Thompson CD, Lubkowski J, Lu W, Lowy DR et al (2006) Human alpha-defensins block papillomavirus infection. Proc Natl Acad Sci U S A 103(5):1516–1521

Bucki R, Sostarecz AG, Byfield FJ, Savage PB, Janmey PA (2007) Resistance of the antibacterial agent ceragenin CSA-13 to inactivation by DNA or F-actin and its activity in cystic fibrosis sputum. J Antimicrob Chemother 60(3):535–545

Bucki R, Leszczynska K, Namiot A, Sokolowski W (2010) Cathelicidin LL-37: a multitask antimicrobial peptide. Arch Immunol Ther Exp 58(1):15–25

Byfield FJ, Kowalski M, Cruz K, Leszczynska K, Namiot A, Savage PB et al (2011) Cathelicidin LL-37 increases lung epithelial cell stiffness, decreases transepithelial permeability, and prevents epithelial invasion by Pseudomonas aeruginosa. J Immunol 187(12):6402–6409

Chen CI, Schaller-Bals S, Paul KP, Wahn U, Bals R (2004) Beta-defensins and LL-37 in bronchoalveolar lavage fluid of patients with cystic fibrosis. J Cyst Fibros 3(1):45–50

Cheng M, Ho S, Yoo JH, Tran DH, Bakirtzi K, Su B et al (2015) Cathelicidin suppresses colon cancer development by inhibition of cancer associated fibroblasts. Clin Exp Gastroenterol 8:13–29

Cherkasov A, Hilpert K, Jenssen H, Fjell CD, Waldbrook M, Mullaly SC et al (2009) Use of artificial intelligence in the design of small peptide antibiotics effective against a broad spectrum of highly antibiotic-resistant superbugs. ACS Chem Biol 4(1):65–74

Chertov O, Michiel DF, Xu L, Wang JM, Tani K, Murphy WJ et al (1996) Identification of defensin-1, defensin-2 and CAP37/azurocidin as T-cell chemoattractant proteins released from interleukin-8-stimulated neutrophils. J Biol Chem 271:2935–2940

Choi KY, Mookherjee N (2012) Multiple immune-modulatory functions of cathelicidin host defense peptides. Front Immunol 3:149

Choi KY, Chow LN, Mookherjee N (2012) Cationic host defence peptides: multifaceted role in immune modulation and inflammation. J Innate Immun 4(4):361–370

Choi KY, Napper S, Mookherjee N (2014) Human cathelicidin LL-37 and its derivative IG-19 regulate interleukin-32-induced inflammation. Immunology 143(1):68–80

Chow LN, Choi KY, Piyadasa H, Bossert M, Uzonna J, Klonisch T et al (2014) Human cathelicidin LL-37-derived peptide IG-19 confers protection in a murine model of collagen-induced arthritis. Mol Immunol 57(2):86–92

Chromek M, Slamova Z, Bergman P, Kovacs L, Podracka L, Ehren I et al (2006) The antimicrobial peptide cathelicidin protects the urinary tract against invasive bacterial infection. Nat Med 12(6):636–641

Chuang CM, Monie A, Wu A, Mao CP, Hung CF (2009) Treatment with LL-37 peptide enhances antitumor effects induced by CpG oligodeoxynucleotides against ovarian cancer. Hum Gene Ther 20(4):303–313

Coffelt SB, Marini FC, Watson K, Zwezdaryk KJ, Dembinski JL, LaMarca HL et al (2009) The pro-inflammatory peptide LL-37 promotes ovarian tumor progression through recruitment of multipotent mesenchymal stromal cells. Proc Natl Acad Sci U S A 106(10):3806–3811

Cowland JB, Johnsen AH, Borregaard N (1995) hCAP-18, a cathelin/pro-bactenecin-like protein of human neutrophil specific granules. FEBS Lett 368(1):173–176

Cullen TW, Schofield WB, Barry NA, Putnam EE, Rundell EA, Trent MS et al (2015) Gut microbiota. Antimicrobial peptide resistance mediates resilience of prominent gut commensals during inflammation. Science 347(6218):170–175

Currie SM, Gwyer Findlay E, McFarlane AJ, Fitch PM, Bottcher B, Colegrave N et al (2016) Cathelicidins have direct antiviral activity against respiratory syncytial virus in vitro and protective function in vivo in mice and humans. J Immunol 196(6):2699–2710

Davidson DJ, Currie AJ, Reid GS, Bowdish DM, MacDonald KL, Ma RC et al (2004) The cationic antimicrobial peptide LL-37 modulates dendritic cell differentiation and dendritic cell-induced T cell polarization. J Immunol 172(2):1146–1156

de Breij A, Riool M, Cordfunke RA, Malanovic N, de Boer L, Koning RI et al (2018) The antimicrobial peptide SAAP-148 combats drug-resistant bacteria and biofilms. Sci Transl Med 10(423):eaan4044

De Smet K, Contreras R (2005) Human antimicrobial peptides: defensins, cathelicidins and histatins. Biotechnol Lett 27(18):1337–1347

De Y, Chen Q, Schmidt AP, Anderson GM, Wang JM, Wooters J et al (2000) LL-37, the neutrophil granule- and epithelial cell-derived cathelicidin, utilizes formyl peptide receptor-like 1 (FPRL1) as a receptor to chemoattract human peripheral blood neutrophils, monocytes, and T cells. J Exp Med 192(7):1069–1074

Deslouches B, Di YP (2017) Antimicrobial peptides with selective antitumor mechanisms: prospect for anticancer applications. Oncotarget 8(28):46635–46651

Donald CD, Sun CQ, Lim SD, Macoska J, Cohen C, Amin MB et al (2003) Cancer-specific loss of beta-defensin 1 in renal and prostatic carcinomas. Lab Investig 83(4):501–505

Doss M, White MR, Tecle T, Hartshorn KL (2010) Human defensins and LL-37 in mucosal immunity. J Leukoc Biol 87(1):79–92

Duplantier AJ, van Hoek ML (2013) The human cathelicidin antimicrobial peptide LL-37 as a potential treatment for polymicrobial infected wounds. Front Immunol 4:143

Epand RM, Vogel HJ (1999) Diversity of antimicrobial peptides and their mechanisms of action. Biochim Biophys Acta 1462(1–2):11–28

Favilli F, Anzilotti C, Martinelli L, Quattroni P, De Martino S, Pratesi F et al (2009) IL-18 activity in systemic lupus erythematosus. Ann N Y Acad Sci 1173:301–309

Fernandez de Caleya R, Gonzalez-Pascual B, Garcia-Olmedo F, Carbonero P (1972) Susceptibility of phytopathogenic bacteria to wheat purothionins in vitro. Appl Microbiol 23(5):998–1000

Fischer N, Sechet E, Friedman R, Amiot A, Sobhani I, Nigro G et al (2016) Histone deacetylase inhibition enhances antimicrobial peptide but not inflammatory cytokine expression upon bacterial challenge. Proc Natl Acad Sci U S A 113(21):E2993–E3001

Frohm M, Agerberth B, Ahangari G, Stahle-Backdahl M, Liden S, Wigzell H et al (1997) The expression of the gene coding for the antibacterial peptide LL-37 is induced in human keratinocytes during inflammatory disorders. J Biol Chem 272(24):15258–15263

Funderburg N, Lederman MM, Feng Z, Drage MG, Jadlowsky J, Harding CV et al (2007) Human -defensin-3 activates professional antigen-presenting cells via toll-like receptors 1 and 2. Proc Natl Acad Sci U S A 104(47):18631–18635

Ganz T (2003a) Defensins: antimicrobial peptides of innate immunity. Nat Rev Immunol 3(9):710–720

Ganz T (2003b) The role of antimicrobial peptides in innate immunity. Integr Comp Biol 43(2):300–304

Ganz T, Selsted ME, Szklarek D, Harwig SS, Daher K, Bainton DF et al (1985) Defensins. Natural peptide antibiotics of human neutrophils. J Clin Invest 76(4):1427–1435

Ghosh D, Porter E, Shen B, Lee SK, Wilk D, Drazba J et al (2002) Paneth cell trypsin is the processing enzyme for human defensin-5. Nat Immunol 3(6):583–590

Gombart AF, Borregaard N, Koeffler HP (2005) Human cathelicidin antimicrobial peptide (CAMP) gene is a direct target of the vitamin D receptor and is strongly upregulated in myeloid cells by 1,25-dihydroxyvitamin D3. FASEB J 19(9):1067–1077

Gordon YJ, Huang LC, Romanowski EG, Yates KA, Proske RJ, McDermott AM (2005) Human cathelicidin (LL-37), a multifunctional peptide, is expressed by ocular surface epithelia and has potent antibacterial and antiviral activity. Curr Eye Res 30(5):385–394

Gronberg A, Mahlapuu M, Stahle M, Whately-Smith C, Rollman O (2014) Treatment with LL-37 is safe and effective in enhancing healing of hard-to-heal venous leg ulcers: a randomized, placebo-controlled clinical trial. Wound Repair Regen 22(5):613–621

Gudmundsson GH, Agerberth B, Odeberg J, Bergman T, Olsson B, Salcedo R (1996) The human gene FALL39 and processing of the cathelin precursor to the antibacterial peptide LL-37 in granulocytes. Eur J Biochem 238(2):325–332

Guo C, Rosoha E, Lowry MB, Borregaard N, Gombart AF (2013) Curcumin induces human cathelicidin antimicrobial peptide gene expression through a vitamin D receptor-independent pathway. J Nutr Biochem 24(5):754–759

Han Q, Wang R, Sun C, Jin X, Liu D, Zhao X et al (2014) Human beta-defensin-1 suppresses tumor migration and invasion and is an independent predictor for survival of oral squamous cell carcinoma patients. PLoS One 9(3):e91867

Hanaoka Y, Yamaguchi Y, Yamamoto H, Ishii M, Nagase T, Kurihara H et al (2016) In vitro and in vivo anticancer activity of human beta-defensin-3 and its mouse homolog. Anticancer Res 36(11):5999–6004

Hancock RE, Lehrer R (1998) Cationic peptides: a new source of antibiotics. Trends Biotechnol 16(2):82–88

Hancock RE, Nijnik A, Philpott DJ (2012) Modulating immunity as a therapy for bacterial infections. Nat Rev Microbiol 10(4):243–254

Hancock RE, Haney EF, Gill EE (2016) The immunology of host defence peptides: beyond antimicrobial activity. Nat Rev Immunol 16(5):321–334

Haney EF, Mansour SC, Hilchie AL, de la Fuente-Nunez C, Hancock RE (2015) High throughput screening methods for assessing antibiofilm and immunomodulatory activities of synthetic peptides. Peptides 71:276–285

Harder J, Bartels J, Christophers E, Schroder JM (2001) Isolation and characterization of human beta -defensin-3, a novel human inducible peptide antibiotic. J Biol Chem 276(8):5707–5713

Harder J, Meyer-Hoffert U, Wehkamp K, Schwichtenberg L, Schroder JM (2004) Differential gene induction of human beta-defensins (hBD-1, -2, -3, and -4) in keratinocytes is inhibited by retinoic acid. J Invest Dermatol 123(3):522–529

Hase K, Eckmann L, Leopard JD, Varki N, Kagnoff MF (2002) Cell differentiation is a key determinant of cathelicidin LL-37/human cationic antimicrobial protein 18 expression by human colon epithelium. Infect Immun 70(2):953–963

Hashimoto S, Uto H, Kanmura S, Sakiyama T, Oku M, Iwashita Y et al (2012) Human neutrophil peptide-1 aggravates dextran sulfate sodium-induced colitis. Inflamm Bowel Dis 18(4):667–675

Heilborn JD, Nilsson MF, Jimenez CI, Sandstedt B, Borregaard N, Tham E et al (2005) Antimicrobial protein hCAP18/LL-37 is highly expressed in breast cancer and is a putative growth factor for epithelial cells. Int J Cancer 114(5):713–719

Hemshekhar M, Anaparti V, Mookherjee N (2016) Functions of cationic host defense peptides in immunity. Pharmaceuticals (Basel) 9(3):40

Hensel JA, Chanda D, Kumar S, Sawant A, Grizzle WE, Siegal GP et al (2011) LL-37 as a therapeutic target for late stage prostate cancer. Prostate 71(6):659–670

Herr C, Beisswenger C, Hess C, Kandler K, Suttorp N, Welte T et al (2009) Suppression of pulmonary innate host defence in smokers. Thorax 64(2):144–149

Hiemstra PS (2015) Parallel activities and interactions between antimicrobial peptides and complement in host defense at the airway epithelial surface. Mol Immunol 68(1):28–30

Hiemstra PS, Amatngalim GD, van der Does AM, Taube C (2016) Antimicrobial peptides and innate lung defenses: role in infectious and noninfectious lung diseases and therapeutic applications. Chest 149(2):545–551

Hilpert K, Volkmer-Engert R, Walter T, Hancock RE (2005) High-throughput generation of small antibacterial peptides with improved activity. Nat Biotechnol 23(8):1008–1012

Hilpert K, Elliott MR, Volkmer-Engert R, Henklein P, Donini O, Zhou Q et al (2006) Sequence requirements and an optimization strategy for short antimicrobial peptides. Chem Biol 13(10):1101–1107

Hirsch T, Jacobsen F, Steinau HU, Steinstraesser L (2008) Host defense peptides and the new line of defence against multiresistant infections. Protein Pept Lett 15(3):238–243

Hoffmann MH, Bruns H, Backdahl L, Neregard P, Niederreiter B, Herrmann M et al (2013) The cathelicidins LL-37 and rCRAMP are associated with pathogenic events of arthritis in humans and rats. Ann Rheum Dis 72(7):1239–1248

Hollox EJ, Huffmeier U, Zeeuwen PL, Palla R, Lascorz J, Rodijk-Olthuis D et al (2008) Psoriasis is associated with increased beta-defensin genomic copy number. Nat Genet 40(1):23–25

Holterman DA, Diaz JI, Blackmore PF, Davis JW, Schellhammer PF, Corica A et al (2006) Overexpression of alpha-defensin is associated with bladder cancer invasiveness. Urol Oncol 24(2):97–108

Hong SA, Kim KH, Lee TJ, Park ES, Kim MK, Myung SC (2017) A role of human beta defensin-1 in predicting prostatic adenocarcinoma in cases of false-negative biopsy. APMIS 125(12):1063–1069

Hoppe T, Kraus D, Novak N, Probstmeier R, Frentzen M, Wenghoefer M et al (2016) Oral pathogens change proliferation properties of oral tumor cells by affecting gene expression of human defensins. Tumour Biol 37(10):13789–13798

Hosoda H, Nakamura K, Hu Z, Tamura H, Reich J, Kuwahara-Arai K et al (2017) Antimicrobial cathelicidin peptide LL37 induces NET formation and suppresses the inflammatory response in a mouse septic model. Mol Med Rep 16(4):5618–5626

Hou M, Zhang N, Yang J, Meng X, Yang R, Li J et al (2013) Antimicrobial peptide LL-37 and IDR-1 ameliorate MRSA pneumonia in vivo. Cell Physiol Biochem 32(3):614–623

Hultmark D, Engstrom A, Bennich H, Kapur R, Boman HG (1982) Insect immunity: isolation and structure of cecropin D and four minor antibacterial components from Cecropia pupae. Eur J Biochem 127(1):207–217

Islam D, Bandholtz L, Nilsson J, Wigzell H, Christensson B, Agerberth B et al (2001) Downregulation of bactericidal peptides in enteric infections: a novel immune escape mechanism with bacterial DNA as a potential regulator. Nat Med 7(2):180–185

Jia J, Zheng Y, Wang W, Shao Y, Li Z, Wang Q et al (2017) Antimicrobial peptide LL-37 promotes YB-1 expression, and the viability, migration and invasion of malignant melanoma cells. Mol Med Rep 15(1):240–248

Joly S, Compton LM, Pujol C, Kurago ZB, Guthmiller JM (2009) Loss of human beta-defensin 1, 2, and 3 expression in oral squamous cell carcinoma. Oral Microbiol Immunol 24(5):353–360

Jones DE, Bevins CL (1992) Paneth cells of the human small intestine express an antimicrobial peptide gene. J Biol Chem 267(32):23216–23225

Jones DE, Bevins CL (1993) Defensin-6 mRNA in human Paneth cells: implications for antimicrobial peptides in host defense of the human bowel. FEBS Lett 315(2):187–192

Jones EA, Kananurak A, Bevins CL, Hollox EJ, Bakaletz LO (2014) Copy number variation of the beta defensin gene cluster on chromosome 8p influences the bacterial microbiota within the nasopharynx of otitis-prone children. PLoS One 9(5):e98269

Joo HS, Fu CI, Otto M (2016) Bacterial strategies of resistance to antimicrobial peptides. Philos Trans R Soc Lond B Biol Sci 371(1695):20150292

Joseph G, Tarnow L, Astrup AS, Hansen TK, Parving HH, Flyvbjerg A et al (2008) Plasma alpha-defensin is associated with cardiovascular morbidity and mortality in type 1 diabetic patients. J Clin Endocrinol Metab 93(4):1470–1475

Kahlenberg JM, Kaplan MJ (2013) Little peptide, big effects: the role of LL-37 in inflammation and autoimmune disease. J Immunol 191(10):4895–4901

Kahlenberg JM, Carmona-Rivera C, Smith CK, Kaplan MJ (2013) Neutrophil extracellular trap-associated protein activation of the NLRP3 inflammasome is enhanced in lupus macrophages. J Immunol 190(3):1217–1226

Kaneda Y, Yamaai T, Mizukawa N, Nagatsuka H, Yamachika E, Gunduz M et al (2009) Localization of antimicrobial peptides human beta-defensins in minor salivary glands with Sjogren's syndrome. Eur J Oral Sci 117(5):506–510

Kaplan CW, Sim JH, Shah KR, Kolesnikova-Kaplan A, Shi W, Eckert R (2011) Selective membrane disruption: mode of action of C16G2, a specifically targeted antimicrobial peptide. Antimicrob Agents Chemother 55(7):3446–3452

Kesting MR, Loeffelbein DJ, Hasler RJ, Wolff KD, Rittig A, Schulte M et al (2009) Expression profile of human beta-defensin 3 in oral squamous cell carcinoma. Cancer Investig 27(5):575–581

Kesting MR, Stoeckelhuber M, Kuppek A, Hasler R, Rohleder N, Wolff KD et al (2012) Human beta-defensins and psoriasin/S100A7 expression in salivary glands: anti-oncogenic molecules for potential therapeutic approaches. BioDrugs 26(1):33–42

Kienhofer D, Hahn J, Schubert I, Reinwald C, Ipseiz N, Lang SC et al (2014) No evidence of pathogenic involvement of cathelicidins in patient cohorts and mouse models of lupus and arthritis. PLoS One 9(12):e115474

Kim JE, Kim HJ, Choi JM, Lee KH, Kim TY, Cho BK et al (2010) The antimicrobial peptide human cationic antimicrobial protein-18/cathelicidin LL-37 as a putative growth factor for malignant melanoma. Br J Dermatol 163(5):959–967

Kim SH, Yang IY, Kim J, Lee KY, Jang YS (2015) Antimicrobial peptide LL-37 promotes antigen-specific immune responses in mice by enhancing Th17-skewed mucosal and systemic immunities. Eur J Immunol 45(5):1402–1413

Kim SH, Kim YN, Jang YS (2017) Cutting edge: LL-37-mediated formyl peptide Receptor-2 signaling in follicular dendritic cells contributes to B cell activation in Peyer's patch germinal centers. J Immunol 198(2):629–633

King AE, Fleming DC, Critchley HO, Kelly RW (2003) Differential expression of the natural antimicrobials, beta-defensins 3 and 4, in human endometrium. J Reprod Immunol 59(1):1–16

Kopfnagel V, Harder J, Werfel T (2013) Expression of antimicrobial peptides in atopic dermatitis and possible immunoregulatory functions. Curr Opin Allergy Clin Immunol 13(5):531–536

Kreuter A, Jaouhar M, Skrygan M, Tigges C, Stucker M, Altmeyer P et al (2011) Expression of antimicrobial peptides in different subtypes of cutaneous lupus erythematosus. J Am Acad Dermatol 65(1):125–133

Kurosaka K, Chen Q, Yarovinsky F, Oppenheim JJ, Yang D (2005) Mouse cathelin-related antimicrobial peptide chemoattracts leukocytes using formyl peptide receptor-like 1/mouse formyl peptide receptor-like 2 as the receptor and acts as an immune adjuvant. J Immunol 174(10):6257–6265

Kuwano K, Tanaka N, Shimizu T, Kida Y (2006) Antimicrobial activity of inducible human beta defensin-2 against mycoplasma pneumoniae. Curr Microbiol 52(6):435–438

Lande R, Ganguly D, Facchinetti V, Frasca L, Conrad C, Gregorio J et al (2011) Neutrophils activate plasmacytoid dendritic cells by releasing self-DNA-peptide

complexes in systemic lupus erythematosus. Sci Transl Med 3(73):73ra19
Larrick JW, Hirata M, Zhong J, Wright SC (1995) Anti-microbial activity of human CAP18 peptides. Immunotechnology 1(1):65–72
Laube DM, Yim S, Ryan LK, Kisich KO, Diamond G (2006) Antimicrobial peptides in the airway. Curr Top Microbiol Immunol 306:153–182
Lehouck A, Mathieu C, Carremans C, Baeke F, Verhaegen J, Van Eldere J et al (2012) High doses of vitamin D to reduce exacerbations in chronic obstructive pulmonary disease. Ann Intern Med 156(2):105–114
Li HN, Barlow PG, Bylund J, Mackellar A, Bjorstad A, Conlon J et al (2009) Secondary necrosis of apoptotic neutrophils induced by the human cathelicidin LL-37 is not proinflammatory to phagocytosing macrophages. J Leukoc Biol 86(4):891–902
Lisanby MW, Swiecki MK, Dizon BL, Pflughoeft KJ, Koehler TM, Kearney JF (2008) Cathelicidin administration protects mice from *Bacillus anthracis* spore challenge. J Immunol 181(7):4989–5000
Luciano N, Valentini V, Calabro A, Elefante E, Vitale A, Baldini C et al (2015) One year in review 2015: Sjogren's syndrome. Clin Exp Rheumatol 33(2):259–271
Mader JS, Mookherjee N, Hancock RE, Bleackley RC (2009) The human host defense peptide LL-37 induces apoptosis in a calpain- and apoptosis-inducing factor-dependent manner involving Bax activity. Mol Cancer Res 7(5):689–702
Malinovschi A, Masoero M, Bellocchia M, Ciuffreda A, Solidoro P, Mattei A et al (2014) Severe vitamin D deficiency is associated with frequent exacerbations and hospitalization in COPD patients. Respir Res 15:131
Mallia P, Footitt J, Sotero R, Jepson A, Contoli M, Trujillo-Torralbo MB et al (2012) Rhinovirus infection induces degradation of antimicrobial peptides and secondary bacterial infection in chronic obstructive pulmonary disease. Am J Respir Crit Care Med 186(11):1117–1124
Mangoni ML, McDermott AM, Zasloff M (2016) Antimicrobial peptides and wound healing: biological and therapeutic considerations. Exp Dermatol 25(3):167–173
Martineau AR, James WY, Hooper RL, Barnes NC, Jolliffe DA, Greiller CL et al (2015) Vitamin D3 supplementation in patients with chronic obstructive pulmonary disease (ViDiCO): a multicentre, double-blind, randomised controlled trial. Lancet Respir Med 3(2):120–130
Meisel JS, Sfyroera G, Bartow-McKenney C, Gimblet C, Bugayev J, Horwinski J et al (2018) Commensal microbiota modulate gene expression in the skin. Microbiome 6(1):20
Melle C, Ernst G, Schimmel B, Bleul A, Thieme H, Kaufmann R et al (2005) Discovery and identification of alpha-defensins as low abundant, tumor-derived serum markers in colorectal cancer. Gastroenterology 129(1):66–73
Merkel D, Rist W, Seither P, Weith A, Lenter MC (2005) Proteomic study of human bronchoalveolar lavage fluids from smokers with chronic obstructive pulmonary disease by combining surface-enhanced laser desorption/ionization-mass spectrometry profiling with mass spectrometric protein identification. Proteomics 5(11):2972–2980
Miles K, Clarke DJ, Lu W, Sibinska Z, Beaumont PE, Davidson DJ et al (2009) Dying and necrotic neutrophils are anti-inflammatory secondary to the release of alpha-defensins. J Immunol 183(3):2122–2132
Mily A, Rekha RS, Kamal SM, Arifuzzaman AS, Rahim Z, Khan L et al (2015) Significant effects of oral phenylbutyrate and vitamin D3 adjunctive therapy in pulmonary tuberculosis: a randomized controlled trial. PLoS One 10(9):e0138340
Miraglia E, Nylen F, Johansson K, Arner E, Cebula M, Farmand S et al (2016) Entinostat up-regulates the CAMP gene encoding LL-37 via activation of STAT3 and HIF-1alpha transcription factors. Sci Rep 6:33274
Molhoek EM, den Hertog AL, de Vries AM, Nazmi K, Veerman EC, Hartgers FC et al (2009) Structure-function relationship of the human antimicrobial peptide LL-37 and LL-37 fragments in the modulation of TLR responses. Biol Chem 390(4):295–303
Mookherjee N, Hancock RE (2007) Cationic host defence peptides: innate immune regulatory peptides as a novel approach for treating infections. Cell Mol Life Sci 64(7–8):922–933
Mookherjee N, Brown KL, Bowdish DM, Doria S, Falsafi R, Hokamp K et al (2006) Modulation of the TLR-mediated inflammatory response by the endogenous human host defense peptide LL-37. J Immunol 176(4):2455–2464
Mookherjee N, Hamill P, Gardy J, Blimkie D, Falsafi R, Chikatamarla A et al (2009) Systems biology evaluation of immune responses induced by human host defence peptide LL-37 in mononuclear cells. Mol BioSyst 5(5):483–496
Muller CA, Markovic-Lipkovski J, Klatt T, Gamper J, Schwarz G, Beck H et al (2002) Human alpha-defensins HNPs-1, -2, and -3 in renal cell carcinoma: influences on tumor cell proliferation. Am J Pathol 160(4):1311–1324
Murakami M, Ohtake T, Dorschner RA, Schittek B, Garbe C, Gallo RL (2002) Cathelicidin anti-microbial peptide expression in sweat, an innate defense system for the skin. J Invest Dermatol 119(5):1090–1095
Murakami M, Lopez-Garcia B, Braff M, Dorschner RA, Gallo RL (2004) Postsecretory processing generates multiple cathelicidins for enhanced topical antimicrobial defense. J Immunol 172(5):3070–3077
Nagaoka I, Tamura H, Hirata M (2006) An antimicrobial cathelicidin peptide, human CAP18/LL-37, suppresses neutrophil apoptosis via the activation of formyl-peptide receptor-like 1 and P2X7. J Immunol 176(5):3044–3052
Nakamura T, Furunaka H, Miyata T, Tokunaga F, Muta T, Iwanaga S et al (1988) Tachyplesin, a class of antimicrobial peptide from the hemocytes of the horseshoe

crab (*Tachypleus tridentatus*). Isolation and chemical structure. J Biol Chem 263(32):16709–16713
Nakatsuji T, Chen TH, Narala S, Chun KA, Two AM, Yun T et al (2017) Antimicrobials from human skin commensal bacteria protect against *Staphylococcus aureus* and are deficient in atopic dermatitis. Sci Transl Med 9(378):eaah4680
Nell MJ, Tjabringa GS, Vonk MJ, Hiemstra PS, Grote JJ (2004) Bacterial products increase expression of the human cathelicidin hCAP-18/LL-37 in cultured human sinus epithelial cells. FEMS Immunol Med Microbiol 42(2):225–231
Nemeth BC, Varkonyi T, Somogyvari F, Lengyel C, Fehertemplomi K, Nyiraty S et al (2014) Relevance of alpha-defensins (HNP1-3) and defensin beta-1 in diabetes. World J Gastroenterol 20(27):9128–9137
Neumann A, Berends ET, Nerlich A, Molhoek EM, Gallo RL, Meerloo T et al (2014) The antimicrobial peptide LL-37 facilitates the formation of neutrophil extracellular traps. Biochem J 464(1):3–11
Nijnik A, Madera L, Ma S, Waldbrook M, Elliott MR, Easton DM et al (2010) Synthetic cationic peptide IDR-1002 provides protection against bacterial infections through chemokine induction and enhanced leukocyte recruitment. J Immunol 184(5):2539–2550
Niyonsaba F, Madera L, Afacan N, Okumura K, Ogawa H, Hancock RE (2013) The innate defense regulator peptides IDR-HH2, IDR-1002, and IDR-1018 modulate human neutrophil functions. J Leukoc Biol 94(1):159–170
Niyonsaba F, Kiatsurayanon C, Chieosilapatham P, Ogawa H (2017) Friends or foes? Host defense (antimicrobial) peptides and proteins in human skin diseases. Exp Dermatol 26(11):989–998
Nizet V, Ohtake T, Lauth X, Trowbridge J, Rudisill J, Dorschner RA et al (2001) Innate antimicrobial peptide protects the skin from invasive bacterial infection. Nature 414(6862):454–457
Obata-Onai A, Hashimoto S, Onai N, Kurachi M, Nagai S, Shizuno K et al (2002) Comprehensive gene expression analysis of human NK cells and CD8(+) T lymphocytes. Int Immunol 14(10):1085–1098
Okrent DG, Lichtenstein AK, Ganz T (1990) Direct cytotoxicity of polymorphonuclear leukocyte granule proteins to human lung-derived cells and endothelial cells. Am Rev Respir Dis 141(1):179–185
Ong PY, Ohtake T, Brandt C, Strickland I, Boguniewicz M, Ganz T et al (2002) Endogenous antimicrobial peptides and skin infections in atopic dermatitis. N Engl J Med 347(15):1151–1160
Ostaff MJ, Stange EF, Wehkamp J (2013) Antimicrobial peptides and gut microbiota in homeostasis and pathology. EMBO Mol Med 5(10):1465–1483
Pace E, Ferraro M, Minervini MI, Vitulo P, Pipitone L, Chiappara G et al (2012) Beta defensin-2 is reduced in central but not in distal airways of smoker COPD patients. PLoS One 7(3):e33601
Paone G, Conti V, Vestri A, Leone A, Puglisi G, Benassi F et al (2011) Analysis of sputum markers in the evaluation of lung inflammation and functional impairment in symptomatic smokers and COPD patients. Dis Markers 31(2):91–100
Park K, Elias PM, Oda Y, Mackenzie D, Mauro T, Holleran WM et al (2011) Regulation of cathelicidin antimicrobial peptide expression by an endoplasmic reticulum (ER) stress signaling, vitamin D receptor-independent pathway. J Biol Chem 286(39):34121–34130
Persson LJ, Aanerud M, Hardie JA, Miodini Nilsen R, Bakke PS, Eagan TM et al (2017) Antimicrobial peptide levels are linked to airway inflammation, bacterial colonisation and exacerbations in chronic obstructive pulmonary disease. Eur Respir J 49(3):1601328
Pezzulo AA, Tang XX, Hoegger MJ, Abou Alaiwa MH, Ramachandran S, Moninger TO et al (2012) Reduced airway surface pH impairs bacterial killing in the porcine cystic fibrosis lung. Nature 487(7405):109–113
Phan TK, Lay FT, Poon IK, Hinds MG, Kvansakul M, Hulett MD (2016) Human beta-defensin 3 contains an oncolytic motif that binds PI(4,5)P2 to mediate tumour cell permeabilisation. Oncotarget 7(2):2054–2069
Piyadasa H, Hemshekhar M, Altieri A, Basu S, van der Does AM, Halayko AJ et al (2018a) Immunomodulatory innate defence regulator (IDR) peptide alleviates airway inflammation and hyper-responsiveness. Thorax. https://doi.org/10.1136/thoraxjnl-2017-210739. [Epub ahead of print]
Piyadasa H, Hemshekhar M, Carlsten C, Mookherjee N (2018b) Inhaled diesel exhaust decreases the antimicrobial peptides alpha-Defensin and S100A7 in human bronchial secretions. Am J Respir Crit Care Med 197(10):1358–1361
Powers JP, Hancock RE (2003) The relationship between peptide structure and antibacterial activity. Peptides 24(11):1681–1691
Quayle AJ, Porter EM, Nussbaum AA, Wang YM, Brabec C, Yip KP et al (1998) Gene expression, immunolocalization, and secretion of human defensin-5 in human female reproductive tract. Am J Pathol 152(5):1247–1258
Radic M (2014) Clearance of apoptotic bodies, NETs, and biofilm DNA: implications for autoimmunity. Front Immunol 5:365
Raqib R, Sarker P, Mily A, Alam NH, Arifuzzaman AS, Rekha RS et al (2012) Efficacy of sodium butyrate adjunct therapy in shigellosis: a randomized, double-blind, placebo-controlled clinical trial. BMC Infect Dis 12:111
Rehaume LM, Hancock RE (2008) Neutrophil-derived defensins as modulators of innate immune function. Crit Rev Immunol 28(3):185–200
Rekha RS, Rao Muvva SS, Wan M, Raqib R, Bergman P, Brighenti S et al (2015) Phenylbutyrate induces LL-37-dependent autophagy and intracellular killing of Mycobacterium tuberculosis in human macrophages. Autophagy 11(9):1688–1699
Ren SX, Cheng AS, To KF, Tong JH, Li MS, Shen J et al (2012) Host immune defense peptide LL-37 activates caspase-independent apoptosis and suppresses colon cancer. Cancer Res 72(24):6512–6523

Rivas-Santiago B, Castaneda-Delgado JE, Rivas Santiago CE, Waldbrook M, Gonzalez-Curiel I, Leon-Contreras JC et al (2013) Ability of innate defence regulator peptides IDR-1002, IDR-HH2 and IDR-1018 to protect against Mycobacterium tuberculosis infections in animal models. PLoS One 8(3):e59119

Rodriguez-Jimenez FJ, Krause A, Schulz S, Forssmann WG, Conejo-Garcia JR, Schreeb R et al (2003) Distribution of new human beta-defensin genes clustered on chromosome 20 in functionally different segments of epididymis. Genomics 81(2):175–183

Rohde G, Message SD, Haas JJ, Kebadze T, Parker H, Laza-Stanca V et al (2014) CXC chemokines and antimicrobial peptides in rhinovirus-induced experimental asthma exacerbations. Clin Exp Allergy 44(7):930–939

Roudi R, Syn NL, Roudbary M (2017) Antimicrobial peptides as biologic and immunotherapeutic agents against cancer: a comprehensive overview. Front Immunol 8:1320

Salzman NH, Hung K, Haribhai D, Chu H, Karlsson-Sjoberg J, Amir E et al (2010) Enteric defensins are essential regulators of intestinal microbial ecology. Nat Immunol 11(1):76–83

Saraheimo M, Forsblom C, Pettersson-Fernholm K, Flyvbjerg A, Groop PH, Frystyk J (2008) Increased levels of alpha-defensin (−1, −2 and −3) in type 1 diabetic patients with nephropathy. Nephrol Dial Transplant 23(3):914–918

Schauber J, Svanholm C, Termen S, Iffland K, Menzel T, Scheppach W et al (2003) Expression of the cathelicidin LL-37 is modulated by short chain fatty acids in colonocytes: relevance of signalling pathways. Gut 52(5):735–741

Schauber J, Iffland K, Frisch S, Kudlich T, Schmausser B, Eck M et al (2004) Histone-deacetylase inhibitors induce the cathelicidin LL-37 in gastrointestinal cells. Mol Immunol 41(9):847–854

Scott MG, Vreugdenhil AC, Buurman WA, Hancock RE, Gold MR (2000) Cutting edge: cationic antimicrobial peptides block the binding of lipopolysaccharide (LPS) to LPS binding protein. J Immunol 164(2):549–553

Scott MG, Davidson DJ, Gold MR, Bowdish D, Hancock RE (2002) The human antimicrobial peptide LL-37 is a multifunctional modulator of innate immune responses. J Immunol 169(7):3883–3891

Scott MG, Dullaghan E, Mookherjee N, Glavas N, Waldbrook M, Thompson A et al (2007) An anti-infective peptide that selectively modulates the innate immune response. Nat Biotechnol 25(4):465–472

Seil MN, Nagant C, Dehaye J-P, Vandenbranden M, Lensink MF (2010) Spotlight on human LL-37, an immunomodulatory peptide with promising cell-penetrating properties. Pharmaceuticals (Basel) 3(11):3435–3460

Sharma S, Verma I, Khuller GK (2001) Therapeutic potential of human neutrophil peptide 1 against experimental tuberculosis. Antimicrob Agents Chemother 45(2):639–640

Sieprawska-Lupa M, Mydel P, Krawczyk K, Wojcik K, Puklo M, Lupa B et al (2004) Degradation of human antimicrobial peptide LL-37 by *Staphylococcus aureus*-derived proteinases. Antimicrob Agents Chemother 48(12):4673–4679

Sierra JM, Fuste E, Rabanal F, Vinuesa T, Vinas M (2017) An overview of antimicrobial peptides and the latest advances in their development. Expert Opin Biol Ther 17(6):663–676

Simmaco M, Mignogna G, Barra D, Bossa F (1993) Novel antimicrobial peptides from skin secretion of the European frog *Rana esculenta*. FEBS Lett 324(2):159–161

Simmaco M, Mignogna G, Canofeni S, Miele R, Mangoni ML, Barra D (1996) Temporins, antimicrobial peptides from the European red frog *Rana temporaria*. Eur J Biochem 242(3):788–792

Simmaco M, Kreil G, Barra D (2009) Bombinins, antimicrobial peptides from Bombina species. Biochim Biophys Acta 1788(8):1551–1555

Sorensen OE, Follin P, Johnsen AH, Calafat J, Tjabringa GS, Hiemstra PS et al (2001) Human cathelicidin, hCAP-18, is processed to the antimicrobial peptide LL-37 by extracellular cleavage with proteinase 3. Blood 97(12):3951–3959

Sorensen OE, Cowland JB, Theilgaard-Monch K, Liu L, Ganz T, Borregaard N (2003) Wound healing and expression of antimicrobial peptides/polypeptides in human keratinocytes, a consequence of common growth factors. J Immunol 170(11):5583–5589

Steiner H, Hultmark D, Engstrom A, Bennich H, Boman HG (1981) Sequence and specificity of two antibacterial proteins involved in insect immunity. Nature 292(5820):246–248

Steinmann J, Halldorsson S, Agerberth B, Gudmundsson GH (2009) Phenylbutyrate induces antimicrobial peptide expression. Antimicrob Agents Chemother 53(12):5127–5133

Steinstraesser L, Hirsch T, Schulte M, Kueckelhaus M, Jacobsen F, Mersch EA et al (2012) Innate defense regulator peptide 1018 in wound healing and wound infection. PLoS One 7(8):e39373

Sthoeger ZM, Bezalel S, Chapnik N, Asher I, Froy O (2009) High alpha-defensin levels in patients with systemic lupus erythematosus. Immunology 127(1):116–122

Stone VN, Xu P (2017) Targeted antimicrobial therapy in the microbiome era. Mol Oral Microbiol 32(6):446–454

Suarez-Carmona M, Hubert P, Delvenne P, Herfs M (2015) Defensins: "simple" antimicrobial peptides or broad-spectrum molecules? Cytokine Growth Factor Rev 26(3):361–370

Sun J, Furio L, Mecheri R, van der Does AM, Lundeberg E, Saveanu L et al (2015) Pancreatic beta-cells limit autoimmune diabetes via an immunoregulatory antimicrobial peptide expressed under the influence of the gut microbiota. Immunity 43(2):304–317

Supanchart C, Thawanaphong S, Makeudom A, Bolscher JG, Nazmi K, Kornak U et al (2012) The antimicrobial

peptide, LL-37, inhibits in vitro osteoclastogenesis. J Dent Res 91(11):1071–1077
Tewary P, de la Rosa G, Sharma N, Rodriguez LG, Tarasov SG, Howard OM et al (2013) Beta-Defensin 2 and 3 promote the uptake of self or CpG DNA, enhance IFN-alpha production by human plasmacytoid dendritic cells, and promote inflammation. J Immunol 191(2):865–874
Tjabringa GS, Ninaber DK, Drijfhout JW, Rabe KF, Hiemstra PS (2006) Human cathelicidin LL-37 is a chemoattractant for eosinophils and neutrophils that acts via formyl-peptide receptors. Int Arch Allergy Immunol 140(2):103–112
Tomasinsig L, Zanetti M (2005) The cathelicidins – structure, function and evolution. Curr Protein Pept Sci 6(1):23–34
Tongaonkar P, Golji AE, Tran P, Ouellette AJ, Selsted ME (2012) High fidelity processing and activation of the human alpha-defensin HNP1 precursor by neutrophil elastase and proteinase 3. PLoS One 7(3):e32469
Turner-Brannen E, Choi KY, Lippert DN, Cortens JP, Hancock RE, El-Gabalawy H et al (2011) Modulation of interleukin-1beta-induced inflammatory responses by a synthetic cationic innate defence regulator peptide, IDR-1002, in synovial fibroblasts. Arthritis Res Ther 13(4):R129
Uraki S, Sugimoto K, Shiraki K, Tameda M, Inagaki Y, Ogura S et al (2015) Corrigendum: human beta-defensin-3 inhibits migration of colon cancer cells via downregulation of metastasis-associated 1 family, member 2 expression. Int J Oncol 46(4):1858
Valore EV, Ganz T (1992) Posttranslational processing of defensins in immature human myeloid cells. Blood 79(6):1538–1544
van der Does AM, Beekhuizen H, Ravensbergen B, Vos T, Ottenhoff THM, van Dissel JT et al (2010) LL-37 directs macrophage differentiation toward macrophages with a Proinflammatory signature. J Immunol 185(3):1442–1449
Van Wetering S, Mannesse-Lazeroms SPG, Van Sterkenburg MAJA, Daha MR, Dijkman JH, Hiemstra PS (1997) Effect of defensins on IL-8 synthesis in airway epithelial cells. Am J Physiol (Lung Cell Mol Physiol) 272(16):L888–LL96
Varoga D, Paulsen FP, Kohrs S, Grohmann S, Lippross S, Mentlein R et al (2006) Expression and regulation of human beta-defensin-2 in osteoarthritic cartilage. J Pathol 209(2):166–173
Varoga D, Klostermeier E, Paulsen F, Wruck C, Lippross S, Brandenburg LO et al (2009) The antimicrobial peptide HBD-2 and the toll-like receptors-2 and -4 are induced in synovial membranes in case of septic arthritis. Virchows Arch 454(6):685–694
von Haussen J, Koczulla R, Shaykhiev R, Herr C, Pinkenburg O, Reimer D et al (2008) The host defence peptide LL-37/hCAP-18 is a growth factor for lung cancer cells. Lung Cancer 59(1):12–23
Vordenbaumen S, Fischer-Betz R, Timm D, Sander O, Chehab G, Richter J et al (2010) Elevated levels of human beta-defensin 2 and human neutrophil peptides in systemic lupus erythematosus. Lupus 19(14):1648–1653
Vragniau C, Hubner JM, Beidler P, Gil S, Saydaminova K, Lu ZZ et al (2017) Studies on the interaction of tumor-derived HD5 alpha defensins with adenoviruses and implications for oncolytic adenovirus therapy. J Virol 91(6):e02030–e02016
Wan M, van der Does AM, Tang X, Lindbom L, Agerberth B, Haeggstrom JZ (2014) Antimicrobial peptide LL-37 promotes bacterial phagocytosis by human macrophages. J Leukoc Biol 95(6):971–981
Wang K, Wang JH, Baskaran H, Wang R, Jurevic R (2012) Effect of human beta-defensin-3 on head and neck cancer cell migration using micro-fabricated cell islands. Head Neck Oncol 4:41
Wang G, Mishra B, Epand RF, Epand RM (2014) High-quality 3D structures shine light on antibacterial, anti-biofilm and antiviral activities of human cathelicidin LL-37 and its fragments. Biochim Biophys Acta 1838(9):2160–2172
Wang G, Li X, Wang Z (2016) APD3: the antimicrobial peptide database as a tool for research and education. Nucleic Acids Res 44(D1):D1087–D1093
Weber G, Chamorro CI, Granath F, Liljegren A, Zreika S, Saidak Z et al (2009) Human antimicrobial protein hCAP18/LL-37 promotes a metastatic phenotype in breast cancer. Breast Cancer Res: BCR 11(1):R6
Welling MM, Hiemstra PS, van den Barselaar MT, Paulusma-Annema A, Nibbering PH, Pauwels EK et al (1998) Antibacterial activity of human neutrophil defensins in experimental infections in mice is accompanied by increased leukocyte accumulation. J Clin Invest 102(8):1583–1590
Wilson CL, Ouellette AJ, Satchell DP, Ayabe T, Lopez-Boado YS, Stratman JL et al (1999) Regulation of intestinal alpha-defensin activation by the metalloproteinase matrilysin in innate host defense. Science 286(5437):113–117
Winter J, Kraus D, Reckenbeil J, Probstmeier R (2016) Oncogenic relevant defensins: expression pattern and proliferation characteristics of human tumor cell lines. Tumour Biol 37(6):7959–7966
Woo JS, Jeong JY, Hwang YJ, Chae SW, Hwang SJ, Lee HM (2003) Expression of cathelicidin in human salivary glands. Arch Otolaryngol Head Neck Surg 129(2):211–214
Wu WK, Sung JJ, To KF, Yu L, Li HT, Li ZJ et al (2010) The host defense peptide LL-37 activates the tumor-suppressing bone morphogenetic protein signaling via inhibition of proteasome in gastric cancer cells. J Cell Physiol 223(1):178–186
Xu WD, Zhang M, Feng CC, Yang XK, Pan HF, Ye DQ (2013) IL-32 with potential insights into rheumatoid arthritis. Clin Immunol 147(2):89–94
Yamaguchi Y, Nagase T, Makita R, Fukuhara S, Tomita T, Tominaga T et al (2002) Identification of multiple novel epididymis-specific beta-defensin isoforms in humans and mice. J Immunol 169(5):2516–2523
Yamasaki K, Schauber J, Coda A, Lin H, Dorschner RA, Schechter NM et al (2006) Kallikrein-mediated prote-

olysis regulates the antimicrobial effects of cathelicidins in skin. FASEB J 20(12):2068–2080
Yang D, Chertov O, Bykovskaia SN, Chen Q, Buffo MJ, Shogan J et al (1999) Beta-defensins: linking innate and adaptive immunity through dendritic and T cell CCR6. Science 286(5439):525–528
Yang D, Chen Q, Schmidt AP, Anderson GM, Wang JM, Wooters J et al (2000) LL-37, the neutrophil granule- and epithelial cell-derived cathelicidin, utlizes formyl peptide receptor-like 1 (FPRL1) as a receptor to chemoattract human peripheral blood neutrophils, monocytes, and T cells. J Exp Med 192(7):1069–1074
Yeung AT, Gellatly SL, Hancock RE (2011) Multifunctional cationic host defence peptides and their clinical applications. Cell Mol Life Sci 68(13):2161–2176
Yim S, Dhawan P, Ragunath C, Christakos S, Diamond G (2007) Induction of cathelicidin in normal and CF bronchial epithelial cells by 1,25-dihydroxyvitamin D(3). J Cyst Fibros 6(6):403–410
Yin J, Yu FS (2010) LL-37 via EGFR transactivation to promote high glucose-attenuated epithelial wound healing in organ-cultured corneas. Invest Ophthalmol Vis Sci 51(4):1891–1897
Zasloff M (1987) Magainins, a class of antimicrobial peptides from Xenopus skin: isolation, characterization of two active forms, and partial cDNA sequence of a precursor. Proc Natl Acad Sci U S A 84(15):5449–5453
Zhang Z, Cherryholmes G, Shively JE (2008) Neutrophil secondary necrosis is induced by LL-37 derived from cathelicidin. J Leukoc Biol 84(3):780–788
Zhang Z, Cherryholmes G, Chang F, Rose DM, Schraufstatter I, Shively JE (2009) Evidence that cathelicidin peptide LL-37 may act as a functional ligand for CXCR2 on human neutrophils. Eur J Immunol 39(11):3181–3194
Zhao Y, Chen F, Wu W, Sun M, Bilotta AJ, Yao S et al (2018) GPR43 mediates microbiota metabolite SCFA regulation of antimicrobial peptide expression in intestinal epithelial cells via activation of mTOR and STAT3. Mucosal Immunol 11(3):752–762

Part IV

Towards Clinical Applications

Selectivity of Antimicrobial Peptides: A Complex Interplay of Multiple Equilibria

11

Sara Bobone and Lorenzo Stella

Abstract

Antimicrobial peptides (AMPs) attack bacterial membranes selectively, killing microbes at concentrations that cause no toxicity to the host cells. This selectivity is not due to interaction with specific receptors but is determined by the different lipid compositions of the membranes of the two cell types and by the peculiar physicochemical properties of AMPs, particularly their cationic and amphipathic character. However, the available data, including recent studies of peptide-cell association, indicate that this picture is excessively simplistic, because selectivity is modulated by a complex interplay of several interconnected phenomena. For instance, conformational transitions and self-assembly equilibria modulate the effective peptide hydrophobicity, the electrostatic and hydrophobic contributions to the membrane-binding driving force are nonadditive, and kinetic processes can play an important role in selective bacterial killing in the presence of host cells. All these phenomena and their bearing on the final activity and toxicity of AMPs must be considered in the definition of design principles to optimize peptide selectivity.

Keywords

Antimicrobial peptides · Host defense peptides · Selectivity · Toxicity · Peptide-membrane association · Aggregation · Hydrophobicity · Amphipathicity

11.1 Introduction

The scientific and medical interest for antimicrobial peptides (AMPs), short peptides produced by most organisms as part of their innate immune defenses, derives from their wide-spectrum bactericidal properties and their possible application to fight drug-resistant bacteria. However, in view of clinical applications, the absence of significant toxicity is almost as important as a good activity. In this respect, one of the appealing properties of many AMPs is their cell selectivity, i.e., the ability to kill bacterial cells at concentrations significantly lower than those causing damage to cells of the host organism, at least in in vitro tests. Still, potential toxicity is commonly listed as one of the challenges limiting the clinical application of AMPs as systemic drugs (Hancock and Sahl 2006; Eckert 2011; Yeung et al. 2011; Seo et al. 2012; Carneiro et al. 2015; Pachón-Ibáñez et al. 2017), and therefore, several research efforts are devoted to understand and further improve AMP selectivity.

S. Bobone · L. Stella (✉)
Department of Chemical Science and Technologies, University of Rome Tor Vergata, Rome, Italy
e-mail: stella@stc.uniroma2.it

K. Matsuzaki (ed.), *Antimicrobial Peptides*, Advances in Experimental Medicine and Biology 1117,
https://doi.org/10.1007/978-981-13-3588-4_11

This chapter discusses AMP selectivity, the origin and the structural determinants of this property, the design strategies available to improve it, and the results of recent studies on the quantitative determination of peptide-cell association. Overall, the available data indicate that selectivity is the result of a complex interplay of several interconnected phenomena, including peptide association to target and host cells, peptide conformational equilibria, and AMP aggregation. Any modification to the peptide sequence and structure necessarily affects all of these processes, which must therefore be fully understood and considered in the rational design of new peptide or peptidomimetic molecules with improved selectivity properties. Our attempt is to give a critical overview of the available evidences, in order to provide a rationale for future efforts in this area. To this end, we have strived to derive, whenever possible, generalizations of the findings reported in the literature, but we have to stress from the beginning that the presence of exceptions to every rule is the norm, in such a diverse set as AMPs, also as a consequence of the complications mentioned above.

The different aspects of AMP selectivity have last been reviewed by Matsuzaki in 2009 (Matsuzaki 2009). Selectivity or toxicity has often been considered in general review articles on AMPs (Alba et al. 2012; Teixeira et al. 2012; Oddo and Hansen 2017; Hollmann et al. 2018). Some reviews have summarized our current knowledge on the structural determinants of AMP activity and selectivity (Takahashi et al. 2010; Huang et al. 2010; Strömstedt et al. 2010; Tossi 2011; Ruiz et al. 2014; Ebenhan et al. 2014a). Finally, for a recent discussion on how the interaction of AMPs with target and host cells determines their selectivity, see Savini et al. (2018).

As illustrated in other chapters of this book, AMPs have multiple functions, including anticancer, antifungal, and antiviral activities. For the sake of brevity and simplicity, in this chapter, we will essentially limit ourselves to discuss selectivity for bacterial versus host cells. Selectivity of anticancer peptides has been reviewed by Phoenix and coworkers (2012) and Harris et al. (2013).

11.2 AMPs Are Selective for Microbial Cells

AMPs have been isolated from natural sources based on their antimicrobial activity. The minimum inhibitory concentration (MIC, i.e., the lowest concentration of antimicrobial agent that inhibits the visible growth of a microorganism) (Wiegand et al. 2008) or the minimum bactericidal concentration (MBC, i.e., the minimal drug dosage killing at least 99.9% of the bacterial cells) (Lorian 2005) for AMPs is usually in the low μM range (Giacometti et al. 1998). AMPs are typically bactericidal, and therefore, the MIC and MBC values are usually similar (Giacometti et al. 1998).

The active concentration of a bioactive, therapeutically useful molecule must be much lower than the concentration causing toxic effects to the host cells. This property is quantified by the therapeutic index (TI), i.e., the ratio of the active concentration to the toxic concentration (see the legend to Table 11.1 for a detailed definition of these parameters). In the case of AMPs, whose main mechanism of bactericidal action is membranolytic (as discussed in Sect. 11.3), toxicity is most commonly assessed by measuring the lysis of erythrocytes (Fig. 11.1). Table 11.1 summarizes some TI values of natural and artificial AMPs, which are typically in the range 10–1000. However, it should be considered that unfortunately a strong variability is present in the literature regarding the definition of the toxic concentration, because different thresholds of lysed red blood cells (RBCs) are utilized to define the minimum hemolytic concentration (MHC), ranging from barely detectable to full hemolysis (see references cited in Table 11.1) (Bacalum and Radu 2015). In addition, MIC values depend on the specific strains tested in the assay.

When toxicity is assayed on other human cells, the results are generally not very different from those obtained using hemolysis (Table 11.2), but combining the two toxicity tests obviously

Table 11.1 Therapeutic index (TI) of natural and artificial AMPs

Name	Sequence	TI	Calculated as	References
Natural Peptides				
Temporin-L	FVQWFSKFLGRIL-NH2	**1**	HC_{50}/MIC (Gram + &–)	Mangoni (2011)
Magainin-1	GIGKFLHSAGKFGKAFVGEIMKS	**2.5–4**	HC_5/MIC (*E. coli*)	Bacalum and Radu (2015)
PMAP-36	GRFRRLRKKTRKRLKKIGKVLKWIPPIVGSIPLGCG-NH2	**3**	HC_5/MIC (Gram + &– and *C. albicans*)	Lyu et al. (2016)
Mastoparan X	INWKGIAAMAKKLL-NH2	**3.4**	HC_5/MIC (*E. coli*)	Henriksen et al. (2014)
Arenicin-1	RWCVYAYVRVRGVLVRYRRCW	**5**	HC_{50}/MIC (Gram + &–)	Panteleev et al. (2015)
Polistes Mastoparan	VDWKKIGQHILSVL-NH2	**6**	HC_5/MIC (*E. coli*)	Bacalum and Radu (2015)
Magainin 2B	GIGKFLHAAKKFAKAFVAEIMNS	**9**	HC_{50}/IC_{50} (*B. anthracis*)	Dawson et al (2011)
Dermaseptin 1	ALWKTMLKKLGTMALHAGKAALGAAADTISQGTQ	**0.83–7**	HC_5/MIC (*E. coli*)	Bacalum and Radu (2015)
LL-37	LLGDFFRKSKEKIGKEFKRIVQRIKDFLRNLVPRTES	**12**	HC_5/MIC (*E. coli, S. aureus*)	Luo et al. (2017)
Indolicidin	ILPWKWPWWPWRR-NH2	**13** **4–11**	HC_{50}/MIC (Gram + &–) HC_5/MIC (*E. coli*)	Nan et al. (2009) Bacalum and Radu (2015)
Gramicidin-S	VOLFPVOLFP	**1–18** **2**	HC_{50}/MIC (Gram + &–) HC_{100}/MIC (Gram + &–)	Swierstra et al. (2016) Kondejewski (1999)
XT-7	GLLGPLLKIAAKVGSNLL	**18**	HC_{50}/MIC (*E. coli*)	Kamech et al. (2012)
PGLa	GMASKAGAIAGKIAKVALKAL-NH2	**24**	HC_{10}/MIC (Gram + &–)	Strandberg et al. (2007)
Tachyplesin I	KWCFRVCYRGICYRRCR-NH2	**2–24**	HC_5/MIC (*E. coli*)	Bacalum and Radu (2015)
SMAP-29	RGLRRLGRKIAHGVKKYGPTVLRIIRIAG	**26**	HC_{50}/IC_{50} (*B. anthracis*)	Dawson and Liu (2011)
Ascaphin-8	GFKDLLKGAAKALVKTVLF	**37**	HC_{50}/MIC (*E. coli*)	Kamech et al. (2012)
Papillosin	GFWKKVGSAAWGGVKAAAKGAAVGGLNALAKHIQ	**34**	HC_{50}/IC_{50} (*B. anthracis*)	Dawson and Liu (2011)
PMAP-23	RIIDLLWRVRRPQKPKFVTVWVR	**23** **>57**	HC_5/MIC (Gram+ &–) $HC_{=0}$/MIC (Gram + &–)	Velduhizen (2017) Kang et al. (1999)
Ascaphin-1	GIRDVLKGAAKAFVKTVAGHIAN-NH2	**>60**	HC_{50}/MIC (*E. coli*)	Juretic et al. (2009)
Magainin 2	GIGKFLHSAKKFGKAFVGEIMNS	**3–82**	HC_5/MIC (*E. coli*)	Bacalum and Radu (2015)
Cecropin B	KWKVFKKIEKMGRNIRNGIVKAGPAIAVLGEAKAL-NH2	**20–130**	HC_5/MIC (*E. coli*)	Bacalum and Radu (2015)
Cecropin A	KWKLFKKIEKVGQNIRDGIIKAGPAVAVVGQATQIAK-NH2	**177–579**	HC_5/MIC (*E. coli*)	Bacalum and Radu (2015)

(continued)

Table 11.1 (continued)

Modified analogues of natural peptides				
Mastoparan-X Ala1	ANWKGIAAMAKKLL-NH_2	**7**	HC_5/MIC (*E. coli*)	Henriksen et al. (2014)
PMAP-36 T115	TRKRLKKIGKVLKWI-NH_2	**29**	HC_5/MIC (Gram + &– and *C. albicans*)	Lyu et al. (2016)
Gramicidin GS14K4	VKLkVyPLKVKLyP	**31**	HC_{100}/MIC (Gram + &–)	Kondejewski (1999)
D-Piscidin1 19K	ffhhifrgkvhvgktihrlvtg-NH_2	**33**	HC_{50}/MIC (*A. baumannii*)	Jiang (2014)
Arenicin-1 V8R	RWCVYAYRRVRGVLVRYRRCW	**80**	HC_{50}/MIC (Gram + &–)	Panteleev et al. (2015)
[K2, K16] XT-7	GKLGPLLKIAAKVGSKLL	**>130**	HC_{50}/MIC (*E. coli*)	Kamech et al. (2012)
Indolicidin-A7	ILKWKWKWWKWRR-NH_2	**190**	HC_{50}/MIC (Gram + &–)	Nan et al. (2009)
D-Dermaseptin S4 L7K,A14K	alwmtlkkkvlkakakalnavlvgana-NH_2	**219**	HC_{50}/MIC (*A. baumannii*)	Jiang (2014)
[I2, K19] ascaphin-8	GIKDLLKGAAKALVKTVLK	**>480**	HC_{50}/MIC (*E. coli*)	Kamech et al. (2012)
Gramicidin V3/A3	AKLkAyPLKAKLyP	**520**	HC_{100}/MIC (*C. xerosis*)	Kondejewski (2002)
	Designed peptides			
V13K	Ac-KWKSFLKTFKSAKKTVLHTALKAISS-NH_2	**163**	$HC_{>0}$/MIC (Gram + &–)	Chen et al. (2005)
P5	KWKKLLKKPLLKKLLKKL-NH_2	**>150**	$HC_{=0}$/MIC (Gram + &–)	Park et al. (2003)
Pep-1-K	KKTWWKTWWTKWSQPKKKRKV	**174**	$HC_{>0}$/MIC (Gram + &–)	Zhu et al. (2009)
PK-12-KKP	KKPWWKPWWPKWKK	**200**	$HC_{>0}$/MIC (Gram + &–)	Zhu et al. (2009)
D16	Ac-klksllktlskakkkklktllkalsk-NH_2	**890** **3355**	HC_{50}/MIC (*P. aeruginosa*) HC_{50}/MIC (*A. baumannii*)	Jiang (2011)

The TI is defined as the ratio of hemolytic to inhibitory peptide concentration. HC_x is defined as the peptide concentration causing the *x*% of hemolysis. $HC_{>0}$ is the minimal peptide concentration that produces detectable hemolysis; $HC_{=0}$ is the highest peptide concentration that causes no detectable release of hemoglobin. HC_{100} is the minimal concentration causing total lysis. MIC is defined as the minimum concentration that inhibits bacterial growth; when calculated on more than one strain, the geometric mean of the values is reported. Data presented as a range result from different MIC values reported in the literature for the same peptide. –NH_2 in the peptide sequence indicates amidation at the C-terminus, while Ac- indicates acylation at the N-terminus. *O* is pyrrolysine. Amino acids are colored in blue, red, orange, and green based on their cationic, anionic, polar, and hydrophobic character, respectively. Lowercase letters indicate D-amino acids

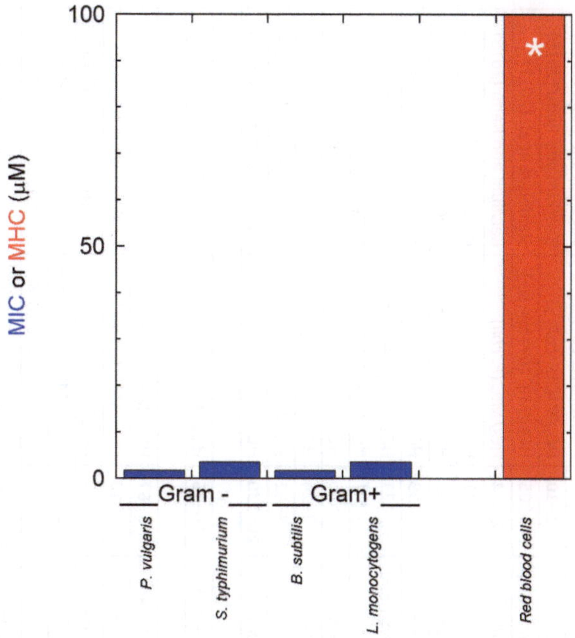

Fig. 11.1 Selective cytotoxicity in in vitro assays. Minimal inhibitory concentrations against different Gram- and Gram+ bacterial strains and minimal hemolytic concentration for the designed artificial AMP P5. The asterisk indicates that no hemolysis was observed for P5 in the peptide concentration range investigated (up to 100 μM). Adapted, with permission, from research originally published in Bobone et al. 2013, published by the European Peptide Society and John Wiley & Sons, Ltd.

provides a clearer picture of the selectivity of AMPs (Bacalum and Radu 2015).

AMP activity and toxicity are usually measured in separate assays performed under rather different conditions (for instance, regarding cell density, see Sect. 11.9) (Matsuzaki 2009). We have argued that experiments on bacterial and human cells in co-culture would provide a more stringent test of peptide selectivity (Savini et al. 2017, 2018). However, this approach has been employed only in a few cases. These studies, discussed in detail in Sect. 9.3, demonstrated that AMPs are selective even when acting on bacteria co-cultured with mammalian cells (Fig. 11.2a).

Several evidences indicate that AMPs are selective also in vivo. For instance, a large body of studies starting in 1999 (Welling) has shown that radiolabeled (Lupetti et al. 2003; Brouwer et al. 2008; Akhtar et al. 2012; Ebenhan 2014a) or fluorescent (Akram et al. 2015) AMPs can be used to image infections in vivo and can even discriminate between infection and inflammation, thanks to their specific binding to bacterial cells (Welling et al. 2000) (Fig. 11.2b, c). Among the peptides used for this purpose, there are defensins, cathelicidins, lactoferricins, histatins, artificial peptoids, and particularly sequences derived from ubiquicidin (Lupetti et al. 2003; Brouwer et al. 2008; Akhtar et al. 2012; Ebenhan 2014a, b; Dutta et al. 2017; Lei et al. 2018). Several imaging studies employed ubiquicidin 29–41 (Meléndez-Alafort et al. 2004; Akhtar et al. 2005; Vallejo et al. 2008; Gandomkar et al. 2009; de Murphy et al. 2010; Assadi et al. 2011; Ostovar et al. 2013; Saeed et al. 2013; Kahrom et al. 2014; Ebenhan et al. 2018; Bhatt et al. 2018), which has moderate activity and selectivity in the standard assays (MIC 40 μM, TI > 5) (Brouwer et al. 2006; Lupetti et al. 2008) but accumulates at the site of infection. For instance, one study reported overall values of sensitivity, specificity, and accuracy for infection detection of 100%, 80%, and 94% (Akhtar et al. 2005). Analogues of the ubiquicidin peptide have been used also for targeted

Table 11.2 Comparison of TI values determined with RBCs or with other eukaryotic cells

Name	Sequence	TI (cells)	Calculated as	TI (RBCs)	Calculated as	References
Natural peptides						
Magainin2-NH_2	GIGKFLHSAKKFGKAFVGEIMNS-NH_2	6	LC_{99} (3T3)/MIC (Gram+ &–)	>130	HC_{100} /MIC (Gram+ &–)	Javadpour et al. (1996)
Polistes Mastoparan	VDWKKIGQHILSVL-NH2	7	LC_{50}(PBMC)MIC (*E. coli*)	6	HC_5/MIC (*E. coli*)	Bacalum and Radu (2015)
Dermaseptin 1	ALWKTMLKKLGTMALHAGKAALGAAADTISQGTQ	1–7	LC_{50} (PBMC)/MIC (*E. coli*)	1–10	HC_5/MIC (*E. coli*)	Bacalum and Radu (2015)
Lasioglossin III	VNWKKILGKIIKVVK-NH2	>8	LC_{50} (HUVEC)/MIC (Gram+ &–)	>105	HC_{50}/MIC (Gram+ &–)	Slaninová et al. (2012)
Magainin I	GIGKFLHSAGKFGKAFVGEIMKS	7–12	LC_{50} (PBMC)/MIC (*E. coli*)	2.5–4	HC_5/MIC (*E. coli*)	Bacalum and Radu (2015)
Indolicidin	ILPWKWPWWPWRR-NH_2	6–17	LC_{50} (PBMC)/MIC (*E. coli*)	4–11	HC_5/MIC (*E. coli*)	Bacalum and Radu (2015)
BMAP-28	GGLRSLGRKILRAWKKYGPIIVPIIRIG	17	LC_{95} (HN)/MIC(Gram+ &–)	55	HC_{90}/MIC (Gram+ &–)	Skerlavaj et al. (1996)
BMAP-27	GRFKRFRKKFKKLFKKLSPVIPLLHLG	18	LC_{80} (HN)/MIC (Gram+ &–)	59	HC_{30}/MIC (Gram+ &–)	Skerlavaj et al. (1996)
PMAP-23	RIIDLLWRVRRPQKPKFVTVWVR	23	LC_{30} (IPEC-J2)/MIC (Gram+ &–)	23 >57	HC_5/MIC (Gram+ &–) $HC_{=0}$/MIC (Gram + &–)	Velduhizen (2017) and Kang et al. (1999)
Cecropin B-NH_2	KWKVFKKIEKMGRNIRNGIVKAGPAIAVLGEAKAL-NH_2	29	LC_{99} (3T3)/MIC (Gram+ &–)	>86	HC_{100} /MIC (Gram+ &–)	Javadpour et al. (1996)
Tachyplesin I	KWCFRVCYRGICYRRCR-NH_2	3–30	LC_{50} (PBMC)/MIC (*E. coli*)	2–24	HC_5/MIC (*E. coli*)	Bacalum and Radu (2015)
Magainin2	GIGKFLHSAKKFGKAFVGEIMNS	3–82	LC_{50} (PBMC)/MIC (*E. coli*)	3–82	HC_5/MIC (*E. coli*)	Bacalum and Radu (2015)
Cecropin A	KWKLFKKIEKVGQNIRDGIIKAGPAVAVVGQATQIAK-NH_2	27–180	LC_{50} (PBMC)/MIC (*E. coli*)	20–130	HC_5/MIC (*E. coli*)	Bacalum and Radu (2015)
Modified analogues of natural peptides						
A(A1R A8R I17K)	RIGSILGRLAKGLPTLKSWIKNR-NH_2	1.5–3	LC_{50} (L929)/MIC (Gram+ &–)	10–16	HC_{10}/MIC (Gram+ &–)	Zhang (2016)
Designed peptides						
D-LAK120 AP13	kklalalakkwlplakklalalakk-NH_2	4	LC_{50} (RAW 264.7)/MIC (Gram–)	67	HC_{50}/MIC (Gram–)	Vermeer (2012)
WK12	KWWKWWKKWWKK	>10	LC_{20} (PBMC)/MIC (Gram–)	>10	HC_{10}/MIC (Gram–)	Deslouches et al. (2016)
WR12	RWWRWWRRWWRR	>20	LC_{20} (PBMC)/MIC (Gram–)	>20	HC_{10}/MIC (Gram–)	Deslouches et al. (2016)
$(KLAKKLA)_2$	KLAKKLAKLAKKLA	>45	LC_{99} (3T3)/MIC (Gram+ &–)	>125	HC_{100} /MIC (Gram+ &–)	Javadpour et al. (1996)
$(KLAKLAK)_2$	KLAKLAKKLAKLAK	>86	LC_{99} (3T3)/MIC (Gram+ &–)	>125	HC_{100} /MIC (Gram+ &–)	Javadpour et al. (1996)

LC is the lethal peptide concentration. LC_x is defined as the peptide concentration killing *x%* of the cells. For HC and MIC definitions and for the color code used for the sequences, please see Table 11.1. When data are presented as a range, they result from different MIC values reported in the literature for the same peptide. *HBRC* human red blood cells, *PBMC* peripheral blood mononuclear cells ($3 \cdot 10^6$/mL), *3T3 cells* murine fibroblast cell line ($2 \cdot 10^5$/mL), *HUVEC cells* human umbilical vein endothelial cells ($2 \cdot 10^4$/mL), *HN* human neutrophils ($4 \cdot 10^6$/mL), *IPEC-J2* porcine intestinal epithelial cells ($1.5 \cdot 10^5$/mL), *L929* mouse fibroblast cells ($1.5 \cdot 10^4$/mL), *RAW 264.7* murine macrophage cells ($3 \cdot 10^5$/mL). The density of RBCs used in the hemolytic activity assays (when specified in the original reference) ranged from 10^6 to 10^9 cells/mL. Lowercase letters indicate D-amino acids

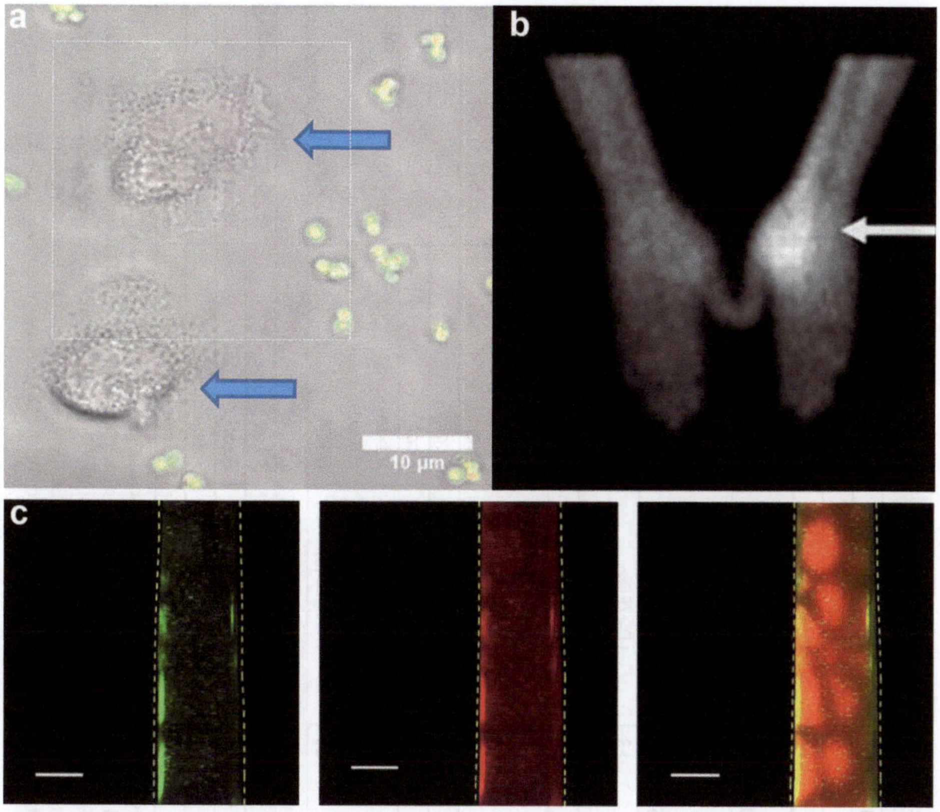

Fig. 11.2 Selective targeting of bacterial cells by AMPs in vitro and in vivo. (**a**) Optical and fluorescence microscopy image of a labeled ubiquicidin analogue (visible by the green fluorescence) selectively binding to *S. aureus* bacteria, in co-culture with isolated human neutrophils (blue arrows) Adapted, with permission, from research originally published in Akram et al. 2015, published by The Royal Society of Chemistry. (**b**) Positron emission tomography image of a patient with an infection in the left hand (indicated by the arrow), traced with a radiolabeled ubiquicidin analogue. No significant peptide uptake in the contralateral hand was noted. The image was obtained 30 min after tracer administration (Reproduced, with permission, from research originally published in Akhtar et al. 2012). (**c**) Visualization of in vivo targeting of human α-defensin 5 (HD5) toward *E. coli* cells. The mesenteric vein was imaged intravitally in mice by two-photon laser scanning microscopy, 30 min after injection of *E.coli* cells expressing a green fluorescent protein (visualized in the left panel), and treatment with HD5 labeled with a red fluorescent probe (imaged in the center panel). Colocalization is demonstrated by the overlapped images (right panel). Scale bars, 50 µm. Adapted, with permission, from Lei et al. 2018. Copyright (2018) American Chemical Society

delivery of traditional antibiotics to the infection site (Chen et al. 2015).

Although in vivo studies of the activity of AMPs abound, similar investigations characterizing their toxicity are more sparse (Mahlapuu et al. 2016). Some TIs derived from animal studies are summarized in Table 11.3, and the range of values is similar to that obtained in vitro.

The selectivity of AMPs for bacterial cells is demonstrated also by the fact that these peptides have been exploited in sensing elements that can detect infection (Mannoor et al. 2010; Shriver-Lake et al. 2012; Silva et al. 2014; Hoyos-Nogués et al. 2018), even in whole blood (Shi et al. 2017) or other complex biological samples (Qiao et al. 2017).

Overall, the results collected in the literature support an interesting selectivity of AMPs for target versus host cells. The origin of this property is necessarily related to the mechanism of action of AMPs.

Table 11.3 TI values obtained from in vivo studies

Name	Sequence	TI	Calculated as	Bacterial strain	References
Natural peptides					
P3	VNFKLLSHSLLVTLASHL	**3**	LD_{50}/ED_{60}	*E.coli*	Zhang (2015)
OH-CATH30	KFFKKLKNSVKKRAKKFFKKPRVIGVSIPF	**12**	LD_{50}/ED_{80}	*E.coli*	Li (2012)
BMAP-27	GRFKRFRKKFKKLFKKLSPVIPLLHLG	**47–55**	LD_{50}/ED_{100}	*P.aeruginosa*	Benincasa (2003)
		190–220	LD_{50}/ED_{100}	*E.coli*	
BMAP-28	GGLRSLGRKILRAWKKYGPIIVPIIRIG	**24–27**	LD_{50}/ED_{100}	*E.coli*	Benincasa (2003)
		47–55	LD_{50}/ED_{100}	*S.aureus*	
Modified analogues of natural peptides					
HD5-myr	ATCYCRTGRCATRESLSGVCEISGRLYRLCCR-myr	**>3**	$LD_{=0}/ED_{90}$	*E. coli*	Lei (2018)
RN7-IN8	FLGGLIKWPWWPWRR-NH_2	**4**	LD_{100}/ED_{50}	*S. pneumoniae*	Jindal (2017)
JH3	RRFKLLSHSLLVTLASHL	**4.5**	LD_{50}/ED_{90}	*E.coli*	Zhang (2015)
TP3	FIHHIIGGLFSVGKHIHSLIHGH	**7**	MTD/ED_{60}	*A.baumannii*	Pan (2015)
		7	MTD/ED_{70}	*K pneumoniae*	Pan (2015)
		>24	MTD/ED_{100}	MRSA	Huang (2015a)
D-OH-CATH30	kffkklknsvkkrakkffkkprvigvsipf	**8**	LD_{50}/ED_{100}	*E.coli*	Li (2012)
OH-CM6	KFFKKLKKAVKKGFKKFAKV	**10**	LD_{50}/ED_{70}	*E.coli*	Li (2012)
TP4	FIHHIIGGLFSAGKAIHRLIRRRRR	**20**	MTD/ED_{90}	*A.baumannii*	Pan (2015)
		20	MTD/ED_{90}	*K pneumoniae*	Pan (2015)
		>24	MTD/ED_{100}	MRSA	Huang (2015b)
Designed peptides					
A3-APO	(Chex-RPEKPRPYLPRPRPPRPVR)$_2$-Dab-NH_2	**2.5**	LD_{50}/ED_{100}	*E.coli*	Szabo (2010)
Onc72	VDKPPYLPRPRPPROIYNO-NH_2	**>20**	LD_{50}/ED_{50}	*E.coli*	Knappe (2012)
(LLKK)$_2$C	LLKKLLKKC	**28**	LD_{50}/ED_{50}	*A.baumannii*	Huang (2012)
C(LLKK)$_2$C	CLLKKLLKKC	**34**	LD_{50}/ED_{50}	*A.baumannii*	Huang (2012)

TI was calculated as the ratio of the toxic or lethal dose to the effective dose of peptides in mice. LD_x is the minimum dose that was lethal for at least x% of animals; $LD_{=0}$ is the highest peptide concentration that caused no deaths; MTD is minimum dose that caused toxicity (narrowing of the eyes was taken as symptom), MLD is the minimum lethal dose, and ED_x is the effective dose, i.e., the minimum dose causing the survival of at least x% of the animals. Bacteria and peptides were injected intraperitoneally, except for Huang (2015a, b) and Pan (2015), where toxicity tests were performed by intramuscular injection. -myr indicates myristoylation at the C-terminus. *Chex* 1-amino-cyclohexane carboxylic acid, *Dab* 2,4-diamino-butyric acid, *O* pyrrolysine. MRSA is methicillin-resistant *S. aureus*. For the color code used for the sequences, please see Table 11.1. Lowercase letters indicate D-amino acids

11.3 Cellular Membranes Are the Main Target of AMPs

Selectivity is not surprising when a biomolecule associates to a specific receptor or protein (Le Joncour and Laakkonen 2018). However, this is not the case for most AMPs. In general, natural AMPs and their enantiomers comprising all D-amino acids have a comparable antimicrobial activity (Fig. 11.3), while any interaction of a peptide with a protein, due to the chirality of both systems, would be favored for one enantiomer over the other. As is often the case for AMPs, exceptions to this rule have been reported (Otvos et al. 2000; Bulet and Stocklin 2005; de la Fuente-Núñez et al. 2015), showing that the mechanism of action of a minority of AMPs could be receptor mediated. On the other hand, microbiological assays of membrane permeability and microscopic imaging of bacteria treated with AMPs clearly show that cell membranes are damaged and that, as a consequence, transmembrane gradients are dissipated (Tiozzo et al. 1998; Arcidiacono et al. 2009; Hartmann et al. 2010; Agrawal and Weisshaar 2018) (Fig. 11.4). Usually, membrane perturbation and bacterial killing are correlated, further supporting membrane disruption as the main bactericidal mechanism. However, also in this case, exceptions exist, indicating that a subclass of AMPs might act through different killing mechanisms (He et al. 2014; Friedrich et al. 2000). Finally, it is worth mentioning that AMPs usually are able to perturb the permeability of artificial membranes, comprising only phospholipids (Fig. 11.4c) (Orioni et al. 2009; Bocchinfuso et al. 2011; Braun et al. 2017; Savini et al. 2018). This observation demonstrates that the membrane-perturbing activity is purely the result of physicochemical interaction between the peptide and the lipid bilayer, and not the consequence of some biological process.

Overall, literature data clearly demonstrate that membrane perturbation is the main mecha-

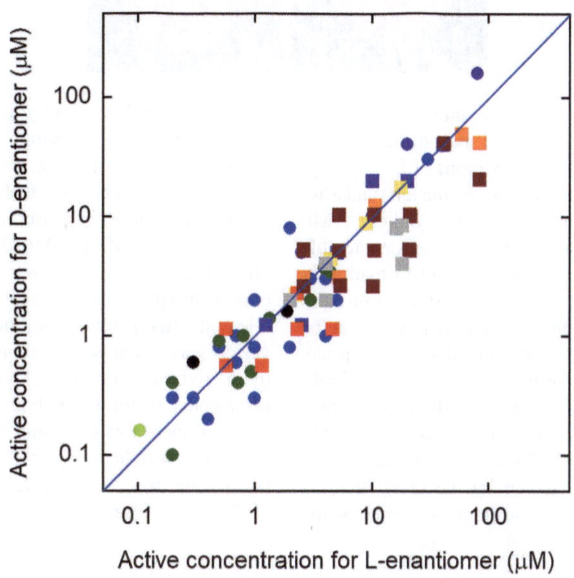

Fig. 11.3 Activity of natural AMPs and their enantiomeric analogues. Comparison of the antibacterial activities (circles for MBC, squares for MIC) of enantiomeric peptides. Data refer to magainin 2 (Bessalle et al. 1990) (violet), cecropin A (Wade et al. 1990) (blue), melittin (Juvvadi et al. 1996) (dark green), LL-37 (Dean et al. 2011) (light green), KSLK (Hong et al. 1999) (yellow), temporin A (Wade et al. 2000) (orange; data refer to IC_{50} values), camel 48 (Oh et al. 2000) (red), V681 and analogues (Chen et al. 2006) (dark red; data refer to LC_{50} values), lactoferricin B analogues (Wakabayashi et al. 1999) (silver), and cecropin B (Bland et al. 2001) (black). The blue line is the diagonal of the plot (corresponding to identical activity for D and L enantiomers) and not a fit. A version of this figure with a more limited set of data has been published previously (Savini et al. 2018)

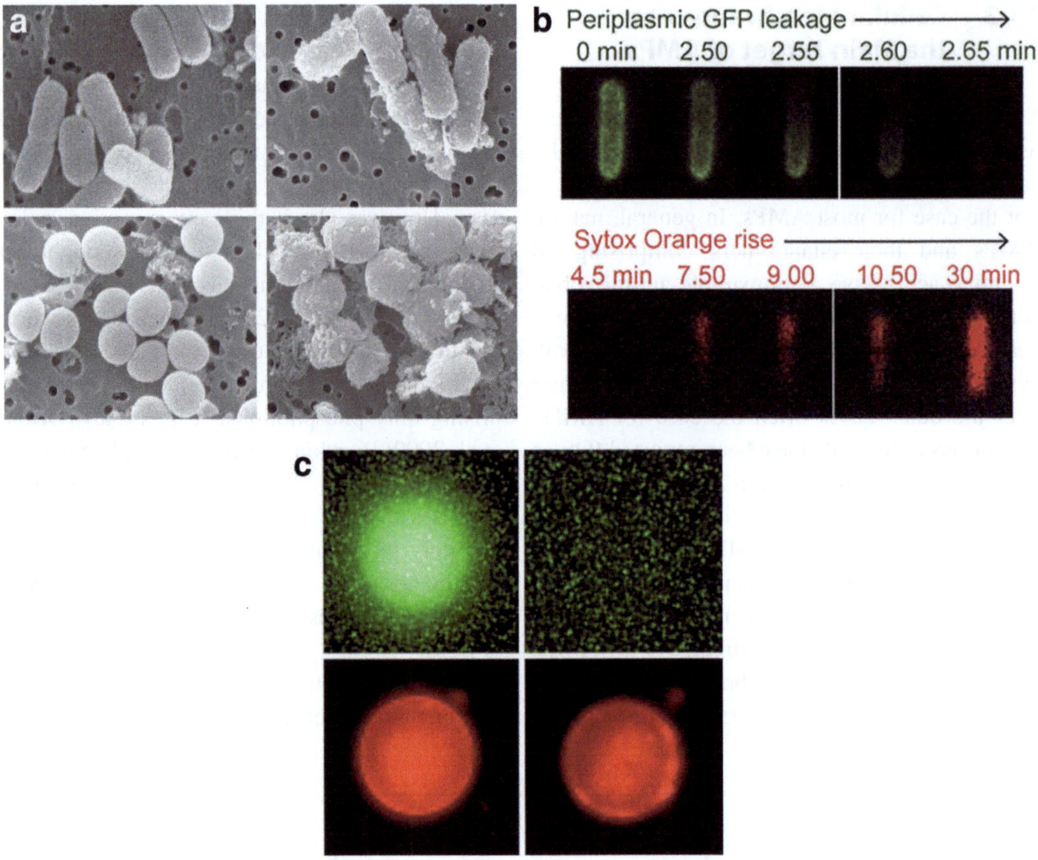

Fig. 11.4 Fluorescence and electron microscopy images of the effects of AMPs on bacterial and artificial membranes. (**a**) Scanning electron micrographs of *E. coli* (top) and *S. aureus* (bottom) before (images on the left) and after (images on the right) treatment with the synthetic AMP PGYa (30 min, 10 μM). The images show a considerable roughening of the bacterial membranes and formation of blebs on the cell surface, in contrast to the smooth surfaces of untreated bacteria, providing a strong indication that the membrane is being considerably altered by the peptide. Adapted, with permission, from research originally published in Tiozzo et al. 1998 © Elsevier. (**b**) Fluorescence microscopy images of an *E. coli* cell attacked by the AMP cecropin A (0.5 μM). The leakage of periplasmic green fluorescent protein (GFP), shown by the green fluorescence, indicates perturbation of the outer membrane, while uptake of the DNA stain Sytox Orange (red fluorescence) demonstrates pore formation in the plasma membrane (Adapted, with permission, from research originally published in Agrawal 2018 © Elsevier). (**c**) Fluorescence microscopy images of perturbation of a giant unilamellar vesicle by the AMP PMAP-23. The top panels report the green fluorescence emission from carboxyfluorescein molecules entrapped inside the GUV, which were completely released after peptide addition (right). By contrast, the vesicle was still present after peptide addition, as indicated by the red fluorescence of rhodamine-labeled phospholipids located in the GUV bilayer (bottom panels). Taken together, these images demonstrate pore formation by the AMP. The vesicle diameter is about 20 μm. Adapted, with permission, from research originally published in Orioni et al. 2009 © Elsevier

nism of direct bacterial killing for most AMPs. Even for those AMPs that act through a different antibacterial mechanism (Nicolas 2009; Otvos 2017), the cell envelope is the first cell component that the peptides encounter, and they have to cross the extracellular membrane (when present) and the cell wall to reach the plasma membrane and eventually the cell interior.

Incidentally, the fact that AMPs target microbial membranes determines their broad-spectrum activity, their bactericidal, rather than bacteriostatic, mechanism of action and also the higher difficulty for bacteria in developing resistance against them (compared to resistance against conventional antibiotics acting on a protein target) (Perron et al. 2006; Otvos 2017).

11.4 Bacterial and Host Cells Have Different Membrane Structure and Composition

If membranes are the target, then it is conceivable that selectivity arises from a difference in membrane composition of the various cell types. Indeed, bacterial and eukaryotic cells have very different cell envelopes (Wang 2017). Bacteria can be divided into Gram-positive and Gram-negative, depending on whether they are colored by the Gram stain or not. This assay reflects differences in the composition of the cell envelope. In both cases, the plasma membrane is surrounded by a cell wall. However, in Gram-positive bacteria, this is formed by a thick peptidoglycan and lipoteichoic acid layer (40–80 nm). By contrast, in Gram negatives, a thin peptidoglycan layer (8 nm thick) is contained in a second (outer) membrane, with asymmetric composition: phospholipids are the main components of the inner leaflet, while the outer layer is mainly formed by lipopolysaccharides (LPS). On the other hand, eukaryotic cells only have the plasma membrane, with asymmetric lipid composition in the two leaflets of the bilayer (Fig. 11.5).

In addition to the different structures of the cell envelope, important differences are present in the lipid composition of the cellular membranes. Tables 11.4 and 11.5 summarize the lipid content of bacterial and RBC membranes.

Bacterial membranes contain a significant fraction of negatively charged lipids: in Gram negatives, both membranes contain phosphatidylglycerol (PG, ~20% overall) and cardiolipin (CL, ~5% overall); in Gram-positive bacteria, the content of anionic lipids is much higher, again with PG and CL being the most important components (Malanovic and Lohner 2016). However, the membranes of these cells can also contain the positively charged L-lysyl-PG (LPG). In both cases, the main zwitterionic component is phosphatidylethanolamine (PE), and no sterols are present. Of course, the values for the composition reported in Table 11.4 are only approximate, since they change with the specific strain and growth conditions. In addition, lipid composition is not homogeneous over the cell surface (Renner and Weibel 2011; Oliver et al. 2014).

Human cells contain cholesterol and have no anionic phospholipids in the outer leaflet of their cell membrane. Some negatively charged glycolipids, such as gangliosides, are present on the cell surface (Miyazaki et al. 2012), but they are minor components in most cell membranes (with the exception of nerve cells) (Storch and Kleinfeld 1985). These properties are exemplified by RBCs, which are commonly used to test toxicity and selectivity (Table 11.5). For eukaryotes, the main zwitterionic components are phosphatidylcholine (PC), sphingomyelin (SM), and PE.

Overall, we can translate these differences in lipid composition in distinct physicochemical properties. Bacterial membranes contain more anionic lipids in the outer surface of their bilayers than eukaryotic cells. This difference combines

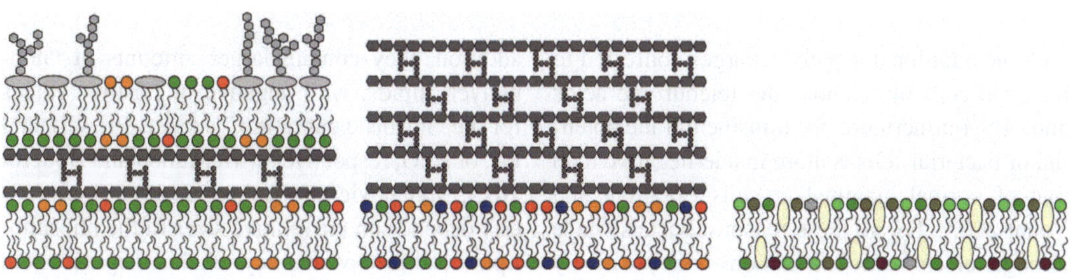

Fig. 11.5 Schematic depiction of the structure of the cellular envelope in different cell types. The three panels, from left to right, schematize the structure of the cellular envelope in Gram- and Gram+ bacteria and in human cells, respectively. Proteins, glycolipids, and lipoteichoic acids (in Gram+ bacteria) have been omitted, for the sake of clarity. LPS, light gray; peptidoglycan, dark gray; PC, light green; PE, dark green; SM, light brown; PI, light gray; PG, red; PS, dark red; CL, orange; L-lysyl PG, blue; cholesterol, beige

Table 11.4 Phospholipid composition of the membranes of Gram-negative (name highlighted in red) and Gram-positive (name highlighted in blue) bacteria

Phospholipid composition of bacterial membranes					
	PE	PG	CL	PA	L-lysyl PG
Total charge	0	−1	−2 (−1)	−1	+1
Intrinsic curvature	−	0	−	−	+
E. coli (both membranes) (Ames 1968) (Raetz 1986) (Morein et al. 1996) (Rowlett et al. 2017)	 69 75–85 79 78	 19 10–20 17 12	 6.5 5–15 4 6	 <3	
E. coli (cell membrane) (Morein et al. 1996)	 75	 19	 6		
S. typhimurium (both membranes) (Osborn 1972) (Ames 1968)	 86 78	 13 18	 1 3		
S. typhimurium (cell membrane) (Osborn 1972)	 76	 21	 3		
B. subtilis (Op den Kamp et al. 1969)* (Bishop et al. 1967)**	 30 34	 36 49	 12 11		 22
S. aureus (Hayami 1979): (strain Newman) (strain Tazaki)		 49 47	 13 10	 0.6 1	 34 38

Phospholipid composition data are expressed as molar percentages. In the case of Hayami et al. (1979), data were calculated converting the % of phosphorus to molar % by considering two P atoms per CL molecule and one for the other lipids. In the case of Osborn et al. (1972), data were calculated converting the % of [2-^{3}H] glycerol to molar %, by considering three ^{3}H atoms per CL molecule, two per PG, and one for the other lipids. Data on total charge at physiologic pH and intrinsic curvature were taken from Marsh (1990), McMahon and Boucrot (2015), Malanovic and Lohner (2016), and Boyd et al. (2017)

PE phosphatidylethanolamine, *PG* phosphatidylglycerol, *CL* cardiolipin, *PA* phosphatidic acid, *L-lysyl PG* L-lysyl phosphatidylglycerol

*Indicates a growth condition without glucose and sulfate

**Indicates that 6% of lipo-amino acids were also recovered.

with the additional negative charges conferred to bacterial cells by teichoic and teichuronic acids and LPS. Furthermore, the transmembrane potential of bacterial cells is more inside-negative than that of normal mammalian cells (Yeaman and Yount 2003). For all these reasons, bacteria have stronger electrostatic interactions with positively charged molecules than eukaryotic cells. Another difference is that bacterial membranes are more disordered and less well packed than those of eukaryotes, due to the lack of cholesterol. In addition, they contain larger amounts of "non-bilayer" lipids, with negative or positive values for the "intrinsic curvature," such as PE, CL, and PA, or LPG, respectively (McMahon and Boucrot 2015; Malanovic and Lohner 2016). This property depends on the relative sizes of the phospholipid head-groups and acyl chains. Lipids where the cross-sectional area occupied by head-groups and tails is similar (e.g., PC, PG, PS) are said to have a cylindrical shape and pack well in locally flat bilayer structures (zero intrinsic curvature).

Table 11.5 Phospholipid and cholesterol content of human erythrocyte membrane

Phospholipid composition of human RBC membranes						
	PC	**PE**	**SM**	**PI**	**PS**	**PA**
Total charge	0	0	0	−1	−1	−1 (−2)
Intrinsic curvature	0	−	0	+	0	−
Both leaflets						
(Dodge and Phillips 1967)	29.2±1.5	27.5±1.5	25.4±1.4	0.6±0.5	14.8±1.7	1.1±0.5
(Broekhuyse 1969)	28.3±2.1	26.7±1.0	25.8±1.7	1.9±0.6	12.7±1.3	
(White 1973)	34.7	28.0	20.1		14.3	
(Verkeleij 1973)	28	26	24		13	
(Van Meer 1981)	29.5	25.9	25.3		12.2	
Outer leaflet						
(Verkeleij 1973)	42	10	40		0	
(Virtanen et al. 1998)	44.8	11.1	42.1			
Inner leaflet						
(Verkeleij 1973)	14	42	8		26	
(Virtanen et al. 1998)	14.0	43.9	9.1	1.2	29.6	2.2
Asymmetry of distribution						
(% in the outer leaflet)						
(Verkleij 1973)	76	20	82		0	
(Zwaal et al. 1973)	62		83			
(Gordesky and Marinetti 1973)		15			0	
(Zwaal et al. 1975)	75					
(Gordesky et al. 1975)		33			0	
(Van Meer 1981)	78	20	80			
(Bütikofer et al. 1990)				24		
(Gascard et al. 1991)				20		
Cholesterol content as cholesterol/phospholipid molar ratio						
(Cooper 1975)	0.95					
(Ballas and Krasnow 1980)	0.75					
(Chabanel 1983)	0.80					

Phospholipid composition data are expressed as molar percentages. Data from Verkeleij (1973) were derived from a figure in the cited reference. Data on total charge at physiologic pH and intrinsic curvature were taken from Marsh (1990) and McMahon and Boucrot (2015)

PC phosphatidylcholine, *PE* phosphatidylethanolamine, *SM* sphingomyelin, *PI* phosphatidylinositol, *PS* phosphatidylserine, *PA* phosphatidic acid

By contrast, lipids where the head-group is smaller than the tails (e.g., PE or PA) favor concave shapes of the monolayer (negative curvature). The opposite is true for lipids with comparatively larger polar heads (e.g., LPG) which have a positive curvature (Koller and Lohner 2014).

The differences in lipid composition and in physical properties between bacterial and human cell membranes are considered to be the origin of AMP selectivity. Similar considerations on membrane composition (particularly regarding the content of anionic lipids and sterols) have been proposed to explain the selectivity for cancer cells (Hoskin and Ramamoorthy 2008; Schweizer 2009; Phoenix et al. 2012; Gaspar et al. 2013), fungi (van der Weerden et al. 2013; Rautenbach et al. 2016), protozoa (Rivas et al. 2009), and enveloped viruses (Aloia et al. 1993; Findlay et al. 2013), since in all cases the lipid distribution is different from that of a normal eukaryotic cell.

11.5 Lipid Composition Determines the Affinity of AMPs for Lipid Bilayers

The hypothesis of a selectivity based on differences in lipid composition has been tested by studying the interaction of AMPs with model membranes mimicking the composition of the natural bilayers. With liposomes, it is possible to vary the lipid composition at will and to measure both peptide-membrane association and peptide-induced membrane permeability (Bocchinfuso et al. 2011; Savini et al. 2018). The role of various membrane properties in AMP selectivity is summarized in the following sections.

11.5.1 Membrane Charge

In model membranes, the presence of anionic lipids increases peptide association to the bilayer and, as a consequence, peptide-induced leakage (Matsuzaki et al. 1989, 1995; Gazit et al. 1995; Abraham et al. 2005; Sood et al. 2008; Russell et al. 2010; Bobone et al. 2013; Golbek et al. 2017; Maturana et al. 2017). On the other hand, the positively charged lipid lysyl-PG, present in Gram+ bacteria, inhibits AMP activity (Nishi et al. 2004; Andra et al. 2011). These findings are a straightforward consequence of electrostatic interaction of the membranes with the positively charged AMPs (see Sect. 11.6). Regarding the anionic gangliosides present in the outer leaflet of eukaryotic membranes, Matsuzaki and coworkers (2012) demonstrated that, although their acidic moieties favor the association of AMPs to model membranes, this interaction does not lead to strong membrane perturbation, since the peptides remain trapped in the sugar region.

11.5.2 Cholesterol Content

Several studies also reported an AMP inhibitory effect of cholesterol. For instance, the presence of cholesterol inhibits the membrane-perturbing activity of magainin, pardaxin, LL-37, temporin L, human defensin HNP1, and other AMPs (Matsuzaki et al. 1995; Tytler et al. 1995; Hallock et al. 2002; Sood et al. 2008; Sood and Kinnunen 2008; Gonçalves et al. 2012; Verly et al. 2008; Wu et al. 2010; McHenry et al. 2012). The membrane-ordering effects of cholesterol in fluid bilayers are well established: insertion of the rigid ring structure of the sterol limits the possibility for trans-gauche isomerization for adjacent phospholipid tails, leading to an increase in bilayer order, packing, thickness, and rigidity (Henriksen et al. 2006; Mouritsen and Zuckermann 2004). All these effects could contribute to reduce peptide binding and membrane perturbation (McIntosh et al. 2002). However, the relevance of cholesterol for AMP selectivity has been recently questioned. While all the investigations listed above were performed on simple lipid mixtures, a comprehensive study by Ramamoorthy and coworkers on more realistic lipid compositions showed that cholesterol's protective effect against AMPs does not occur in lipid systems containing raft domains and presenting phase separation (McHenry et al. 2012; Brender et al. 2012). There are examples where the activity of AMPs is not affected by the presence of cholesterol even in simple lipid mixtures (Bobone et al. 2013). On the other hand, Matsuzaki et al. (1995) demonstrated an AMP-inhibiting effect of cholesterol in real cells, by artificially varying the cholesterol content of RBCs.

It is worth mentioning that the inhibitory effect is specific of cholesterol, while ergosterol, present in fungal membranes, does not appear to inhibit peptide binding and activity to the same extent, in agreement with the specific activity of antifungal peptides (Sood and Kinnunen 2008; Gonçalves et al. 2012) and with the comparatively smaller effects of ergosterol on membrane order (Henriksen et al. 2006).

11.5.3 Intrinsic Curvature

The situation is less clear regarding the effect of the presence of negative curvature lipids (PE) in bacterial membranes. PE has been shown to inhibit pore formation by magainin, melittin, alamethicin, PMAP-23, and mastoparan X (Matsuzaki et al. 1998; Allende et al. 2005; Lee et al. 2005; Bobone et al. 2012). On the other hand, the activity of some AMPs is favored by the presence of PE (Schröder-Borm et al. 2003; Epand et al. 2006; Leite et al. 2015). Inhibition of pore formation by PE can be understood by considering that the peptides act by inserting in the head-group region of the membrane, thus imposing a positive curvature strain, which is released after a threshold of membrane-bound peptide concentration is reached, through the formation of membrane defects or pores. The presence of lipids with negative intrinsic curvature would counteract this mechanism (Matsuzaki et al. 1998, Lee et al. 2005). Similar considerations, on the other hand, suggest that PE can favor membrane binding of AMPs, by reducing the intrinsic curvature strain needed for peptide insertion in the polar region of the bilayer; an increased binding to PE-containing membranes has been reported for some AMPs (Schröder-Borm et al. 2003; Phoenix et al. 2015). However, reasoning only in terms of intrinsic curvature might be misleading. For instance, the peculiar lipid-lipid interactions made possible by the structure of PE could inhibit peptide insertion: this phospholipid contains a primary amine (lacking in PC), which allows it to form strong hydrogen bonds with phosphate or CO groups in other lipids (Lewis and McElhaney 2005). This H-bond network is responsible for the melting temperatures of PE lipids being higher than those of their corresponding PC analogues (Lewis and McElhaney 2005). This difference might lead to an increase in the energy needed to insert a peptide in the bilayer, or to open a pore, in the presence of PE. Finally, an additional mechanism invoked to explain different activities on vesicles lacking or containing PE involves peptide-induced formation of lipid domains (Epand et al. 2006). Overall, these considerations can explain why different final effects are observed on the membrane-perturbing activity of AMPs, depending on which of the various phenomena predominates in each specific case. In any case, it is difficult to ascribe a well-defined role in AMP selectivity to the PE content.

11.6 Thermodynamics of Peptide-Membrane Association

In principle, selectivity for different membrane compositions could result from two effects. AMPs could have a higher affinity for bacterial membranes than for human bilayers, or they could be more effective in perturbing the former, once inserted (Wimley and Hristova 2011). The data on model membranes presented above clearly indicate that differential binding is an important aspect of AMP selectivity.

AMPs are usually short (about 10–50 residues in length), and their sequences and structures have no common features, except for the cationic charge (most AMPs fall in the range of +2 to +4 e), and amphipathic character, with an overall content of about 50% hydrophobic residues (Wang 2017). The role of these properties is easily rationalized: charge imparts selectivity toward bacterial versus eukaryotic membranes, and apolar residues provide a hydrophobic driving force for binding and insertion into membranes, leading to perturbation of bilayer integrity. Indeed, peptides interacting only electrostatically usually do not cause significant membrane leakage, because their depth of insertion is too shallow (Wimley 2010a).

11.6.1 Hydrophobic and Electrostatic Driving Forces Are Nonadditive

Different treatments are used in the literature to describe peptide-membrane interactions, and comprehensive reviews are available on this topic (White and Wimley 1999; Wieprecht and Seelig 2002; Simon and McIntosh 2002; Santos et al. 2003; Seelig 2004; Wimley 2010a). Here we will just briefly mention that, since peptide-membrane association does not have a specific stoichiometry, it is not correctly described by a binding equilibrium, and it is better treated as a partition equilibrium between the water and the membrane phase (White and Wimley 1999; Wieprecht and Seelig 2002; Santos et al. 2003; Wimley 2010a). In this view, the main effect of Coulombic interactions can be described as an increase in local peptide concentration in the vicinity of the bilayer, according to Gouy-Chapman theory (Beschiaschvili and Seelig 1990; Wieprecht and Seelig 2002; Seelig 2004). However, the thermodynamic contributions of electrostatic and hydrophobic effects to the driving force of peptide water-membrane partition are not simply additive (Ladokhin and White 2001). This finding is due mainly to the different depths of polar and aliphatic moieties of phospholipids in the bilayer: charged groups are located on the surface of the membrane, well separated from the hydrocarbon core, and the physicochemical properties of the bilayer vary steeply in the head-group region (Wimley 2010a). As a consequence, the depth of insertion of a peptide in the membrane is determined by the interplay between hydrophobic effect and Coulombic forces (Wimley 2010a). Highly charged, hydrophilic molecules sit on the membrane surface, and strongly hydrophobic peptides insert into the hydrocarbon core, while cationic, amphipathic peptides are located at an intermediate position, which depends on their specific properties (Bocchinfuso et al. 2009; Farrotti et al. 2015). In turn, the depth of insertion in the bilayer modulates the intensity of electrostatic and hydrophobic contributions: strongly simplifying, one could say that the position of a peptide in the membrane determines the average distance between the peptide and the charged lipid moieties and the degree of insertion of the peptide in the water-free hydrocarbon core. An increase in peptide hydrophobicity ultimately reduces the effect of electrostatic interactions; on the other hand, augmenting the Coulombic forces diminishes the hydrophobic contribution to the binding free energy (Ladokhin and White 2001).

11.6.2 Multiple Interconnected Equilibria Modulate Peptide Activity and Selectivity

Several artificial peptides were designed having the required characteristics of cationic charge and amphipathic character. However, in many cases, such peptides turned out to be highly toxic (Dathe et al. 1996; Cornut et al. 1994; Bobone et al. 2013). These findings demonstrated that a cationic charge is not sufficient for specificity and provided a first indication of the complexities of peptide-membrane interaction discussed above. Peptides in solution can assume different conformations and aggregation states, and once membrane-bound, they can change conformation, orientation, insertion depth, and aggregation state (Fig. 11.6). All these phenomena are regulated by interconnected equilibria, and therefore they contribute in determining the final membrane-perturbing activity (Stella et al. 2004; Mazzuca et al. 2005; Gatto et al. 2006; Bobone et al. 2013). Every modification in peptide properties can affect all these processes (Gatto et al. 2006). In our opinion, this is the reason why the rational design of peptides with improved selectivity has met with limited success, and it has progressed through a trial-and-error process. Even so, several useful principles for the optimization of AMP selectivity have been defined.

11.7 Selectivity of AMPs Is Determined by Their Physicochemical Properties

Helical peptides are the most abundant and best-characterized class of AMPs. Investigations on the structural determinants of AMP selectivity have mostly focused on this type of peptides. They are usually disordered in solution but attain a helical conformation when membrane-bound, with a spatially amphipathic distribution of the side chains, where most of the hydrophobic residues face toward the membrane center and the polar and charged residues are oriented toward the water phase. From a physicochemical point of view, they can be characterized by several parameters, such as charge, hydrophobicity, or amphipathicity. In addition to the considerations discussed in the previous section, on the multiple processes involved in peptide-membrane interactions, investigations on the role of each of these parameters are complicated by the fact that varying the sequence by a single

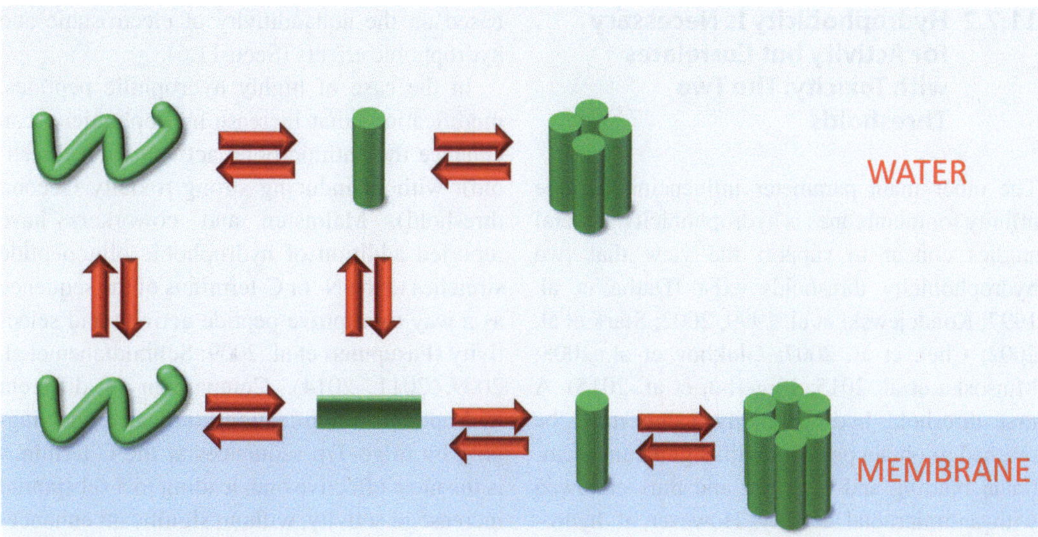

Fig. 11.6 Schematic depiction of the phenomena involved in peptide-membrane interaction. In addition to water-membrane partition, conformational, orientational, and aggregation equilibria are at play in determining the final peptide activity and selectivity

amino acid substitution usually causes a variation in multiple physicochemical properties of the peptides (Wieprecht et al. 1997a; Dathe et al. 2001). For instance, inserting an additional cationic residue does not vary only the peptide charge but also its hydrophobicity and amphipathicity. However, some systematic studies have been performed where the authors tried to compare peptide sequences where multiple substitutions were inserted to cause the significant variation of one parameter only, while the others were kept as constant as possible (Dathe et al. 1997, 2001, 2002; Wieprecht et al. 1997a, b, c; Dathe and Wieprecht 1999; Giangaspero et al. 2001; Zelezetsky et al. 2005; Zelezetsky and Tossi 2006).

11.7.1 Cationic Charges Favor Selectivity

Based on the results on the importance of anionic lipids for the membrane activity of AMPs, it is not surprising that a positive correlation between peptide positive charge and antimicrobial activity and selectivity has often been described (Bessalle et al. 1992; Matsuzaki et al. 1997; Dathe et al. 2001; Giangaspero et al. 2001; Zelezetsky and Tossi 2006; Bobone et al. 2011). However, several studies reported that increasing cationicity above a certain level (+5, +8, or +9 depending on the specific case) is not beneficial and might even cause a decrease in activity or selectivity (Dathe et al. 2001; Giangaspero et al. 2001; Zelezetsky and Tossi 2006; Jiang et al. 2008). This last finding might be due to an overly shallow insertion of the peptide in the bilayer (Wimley 2010a) and to the nonadditivity of electrostatic and hydrophobic effects, discussed in Sect. 11.6.

One of the ways to increase the total positive charge of the peptide is C-terminal amidation, which is frequent in natural sequences and has the additional advantage of reducing susceptibility to proteolytic degradation (Huang et al. 2010; Mura et al. 2016). However, this approach to increase peptide selectivity is not generally valid, possibly because it also affects the stability of helical conformations in solution (Dennison et al. 2009) (see Sect. 11.8).

11.7.2 Hydrophobicity Is Necessary for Activity but Correlates with Toxicity: The Two Thresholds

The other main parameter influencing peptide affinity for membranes is hydrophobicity. Several studies concur to support the view that two hydrophobicity thresholds exist (Dathe et al. 1997; Kondejewski et al. 1999, 2002; Stark et al. 2002; Chen et al. 2007; Glukhov et al. 2008; Mojsoska et al. 2015; Uggerhøj et al. 2015). A first threshold hydrophobicity value must be reached to obtain peptides with significant membrane binding and insertion and thus endowed with antimicrobial activity. However, if hydrophobicity surpasses a second, higher threshold, toxicity is observed, because binding to neutral membranes becomes significant. The difference between these two thresholds is due to the electrostatic contributions to peptide binding to bacterial membranes. Therefore, an optimal range of hydrophobicity values exists, in which peptides exhibit antimicrobial activity, but no significant toxicity. Above a third, even higher threshold, activity decreases, because of peptide aggregation and lack of solubility (Gatto et al. 2006; Chen et al. 2007; Chu-Kung et al. 2010; Wimley 2010a). It is difficult to provide quantitative values for these thresholds, since different hydrophobicity scales are used in the literature. Just as an example, Deber and coworkers identified values of 0.4 and approximately 2 in the Liu-Deber scale for the activity and toxicity thresholds, respectively, for the hydrophobicity of the core segment of a series of model peptides (Glukhov et al. 2008).

It is interesting to note that hydrophobicity affects binding to neutral membranes more than to charged bilayers and hemolysis more than bactericidal activity (Wieprecht et al. 1997a; Dathe et al. 2002). The rationale underlying this finding is not immediately obvious, since the hydrophobic driving force is present for both membrane types, and therefore any variation in hydrophobicity should affect both antimicrobial activity and toxicity to the same extent. The experimental observations can be explained based on the nonadditivity of electrostatic and hydrophobic effects (Sect. 11.6).

In the case of highly hydrophilic peptides, modifications that increase hydrophobicity can enhance the antimicrobial activity (first threshold), without inducing strong toxicity (second threshold). Malmsten and coworkers have reported addition of hydrophobic oligopeptide stretches to the N- or C-terminus of the sequence as a way to improve peptide activity and selectivity (Pasupuleti et al. 2009; Schmidtchen et al. 2009, 2011, 2014). Comparison of different hydrophobic modifications indicated that tagging by oligo-Trp sequences at the C-terminus is the most effective one, leading to a substantial increase in activity, without significant enhancement of toxicity. Trp residues have peculiar properties, since they are known to have an affinity for membrane interfaces, thanks to their ability to interact both with hydrophobic moieties and with charged groups (through cation-aromatic interactions) (Yau et al. 1998). It has been speculated that the specificity-enhancing effect of Trp might be linked to the difficulty of inserting such a bulky residue in the tightly packed, cholesterol-containing membranes of eukaryotes (Pasupuleti et al. 2009; Schmidtchen et al. 2009, 2011, 2014). However, preferential interaction of Trp with cholesterol has also been hypothesized (de Kruijff 1990), although it is disputed (Holt et al. 2008), and in some cases, introduction of Trp residues has been linked to enhanced peptide toxicity (Oddo and Hansen 2017; Matsuzaki et al. 1997). Another common approach to increase the hydrophobicity of highly hydrophilic peptides is lipidation (Gatto et al. 2006). Shai's group demonstrated that highly polar peptides, originally devoid of antimicrobial activity, can become antimicrobial, but not toxic, after this modification (Avrahami and Shai 2004; Malina and Shai 2005; Makovitzki et al. 2006, 2008). However, in other cases, lipidation led to strong toxicity (Chu-Kung et al. 2004; Laverty et al. 2010), or even to loss of activity, when it compromised peptide solubility (Toniolo et al. 1996; Gatto et al. 2006; Chu-Kung et al. 2010). These findings highlight the fine-tuning of AMP hydrophobicity needed for optimal activity and selectivity properties.

11.7.3 Excessive Amphipathicity Causes Toxicity

The total quantities of charged and hydrophobic residues provide only a very rough measure of peptide properties, since also their position in the sequence and structure are obviously important. Amphipathicity measures the degree of asymmetry in the distribution of polar and hydrophobic residues. This property can be quantified by the hydrophobic moment. This quantity is usually defined assuming an ideal helical structure and summing the vectors indicating the position of each residue with respect to the helix axis, multiplied by their respective hydrophobicity values (in analogy with the definition of an electric dipole). To compare sequences of different lengths, the mean hydrophobic moment can be obtained by normalizing for the number of amino acids (Eisenberg et al. 1982; Phoenix and Harris 2002). Peptide amphipathicity is a very important parameter for determining the free energy of membrane binding (Fernández-Vidal et al. 2007). As early as 1981, De Grado demonstrated that amphipathicity is sufficient to induce lytic activity in a helical peptide (De Grado et al. 1981). The specific value of the hydrophobic moment becomes particularly important for selectivity in an intermediate range of hydrophobicity values, when the hydrophilic or hydrophobic components of the peptide do not predominate in determining its behavior (Dathe and Wieprecht 1999; Dathe et al. 2002). Similar to what has been reported for hydrophobicity, an increased hydrophobic moment affects the activity on neutral membranes more than that on charged bilayers (Wieprecht et al. 1997b; Dathe et al. 2002). Increasing amphipathicity above a critical threshold results in strong interaction with neutral membranes, leading to toxicity (Wieprecht et al. 1997b; Dathe and Wieprecht 1999; Fernández-Vidal et al. 2007; Kindrachuk and Napper 2010). As discussed above for hydrophobicity, also in this case, it is difficult to provide a quantitative, generally valid value for this threshold.

Another measure of the distribution of polar and hydrophobic residues is the angle subtended by the polar face of the amphipathic helix, again assuming an ideal conformation and looking along the helix axis (Uematsu and Matsuzaki 2000). The available data on the role of this property in selectivity are limited, but a comprehensive study by Dathe and coworkers (2002) provided some indications. As discussed above, hydrophobicity and hydrophobic moment mostly affect the affinity for neutral membranes and thus the toxic activity. By contrast, in model membranes, the polar angle affects AMP ability to perturb the bilayer, after membrane binding: in charged bilayers, peptide-induced membrane leakage decreases with increasing polar angle, while it is essentially unaffected in neutral membranes (Dathe et al. 2002). However, the effects of the polar angle in cellular assays of activity and toxicity are more limited (Dathe et al. 2002).

11.8 Conformational and Aggregation Equilibria Play an Important Role in Membrane Selectivity: The Concept of Effective Hydrophobicity

All the considerations reported in the previous section are based on hydrophobicity values determined from the peptide amino acid composition and on amphipathicity calculated assuming an ideal helical conformation. In addition, a monomeric peptide state is always considered. However, as discussed in Sect. 11.6, peptides in solution and in the membrane attain specific ensembles of conformations, which can deviate significantly from an ideal alpha-helix. In addition, amphipathic peptides have a strong tendency to aggregate (Fig. 11.6). Conformational equilibria and self-assembly affect the degree to which the hydrophobic moieties of AMPs are exposed to the water phase and therefore modulate the hydrophobic driving force for membrane binding (Bobone et al. 2013). Similarly, water-membrane partition is affected by the peptide conformation, orientation, and depth of insertion in the bilayer. Based on these considerations, in our opinion, AMP selectivity is not determined by the "ideal" peptide hydrophobicity or amphipathicity but by what we call "effective" hydrophobicity and amphipathicity, i.e., the value these parameters assume in the actual conformation and aggregation state attained by the peptide in solution and in the bilayer (Bobone et al. 2013; Uggerhøj et al. 2015).

If peptide conformation and aggregation influence peptide hydrophobicity, the opposite is also true: high hydrophobicity and amphipathicity values favor a stable secondary structure by allowing the formation of intramolecular interactions between apolar residues (Fernández-Vidal et al. 2007). Peptide structure is influenced by self-assembly processes, too (Sal-Man et al. 2002). Therefore, in order to fully understand the determinants of peptide selectivity, conformational and self-assembly equilibria should be considered.

11.8.1 Helicity Correlates with Toxicity

A correlation between peptide helicity and toxicity has been reported in many studies (Tossi et al. 2000; Giangaspero et al. 2001; Zelezetsky et al. 2005; Chen et al. 2005; Khandelia and Kaznessis 2006; Zhang et al. 2011; Mangoni et al. 2011; Chapuis et al. 2012; Bobone et al. 2013; Cherry et al. 2014). In addition, helix-destabilizing Gly or Pro residues are often present close to the center of the sequence of natural, selective AMPs that attain a helical conformation in membranes (Tossi et al. 2000, Bobone et al. 2013). These amino acids are important for peptide selectivity, since their deletion, substitution, or insertion significantly affects toxicity, through the perturbation of the secondary structure (Thennarasu and Nagaraj 1996; Zhang et al. 1999; Shin et al. 2001; Yang et al. 2002, 2006a, b; Lee et al. 2004, 2007; Song et al. 2004; Carotenuto et al. 2008; Bobone et al. 2013; Wang et al. 2015). An increase in selectivity with a reduction in helical structure has been reported for other helix-breaking strategies, such as the insertion of D-amino acids (Shai and Oren 1996, 2001; Oren and Shai 1997; Papo et al. 2002; Chen et al. 2005; Zhu et al. 2007c; Kaminski and Feix 2011; Nan et al. 2012; Huang et al. 2014) or peptoid residues (N-substituted glycines, which lack a H-bonding proton on the N backbone atom and comprise a flexible main-chain methylene group) (Song et al. 2005; Zhu et al. 2007a, b; Kim et al. 2010). Incidentally, these non-proteinogenic residues have the added advantage of reducing peptide susceptibility to proteolysis (Papo et al. 2002; Kim et al. 2010).

The correlation between helicity and toxicity was tentatively explained by proposing that a stable helical conformation enhances the peptide propensity to aggregate (Kindrachuk and Napper 2010; Vermeer et al. 2012): in helical amphipathic peptides, the hydrophobic face of the helix is totally exposed to the aqueous phase, and therefore aggregation is hydrophobically favored. Aggregation, in turn, would inhibit crossing of the LPS layer and cell wall and thus access to the plasma membrane of bacteria (see below). However, helicity normally affects toxicity, rather than antibacterial activity. In addition, in the peptides we investigated, aggregation was significant only at concentrations higher than the membrane-perturbing values, and therefore it was not relevant for activity (Bobone et al. 2013). Probably a higher tendency to aggregate and an enhanced toxicity are just two independent consequences of the hydrophobicity induced by a stable helical structure, but the lack of selectivity in helical peptides is not caused by peptide aggregation (see also below). Based on a systematic study in which a central proline residue was moved along the sequence or deleted, we obtained data supporting an alternative explanation (Bobone et al. 2013). In a perfectly helical, amphipathic structure, the apolar residues are completely exposed on the hydrophobic face of the helix. Even though short peptides are often unstructured in water, helical conformations can be at least partially populated, also thanks to the stabilization due to the interaction between hydrophobic side chains aligned along the helix (a motif often called "leucine zipper" (Asthana et al. 2004)). White and coworkers reported a strong correlation between the amphiphilicity of a peptide sequence in an ideal helical conformation and both the degree of helicity in solution and the affinity for neutral membranes (Fernández-Vidal et al. 2007). Destabilization of the helical conformation allows the peptide to fold onto itself, hiding the apolar side chains from the water phase, and reducing the effective hydrophobicity of the peptide, and thus the driving force for binding neutral membranes (Bobone et al. 2013; Büttner et al. 1992). Destabilization of the helix also increases the entropic cost of membrane binding, since association to the bilayer is normally followed by peptide structuring (Zelezetsky et al. 2005). Interestingly, once membrane-bound, the helix-destabilizing modifications do not preclude the attainment of an amphipathic helical conformation (Bobone et al. 2013; Orioni et al. 2009; Oren and Shai 2000). Therefore, variations in the free energy of membrane binding are determined essentially by changes in the solution conformation. Unstructured conformations would inhibit binding to neutral membranes but would affect only marginally the affinity for charged bilayers, leading to enhanced selectivity. Related to this interpretation is the concept of position-dependent hydrophobicity: hydrophobic residues in unstructured regions of the peptide contribute to the effective hydrophobicity and to toxicity less than those in helical segments (Tachi et al. 2002).

The effective, conformation-dependent hydrophobicity can be calculated from peptide structures (Gaillard et al. 1994), but it can also be determined experimentally. Reversed-phase chromatography retention times have resulted to be an accurate measure of the effective hydrophobicity of peptides (Krause et al. 1995; Zhou et al. 1990; Kim et al. 2005). Interestingly, strong correlation between RP-HPLC retention times and the hemolytic activity of AMPs has been reported (Blondelle and Houghten 1991; Kondejewski et al. 1999; Tachi et al. 2002).

11.8.2 Imperfect Amphipathicity Optimizes Selectivity

Another idea related to effective hydrophobicity is imperfect amphipathicity. Insertion of a polar/charged residue in the hydrophobic face of an amphipathic helix has proven to be a reliable method to increase AMP selectivity (Asthana et al. 2004; Chen et al. 2005; Ahmad et al. 2006, 2009a, b; Hawrani et al. 2008; Pandey et al. 2010, 2011; Jiang et al. 2011, 2014, 2018; Son et al. 2013; Dalzini et al. 2016; Zhang et al. 2016). Hodges and coworkers even termed these misplaced polar residues "specificity determinants" (Jiang et al. 2011, 2014, 2018), even though exceptions to the selectivity-improving effect of this approach have been reported (Wang et al. 2018). Interestingly, an imperfectly amphipathic structure is a common property of many natural, selective AMPs (Wimley 2010b; Orioni et al. 2009). Again, this approach reduces the hydrophobic driving force for binding neutral membranes. On the other hand, antimicrobial activity is usually not affected significantly by these changes: in charged bilayers, binding takes place all the same, thanks to the electrostatic attraction; once membrane-bound, the peptide is able to attain a (possibly distorted) helical conformation, as demonstrated by spectroscopic and simulative studies (Hawrani et al. 2008; Orioni et al. 2009). In the bilayer, imperfect amphipathicity will contribute to membrane disruption, by driving some polar head-groups in the hydrophobic core of the membrane, as we observed for PMAP-23 (Orioni et al. 2009) (Fig. 11.7). This type of membrane activity has been termed "interfacial activity" by Wimley (2010b). Finally, it is worth mentioning that we recently observed an effect of imperfect amphipathicity on toxicity also in the case of peptidomimetic antimicrobial molecules (Konai et al. 2018). Two small amphipathic, cationic molecules were characterized by the same compositional hydrophobicity but had very different selectivity. By combining molecular dynamics simulations and RP-HPLC retention times, we demonstrated that this was due to imperfect amphipathicity and lower effective hydrophobicity of the selective analogue compared to the toxic compound.

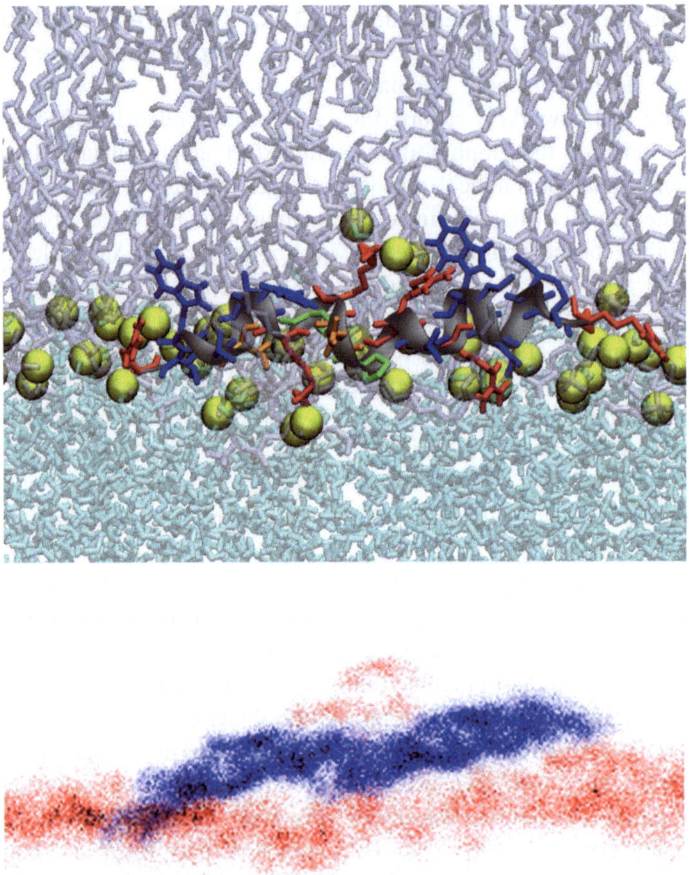

Fig. 11.7 Interfacial activity of an imperfectly amphipathic AMP. Effects of the imperfectly amphipathic AMP PMAP-23 on the structure of a lipid bilayer, as observed in MD simulations. Two charged residues located on the hydrophobic side of the helix drive three phospholipid head-groups and some water molecules into the hydrophobic core of the membrane. Water is represented in cyan, phospholipids in gray, and phospholipids' phosphorus atoms as yellow spheres. The peptide backbone is shown in gray, charged side chains in red, polar amino acids in orange, apolar residues in blue, and prolines in green. The lipid composition was POPG/POPC (1:3 mol/mol).The bottom panel reports the density map of the lipid phosphorus atoms (red) and of the peptide backbone atoms (blue). Adapted, with permission, from research originally published in Orioni et al. 2009 © Elsevier

11.8.3 Effects of Peptide Aggregation in the Aqueous Phase on Activity and Toxicity Are System Dependent

AMPs, due to their amphipathic nature, are susceptible to aggregation in water (Tian et al. 2015). Some peptides oligomerize through the interaction of the apolar sides of their amphipathic helices (Oren et al. 1999; Asthana et al. 2004; Raimondo et al. 2005; Ahmad et al. 2006), while others form micellar structures (Liu et al. 2009; Wang et al. 2010; Joshi et al. 2015; Lin and Grossfield 2015; Haney et al. 2017; Lei et al. 2018), fibrils (Tu et al. 2007; Chen et al. 2010; Chen and Liang 2013; Shankar et al. 2013; Chairatana and Nolan 2014; Ravi et al. 2015), or even hydrogels (Veiga et al. 2012; McCloskey et al. 2014; Haney et al. 2017). Often the aggregates disassemble into monomers once membrane-bound (Ghosh et al. 1997). The critical concentration for self-assembly can vary significantly from one specific case to the other, also depending on the experimental conditions and particularly on salt concentration.

The results of experimental and theoretical studies on the effects of aggregation on peptide activity and selectivity are extremely contradictory. Both negative (Feder et al. 2000; Kustanovich et al. 2002; Chen et al. 2006, 2007; Daschbach et al. 2012; Lin and Grossfield 2015; Farrotti et al. 2017; Haney et al. 2017; Bagheri et al. 2018; Zou et al. 2018) and positive (Sal-Man et al. 2002; Avrahami and Shai 2002; Liu et al. 2009; Chen et al. 2010; Joshi et al. 2015; Ravi et al. 2015; Lei et al. 2018) correlations between aggregation and activity have been reported, as well as lack of activity changes following aggregation (Chen and Liang 2013). Similarly, some studies found that toxicity was not significantly affected by aggregation (Lei et al. 2018), while others reported an increase (Chen and Liang 2013; Lin and Grossfield 2015) or a decrease (Kustanovich et al. 2002; Chen et al. 2006; Chen et al. 2007) in selectivity upon self-assembly. Shankar et al. (2013) suggested that toxicity of self-assembled lipopeptides depends on the specific structure of the fibrillar aggregates.

The reported discrepancies are most likely due to the fact that peptide aggregation is usually controlled by varying the peptide properties or by modulating electrostatic interactions by changing the ionic strength of the solution. It is therefore difficult to discriminate between the direct effects of these changes (e.g., an increase in hydrophobicity) and the consequence of the variations they induce in aggregation. One approach to solve this problem is covalent linking of the monomers (Sal-Man et al. 2002; Dempsey et al. 2003), but it does not exactly mimic self-assembly driven by hydrophobic interactions.

Thermodynamic considerations on the interconnected equilibria involved in AMP activity indicate a possible positive role of aggregation in enhancing peptide selectivity. Aggregation, which is hydrophobically driven, reduces the effective peptide hydrophobicity by hiding the apolar moieties in the molecule from the aqueous phase. As a consequence, the hydrophobic driving force for membrane binding is reduced in the aggregates (Stella et al. 2004; Mazzuca et al. 2005; Gatto et al. 2006; Chen et al. 2007; Chu-Kung et al. 2010; Farrotti et al. 2017). Considering the various hydrophobicity thresholds discussed above, aggregation could therefore lead to a reduced toxicity. At the same time, preassembly of AMPs causes a local release of a high concentration at a single site in the membrane, and this could cause higher activity (Ravi et al. 2015). In addition, computational studies suggested that self-assembly could lead to membrane selectivity also by affecting the kinetics of membrane binding (Lin and Grossfield 2015): binding to host mammalian membranes will be slow and inefficient as long as the lipopeptides are micellized in solution, while binding to the bacterial surface will still be efficient, thanks to electrostatic interactions and to the higher fluidity of the membrane. On the other hand, in cellular assays, the large size of the aggregates, compared to monomers, could impair selectivity: preassembled AMPs might be unable to cross the LPS layer or the cell wall and thus to reach the plasma membrane of bacteria. At the same time, they would still be able to interact with the "naked" membrane of host cells (Oren and Shai 2000; Kustanovich et al. 2002; Sal-Man et al. 2002; Mangoni and Shai 2009).

Discussing aggregation, it is important to note that this phenomenon reduces susceptibility to proteolytic degradation and affects the pharmacokinetics and pharmacodynamics in vivo (Raimondo et al. 2005; Tu et al. 2007; Chen and Liang 2013; Lei et al. 2018). It is also worth mentioning that human α-defensin 6 (HD6) has negligible direct killing activity but prevents infections by self-assembling into a network of fibrils that capture pathogens and thus contrast microbial invasion (Chairatana and Nolan 2014).

11.9 AMP Binding to Cells

As discussed in the previous sections, AMP selectivity is usually interpreted essentially on the basis of the different affinities observed in liposome studies for bilayers mimicking the membranes of bacteria or eukaryotes. However, quite surprisingly, peptide affinities toward the two types of cells are largely uncharacterized. In addition, if peptide activity is modulated by a cell-binding equilibrium, it should depend on the density of cells, but antimicrobial activity and toxicity assays are usually carried out using standardized, fixed cell densities, which are not necessarily representative of the cell concentrations present in a typical infection site (Savini et al. 2018). Finally, bactericidal and hemolytic activities are routinely determined in separate assays, but when the two cell populations are present at the same time, they compete for peptide association. All these aspects have received limited attention, until quite recently. Biophysical studies on model membranes allow the determination of both membrane-binding and bilayer-perturbing activities, while microbiological studies usually report activities only in terms of total peptide concentration. We recently reviewed the few studies that are trying to apply to cellular experiments the same quantitative approaches normally used with model systems (Savini et al. 2018). Here, only the aspects relevant to AMP selectivity are summarized.

11.9.1 AMPs Have a Higher Affinity for Bacterial Than for Eukaryotic Cells

Only a handful of studies reported data on AMP binding to bacterial and eukaryotic cells. As soon as 1988, Bruce Merrifield and his group (Steiner et al. 1988) measured binding of cecropin A and some of its analogues to *Escherichia coli*, *B. megaterium*, *B. thuringiensis*, and *P. aeruginosa* cells and to erythrocytes. While binding to the bacteria was significant (between 70% and 80% for the natural peptide, under the conditions studied), no detectable association was observed for RBCs, at a cell density corresponding to a membrane area similar to that present in the experiments with bacteria. Welling et al. (2000) measured the binding of defensins 1–3, ubiquicidin, and human lactoferrin to bacteria and activated murine peritoneal leucocytes. In the presence of the same cell density ($2 \cdot 10^7$ cell/mL), the peptides bound 5–500 times more efficiently to bacteria than to mammalian cells, even though the latter are much bigger. Similarly, Ferro-Flores et al. (2003) reported that, in the presence of $2 \cdot 10^7$ cell/mL, an ubiquicidin analogue was 35% bound in the case of bacteria (*S. aureus*) while less than 4% in the case of human tumor cell lines LS174T and ACHN (which, again, are significantly bigger than bacteria). Comparable results have been reported for two ubiquicidin analogues (approximately 45–100% binding to *S. aureus* while only 10% to leukocytes, in the presence of $2 \cdot 10^5$ cell/mL), although in this case, selectivity was surprisingly observed also for an anionic peptide used as negative control (Ebenhan et al. 2014b). Wimley and coworkers (Starr et al. 2016) measured the binding of the artificial AMP ARVA to *E. coli*, *S. aureus*, and RBCs. In all cases, association to bacteria was more favorable than to RBCs. After accounting for the differences in cell size, the authors estimated that the affinity for bacterial membranes was more than two orders of magnitude higher than for erythrocytes. Finally, an analogue of LL37 was reported to bind *E. coli*, *S. aureus*, and *M. smegmatis*, but not to hepatic cells, under conditions of comparable cell numbers (Dutta et al. 2017). Overall, these data indicate that the differential affinity routinely observed with model bilayers is present also for the membranes of real cells.

11.9.2 Activity and Toxicity Are Cell Density Dependent

Another aspect that has remained essentially uncharacterized until very recently is whether the activities of AMPs depend on the density of cells present in the assays. Based on a partition equilibrium, the fraction of membrane-bound peptide obviously depends on the concentration of cells in the sample. Therefore, it is to be expected that MIC/MBC/MHC values depend on the concentration of cells used in the assays. In broth dilution assays of antimicrobial activity, the recommended value for the initial cell density (inoculum) is $5 \cdot 10^5$ cell/mL (Patel et al. 2012), which was selected for minimizing false-positive and false-negative results in the clinical practice (Wiegand et al. 2008). However, bacterial cell densities in clinically relevant infections range from 1 to 10^9 cell/mL. Similarly, hemolytic activity assays are normally performed with $5 \cdot 10^8$ cell/mL, which is 1/10 of the cell density in whole blood (Savini et al. 2018 and references therein). Matsuzaki (2009) pointed out that the cell densities in the two assays are very different, also considering that the membrane area of an erythrocyte is approximately ten times bigger than that of a typical bacterium. Therefore, he wondered if TI values such as those reported in Table 11.1 are an experimental artifact due simply to the fact that more peptide is probably needed to kill a higher number of bigger cells.

In the case of traditional antibiotics, it is well known that the MIC often depends on the size of the bacterial inoculum ("inoculum effect"). By contrast, in the case of AMPs, this possible dependence has been investigated only in very few studies (Savini et al. 2018). In the 1990s, Levison et al. (1993) reported that the bactericidal activity of magainins against *P. aeruginosa* was inoculum dependent above $3 \cdot 10^5$ cell/mL, but it did not vary if the inoculum was reduced below this value. Similarly, Jones et al. (1994) observed an inoculum effect for lactoferricin B against *E. coli*, with a plateau at inoculum densities below 10^6 cell/mL. Ulrich and coworkers measured MIC values for gramicidin S and PGLa at two cell density values and observed a cell density dependence (Hartmann et al. 2010). More recently, we measured the MBC values for a fluorescent analogue of PMAP-23 in the presence of different *E. coli* cell densities (Savini et al. 2017). Also in our case, the MBC increased with inoculum size but reached a plateau at densities of $5 \cdot 10^6$ cell/mL and below. A similar trend was reported by Poon and coworkers for pexiganan (Jepson et al. 2016). Finally, a recent study reported an inoculum effect also for LL-37 (Snoussi et al. 2018).

Overall, these studies show that AMP activity is strongly dependent on the density of cells in the assay, with a linear (Savini et al. 2017) or sublinear (Jepson et al. 2016) trend. If a threshold concentration must be reached in the membrane to form pores (Melo et al. 2009), it is obvious that more membrane-bound peptide molecules are needed to kill a higher number of cells (Savini et al. 2017, 2018). Therefore, under conditions of relatively high cell densities, where the peptide is completely bound to the cells, a strong cell density dependence of the activity is expected. Data by Jepson et al. (2016) and Snoussi et al. (2018) indicate that this effect might be due also to peptide sequestration by strong binding to killed bacterial cells. We tentatively explained the plateau in the low cell density regime as due to the cell-binding equilibrium. At low cell densities, some of the peptide remains free in solution, and this fraction increases with decreasing concentrations of cells. As a consequence, the two effects (when there are less cells to kill, also a lower fraction of peptide is cell-bound) cancel each other, leading to a plateau in the total peptide concentration needed in the sample to kill the bacteria (Savini et al. 2017, 2018).

Interestingly, we observed a cell density dependence also for the hemolytic activity, with a plateau at densities below 10^7 cell/mL (Savini et al. 2017).

11.9.3 Competition for Cell Binding in Co-culture Experiments Might Be Regulated by Kinetic Phenomena

Since both antimicrobial activity and toxicity seem to depend on the density of cells, the effective selectivity also depends on the value of this parameter used in the two assays (MIC and MHC). The quantity of peptide that binds to a type of cell or to the other is determined by the respective affinities but also by the concentration of cells of each type. In the end, the peptide is active and/or toxic if the respective threshold of bound peptide needed for membrane perturbation is reached. In principle, under conditions where the host cells are in large excess (as in the case of systemic treatment of an infection), lack of toxicity is expected even if the affinities for the two cell types are similar (Matsuzaki 2009; Savini et al. 2017).

A limited number of studies have tested peptide activity and toxicity in assays where both bacteria and mammalian cells were present. Mor and coworkers showed that AMPs bound to RBCs are able to transfer to microbial cells, exerting their activity (a phenomenon they termed "affinity-driven molecular transfer") (Feder et al. 2001). Derivatives of lentivirus lytic peptides killed *P. aeruginosa* bacteria interacting with cultured human airway epithelial cells, at peptide concentrations that only moderately affected the cell monolayer (Phadke et al. 2003). Fluorescence microscopy images showed that in a co-culture of *S. aureus* and human cells (endothelial cells or neutrophils), AMPs concentrate on the bacterial cells (Matsuzaki 2009; Akram et al. 2015) (Fig. 11.2a). Chen and Liang (2013) showed that the artificial AMP CL-1 is able to kill selectively *S. aureus* in co-culture with human cells, and Malmsten and coworkers reported activity without hemoliticity in bacteria-supplemented blood for an engineered AMP (Schmidtchen et al. 2011). A striking evidence of selectivity is provided by the fact that some AMPs (e.g., LL-37) are able to kill *S. aureus* bacteria internalized into mammalian cells (Noore et al. 2012). Selectivity in co-culture has been reported also for AMP-inspired systems, such as peptidomimetics, cationic peptidopolysaccharides, and peptide hydrogels (Salick et al. 2007; Li et al. 2012; Konai et al. 2018). Some co-culture data have been reported also for antifungal, antiprotozoan, and anticancer activities. For instance, an analogue of the antifungal peptide PAF26 concentrated on fungal cells in co-culture with human lung epithelial cells (Mendive-Tapia et al. 2016); an artificial anticancer peptide concentrated in cancer cells, co-cultured with primary cells (Chen et al. 2014). Some AMPs (e.g., dermaseptins or NK-lysin) are able to kill protozoan parasites such as *P. falciparum* or *T. cruzi* inside human cells, disrupting the plasma membrane of the intracellular parasites without harming that of the host (Ghosh et al. 1997; Krugliak et al. 2000; Jacobs et al. 2003; Gelhaus et al. 2008).

Based on association equilibria, when the two cell types are present at the same time, competition for peptide binding should take place: the antimicrobial activity and/or the toxicity of the peptides should be inhibited by sequestration of a fraction of the peptide molecules due to binding to the other cell population. Very surprisingly, we observed that this is not the case (Savini et al. 2017). We measured both bacterial killing and hemolysis for analogues of PMAP-23 and esculentin, in a mixed population of *E. coli* and erythrocytes, and compared these results with the traditional assays performed on the two cell types separately (Fig. 11.8). The activities on bacteria and on RBCs were essentially unaffected by the presence of the other cell population, in contradiction with the predictions based on binding equilibria. Data from Wimley's lab provide further support to the conclusion that out of equilibrium, kinetic phenomena are at play when two cell populations are present at the same time. They showed that the results of such experiments depend on the order of addition of the different components: like in our case, no change in the MIC was observed when AMPs were added to a mixture of bacteria and RBCs or to bacteria alone. However, the antimicrobial activity was significantly inhibited when the peptide was incubated with RBCs first, and then both were added to the bacterial culture (Starr et al. 2016) (Fig. 11.8). Actually, while equilibrium processes can be invoked before the cell membranes are perturbed, it is easy to realize that this approach is too simplistic in the case of bilayer disruption, which allows access to multiple additional binding targets (e.g., inside the cell) (Snoussi et al. 2018).

Overall, quantitative measurements of peptide interactions with cells confirmed that AMPs have a higher affinity for bacterial than for host cells. Experiments performed with varying cell densities indicated that both activity and toxicity depend on this parameter (even though a plateau is observed at low cell densities) and that therefore the measured selectivity depends on the specific conditions of the experiments. Finally, experiments with mixed bacterial and eukaryotic cells showed that, contrary to expectations based on equilibrium considerations, competition for peptide binding does not lead to a loss in activity.

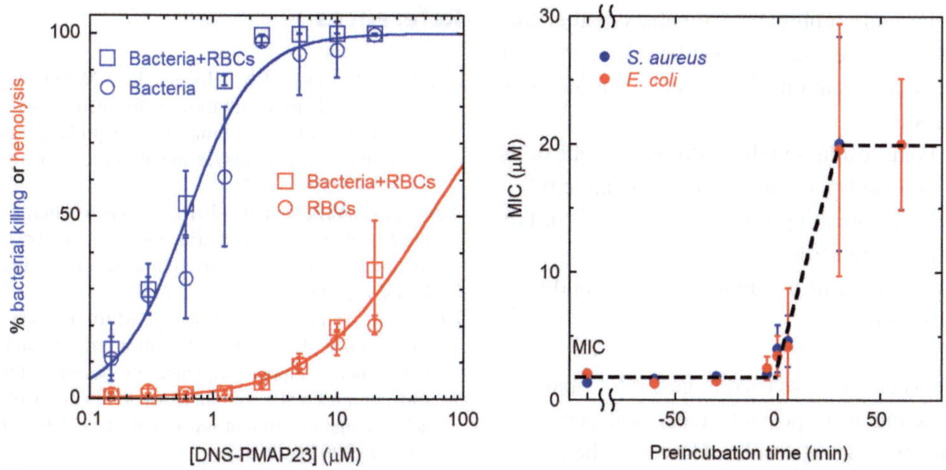

Fig. 11.8 Antimicrobial and hemolytic activity in assays with both bacterial and erythrocytes. Left panel, bactericidal (blue) and hemolytic (red) activities of the AMP DNS-PMAP23 in the presence of both bacteria and erythrocytes (squares) or of one cell type only (circles). Both activities are only slightly affected by the presence of the other cell population. 4.5 × 10[7] *E. coli* cells/mL, 4.5 × 10[8] RBCs/mL. (Reproduced with permission from Savini et al. 2017 https://pubs.acs.org/doi/abs/10.1021/acschembio.6b00910 (Copyright 2017, American Chemical Society)). Further permissions related to the material excerpted should be directed to the ACS. Right panel: effect of incubation time on MIC values in assays with both bacteria and erythrocytes. The artificial AMP ARVA-D was tested against *E. coli* (red) and *S. aureus* (blue) under various conditions. "MIC" represents measurements done in the absence of RBCs. All other experiments include 10[9] human RBC/mL. Time zero represents the experiments in which RBC and bacteria were first mixed, followed by peptide addition, i.e., no preincubation with either cell type. Negative times represent peptide preincubation with bacteria before the addition of RBCs. Positive times represent peptide preincubation with RBCs, followed by addition of bacteria. Points plotted at 20 μM had MIC values ≥20 μM. Significant inhibition of peptide antimicrobial activity due to the presence of RBCs was observed only in the case of preincubation with erythrocytes. Proteolytic degradation effects can be ruled out, since ARVA-D is a peptide comprising all D-amino acids. (Adapted with permission from Starr 2016 (Copyright 2016, American Chemical Society))

This finding definitely warrants further co-culture studies.

11.10 Concluding Remarks

A large body of studies has been devoted to characterize, understand, and improve the selectivity of AMPs. These data support the view that selectivity arises due to the different lipid composition of bacterial and host cell membranes. AMPs are able to discriminate between the two types of bilayers, thanks to their physicochemical properties. We can summarize here the main guidelines for optimization of peptide selectivity:

- Increasing the cationic charge, by C-terminal amidation, substitution of anionic residues, or insertion of cationic amino acids in the polar side of the peptide structure, leads to a better selectivity. However, an excessive increase in positive charge, above a threshold that depends on each specific case, might be ineffective or even detrimental.
- Reducing hydrophobicity, amphipathicity, and helicity is an effective strategy. These properties are necessary for activity, since they are responsible for the hydrophobic driving force for membrane binding and for insertion in the bilayer. However, several studies have demonstrated that they affect toxicity more than activity. This finding is probably a consequence of the nonadditivity of electrostatic and hydrophobic effects. Therefore, an intermediate range of hydrophobicity and amphipathicity values optimizes selectivity.
- The real determinants of selectivity are the effective, conformation-dependent, hydropho-

bicity and amphipathicity values (rather than the parameters determined based on an idealized conformation). They can be optimized by introducing:
  - Polar residues in the hydrophobic side of the peptide helix (imperfect amphipathicity)
  - Helix-breaking residues, such as Pro, Gly, D-amino acids, or peptoids
- Aggregation in the aqueous phase modulates selectivity, too.

In this chapter, we tried to generalize and simplify as much as possible the results collected over many years of studies. However, the readers that had the patience to follow us until here must have realized that the literature on AMP selectivity is full of exceptions and contradictions. Peptide association to bacterial and host cells is modulated (in a nonadditive way) by electrostatic and hydrophobic interactions. The effective peptide hydrophobicity is determined by peptide conformation and aggregation state. Therefore, AMP selectivity is finely regulated by interconnected binding, aggregation and conformational equilibria. Any variation in peptide property will affect all of these phenomena. Therefore, in order to predict, or at least to understand, the effect of peptide modifications on the final selectivity, all the possible processes involved in peptide behavior in the aqueous and membrane phases must be considered.

We still do not fully understand what happens when AMPs act on bacteria and human cells together. Recent data indicate that in this case even the complex scenario outlined above is overly simplified, since kinetic phenomena probably have to be taken into account. Further studies in this area are definitely warranted and hold the promise to provide a better understanding of AMP selectivity.

Acknowledgments The authors gratefully thank Dr. F. Savini and Dr. A. Papi for their help with Figs. 11.3 and 11.5. Research in our lab is currently supported by the Italian Ministry for Education, University and Research (grant PRIN 20157WW5EH_007), and by the Italian Association for Cancer Research (AIRC grant IG 2016 19171).

References

Abraham T, Lewis RN, Hodges RS, McElhaney RN (2005) Isothermal titration calorimetry studies of the binding of the antimicrobial peptide gramicidin S to phospholipid bilayer membranes. Biochemistry 44(33):11279–11285

Agrawal A, Weisshaar JC (2018) Effects of alterations of the *E. coli* lipopolysaccharide layer on membrane permeabilization events induced by Cecropin A. Biochim Biophys Acta 1860(7):1470–1479

Ahmad A, Yadav SP, Asthana N, Mitra K, Srivastava SP, Ghosh JK (2006) Utilization of an amphipathic leucine zipper sequence to design antibacterial peptides with simultaneous modulation of toxic activity against human red blood cells. J Biol Chem 281(31):22029–22038

Ahmad A, Asthana N, Azmi S, Srivastava RM, Pandey BK, Yadav V, Ghosh JK (2009a) Structure–function study of cathelicidin-derived bovine antimicrobial peptide BMAP-28: design of its cell-selective analogs by amino acid substitutions in the heptad repeat sequences. Biochim Biophys Acta 1788(11):2411–2420

Ahmad A, Azmi S, Srivastava RM, Srivastava S, Pandey BK, Saxena R, Bajpai VK, Ghosh JK (2009b) Design of nontoxic analogues of cathelicidin-derived bovine antimicrobial peptide BMAP-27: the role of leucine as well as phenylalanine zipper sequences in determining its toxicity. Biochemistry 48(46):10905–10917

Akhtar MS, Qaisar A, Irfanullah J, Iqbal J (2005) Antimicrobial peptide ^{99m}Tc-ubiquicidin 29-41 as human infection-imaging agent: clinical trial. J Nucl Med 46(4):567–573

Akhtar MS, Imran MB, Nadeem MA, Shahid A (2012) Antimicrobial peptides as infection imaging agents: better than radiolabeled antibiotics. Int J Pept. https://doi.org/10.1155/2012/965238

Akram AR, Avlonitis N, Lilienkampf A, Perez-Lopez AM, McDonald N, Chankeshwara SV, Scholefield E, Haslett C, Bradley M, Dhaliwal K (2015) A labelled-ubiquicidin antimicrobial peptide for immediate in situ optical detection of live bacteria in human alveolar lung tissue. Chem Sci 6(12):6971–6979

Alba A, López-Abarrategui C, Otero-González AJ (2012) Host defense peptides: an alternative as antiinfective and immunomodulatory therapeutics. Biopolymers 98(4):251–267

Allende D, Simon SA, McIntosh TJ (2005) Melittin-induced bilayer leakage depends on lipid material properties: evidence for toroidal pores. Biophys J 88(3):1828–1837

Aloia RC, Tian H, Jensen FC (1993) Lipid composition and fluidity of the human immunodeficiency virus envelope and host cell plasma membranes. Proc Natl Acad Sci U S A 90(11):5181–5185

Ames GF (1968) Lipids of Salmonella typhimurium and *Escherichia coli*: structure and metabolism. J Bacteriol 95(3):833–843

Andra J, Goldmann T, Ernst CM, Peschel A, Gutsmann T (2011) Multiple peptide resistance factor (MprF)-mediated resistance of *Staphylococcus aureus* against antimicrobial peptides coincides with a modulated peptide interaction with artificial membranes comprising lysyl-phosphatidylglycerol. J Biol Chem 286(21):18692–18700

Arcidiacono S, Soares JW, Meehan AM, Marek P, Kirby R (2009) Membrane permeability and antimicrobial kinetics of cecropin P1 against *Escherichia coli*. J Pept Sci 15(6):398–403

Assadi M, Vahdat K, Nabipour I, Sehhat MR, Hadavand F, Javadi H, Tavakoli A, Saberifard J, Kalantarhormozi MR, Zakani A, Eftekhari M (2011) Diagnostic value of 99mTc-ubiquicidin scintigraphy for osteomyelitis and comparisons with 99mTc-methylene diphosphonate scintigraphy and magnetic resonance imaging. Nucl Med Commun 32(8):716–723

Asthana N, Yadav SP, Ghosh JK (2004) Dissection of antibacterial and toxic activity of Melittin: a leucine zipper motif plays a crucial role in determining its hemolytic activity but not antibacterial activity. J Biol Chem 279(53):55042–55050

Avrahami D, Shai Y (2002) Conjugation of a magainin analogue with lipophilic acids controls hydrophobicity, solution assembly, and cell selectivity. Biochemistry 41(7):2254–2263

Avrahami D, Shai Y (2004) A new group of antifungal and antibacterial lipopeptides derived from non-membrane active peptides conjugated to palmitic acid. J Biol Chem 279(13):12277–12285

Bacalum M, Radu M (2015) Cationic antimicrobial peptides cytotoxicity on mammalian cells: an analysis using therapeutic index integrative concept. Int J Pept Res Ther 21:47–55

Bagheri M, Amininasab M, Dathe M (2018) Arg/Trp-rich cyclic α/β-antimicrobial peptides: the roles of H-bonding and hydrophobic/hydrophilic solvent accessible surface areas upon the activity and membrane selectivity. Chem Eur J. 24(53):14242–14253

Ballas SK, Krasnow SH (1980) Structure of erythrocyte membrane and its transport functions. Ann Clin Lab Sci 10(3):209–219

Beschiaschvili G, Seelig J (1990) Melittin binding to mixed phosphatidylglycerol/phosphatidylcholine membranes. Biochemistry 29(1):52–58

Bessalle R, Kapitkovsky A, Gorea A, Shalit I, Fridkin M (1990) All-D-magainin: chirality, antimicrobial activity and proteolytic resistance. FEBS Lett 274(1–2):151–155

Bessalle R, Haas H, Goria A, Shalit I, Fridkin M (1992) Augmentation of the antibacterial activity of magainin by positive-charge chain extension. Antimicrob Agents Chemother 36(2):313–317

Bhatt J, Mukherjee A, Shinto A, Karuppusamy KK, Korde A, Kumar M, Sarma HD, Repaka K, Dash A (2018) Gallium-68 labeled Ubiquicidin derived octapeptide as a potential infection imaging agent. Nucl Med Biol 62–63:47–53

Bishop DG, Rutberg L, Samuelsson B (1967) The chemical composition of the cytoplasmic membrane of Bacillus subtilis. Eur J Biochem 2(4):448–453

Bland JM, De Lucca AJ, Jacks TJ, Vigo CB (2001) All-D-cecropin B: synthesis, conformation, lipopolysaccharide binding, and antibacterial activity. Mol Cell Biochem 218(1–2):105–111

Blondelle SE, Houghten RA (1991) Hemolytic and antimicrobial activities of the twenty-four individual omission analogs of melittin. Biochemistry 30(19):4671–4678

Bobone S, Piazzon A, Orioni B, Pedersen JZ, Nan YH, Hahm KS, Shin SH, Stella L (2011) The thin line between cell-penetrating and antimicrobial peptides: the case of Pep-1 and Pep-1-K. J Pept Sci 17(5):335–341

Bobone S, Roversi D, Giordano L, De Zotti M, Formaggio F, Toniolo C, Park Y, Stella L (2012) The lipid dependence of antimicrobial peptide activity is an unreliable experimental test for different pore models. Biochemistry 51(51):10124–10126

Bobone S, Bocchinfuso G, Park Y, Palleschi A, Hahm KS, Stella L (2013) The importance of being kinked: role of Pro residues in the selectivity of the helical antimicrobial peptide P5. J Pept Sci 19(12):758–769

Bocchinfuso G, Palleschi A, Orioni B, Grande G, Formaggio F, Toniolo C et al (2009) Different mechanisms of action of antimicrobial peptides: insights from fluorescence spectroscopy experiments and molecular dynamics simulations. J Pept Sci Off Publ Eur Pept Soc 15(9):550–558

Bocchinfuso G, Bobone S, Mazzuca C, Palleschi A, Stella L (2011) Fluorescence spectroscopy and molecular dynamics simulations in studies on the mechanism of membrane destabilization by antimicrobial peptides. Cell Mol Life Sci 68(13):2281–2301

Boyd KJ, Alder NN, May ER (2017) Buckling under pressure: curvature-based lipid segregation and stability modulation in cardiolipin-containing bilayers. Langmuir 33(27):6937–6946

Braun S, Pokorná S, Šachl R, Hof M, Heerklotz H, Hoernke M (2017) Biomembrane permeabilization: statistics of individual leakage events harmonize the interpretation of vesicle leakage. ACS Nano 12(1):813–819

Brender JR, McHenry AJ, Ramamoorthy A (2012) Does cholesterol play a role in the bacterial selectivity of antimicrobial peptides? Front Immunol 3:195

Broekhuyse RM (1969) Quantitative two-dimensional thin-layer chromatography of blood phospholipids. Clin Chim Acta 23(3):457–461

Brouwer CP, Bogaards SJ, Wulferink M, Velders MP, Welling MM (2006) Synthetic peptides derived from human antimicrobial peptide ubiquicidin accumulate at sites of infections and eradicate (multi-drug

resistant) *Staphylococcus aureus* in mice. Peptides 27(11):2585–2591
Brouwer CPJM, Sarda-Mantel L, Meulemans A, Guludec DL, Welling MM (2008) The use of technetium-99m radiolabeled human antimicrobial peptides for infection specific imaging. Mini-Rev Med Chem 8(10):1039–1052
Bulet P, Stocklin R (2005) Insect antimicrobial peptides: structures, properties and gene regulation. Protein Pept Lett 12(1):3–11
Bütikofer P, Lin ZW, Chiu DT, Lubin B, Kuypers FA (1990) Transbilayer distribution and mobility of phosphatidylinositol in human red blood cells. J Biol Chem 265(27):16035–16038
Büttner K, Blondelle SE, Ostresh JM, Houghten RA (1992) Perturbation of peptide conformations induced in anisotropic environments. Biopolymers 32(6):575–583
Carneiro VA, Duarte HS, Prado MGV, Silva ML, Teixeira M, dos Santos YM, Vasconcelos IB, Cunha RMS (2015) Antimicrobial peptides: from synthesis to clinical perspectives. In: The battle against microbial pathogens: basic science, technological advances and educational programs, 1st edn. Formatex Research Center, Spain, pp 81–90
Carotenuto A, Malfi S, Saviello MR, Campiglia P, Gomez-Monterrey I, Mangoni ML, Marcellini Hercolani Gaddi L, Novellino E, Grieco P (2008) A different molecular mechanism underlying antimicrobial and hemolytic actions of temporins A and L. J Med Chem 51(8):2354–2362
Chairatana P, Nolan EM (2014) Molecular basis for self-assembly of a human host-defense peptide that entraps bacterial pathogens. J Am Chem Soc 136(38):13267–13276
Chapuis H, Slaninová J, Bednárová L, Monincová L, Buděšínský M, Čeřovský V (2012) Effect of hydrocarbon stapling on the properties of α-helical antimicrobial peptides isolated from the venom of hymenoptera. Amino Acids 43(5):2047–2058
Chen L, Liang JF (2013) Peptide fibrils with altered stability, activity, and cell selectivity. Biomacromolecules 14(7):2326–2331
Chen Y, Mant CT, Farmer SW, Hancock REW, Michael L, Vasil ML, Hodges RS (2005) Rational design of alpha-helical antimicrobial peptides with enhanced activities and specificity/therapeutic index. J Biol Chem 280:12316–12329
Chen Y, Vasil AI, Rehaume L, Mant CT, Burns JL, Vasil ML, Hancock RE, Hodges RS (2006) Comparison of biophysical and biologic properties of α-helical enantiomeric antimicrobial peptides. Chem Biol Drug Des 67(2):162–173
Chen Y, Guarnieri MT, Vasil AI, Vasil ML, Mant CT, Hodges RS (2007) Role of peptide hydrophobicity in the mechanism of action of α-helical antimicrobial peptides. Antimicrob Agents Chemother 51(4):1398–1406
Chen C, Pan F, Zhang S, Hu J, Cao M, Wang J, Xu H, Zhao X, Lu JR (2010) Antibacterial activities of short designer peptides: a link between propensity for nanostructuring and capacity for membrane destabilization. Biomacromolecules 11(2):402–411
Chen C, Hu J, Zeng P, Pan F, Yaseen M, Xu H, Lu JR (2014) Molecular mechanisms of anticancer action and cell selectivity of short α-helical peptides. Biomaterials 35(5):1552–1561
Chen H, Liu C, Chen D, Madrid K, Peng S, Dong X, Zhang M, Gu Y (2015) Bacteria-targeting conjugates based on antimicrobial peptide for bacteria diagnosis and therapy. Mol Pharm 12(7):2505–2516
Cherry MA, Higgins SK, Melroy H, Lee HS, Pokorny A (2014) Peptides with the same composition, hydrophobicity, and hydrophobic moment bind to phospholipid bilayers with different affinities. J Phys Chem B 118(43):12462–12470
Chu-Kung AF, Bozzelli KN, Lockwood NA, Haseman JR, Mayo KH, Tirrell MV (2004) Promotion of peptide antimicrobial activity by fatty acid conjugation. Bioconjug Chem 15(3):530–535
Chu-Kung AF, Nguyen R, Bozzelli KN, Tirrell M (2010) Chain length dependence of antimicrobial peptide-fatty acid conjugate activity. J Colloid Interface Sci 345(2):160–167
Cornut I, Büttner K, Dasseux JL, Dufourcq J (1994) The amphipathic α-helix concept: application to the de novo design of ideally amphipathic Leu, Lys peptides with hemolytic activity higher than that of melittin. FEBS Lett 349(1):29–33
Dalzini A, Bergamini C, Biondi B, De Zotti M, Panighel G, Fato R, Peggion C, Bortolus M, Maniero AL (2016) The rational search for selective anticancer derivatives of the peptide trichogin GA IV: a multi-technique biophysical approach. Sci Rep 6:24000
Daschbach MM, Negin S, You L, Walsh M, Gokel GW (2012) Aggregation and supramolecular membrane interactions that influence anion transport in tryptophan-containing synthetic peptides. Chem Eur J 18(24):7608–7623
Dathe M, Wieprecht T (1999) Structural features of helical antimicrobial peptides: their potential to modulate activity on model membranes and biological cells. Biochim Biophys Acta 1462(1):71–87
Dathe M, Schümann M, Wieprecht T, Winkler A, Beyermann M, Krause E, Matsuzatki K, Murase O, Bienert M (1996) Peptide helicity and membrane surface charge modulate the balance of electrostatic and hydrophobic interactions with lipid bilayers and biological membranes. Biochemistry 35(38):12612–12622
Dathe M, Wieprecht T, Nikolenko H, Handel L, Maloy WL, MacDonald DL, Beyermann M, Bienert M (1997) Hydrophobicity, hydrophobic moment and angle subtended by charged residues modulate antibacterial and haemolytic activity of amphipathic helical peptides. FEBS Lett 403(2):208–212
Dathe M, Nikolenko H, Meyer J, Beyermann M, Bienert M (2001) Optimization of the antimicrobial activity of magainin peptides by modification of charge. FEBS Lett 501(2–3):146–150

Dathe M, Meyer J, Beyermann M, Maul B, Hoischen C, Bienert M (2002) General aspects of peptide selectivity towards lipid bilayers and cell membranes studied by variation of the structural parameters of amphipathic helical model peptides. Biochim Biophys Acta 1558(2):171–186

Dawson RM, Liu CQ (2011) Analogues of peptide SMAP-29 with comparable antimicrobial potency and reduced toxicity. Int J Antimicrob Agents 37(5):432–437

Dawson RM, Fox MA, Atkins HS, Liu CQ (2011) Potent antimicrobial peptides with selectivity for *Bacillus anthracis* over human erythrocytes. Int J Antimicrob Agents 38(3):237–242

de Kruijff B (1990) Cholesterol as a target for toxins. Biosci Rep 10(2):127–130

de la Fuente-Núñez C, Reffuveille F, Mansour SC, Reckseidler-Zenteno SL, Hernández D, Brackman G, Coenye T, Hancock REW (2015) D-enantiomeric peptides that eradicate wild-type and multidrug-resistant biofilms and protect against lethal Pseudomonas aeruginosa infections. Chem Biol 22(2):196–205

de Murphy CA, Gemmel F, Balter J (2010) Clinical trial of specific imaging of infections. Nucl Med Commun 31(8):726–733

Dean SN, Bishop BM, Van Hoek ML (2011) Susceptibility of Pseudomonas aeruginosa biofilm to alpha-helical peptides: D-enantiomer of LL-37. Front Microbiol 2:128

DeGrado WF, Kezdy FJ, Kaiser ET (1981) Design, synthesis, and characterization of a cytotoxic peptide with melittin-like activity. J Am Chem Soc 103(3):679–681

Dempsey CE, Ueno S, Avison MB (2003) Enhanced membrane permeabilization and antibacterial activity of a disulfide-dimerized magainin analogue. Biochemistry 42(2):402–409

Dennison SR, Harris F, Bhatt T, Singh J, Phoenix DA (2009) The effect of C-terminal amidation on the efficacy and selectivity of antimicrobial and anticancer peptides. Mol Cell Biochem 332(1–2):43–50

Deslouches B, Hasek ML, Craigo JK, Steckbeck JD, Montelar RC (2016) Comparative functional properties of engineered cationic antimicrobial peptides consisting exclusively of tryptophan and either lysine or arginine. J Med Microbiol 65(6):554–565

Dodge JT, Phillips GB (1967) Composition of phospholipids and of phospholipid fatty acids and aldehydes in human red cells. J Lipid Res 8(6):667–675

Dutta J, Baijnath S, Somboro AM, Nagiah S, Albericio F, de la Torre BG, Marjanovic-Painter B, Zeevaart JR, Sathekge M, Kruger HG, Chuturgoon A, Naicker T, Ebenhan T, Govender T (2017) Synthesis, in vitro evaluation, and 68Ga-radiolabeling of CDP1 toward PET/CT imaging of bacterial infection. Chem Biol Drug Des 90(4):572–579

Ebenhan T, Gheysens O, Kruger HG, Zeevaart JR, Sathekge MM (2014a) Antimicrobial peptides: their role as infection-selective tracers for molecular imaging. Biomed Res Int. https://doi.org/10.1155/2014/867381

Ebenhan T, Chadwick N, Sathekge MM, Govender P, Govender T, Kruger HG, Marjanovic-Painter B, Zeevaart JR (2014b) Peptide synthesis, characterization and 68Ga-radiolabeling of NOTA-conjugated ubiquicidin fragments for prospective infection imaging with PET/CT. Nucl Med Biol 41(5):390–400

Ebenhan T, Sathekge M, Lenagana T, Koole M, Gheysens O, Govender T, Zeevaart JR (2018) 68Ga-NOTA-functionalized ubiquicidin: cytotoxicity, biodistribution, radiation dosimetry and first-in-human positron emission tomography/computed tomography imaging of infections. J Nucl Med 59(2):334–339

Eckert R (2011) Road to clinical efficacy: challenges and novel strategies for antimicrobial peptide development. Future Microbiol 6(6):635–651

Eisenberg D, Weiss RM, Terwilliger TC (1982) The helical hydrophobic moment: a measure of the amphiphilicity of a helix. Nature 299(5881):371–374

Epand RF, Schmitt MA, Gellman SH, Epand RM (2006) Role of membrane lipids in the mechanism of bacterial species selective toxicity by two α/β-antimicrobial peptides. Biochim Biophys Acta 1758(9):1343–1350

Farrotti A, Bocchinfuso G, Palleschi A, Rosato N, Salnikov ES, Voievoda N, Bechinger B, Stella L (2015) Molecular dynamics methods to predict peptide locations in membranes: LAH4 as a stringent test case. Biochim Biophys Acta 1848(2):581–592

Farrotti A, Conflitti P, Srivastava S, Ghosh JK, Palleschi A, Stella L, Bocchinfuso G (2017) Molecular dynamics simulations of the host defense peptide temporin L and its Q3K derivative: an atomic level view from aggregation in water to bilayer perturbation. Molecules 22(7):1235

Feder R, Dagan A, Mor A (2000) Structure-activity relationship study of antimicrobial dermaseptin S4 showing the consequences of peptide oligomerization on selective cytotoxicity. J Biol Chem 275(6):4230–4238

Feder R, Nehushtai R, Mor A (2001) Affinity driven molecular transfer from erythrocyte membrane to target cells. Peptides 22(10):1683–1690

Fernández-Vidal M, Jayasinghe S, Ladokhin AS, White SH (2007) Folding amphipathic helices into membranes: amphiphilicity trumps hydrophobicity. J Mol Biol 370(3):459–470

Ferro-Flores G, de Murphy CA, Pedraza-López M, Meléndez-Alafort L, Zhang YM, Rusckowski M, Hnatowich DJ (2003) In vitro and in vivo assessment of 99mTc-UBI specificity for bacteria. Nucl Med Biol 30(6):597–603

Findlay EG, Currie SM, Davidson DJ (2013) Cationic host defence peptides: potential as antiviral therapeutics. BioDrugs 27(5):479–493

Friedrich CL, Moyles D, Beveridge TJ, Hancock RE (2000) Antibacterial action of structurally diverse cationic peptides on gram-positive bacteria. Antimicrob Agents Chemother 44(8):2086–2092

Gaillard P, Carrupt PA, Testa B, Boudon A (1994) Molecular lipophilicity potential, a tool in 3D QSAR: method and applications. J Comput Aided Mol Des 8(2):83–96

Gandomkar M, Najafi R, Shafiei M, Mazidi M, Goudarzi M, Mirfallah SH, Ebrahimi F, Heydarpor HR, Abdie N (2009) Clinical evaluation of antimicrobial peptide [99mTc/Tricine/HYNIC0] ubiquicidin 29–41 as

a human-specific infection imaging agent. Nucl Med Biol 36(2):199–205
Gascard P, Tran D, Sauvage M, Sulpice JC, Fukami K, Takenawa T, Claret M, Giraud F (1991) Asymmetric distribution of phosphoinositides and phosphatidic acid in the human erythrocyte membrane. Biochim Biophys Acta 1069(1):27–36
Gaspar D, Veiga AS, Castanho MA (2013) From antimicrobial to anticancer peptides. A review. Front Microbiol 4:294
Gatto E, Mazzuca C, Stella L, Venanzi M, Toniolo C, Pispisa B (2006) Effect of peptide lipidation on membrane perturbing activity: a comparative study on two trichogin analogues. J Phys Chem B 110(45):22813–22818
Gazit E, Boman A, Boman HG, Shai Y (1995) Interaction of the mammalian antibacterial peptide cecropin P1 with phospholipid vesicles. Biochemistry 34(36):11479–11488
Gelhaus C, Jacobs T, Andrä J, Leippe M (2008) The antimicrobial peptide NK-2, the core region of mammalian NK-lysin, kills intraerythrocytic *Plasmodium falciparum*. Antimicrob Agents Chemother 52(5):1713–1720
Ghosh JK, Shaool D, Guillaud P, Cicéron L, Mazier D, Kustanovich I, Shai Y, Mor A (1997) Selective cytotoxicity of dermaseptin S3 toward intraerythrocytic plasmodium falciparum and the underlying molecular basis. J Biol Chem 272(50):31609–31616
Giacometti A, Cirioni O, Greganti G, Quarta M, Scalise G (1998) In vitro activities of membrane-active peptides against gram-positive and gram-negative aerobic bacteria. Antimicrob Agents Chemother 42(12):3320–3324
Giangaspero A, Sandri L, Tossi A (2001) Amphipathic α helical antimicrobial peptides. A systematic study of the effects of structural and physical properties on biological activity. Eur J Biochem 268(21):5589–5600
Glukhov E, Burrows LL, Deber CM (2008) Membrane interactions of designed cationic antimicrobial peptides: the two thresholds. Biopolymers 89(5):360–371
Golbek TW, Franz J, Elliott Fowler J, Schilke KF, Weidner T, Baio JE (2017) Identifying the selectivity of antimicrobial peptides to cell membranes by sum frequency generation spectroscopy. Biointerphases 12(2):02D406
Gonçalves S, Abade J, Teixeira A, Santos NC (2012) Lipid composition is a determinant for human defensin HNP1 selectivity. Biopolymers 98(4):313–321
Gordesky SE, Marinetti GV (1973) The asymmetric arrangement of phospholipids in the human erythrocyte membrane. Biochem Biophys Res Commun 50(4):1027–1031
Gordesky SE, Marinetti GV, Love R (1975) The reaction of chemical probes with the erythrocyte membrane. J Membr Biol 20(1–2):111–132
Hallock KJ, Lee DK, Omnaas J, Mosberg HI, Ramamoorthy A (2002) Membrane composition determines pardaxin's mechanism of lipid bilayer disruption. Biophys J 83(2):1004–1013
Hancock RE, Sahl HG (2006) Antimicrobial and host-defense peptides as new anti-infective therapeutic strategies. Nat Biotechnol 24(12):1551
Haney EF, Wu BC, Lee K, Hilchie AL, Hancock RE (2017) Aggregation and its influence on the immunomodulatory activity of synthetic innate defense regulator peptides. Cell Chem Biol 24(8):969–980
Harris F, Dennison SR, Singh J, Phoenix DA (2013) On the selectivity and efficacy of defense peptides with respect to cancer cells. Med Res Rev 33(1):190–234
Hartmann M, Berditsch M, Hawecker J, Ardakani MF, Gerthsen D, Ulrich AS (2010) Damage of the bacterial cell envelope by antimicrobial peptides gramicidin S and PGLa as revealed by transmission and scanning electron microscopy. Antimicrob Agents Chemother 54(8):3132–3142
Hawrani A, Howe RA, Walsh TR, Dempsey CE (2008) Origin of low mammalian cell toxicity in a class of highly active antimicrobial amphipathic helical peptides. J Biol Chem 283(27):18636–18645
Hayami M, Okabe A, Kariyama R, Abe M, Kanemasa Y (1979) Lipid composition of Staphylococcus aureus and its derived L-forms. Microbiol Immunol 23(6):435–442
He J, Krauson AJ, Wimley WC (2014) Toward the de novo design of antimicrobial peptides: lack of correlation between peptide permeabilization of lipid vesicles and antimicrobial, cytolytic, or cytotoxic activity in living cells. Biopolymers 102(1):1–6
Henriksen J, Rowat AC, Brief E, Hsueh YW, Thewalt JL, Zuckermann MJ, Ipsen JH (2006) Universal behavior of membranes with sterols. Biophys J 90(5):1639–1649
Henriksen JR, Etzerodt T, Gjetting T, Andresen TL (2014) Side chain hydrophobicity modulates therapeutic activity and membrane selectivity of antimicrobial peptide mastoparan-X. PLoS One 9:e91007
Hollmann A, Martinez M, Maturana P, Semorile LC, Maffia PC (2018) Antimicrobial peptides: interaction with model and biological membranes and synergism with chemical antibiotics. Front Chem 6:204
Holt A, de Almeida RF, Nyholm TK, Loura LM, Daily AE, Staffhorst RW, Rijkers DTS, Koeppe RE II, Prieto M, Killian JA (2008) Is there a preferential interaction between cholesterol and tryptophan residues in membrane proteins? Biochemistry 47(8):2638–2649
Hong SY, Oh JE, Lee KH (1999) Effect of D-amino acid substitution on the stability, the secondary structure, and the activity of membrane-active peptide. Biochem Pharmacol 58(11):1775–1780
Hoskin DW, Ramamoorthy A (2008) Studies on anticancer activities of antimicrobial peptides. Biochim Biophys Acta 1778(2):357–375
Hoyos-Nogués M, Gil FJ, Mas-Moruno C (2018) Antimicrobial peptides: powerful biorecognition elements to detect bacteria in biosensing technologies. Molecules 23(7):1683

Huang Y, Huang J, Chen Y (2010) Alpha-helical cationic antimicrobial peptides: relationships of structure and function. Protein Cell 1(2):143–152

Huang Y, He L, Li G, Zhai N, Jiang H, Chen Y (2014) Role of helicity of α-helical antimicrobial peptides to improve specificity. Protein Cell 5(8):631–642

Jacobs T, Bruhn H, Gaworski I, Fleischer B, Leippe M (2003) NK-lysin and its shortened analog NK-2 exhibit potent activities against Trypanosoma cruzi. Antimicrob Agents Chemother 47(2):607–613

Javadpour MM, Juban MM, Lo WCJ, Bishop SM, Alberty BJ, Cowell SM, Becker CL, McLaughlin ML (1996) De novo antimicrobial peptides with low mammalian cell toxicity. J Med Chem 39:3107–3113

Jepson AK, Schwarz-Linek J, Ryan L, Ryadnov MG, Poon WC (2016) What is the 'Minimum Inhibitory Concentration' (MIC) of Pexiganan acting on *Escherichia coli*?-A cautionary case study. In: Leake MC (ed) Biophysics of infection. Springer, Cham, pp 33–48

Jiang Z, Vasil AI, Hale JD, Hancock RE, Vasil ML, Hodges RS (2008) Effects of net charge and the number of positively charged residues on the biological activity of amphipathic α-helical cationic antimicrobial peptides. Biopolymers 90(3):369–383

Jiang Z, Vasil AI, Gera L, Vasil M, Hodges RS (2011) Rational design of a-helical antimicrobial peptides to target Gram-negative pathogens, Acinetobacter baumannii and Pseudomonas aeruginosa: utilization of charge, 'specificity determinants,' total hydrophobicity, hydrophobe type and location as design parameters to improve the therapeutic ratio. Chem Biol Drug Des 77(4):225–240

Jiang Z, Vasil AI, Vasil ML, Hodges RS (2014) "Specificity Determinants" improve therapeutic indices of two antimicrobial peptides piscidin 1 and dermaseptin s4 against the Gram-negative pathogens Acinetobacter baumannii and Pseudomonas aeruginosa. Pharmaceuticals 7(4):366–391

Jiang Z, Mant CT, Vasil M, Hodges RS (2018) Role of positively charged residues on the polar and non- polar faces of amphipathic α- helical antimicrobial peptides on specificity and selectivity for Gram-negative pathogens. Chem Biol Drug Des 91(1):75–92

Jones EM, Smart A, Bloomberg G, Burgess L, Millar MR (1994) Lactoferricin, a new antimicrobial peptide. J Appl Bacteriol 77(2):208–214

Joshi S, Dewangan RP, Yar MS, Rawat DS, Pasha S (2015) N-terminal aromatic tag induced self assembly of tryptophan–arginine rich ultra short sequences and their potent antibacterial activity. RSC Adv 5(84):68610–68620

Juretic D, Vukicevic D, Ilic N, Antcheva N, Tossi A (2009) Computational design of highly selective antimicrobial peptides. J Chem Inf Model 49(12):2873–2882

Juvvadi P, Vunnam S, Merrifield RB (1996) Synthetic melittin, its enantio, retro, and retroenantio isomers, and selected chimeric analogs: their antibacterial, hemolytic, and lipid bilayer action. J Am Chem Soc 118(38):8989–8997

Kahrom M, Bahar MM, Jangjoo A, Erfani M, Sadeghi R, Zakavi SR (2014) Poor sensitivity of 99mTc-labeled ubiquicidin scintigraphy in diagnosis of acute appendicitis. Eur Surg 46(4):173–176

Kamech N, Vukičević D, Ladram A, Piesse C, Vasseur J, Bojović V, Simunić J, Juretić D (2012) Improving the selectivity of antimicrobial peptides from anuran skin. J Chem Inf Model 52(12):3341–3351

Kaminski HM, Feix JB (2011) Effects of D-lysine substitutions on the activity and selectivity of antimicrobial peptide CM15. Polymers 3(4):2088–2106

Kang JH, Shin SY, Jang SY, Kim KL, Hahm KS (1999) Effects of tryptophan residues of porcine myeloid antibacterial peptide PMAP-23 on antibiotic activity. Biochem Biophys Res Commun 264(1):281–286

Khandelia H, Kaznessis YN (2006) Molecular dynamics investigation of the influence of anionic and zwitterionic interfaces on antimicrobial peptides' structure: implications for peptide toxicity and activity. Peptides 27(6):1192–1200

Kim S, Kim SS, Lee BJ (2005) Correlation between the activities of α-helical antimicrobial peptides and hydrophobicities represented as RP HPLC retention times. Peptides 26(11):2050–2056

Kim JK, Lee SA, Shin S, Lee JY, Jeong KW, Nan YH, Park YS, Shin SY, Kim Y (2010) Structural flexibility and the positive charges are the key factors in bacterial cell selectivity and membrane penetration of peptoid-substituted analog of Piscidin 1. Biochim Biophys Acta 1798(10):1913–1925

Kindrachuk J, Napper S (2010) Structure-activity relationships of multifunctional host defence peptides. Mini-Rev Med Chem 10(7):596–614

Koller D, Lohner K (2014) The role of spontaneous lipid curvature in the interaction of interfacially active peptides with membranes. Biochim Biophys Acta 1838(9):2250–2259

Konai MM, Samaddar S, Bocchinfuso G, Santucci V, Stella L, Haldar J (2018) Selectively targeting bacteria by tuning the molecular design of membrane-active peptidomimetic amphiphiles. Chem Commun 54(39):4943–4946

Kondejewski LH, Jelokhani-Niaraki M, Farmer SW, Lix B, Kay CM, Sykes BD, Hancock RE, Hodges RS (1999) Dissociation of antimicrobial and hemolytic activities in cyclic peptide diastereomers by systematic alterations in amphipathicity. J Biol Chem 274(19):13181–13192

Kondejewski LH, Lee DL, Jelokhani-Niaraki M, Farmer SW, Hancock REW, Hodges RS (2002) Optimization of microbial specificity in cyclic peptides by modulation of hydrophobicity within a defined structural framework. J Biol Chem 277(1):67–74

Krause E, Beyermann M, Dathe M, Rothemund S, Bienert M (1995) Location of an amphipathic. alpha-Helix in peptides using reversed-phase HPLC retention behavior of D-amino acid analogs. Anal Chem 67(2):252–258

Krugliak M, Feder R, Zolotarev VY, Gaidukov L, Dagan A, Ginsburg H, Mor A (2000) Antimalarial activities

of dermaseptin S4 derivatives. Antimicrob Agents Chemother 44(9):2442–2451
Kustanovich I, Shalev DE, Mikhlin M, Gaidukov L, Mor A (2002) Structural requirements for potent versus selective cytotoxicity for antimicrobial dermaseptin S4 derivatives. J Biol Chem 277(19):16941–16951
Ladokhin AS, White SH (2001) Protein chemistry at membrane interfaces: non-additivity of electrostatic and hydrophobic interactions. J Mol Biol 309(3):543–552
Laverty G, McLaughlin M, Shaw C, Gorman SP, Gilmore BF (2010) Antimicrobial activity of short, synthetic cationic lipopeptides. Chem Biol Drug Des 75(6):563–569
Le Joncour V, Laakkonen P (2018) Seek & destroy, use of targeting peptides for cancer detection and drug delivery. Bioorg Med Chem 26(10):2797–2806
Lee K, Shin SY, Kim K, Lim SS, Hahm KS, Kim Y (2004) Antibiotic activity and structural analysis of the scorpion-derived antimicrobial peptide IsCT and its analogs. Biochem Biophys Res Commun 323(2):712–719
Lee MT, Hung WC, Chen FY, Huang HW (2005) Many-body effect of antimicrobial peptides: on the correlation between lipid's spontaneous curvature and pore formation. Biophys J 89(6):4006–4016
Lee SA, Kim YK, Lim SS, Zhu WL, Ko H, Shin SY, Hahm KS, Kim Y (2007) Solution structure and cell selectivity of piscidin 1 and its analogues. Biochemistry 46(12):3653–3663
Lei R, Hou J, Chen Q, Yuan W, Cheng B, Sun Y, Jin Y, Ge L, Ben-Sasson SA, Chen J, Wang H, Lu W, Fang X (2018) Self-assembling myristoylated human α-defensin 5 as a next-generation nanobiotics potentiates therapeutic efficacy in bacterial infection. ACS Nano 12(6):5284–5296
Leite NB, Aufderhorst-Roberts A, Palma MS, Connell SD, Neto JR, Beales PA (2015) PE and PS lipids synergistically enhance membrane poration by a peptide with anticancer properties. Biophys J 109(5):936–947
Levison ME, Pitsakis PG, May PL, Johnson CC (1993) The bactericidal activity of magainins against Pseudomonas aeruginosa and Enterococcus faecium. J Antimicrob Chemother 32(4):577–585
Lewis RNAH, McElhaney RN (2005) The mesomorphic phase behavior of lipids. In: Yeagle PL (ed) The structure of biological membranes, 2nd edn. CRC Press, Boca Raton, pp 53–120
Li P, Zhou C, Rayatpisheh S, Ye K, Poon YF, Hammond PT, Duan H, Chan-Park MB (2012) Cationic peptidopolysaccharides show excellent broad-spectrum antimicrobial activities and high selectivity. Adv Mater 24(30):4130–4137
Lin D, Grossfield A (2015) Thermodynamics of micelle formation and membrane fusion modulate antimicrobial lipopeptide activity. Biophys J 109(4):750–759
Liu L, Xu K, Wang H, Tan PJ, Fan W, Venkatraman SS, Li L, Yang YY (2009) Self-assembled cationic peptide nanoparticles as an efficient antimicrobial agent. Nat Nanotechnol 4(7):457–463
Lorian V (2005) Antibiotics in laboratory medicine, 5th edn. Lippincott Williams & Wilkins, Philadelphia
Luo Y, McLean DT, Linden GJ, McAuley DF, McMullan R, Lundy FT (2017) The naturally occurring host defense peptide, LL-37, and its truncated mimetics KE-18 and KR-12 have selected biocidal and antibiofilm activities against *Candida albicans*, Staphylococcus aureus, and *Escherichia coli* in vitro. Front Microbiol 8:544
Lupetti A, Welling MM, Pauwels EK, Nibbering PH (2003) Radiolabelled antimicrobial peptides for infection detection. Lancet Infect Dis 3(4):223–229
Lupetti A, Van Dissel JT, Brouwer CP, Nibbering PH (2008) Human antimicrobial peptides' antifungal activity against Aspergillus fumigatus. Eur J Clin Microbiol Infect Dis 27(11):1125–1129
Lyu Y, Yang Y, Lyu X, Dong N, Shan A (2016) Antimicrobial activity, improved cell selectivity and mode of action of short PMAP-36-derived peptides against bacteria and Candida. Sci Rep 6:27258
Mahlapuu M, Håkansson J, Ringstad L, Björn C (2016) Antimicrobial peptides: an emerging category of therapeutic agents. Front Cell Infect Microbiol 6:194
Makovitzki A, Avrahami D, Shai Y (2006) Ultrashort antibacterial and antifungal lipopeptides. Proc Natl Acad Sci U S A 103(43):15997–16002
Makovitzki A, Baram J, Shai Y (2008) Antimicrobial lipopolypeptides composed of palmitoyl di-and tricationic peptides: in vitro and in vivo activities, self-assembly to nanostructures, and a plausible mode of action. Biochemistry 47(40):10630–10636
Malanovic N, Lohner K (2016) Antimicrobial peptides targeting gram-positive bacteria. Pharmaceuticals 9(3):59
Malina A, Shai Y (2005) Conjugation of fatty acids with different lengths modulates the antibacterial and antifungal activity of a cationic biologically inactive peptide. Biochem J 390(3):695–702
Mangoni ML, Shai Y (2009) Temporins and their synergism against Gram-negative bacteria and in lipopolysaccharide detoxification. Biochim Biophys Acta 1788(8):1610–1619
Mangoni ML, Carotenuto A, Auriemma L, Saviello MR, Campiglia P, Gomez-Monterrey I, Malfi S, Marcellini L, Barra D, Novellino E, Grieco P (2011) Structure-activity relationship, conformational and biological studies of Temporin L analogues. J Med Chem 54(5):1298–1307
Mannoor MS, Zhang S, Link AJ, McAlpine MC (2010) Electrical detection of pathogenic bacteria via immobilized antimicrobial peptides. Proc Natl Acad Sci U S A 107(45):19207–19212
Marsh D (1990) CRC handbook of lipid bilayers. CRC Press, Boca Raton
Matsuzaki K (2009) Control of cell selectivity of antimicrobial peptides. Biochim Biophys Acta 1788(8):1687–1692
Matsuzaki K, Harada M, Handa T, Funakoshi S, Fujii N, Yajima H, Miyajima K (1989) Magainin 1-induced leakage of entrapped calcein out of negatively-

charged lipid vesicles. Biochim Biophys Acta 981(1):130–134

Matsuzaki K, Sugishita K, Fujii N, Miyajima K (1995) Molecular basis for membrane selectivity of an antimicrobial peptide, magainin 2. Biochemistry 34(10):3423–3429

Matsuzaki K, Sugishita KI, Harada M, Fujii N, Miyajima K (1997) Interactions of an antimicrobial peptide, magainin 2, with outer and inner membranes of Gram-negative bacteria. Biochim Biophys Acta 1327(1):119–130

Matsuzaki K, Sugishita KI, Ishibe N, Ueha M, Nakata S, Miyajima K, Epand RM (1998) Relationship of membrane curvature to the formation of pores by magainin 2. Biochemistry 37(34):11856–11863

Maturana P, Martinez M, Noguera ME, Santos NC, Disalvo EA, Semorile L, Maffia PC, Hollmann A (2017) Lipid selectivity in novel antimicrobial peptides: implication on antimicrobial and hemolytic activity. Colloids Surf B Biointerfaces 153:152–159

Mazzuca C, Stella L, Venanzi M, Formaggio F, Toniolo C, Pispisa B (2005) Mechanism of membrane activity of the antibiotic trichogin GA IV: a two-state transition controlled by peptide concentration. Biophys J 88(5):3411–3421

McCloskey AP, Gilmore BF, Laverty G (2014) Evolution of antimicrobial peptides to self-assembled peptides for biomaterial applications. Pathogens 3(4):791–821

McHenry AJ, Sciacca MF, Brender JR, Ramamoorthy A (2012) Does cholesterol suppress the antimicrobial peptide induced disruption of lipid raft containing membranes? Biochim Biophys Acta 1818(12):3019–3024

McIntosh TJ, Vidal A, Simon SA (2002) The energetics of peptide-lipid interactions: modulation by interfacial dipoles and cholesterol. In: Peptide-lipid interactions, current topics in membranes, vol 52. Elsevier, Amsterdam, pp 309–338

McMahon HT, Boucrot E (2015) Membrane curvature at a glance. J Cell Sci 128(6):1065–1070

Meléndez-Alafort L, Rodríguez-Cortés J, Ferro-Flores G, De Murphy CA, Herrera-Rodríguez R, Mitsoura E, Martínez-Duncker C (2004) Biokinetics of 99mTc-UBI 29-41 in humans. Nucl Med Biol 31(3):373–379

Melo MN, Ferre R, Castanho MA (2009) Antimicrobial peptides: linking partition, activity and high membrane-bound concentrations. Nat Rev Microbiol 7(3):245–250

Mendive-Tapia L, Zhao C, Akram AR, Preciado S, Albericio F, Lee M, Serrels A, Kielland N, Read ND, Lavilla R, Vendrell M (2016) Spacer-free BODIPY fluorogens in antimicrobial peptides for direct imaging of fungal infection in human tissue. Nat Commun 7:10940

Miyazaki Y, Aoki M, Yano Y, Matsuzaki K (2012) Interaction of antimicrobial peptide magainin 2 with gangliosides as a target for human cell binding. Biochemistry 51(51):10229–10235

Mojsoska B, Zuckermann RN, Jenssen H (2015) Structure-activity relationship study of novel peptoids that mimic the structure of antimicrobial peptides. Antimicrob Agents Chemother 59(7):4112–4120

Morein S, Andersson AS, Rilfors L, Lindblom G (1996) Wild-type *Escherichia coli* cells regulate the membrane lipid composition in a window between gel and non-lamellar structures. J Biol Chem 271(12):6801–6809

Mouritsen OG, Zuckermann MJ (2004) What's so special about cholesterol? Lipids 39(11):1101–1113

Mura M, Wang J, Zhou Y, Pinna M, Zvelindovsky AV, Dennison SR, Phoenix DA (2016) The effect of amidation on the behaviour of antimicrobial peptides. Eur Biophys J 45(3):195–207

Nan YH, Bang JK, Shin SY (2009) Design of novel indolicidin-derived antimicrobial peptides with enhanced cell specificity and potent anti-inflammatory activity. Peptides 30(5):832–838

Nan YH, Lee BJ, Shin SY (2012) Prokaryotic selectivity, anti-endotoxic activity and protease stability of diastereomeric and enantiomeric analogs of human antimicrobial peptide LL-37. Bull Kor Chem Soc 33(9):2883–2889

Nicolas P (2009) Multifunctional host defense peptides: intracellular-targeting antimicrobial peptides. FEBS J 276(22):6483–6496

Nishi H, Komatsuzawa H, Fujiwara T, McCallum N, Sugai M (2004) Reduced content of lysyl-phosphatidylglycerol in the cytoplasmic membrane affects susceptibility to moenomycin, as well as vancomycin, gentamicin, and antimicrobial peptides, in *Staphylococcus aureus*. Antimicrob Agents Chemother 48(12):4800–4807

Noore J, Noore A, Li B (2012) Cationic antimicrobial peptide LL-37 is effective against both extra-and intracellular *staphylococcus aureus*. Antimicrob Agents Chemother 57(3):1283–1290

Oddo A, Hansen PR (2017) Hemolytic activity of antimicrobial peptides. In: Hansen PR (ed) Antimicrobial peptides: methods and protocols. Methods in molecular biology, vol 1548. Humana Press, New York, pp 427–435

Oh H, Hedberg M, Wade D, Edlund C (2000) Activities of synthetic hybrid peptides against anaerobic bacteria: aspects of methodology and stability. Antimicrob Agents Chemother 44(1):68–72

Oliver PM, Crooks JA, Leidl M, Yoon EJ, Saghatelian A, Weibel DB (2014) Localization of anionic phospholipids in *Escherichia coli* cells. J Bacteriol 196(19):3386–3398

Op den Kamp JAF, Redai I, van Deenen LLM (1969) Phospholipid composition of Bacillus subtilis. J Bacteriol 99(1):298–303

Oren Z, Shai Y (1997) Selective lysis of bacteria but not mammalian cells by diastereomers of melittin: structure-function study. Biochemistry 36(7):1826–1835

Oren Z, Shai Y (2000) Cyclization of a cytolytic amphipathic α-helical peptide and its diastereomer: effect on structure, interaction with model membranes, and biological function. Biochemistry 39(20):6103–6114

Oren Z, Lerman JC, Gudmundsson GH, Agerberth B, Shai Y (1999) Structure and organization of the human antimicrobial peptide LL-37 in phospholipid membranes: relevance to the molecular basis for its non-cell-selective activity. Biochem J 341(3):501–513

Orioni B, Bocchinfuso G, Kim JY, Palleschi A, Grande G, Bobone S, Park Y, Kim JI, Hahm KS, Stella L (2009) Membrane perturbation by the antimicrobial peptide PMAP-23: a fluorescence and molecular dynamics study. Biochim Biophys Acta 1788(7):1523–1533

Osborn MJ, Gander JE, Parisi E, Carson J (1972) Mechanism of assembly of the outer membrane of Salmonella typhimurium isolation and characterization of cytoplasmic and outer membrane. J Biol Chem 247(12):3962–3972

Ostovar A, Assadi M, Vahdat K, Nabipour I, Javadi H, Eftekhari M, Assadi M (2013) A pooled analysis of diagnostic value of 99mTc-ubiquicidin (UBI) scintigraphy in detection of an infectious process. Clin Nucl Med 38(6):413–416

Otvos L (2017) Racing on the wrong track. Front Chem 5:42

Otvos L Jr, Bokonyi K, Varga I, Otvos BI, Hoffmann R, Ertl HC, Wade JD, McManus AM, Craik DJ, Bulet P (2000) Insect peptides with improved protease-resistance protect mice against bacterial infection. Protein Sci 9(4):742–749

Pachón-Ibáñez ME, Smani Y, Pachón J, Sánchez-Céspedes J (2017) Perspectives for clinical use of engineered human host defense antimicrobial peptides. FEMS Microbiol Rev 41(3):323–342

Pandey BK, Ahmad A, Asthana N, Azmi S, Srivastava RM, Srivastava S, Verma R, Vishwakarma AL, Ghosh JK (2010) Cell-selective lysis by novel analogues of melittin against human red blood cells and *Escherichia coli*. Biochemistry 49(36):7920–7929

Pandey BK, Srivastava S, Singh M, Ghosh JK (2011) Inducing toxicity by introducing a leucine-zipper-like motif in frog antimicrobial peptide, magainin 2. Biochem J 436(3):609–620

Panteleev P, Bolosov IA, Balandin SV, Ovchinnikova TV (2015) Design of antimicrobial peptide arecinin analogs with improved therapeutic indices. J Pept Sci 21(2):105–113

Papo N, Oren Z, Pag U, Sahl HG, Shai Y (2002) The consequence of sequence alteration of an amphipathic α-helical antimicrobial peptide and its diastereomers. J Biol Chem 277(37):33913–33921

Park Y, Lee DG, Jang SH, Woo EH, Jeong HG, Choi CH, Hahm KS (2003) A Leu-Lys-rich antimicrobial peptide: activity and mechanism. Biochim Biophys Acta 1645(2):172–182

Pasupuleti M, Schmidtchen A, Chalupka A, Ringstad L, Malmsten M (2009) End-tagging of ultra-short antimicrobial peptides by W/F stretches to facilitate bacterial killing. PLoS One 4(4):e5285

Patel JB, Cockerill FR, Bradford PA, Eliopulos GM, Hindler JA, Jenkins SG, Lewis II JS, Limbago B, Miller LA, Nicolau DP, Powell DP, Swenson JM, Traczewski MM, Turnidge JD, Weistein MP, Zimmer BL (2012) Methods for dilution antimicrobial susceptibility tests for bacteria that grow aerobically, approved standard, 10th edn. Clinical and Laboratory Standards Institute CLSI document M07-A10

Perron GG, Zasloff M, Bell G (2006) Experimental evolution of resistance to an antimicrobial peptide. Proc Biol Sci 273(1583):251–256

Phadke SM, Islam K, Deslouches B, Kapoor SA, Stolz DB, Watkins SC, Montelaro RC, Pilewski JM, Mietzner TA (2003) Selective toxicity of engineered lentivirus lytic peptides in a CF airway cell model. Peptides 24(8):1099–1107

Phoenix DA, Harris F (2002) The hydrophobic moment and its use in the classification of amphiphilic structures. Mol Membr Biol 19(1):1–10

Phoenix DA, Dennison SR, Harris F (eds) (2012) Antimicrobial peptides. Wiley, New York

Phoenix DA, Harris F, Mura M, Dennison SR (2015) The increasing role of phosphatidylethanolamine as a lipid receptor in the action of host defence peptides. Prog Lipid Res 59:26–37

Qiao Z, Lei C, Fu Y, Li Y (2017) Rapid and sensitive detection of *E. coli* O157: H7 based on antimicrobial peptide functionalized magnetic nanoparticles and urease-catalyzed signal amplification. Anal Methods 9(35):5204–5210

Raetz CR (1986) Molecular genetics of membrane phospholipid synthesis. Annu Rev Genet 20(1):253–291

Raimondo D, Andreotti G, Saint N, Amodeo P, Renzone G, Sanseverino M, Zocchi I, Molle G, Motta A, Scaloni A (2005) A folding-dependent mechanism of antimicrobial peptide resistance to degradation unveiled by solution structure of distinction. Proc Natl Acad Sci U S A 102(18):6309–6314

Rautenbach M, Troskie AM, Vosloo JA (2016) Antifungal peptides: to be or not to be membrane active. Biochimie 130:132–145

Ravi J, Bella A, Correia AJ, Lamarre B, Ryadnov MG (2015) Supramolecular amphipathicity for probing antimicrobial propensity of host defence peptides. Phys Chem Chem Phys 17(24):15608–15614

Renner LD, Weibel DB (2011) Cardiolipin microdomains localize to negatively curved regions of *Escherichia coli* membranes. Proc Natl Acad Sci U S A 108(15):6264–6269

Rivas L, Luque-Ortega JR, Andreu D (2009) Amphibian antimicrobial peptides and Protozoa: lessons from parasites. Biochim Biophys Acta 1788(8):1570–1581

Rowlett VW, Mallampalli VK, Karlstaedt A, Dowhan W, Taegtmeyer H, Margolin W, Vitrac H (2017) The impact of membrane phospholipid alterations in *Escherichia coli* on cellular function and bacterial stress adaptation. J Bacteriol 199(13). https://doi.org/10.1128/JB.00849-16

Ruiz J, Calderon J, Rondón-Villarreal P, Torres R (2014) Analysis of structure and hemolytic activity relationships of antimicrobial peptides (AMPs). In: Advances in computational biology. Springer, Cham, pp 253–258

Russell AL, Kennedy AM, Spuches AM, Venugopal D, Bhonsle JB, Hicks RP (2010) Spectroscopic and ther-

modynamic evidence for antimicrobial peptide membrane selectivity. Chem Phys Lipids 163(6):488–497
Saeed S, Zafar J, Khan B, Akhtar A, Qurieshi S, Fatima S, Ahmad N, Irfanullah J (2013) Utility of 99mTc-labelled antimicrobial peptide ubiquicidin (29-41) in the diagnosis of diabetic foot infection. Eur J Nucl Med Mol Imaging 40(5):737–743
Salick DA, Kretsinger JK, Pochan DJ, Schneider JP (2007) Inherent antibacterial activity of a peptide-based β-hairpin hydrogel. J Am Chem Soc 129(47):14793–14799
Sal-Man N, Oren Z, Shai Y (2002) Preassembly of membrane-active peptides is an important factor in their selectivity toward target cells. Biochemistry 41(39):11921–11930
Santos NC, Prieto M, Castanho MA (2003) Quantifying molecular partition into model systems of biomembranes: an emphasis on optical spectroscopic methods. Biochim Biophys Acta 1612(2):123–135
Savini F, Luca V, Bocedi A, Massoud R, Park Y, Mangoni ML, Stella L (2017) Cell-density dependence of host-defense peptide activity and selectivity in the presence of host cells. ACS Chem Biol 12(1):52–56
Savini F, Bobone S, Roversi D, Mangoni ML, Stella L (2018) From liposomes to cells: filling the gap between physicochemical and microbiological studies of the activity and selectivity of host-defense peptides. Pept Sci 110(5):e24041
Schmidtchen A, Pasupuleti M, M&oumlrgelin M, Davoudi M, Alenfall J, Chalupka A, Malmsten M (2009) Boosting antimicrobial peptides by hydrophobic oligopeptide end-tags. J Biol Chem 284(26):17584–17594
Schmidtchen A, Ringstad L, Kasetty G, Mizuno H, Rutland MW, Malmsten M (2011) Membrane selectivity by W-tagging of antimicrobial peptides. Biochim Biophys Acta 1808(4):1081–1091
Schmidtchen A, Pasupuleti M, Malmsten M (2014) Effect of hydrophobic modifications in antimicrobial peptides. Adv Colloid Interf Sci 205:265–274
Schröder-Borm H, Willumeit R, Brandenburg K, Andrä J (2003) Molecular basis for membrane selectivity of NK-2, a potent peptide antibiotic derived from NK-lysin. Biochim Biophys Acta 1612(2):164–171
Schweizer F (2009) Cationic amphiphilic peptides with cancer-selective toxicity. Eur J Pharmacol 625(1–3):190–194
Seelig J (2004) Thermodynamics of lipid–peptide interactions. Biochim Biophys Acta 1666(1):40–50
Seo MD, Won HS, Kim JH, Mishig-Ochir T, Lee BJ (2012) Antimicrobial peptides for therapeutic applications: a review. Molecules 17(10):12276–12286
Shai Y, Oren Z (1996) Diastereomers of cytolysins, a novel class of potent antibacterial peptides. J Biol Chem 271(13):7305–7308
Shai Y, Oren Z (2001) From "carpet" mechanism to de-novo designed diastereomeric cell-selective antimicrobial peptides. Peptides 22(10):1629–1641
Shankar SS, Benke SN, Nagendra N, Srivastava PL, Thulasiram HV, Gopi HN (2013) Self-assembly to function: design, synthesis, and broad spectrum antimicrobial properties of short hybrid E-vinylogous lipopeptides. J Med Chem 56(21):8468–8474
Shi X, Zhang X, Yao Q, He F (2017) A novel method for the rapid detection of microbes in blood using pleurocidin antimicrobial peptide functionalized piezoelectric sensor. J Microbiol Methods 133:69–75
Shin SY, Yang ST, Park EJ, Eom SH, Song WK, Kim JI, Lee SH, Lee MK, Lee DG, Hahm KS, Kim Y (2001) Antibacterial, antitumor and hemolytic activities of α-helical antibiotic peptide, P18 and its analogs. J Pept Res 58(6):504–514
Shriver-Lake LC, North SH, Dean SN, Taitt CR (2012) Antimicrobial peptides for detection and diagnostic assays. In: Designing receptors for the next generation of biosensors. Springer, Berlin, pp 85–104
Silva RR, Avelino KY, Ribeiro KL, Franco OL, Oliveira MD, Andrade CA (2014) Optical and dielectric sensors based on antimicrobial peptides for microorganism diagnosis. Front Microbiol 5:443. https://doi.org/10.3389/fmicb.2014.00443
Simon SA, McIntosh TJ (2002) Peptide-lipid interactions, current topics in membranes, vol 52. Elsevier, Amsterdam
Skerlavaj B, Renato Gennaro R, Luigi Bagella L, Laura Merluzzi L, Angela Risso A, Zanetti M (1996) Biological characterization of two novel cathelicidin-derived peptides and identification of structural requirements for their antimicrobial and cell lytic activities. J Biol Chem 271(45):28375–28381
Slaninová J, Mlsová V, Kroupová H, Alán L, Tůmová T, Monincová L, Borovičková L, Fučík V, Ceřovský V (2012) Toxicity study of antimicrobial peptides from wild bee venom and their analogs toward mammalian normal and cancer cells. Peptides 33(1):18–26
Snoussi M, Talledo JP, Del Rosario NA, Ha BY, Kosmrlj A, Taheri-Araghi S (2018) Heterogeneous absorption of antimicrobial peptide LL37 in *Escherichia coli* cells enhances population survivability. eLife 7:e38174
Son M, Lee Y, Hwang H, Hyun S, Yu J (2013) Disruption of interactions between hydrophobic residues on nonpolar faces is a key determinant in decreasing hemolysis and increasing antimicrobial activities of α-helical amphipathic peptides. ChemMedChem 8(10):1638–1642
Song YM, Yang ST, Lim SS, Kim Y, Hahm KS, Kim JI, Shin SY (2004) Effects of L-or D-Pro incorporation into hydrophobic or hydrophilic helix face of amphipathic α-helical model peptide on structure and cell selectivity. Biochem Biophys Res Commun 314(2):615–621
Song YM, Park Y, Lim SS, Yang ST, Woo ER, Park IS, Lee JS, Kim JI, Hahm KS, Kim Y, Shin SY (2005) Cell selectivity and mechanism of action of antimicrobial model peptides containing peptoid residues. Biochemistry 44(36):12094–12106
Sood R, Kinnunen PK (2008) Cholesterol, lanosterol, and ergosterol attenuate the membrane association of LL-37 (W27F) and temporin L. Biochim Biophys Acta 1778(6):1460–1466

Sood R, Domanov Y, Pietiäinen M, Kontinen VP, Kinnunen PK (2008) Binding of LL-37 to model biomembranes: insight into target vs host cell recognition. Biochim Biophys Acta 1778(4):983–996

Stark M, Liu LP, Deber CM (2002) Cationic hydrophobic peptides with antimicrobial activity. Antimicrob Agents Chemother 46(11):3585–3590

Starr CG, He J, Wimley WC (2016) Host cell interactions are a significant barrier to the clinical utility of peptide antibiotics. ACS Chem Biol 11(12):3391–3399

Steiner H, Andreu D, Merrifield RB (1988) Binding and action of cecropin and cecropin analogues: antibacterial peptides from insects. Biochim Biophys Acta 939(2):260–266

Stella L, Mazzuca C, Venanzi M, Palleschi A, Didone M, Formaggio F, Toniolo C, Pispisa B (2004) Aggregation and water-membrane partition as major determinants of the activity of the antibiotic peptide trichogin GA IV. Biophys J 86(2):936–945

Storch J, Kleinfeld AM (1985) The lipid structure of biological membranes. Trends Biochem Sci 10(11):418–421

Strandberg E, Tiltak D, Ieronimo M, Kanithasen N, Wadhwani P, Ulrich AS (2007) Influence of C-terminal amidation on the antimicrobial and hemolytic activities of cationic -helical peptides. Pure Appl Chem 79(4):717–728

Strömstedt AA, Ringstad L, Schmidtchen A, Malmsten M (2010) Interaction between amphiphilic peptides and phospholipid membranes. Curr Opin Colloid Interface Sci 15(6):467–478

Swierstra J, Kapoerchan V, Knijnenburg A, van Belkum A, Overhand M (2016) Structure, toxicity and antibiotic activity of gramicidin S and derivatives. Eur J Clin Microbiol Infect Dis 35(5):763–769

Tachi T, Epand RF, Epand RM, Matsuzaki K (2002) Position-dependent hydrophobicity of the antimicrobial magainin peptide affects the mode of peptide-lipid interactions and selective toxicity. Biochemistry 41(34):10723–10731

Takahashi D, Shukla SK, Prakash O, Zhang G (2010) Structural determinants of host defense peptides for antimicrobial activity and target cell selectivity. Biochimie 92(9):1236–1241

Teixeira V, Feio MJ, Bastos M (2012) Role of lipids in the interaction of antimicrobial peptides with membranes. Prog Lipid Res 51(2):149–177

Thennarasu S, Nagaraj R (1996) Specific antimicrobial and hemolytic activities of 18-residue peptides derived from the amino terminal region of the toxin pardaxin. Protein Eng Des Sel 9(12):1219–1224

Tian X, Sun F, Zhou XR, Luo SZ, Chen L (2015) Role of peptide self-assembly in antimicrobial peptides. J Pept Sci 21(7):530–539

Tiozzo E, Rocco G, Tossi A, Romeo D (1998) Wide-spectrum antibiotic activity of synthetic, amphipathic peptides. Biochem Biophys Res Commun 249(1):202–206

Toniolo C, Crisma M, Formaggio F, Peggion C, Monaco V, Goulard C, Rebuffat S, Bodo B (1996) Effect of N α-acyl chain length on the membrane-modifying properties of synthetic analogs of the lipopeptaibol trichogin GA IV. J Am Chem Soc 118(21):4952–4958

Tossi A (2011) Design and engineering strategies for synthetic antimicrobial peptides. In: Prokaryotic antimicrobial peptides. Springer, New York, pp 81–98

Tossi A, Sandri L, Giangaspero A (2000) Amphipathic, α-helical antimicrobial peptides. Biopolymers 55(1):4–30

Tu Z, Hao J, Kharidia R, Meng XG, Liang JF (2007) Improved stability and selectivity of lytic peptides through self-assembly. Biochem Biophys Res Commun 361(3):712–717

Tytler EM, Anantharamaiah GM, Walker DE, Mishra VK, Palgunachari MN, Segrest JP (1995) Molecular basis for prokaryotic specificity of magainin-induced lysis. Biochemistry 34(13):4393–4401

Uematsu N, Matsuzaki K (2000) Polar angle as a determinant of amphipathic α-helix-lipid interactions: a model peptide study. Biophys J 79(4):2075–2083

Uggerhøj LE, Poulsen TJ, Munk JK, Fredborg M, Sondergaard TE, Frimodt-Moller N, Hansen PR, Wimmer R (2015) Rational design of alpha-helical antimicrobial peptides: do's and don'ts. ChemBioChem 16(2):242–253

Vallejo E, Martinez I, Tejero A, Hernandez S, Jimenez L, Bialostozky D, Sanchez G, Ilarraza H, Ferro-Flores G (2008) Clinical utility of 99mTc-labeled ubiquicidin 29–41 antimicrobial peptide for the scintigraphic detection of mediastinitis after cardiac surgery. Arch Med Res 39(8):768–774

van der Weerden NL, Bleackley MR, Anderson MA (2013) Properties and mechanisms of action of naturally occurring antifungal peptides. Cell Mol Life Sci 70(19):3545–3570

Veiga AS, Sinthuvanich C, Gaspar D, Franquelim HG, Castanho MA, Schneider JP (2012) Arginine-rich self-assembling peptides as potent antibacterial gels. Biomaterials 33(35):8907–8916

Velduhizen EJA, Scheenstra MR, Tjeerdsma-van Bokhoven JLM, Coorens M, Schneider VAF, Bikker FJ, van Dijk A, Haagsman HP (2017) Antimicrobial and immunomodulatory activity of PMAP-23 derived peptides. Protein Pept Lett 24(7):609–616

Verkleij AJ, Zwaal RFA, Roelofsen B, Comfurius P, Kastelijn D, van Deenen LLM (1973) The asymmetric distribution of phospholipids in the human red cell membrane. A combined study using phospholipases and freeze-etch electron microscopy. Biochim Biophys Acta 323(2):178–193

Verly RM, Rodrigues MA, Daghastanli KRP, Denadai AML, Cuccovia IM, Bloch C Jr, Frezard F, Santoro MM, Pilo-Veloso D, Bemquerer MP (2008) Effect of cholesterol on the interaction of the amphibian antimicrobial peptide DD K with liposomes. Peptides 29(1):15–24

Vermeer LS, Lan Y, Abbate V, Ruh E, Bui TT, Wilkinson L, Jumagulova E, Kozlowska J, Patel J, McIntyre CA, Yam WC, Siu GKH, Atkinson RA, Lam JKW,

Bansal SS, Drake AF, Mitchell GH, Mason AJ (2012) Conformational flexibility determines selectivity and Antibacterial, Antiplasmodium, and Anticancer potency of cationic α-helical peptides. J Biol Chem 287(41):34120–34133

Virtanen JA, Cheng KH, Somerharju P (1998) Phospholipid composition of the mammalian red cell membrane can be rationalized by a superlattice model. Proc Natl Acad Sci U S A 95(9):4964–4969

Wade D, Boman A, Wåhlin B, Drain CM, Andreu D, Boman HG, Merrifield RB (1990) All-D amino acid-containing channel-forming antibiotic peptides. Proc Natl Acad Sci U S A 87(12):4761–4765

Wade D, Silberring J, Soliymani R, Heikkinen S, Kilpeläinen I, Lankinen H, Kuusela P (2000) Antibacterial activities of temporin A analogs. FEBS Lett 479(1–2):6–9

Wakabayashi H, Matsumoto H, Hashimoto K, Teraguchi S, Takase M, Hayasawa H (1999) N-Acylated and D enantiomer derivatives of a nonamer core peptide of lactoferricin B showing improved antimicrobial activity. Antimicrob Agents Chemother 43(5):1267–1269

Wang G (ed) (2017) Antimicrobial peptides: discovery, design and novel therapeutic strategies. CABI Publishing, Cambridge, MA

Wang H, Xu K, Liu L, Tan JP, Chen Y, Li Y, Fan W, Wei Z, Sheng J, Yang YY, Li L (2010) The efficacy of self-assembled cationic antimicrobial peptide nanoparticles against *Cryptococcus neoformans* for the treatment of meningitis. Biomaterials 31(10):2874–2881

Wang J, Chou S, Xu L, Zhu X, Dong N, Shan A, Chen Z (2015) High specific selectivity and membrane-active mechanism of the synthetic centrosymmetric α-helical peptides with Gly-Gly pairs. Sci Rep 5:15963

Wang J, Chou S, Yang Z, Yang Y, Wang Z, Song J, Dou X, Shan A (2018) Combating drug-resistant fungi with novel imperfectly amphipathic palindromic peptides. J Med Chem 61(9):3889–3907

Welling MM, Nibbering PH, Paulusma-Annema A, Hiemstra PS, Pauwels EK, Calame W (1999) Imaging of bacterial infections with ^{99m}Tc-labeled human neutrophil peptide-1. J Nucl Med 40(12):2073–2080

Welling MM, Paulusma-Annema A, Balter HS, Pauwels EK, Nibbering PH (2000) Technetium-99m labelled antimicrobial peptides discriminate between bacterial infections and sterile inflammations. Eur J Nucl Med 27(3):292–301

White DA (1973) The phospholipid composition of mammalian tissues. In: Ansell GB, Hawthorne JN, Dawson RMC (eds) Form and function of phospholipids. Elsevier, Amsterdam, pp 441–482

White SH, Wimley WC (1999) Membrane protein folding and stability: physical principles. Annu Rev Biophys Biomol Struct 28(1):319–365

Wiegand I, Hilpert K, Hancock REW (2008) Agar and broth dilution methods to determine the minimal inhibitory concentration (MIC) of antimicrobial substances. Nat Protoc 3(2):163–175

Wieprecht T, Seelig J (2002) Isothermal titration calorimetry for studying interactions between peptides and lipid membranes. Curr Top Membr 52:31–56

Wieprecht T, Dathe M, Beyermann M, Krause E, Maloy WL, MacDonald DL, Bienert M (1997a) Peptide hydrophobicity controls the activity and selectivity of magainin 2 amide in interaction with membranes. Biochemistry 36(20):6124–6132

Wieprecht T, Dathe M, Krause E, Beyermann M, Maloy WL, MacDonald DL, Bienert M (1997b) Modulation of membrane activity of amphipathic, antibacterial peptides by slight modifications of the hydrophobic moment. FEBS Lett 417(1):135–140

Wieprecht T, Dathe M, Epand RM, Beyermann M, Krause E, Maloy WL, MacDonald DL, Bienert M (1997c) Influence of the angle subtended by the positively charged helix face on the membrane activity of amphipathic, antibacterial peptides. Biochemistry 36(42):12869–12880

Wimley WC (2010a) Energetics of peptide and protein binding to lipid membranes. In: Proteins membrane binding and pore formation. Springer, New York, pp 14–23

Wimley WC (2010b) Describing the mechanism of antimicrobial peptide action with the interfacial activity model. ACS Chem Biol 5(10):905–917

Wimley WC, Hristova K (2011) Antimicrobial peptides: successes, challenges and unanswered questions. J Membr Biol 239(1–2):27–34

Wu G, Wu H, Fan X, Zhao R, Li X, Wang S, Ma Y, Shen Z, Xi T (2010) Selective toxicity of antimicrobial peptide S-thanatin on bacteria. Peptides 31(9):1669–1673

Yang ST, Shin SY, Kim YC, Kim Y, Hahm KS, Kim JI (2002) Conformation-dependent antibiotic activity of tritrpticin, a cathelicidin-derived antimicrobial peptide. Biochem Biophys Res Commun 296(5):1044–1050

Yang ST, Lee JY, Kim HJ, Eu YJ, Shin SY, Hahm KS, Kim JI (2006a) Contribution of a central proline in model amphipathic α-helical peptides to self-association, interaction with phospholipids, and antimicrobial mode of action. FEBS J 273(17):4040–4054

Yang ST, Jeon JH, Kim Y, Shin SY, Hahm KS, Kim JI (2006b) Possible role of a PXXP central hinge in the antibacterial activity and membrane interaction of PMAP-23, a member of cathelicidin family. Biochemistry 45(6):1775–1784

Yau WM, Wimley WC, Gawrisch K, White SH (1998) The preference of tryptophan for membrane interfaces. Biochemistry 37(42):14713–14718

Yeaman MR, Yount NY (2003) Mechanisms of antimicrobial peptide action and resistance. Pharmacol Rev 55(1):27–55

Yeung AT, Gellatly SL, Hancock REW (2011) Multifunctional cationic host defence peptides and their clinical applications. Cell Mol Life Sci 68(13):2161

Zelezetsky I, Tossi A (2006) Alpha-helical antimicrobial peptides-using a sequence template to guide structure–activity relationship studies. Biochim Biophys Acta 1758(9):1436–1449

Zelezetsky I, Pacor S, Pag U, Papo N, Shai Y, Sahl HG, Tossi A (2005) Controlled alteration of the shape and conformational stability of α-helical cell-lytic pep-

tides: effect on mode of action and cell specificity. Biochem J 390(1):177–188
Zhang L, Benz R, Hancock REW (1999) Influence of proline residues on the antibacterial and synergistic activities of R-helical peptides. Biochemistry 38(25):8102–8111
Zhang Y, Lu H, Lin Y, Cheng J (2011) Water-soluble polypeptides with elongated, charged side chains adopt ultrastable helical conformations. Macromolecules 44(17):6641–6644
Zhang SK, Song JW, Gong F, Li SB, Chang HY, Xie HM, Gao HW, Tan YX, Ji SP (2016) Design of an α-helical antimicrobial peptide with improved cell-selective and potent anti-biofilm activity. Sci Rep 6:27394
Zhou NE, Mant CT, Hodges RS (1990) Effect of preferred binding domains on peptide retention behavior in reversed-phase chromatography: amphipathic alpha-helices. Pept Res 3(1):8–20
Zhu WL, Hahm K, Shin SY (2007a) Cathelicidin-derived Trp/Pro-rich antimicrobial peptides with lysine peptoid residue (Nlys): therapeutic index and plausible mode of action. J Pept Sci 13(8):529–535
Zhu WL, Song YM, Park Y, Park KH, Yang ST, Kim JI, Park IS, Hahm KS, Shin SY (2007b) Substitution of the leucine zipper sequence in melittin with peptoid residues affects self-association, cell selectivity, and mode of action. Biochim Biophys Acta 1768(6):1506–1517
Zhu WL, Nan YH, Hahm K, Shin SY (2007c) Cell selectivity of an antimicrobial peptide melittin diastereomer with D-amino acid in the leucine zipper sequence. J Biochem Mol Biol 40(6):1090–1094
Zhu WL, Hahm KS, Shin SY (2009) Cell selectivity and mechanism of action of short antimicrobial peptides designed from the cell-penetrating peptide Pep-1. J Pept Sci 15(9):569–575
Zou R, Zhu X, Tu Y, Wu J, Landry MP (2018) Activity of antimicrobial peptide aggregates decreases with increased cell membrane embedding free energy cost. Biochemistry 57(18):2606–2610
Zwaal RFA, Roelofsen B, Colley CM (1973) Localization of red cell membrane constituents. Biochim Biophys Acta 300(2):159–182
Zwaal RFA, Roelsfsen B, Comfurius P, van Deenen LLM (1975) Organization of phospholipids in human red cell membranes as detected by the action of various purified phospholipases. Biochim Biophys Acta 406(1):83–96

Design of Antimicrobial Peptides: Progress Made with Human Cathelicidin LL-37

12

Guangshun Wang, Jayaram Lakshmaiah Narayana, Biswajit Mishra, Yingxia Zhang, Fangyu Wang, Chunfeng Wang, D. Zarena, Tamara Lushnikova, and Xiuqing Wang

Abstract

The incorporation of the innate immune system into humans is essential for survival and health due to the rapid replication of invading microbes and the delayed action of the adaptive immune system. Antimicrobial peptides are important components of human innate immunity. Over 100 such peptides have been identified in various human tissues. Human cathelicidin LL-37 is best studied, and there has been a growing interest in designing new peptides based on LL-37. This chapter describes the alternative processing of the human cathelicidin precursor, protease digestion, and lab cutting of LL-37. Both a synthetic peptide library and structure-based design are utilized to identify the active regions. Although challenging, the determination of the 3D structure of LL-37 enabled the identification of the core antimicrobial region. The minimal region of LL-37 can be function-dependent. We discuss the design and potential applications of LL-37 into antibacterial, antibiofilm, antiviral, antifungal, immune modulating, and anticancer peptides. LL-37 has been engineered into 17BIPHE2, a stable, selective, and potent antimicrobial, antibiofilm, and anticancer peptide. Both 17BIPHE2 and SAAP-148 can eliminate the ESKAPE pathogens and show topical in vivo antibiofilm efficacy. Also discussed are other application strategies, including peptide formulation, antimicrobial implants, and peptide-inducing factors such as vitamin D and sunlight. Finally, we summarize what we learned from peptide design based on human LL-37.

Keywords

Anticancer peptides · Antimicrobial peptides · Antiviral peptides · Cathelicidins · LL-37 · Peptide design

12.1 Introduction

The discovery of penicillin in 1928 is one of the greatest achievements in human history. However, overreliance on antimicrobial drugs against infectious diseases led to a rapid development of resistance in bacteria. Antimicrobial resistance posts a great risk to the human health-care system. To counteract the impacts of emerging multidrug resistance, alternatives are in urgent need. Antimicrobial peptides (AMPs) have been recognized as one such alternative because bacterial resistance development is rare or not yet observed

G. Wang (✉) · J. L. Narayana · B. Mishra · Y. Zhang
F. Wang · C. Wang · D. Zarena · T. Lushnikova
X. Wang
Department of Pathology and Microbiology, College of Medicine, University of Nebraska Medical Center, Omaha, NE, USA
e-mail: gwang@unmc.edu

K. Matsuzaki (ed.), *Antimicrobial Peptides*, Advances in Experimental Medicine and Biology 1117,
https://doi.org/10.1007/978-981-13-3588-4_12

(Zasloff 2002; Boman 2003; Jayaram and Chen 2015; Wang et al. 2015; Mishra et al. 2017b). They are ancient molecules optimized through their coevolution with bacteria over millions of years (Peschel and Sahl 2006; Spencer et al. 2014).

According to the antimicrobial peptide database (APD; http://aps.unmc.edu/AP), naturally occurring AMPs (<100 amino acids) have been discovered in the six life kingdoms, including bacteria, archaea, protists, fungi, plants, and animals (Wang and Wang 2004; Wang et al. 2009, 2016). In invertebrates, innate immune systems are the only defense weapon against microbial infection. The discovery of cecropins, magainins, and defensins in the 1980s laid the foundation for deciphering the pathogen-specific innate immune response pathways (Boman 2003; Lehrer and Lu 2012; Meister et al. 1997; Zasloff 1987). In mammals, several defense mechanisms guard against the threat of infection, ranging from the innate to adaptive immune systems, including skin barriers and physical factors such as urine flow, pH, and ionic composition. In humans, 124 AMPs were identified in various tissues as of June 2018 (Wang et al. 2016). Typical examples include lysozyme, defensins, histatins, cathelicidins, lactoferricin, kinocidins, ribonuclease, and dermcidin (reviewed in Wang 2014). On average, these molecules have a chain length of 44.6 amino acids with a net charge of +4.8. Such properties allow them to adopt unique structures for host defense.

Unlike other animals, there is only one cathelicidin gene in humans encoded on chromosome 3p21.3 (Frohm et al. 1997). Cathelicidins are synthesized as preproproteins with a highly conserved N-terminal domain and a highly variable C-terminal antimicrobial domain. The N-terminal domain usually consists of 94–114 amino acids and shares sequence homology with cathelin, a cysteine protease inhibitor derived from porcine neutrophils, hence the name cathelin-like domain (CLD). At present, there are 113 mature cathelicidin peptides in the APD, ranging from hagfish to humans (Wang et al. 2016). Human cathelicidin LL-37 is one of the best-studied AMPs. It is widely distributed in the human saliva, sperm, skin, gastrointestine, urinary tract, and respiratory airways. LL-37 is expressed by a number of cells, including monocytes, neutrophils, mast cells, stem cells, NK cells, and B and T cells. As an innate immune peptide, it is upregulated upon pathogen invasion or by immune stimuli such as vitamin D and sunlight. LL-37 has a broad-spectrum activity at a micromolar concentration against bacteria, fungi, viruses, and fungi at least in vitro (Durr et al. 2006; Vandamme et al. 2012). The observation that the concentration of LL-37 in certain human tissues is below the minimal inhibitory concentration (MIC) required to kill pathogens, however, led to an emphasis on immune modulation (Scott et al. 2002). It is now accepted that human cathelicidin LL-37 is a moonlighting peptide with multiple functional roles, ranging from antimicrobial to immune regulation. The moonlighting properties of LL-37 provide a basis for its involvement in a variety of human diseases such as infection, diabetes, cancer metastasis, and atherosclerosis (Scott et al. 2002; Vandamme et al. 2012).

This chapter summarizes the discovery, design, and potential applications of new AMPs based on the LL-37 template. Both structure-based design and library screening are covered. It is useful to mention the challenges in structural determination of LL-37 by NMR spectroscopy because of the importance of this structure for a better understanding of the antimicrobial activity of the peptide as well as for structure-based peptide discovery. To facilitate a connection between LL-37 and its fragments, we use the nomenclature and numbering of LL-37 throughout the text. For example, KR-12 (Table 12.1) means a 12-residue peptide corresponding to residues 18–29 of LL-37; it starts with amino acids K18R19 at the N-terminus and ends with R29 at the C-terminus. Subsequent sections are devoted to the potential applications of these peptides in antibiofilm, antiviral, antifungal, anticancer, surface immobilization, and immune modulation studies with a focus on the relationship between intact LL-37 and its fragments. It seems a slightly different region can be utilized for peptide design depending on the targeted pathogen or disease, including laboratory preference. Some LL-37 designer peptides have been shown to have antibiofilm efficacy in animal models.

Table 12.1 Amino acid sequences and physical properties of select peptides derived from LL-37

Name	Amino acid sequence	Net charge	Pho%	HP	Boman index	GRAVY
LL-37	^{1}LLGDFFRKSKEKIGKEFKRIVQRIKDFLRNLVPRTES^{37}	+6	35%	47.79	2.99	−0.72
LL-31	^{1}LLGDFFRKSKEKIGKEFKRIVQRIKDFLRNL^{31}	+6	38%	50.98	2.81	−0.64
RK-31	^{7}RKSKEKIGKEFKRIVQRIKDFLRNLVPRTES^{37}	+7	31%	38.23	3.83	−1.16
KS-30	^{8}KSKEKIGKEFKRIVQRIKDFLRNLVPRTES^{37}	+6	30%	37.39	3.47	−1.05
LL-25	^{1}LLGDFFRKSKEKIGKEFKRIVQRIK^{25}	+6	36%	38.4	2.79	−0.75
RK-25	^{7}RKSKEKIGKEFKRIVQRIKDFLRNL^{31}	+7	32%	38.69	3.81	−1.17
IG-25	^{13}IGKEFKRIVQRIKDFLRNLVPRTES^{37}	+4	36%	39.37	3.08	−0.62
IG-24 (P60)	^{13}IGKEFKRIVQRIKDFLRNLVPRTE^{36}	+4	37%	40.22	3.07	−0.62
OP-145	Acetyl^{13}IGKEFKRIVERIKRFLRELVRPLR^{36}	+6	41%	45.31	3.42	−0.51
LL-23	^{1}LLGDFFRKSKEKIGKEFKRIVQR^{23}	+5	34%	38.61	3.01	−0.84
KR-22	^{12}KIGKEFKRIVQRIKDFLRNLVP^{33}	+5	40%	41.18	2.5	−0.45
GKE-21	^{14}GKEFKRIVQRIKDFLRNLVPR^{34}	+5	38%	40.2	3.3	−0.72
FK-21	^{17}FKRIVQRIKDFLRNLVPRTES^{37}	+4	38%	38.67	3.36	−0.59
KR-20	^{18}KRIVQRIKDFLRNLVPRTES^{37}	+4	35%	37.88	3.68	−0.76
GI-20	13GIKEFKRIVQRIKDFLRNLV^{32}	+4	45%	45.07	2.47	−0.225
IG-19	^{13}IGKEFKRIVQRIKDFLRNL^{31}	+4	42%	41.61	2.82	−0.46
GF-17	G^{17}FKRIVQRIKDFLRNLV^{32}	+5	47%	41.28	2.47	−0.094
FK-16	^{17}FKRIVQRIKDFLRNLV^{32}	+5	50%	40.37	2.69	−0.075
FK-13	^{17}FKRIVQRIKDFLR^{29}	+5	46%	35.69	3.48	−0.44
FK-12	^{17}FKRIVQRIKDFL^{28}	+4	50%	33.54	2.53	−0.1
KR-12	^{18}KRIVQRIKDFLR^{29}	+5	41%	32.11	4.02	−0.71
LL-12	*^{1}LLGDFFRKSKEK^{12}*	*+2*	*33%*	*26.35*	*2.81*	*−0.93*
SK-12	*^{9}SKEKIGKEFKRI^{20}*	*+3*	*25%*	*19.97*	*3.36*	*−1.38*
DF-12	*^{26}DFLRNLVPRTES^{37}*	*0*	*33%*	*26.96*	*3.42*	*−0.67*
RI-10	*^{19}RIVQRIKDFL^{28}*	*+2*	*50%*	*31.1*	*2.78*	*−0.01*

Calculated based on the APD website (http://aps.unmc.edu/AP/prediction/prediciton_main.php). Pho%, hydrophobic content; HP, hydrophobicity calculated using the Fisher peptide-analyzing tool website https://www.thermofisher.com/us/en/home/life-science/protein-biology/peptides-proteins/custom-peptide-synthesis-services/peptide-analyzing-tool.html; Boman index, a term renamed in the APD database based on protein-binding potential (Boman 2003); GRAVY, grand average of hydropathy, which is calculated by summing the hydropathy value of each residue and divided by the length of the peptide (Kyte and Doolittle 1982); underlined residues are either artificial or order swapped

12.2 Processing of the Human Cathelicidin Gene Product

12.2.1 Mature Peptide LL-37

Human cathelicidin is expressed as an 18 kDa precursor protein (hCAP-18). It was discovered in 1995 by three laboratories based on the highly conserved "cathelin" domain (Cowland et al. 1995; Larrick et al. 1995; Agerberth et al. 1995). One of the groups predicted the mature form as FALL-39 by comparison with the pig cathelicidin PR-39, a 39-residue peptide rich in amino acids P and R (Agerberth et al., 1995). The mature form was established as a 37-residue peptide LL-37 after its isolation from granulocytes (Gudmundsson et al. 1996). In neutrophils, hCAP-18 is processed to the antimicrobial peptide LL-37 (a 37-residue peptide starting with two leucines) by extracellular cleavage with proteinase 3 (Sorensen et al. 2001). Thus, the LL-37 pathway is composed of the precursor hCAP-18, the cathelin-like domain, LL-37, and its further cleaved fragments under natural conditions. Interestingly, the precursor of human LL-37 can also be cleaved into alternative forms below.

12.2.2 ALL-38 from the Human Reproductive System

AMPs appear to play a role in sperm fertilization. During sexual intercourse, hCAP-18, along with sperm, is injected into the vagina where it is cleaved into ALL-38 under acidic conditions by gastricsin. ALL-38, with one more alanine at the N-terminus than LL-37, showed similar antibacterial activity (Sorensen et al. 2003).

12.2.3 An Uncharacterized Alternative Form from Fat Cells

A recent exciting discovery is that fat cells also participate in host defense. Adipocytes can release a mature peptide longer than LL-37 to protect against the *S. aureus* infection (Zhang et al. 2015). However, the exact peptide sequence has not been elucidated.

12.2.4 TLN-58 from a Diseased Skin State

It is fascinating that hCAP-18 can also be cleaved upstream of the LL-37 sequence into another longer mature peptide with 58 amino acids (Murakami et al. 2017). TLN-58 is isolated from the lesion vesicle of palmoplantar pustulosis, a diseased skin condition on the palms and soles. Similar to human LL-37, this peptide form can upregulate IL-17C, IL-8, IL-23, IL-1α, and IL-1β mRNA and protein expression in normal human keratinocytes.

These findings provide additional evidence for the idea of one cathelicidin gene and multiple peptides (Wang 2014). It is likely that the single cathelicidin gene can be translated and processed in other manners yet to be discovered. Once a link is established between the cathelicidin form and disease, our detection of a particular form of human cathelicidin may serve as a biomarker for that disease.

12.3 Physical and Structural Basis of LL-37 in Targeting Bacterial Membranes

12.3.1 Bacterial Recognition via Electrostatic Interactions

Cationic LL-37 is active against a broad range of Gram-positive and Gram-negative pathogens (updated in Fig. 12.1). It is proposed that this linear peptide, like other cationic AMPs, targets anionic bacterial membranes via the carpet or toroidal pore model (Oren et al. 1999; Henzler et al. 2003; Lee et al. 2011). The membrane targeting of LL-37 is determined by its sequence property. According to the APD (Wang and Wang 2004), LL-37 (M. Wt. 4493.312 and molecular formula $C_{205}H_{341}N_{59}O_{53}$) has a net charge of +6 (i.e., sum of 6 Lys, 5 Arg, 2 Asp, and 3 Glu),

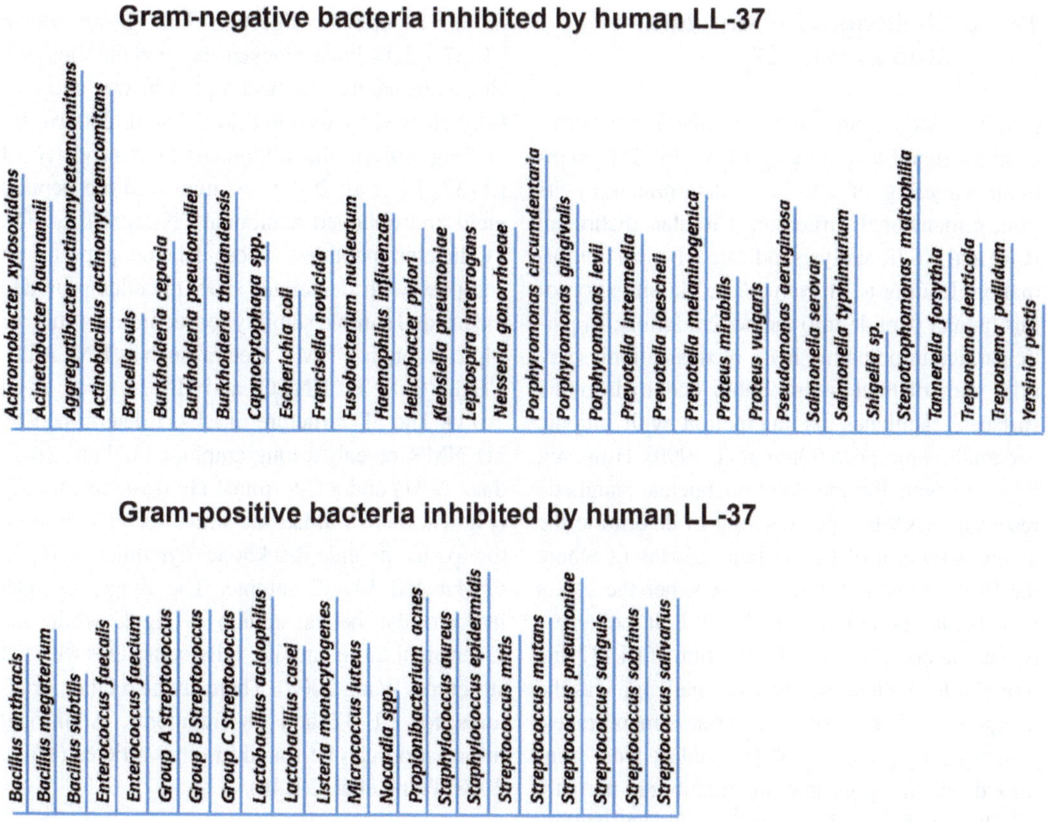

Fig. 12.1 Antibacterial activity of human antimicrobial peptide LL-37. Updated based on Vandamme et al. (2012) and Durr et al. (2006), the antimicrobial peptide database (http://aps.unmc.edu/AP) and the PubMed

allowing it to recognize the negative charge on bacterial surfaces, rather than mammalian cells. While bacterial membranes are rich in anionic PGs, mammalian cell membranes contain mainly zwitterionic phosphorylcholines (PCs) (Wang et al. 2014a, b). This preference of bacterial membranes is supported by in vitro studies using model membranes (Sevcsik et al. 2008).

The negative charge on the bacterial surface originates from a different basis: an outer membrane for Gram-negative bacteria but a cell wall for Gram-positive bacteria. In the outer membranes of Gram-negative bacteria, anionic lipopolysaccharides (LPS) are dominant. They are composed of a polysaccharide moiety and a lipid A. The cell wall of Gram-positive bacteria comprises peptidoglycan and lipoteichoic acids (LTA) that maintain cell integrity. These outer layers confer negative charges to the surface and are important targets for cationic AMPs. Such an electrostatic attraction is the first step of peptide-bacteria interaction. This is because bacterial surface modification to decrease negative charge can reduce peptide activity. For instance, modification of lipid A with glucosamine confers resistance to LL-37 (Shah et al. 2014). In the case of *S. aureus*, knockout of the gene for alanylation of teichoic acid of the cell wall or lysylation of phosphatidylglycerols (PG) in the inner membrane makes the bacterium more susceptible to cationic LL-37 (Saar-Dover et al. 2012; Peschel and Sahl 2006). Likewise, citrullination of LL-37 can make it less positively charged and reduces its LPS neutralization (Koziel et al. 2014). Host Toll-like receptors (TLRs) play an important role in recognizing these bacterial components. TLR2 is required to respond to cell wall preparations of Gram-positive pathogens, whereas TLR4 is involved in recognition of LPS (Takeuchi et al. 1999).

12.3.2 Challenges for Structural Studies of LL-37

Positive charge alone, however, may not be sufficient for membrane binding of AMPs. The membrane targeting of LL-37 is determined by its three-dimensional structure. Circular dichroism (CD) and FT-IR analysis indicate a helical conformation, leading to an immediate classification of this human peptide into the helical family. In particular, an amphipathic helix of this cationic peptide, with distinct hydrophobic and hydrophilic surfaces, facilitates its interaction with anionic bacterial membranes (Oren et al. 1999). However, it has to wait for the solution nuclear magnetic resonance (NMR) spectroscopy to determine the atomic structure of LL-37 that informs us where the helix starts and ends and whether there is a helix break upon membrane binding. It took years of work to complete the 3D structure of LL-37 due to multiple challenges. The first challenge was the complex nature of bacterial membranes. Consequently, solution NMR studies were conducted in the presence of membrane-mimetic micelles such as anionic sodium dodecyl sulfate (SDS). The deuteration of micelles simplified the NMR spectra and allowed us to focus only on the peptide signals. As SDS has a head group different from that of the anionic lipid in bacteria, we also explored the possible use of a series of short-chain PGs for structural studies of AMPs (Wang et al. 2004; Keifer et al. 2004; Wang 2006, 2007, 2008). The second challenge was spectral resolution. The first solution NMR studies revealed the need of 3D NMR for structural determination of LL-37 bound to SDS micelles (Li et al. 2006a). The separation of the overlapped cross peaks onto numerous 2D planes along the ^{15}N or ^{13}C dimension of 3D NMR spectra yielded the needed spectral resolution for a complete assignment of the LL-37 signals and subsequent NOE assignments. The third challenge was the requirement of establishing a bacterial expression system to produce stable isotope-labeled LL-37 required for 3D NMR studies. The toxicity of LL-37 to the expression host made it necessary to express the peptide as a fusion protein. Another challenge was the difficulty to release the recombinant peptide from the fusion protein by enzyme digestion. The aggregation of LL-37 might have blocked the enzyme cleavage site. Fortunately, the fusion protein was successfully cleaved by formic acid (Li et al. 2006b). By making use of the oligomerization property of LL-37 (Li et al. 2007), we improved the peptide yield and obtained additional ^{15}N-labeled or ^{15}N, ^{13}C-labeled peptides. These labeled peptides in complex with deuterated SDS micelles provided the needed sample stability for recording a suite of triple-resonance NMR spectra, such as HNCACB, CBCA(CO)NH, HNCO, and HNCA (Kay et al. 2011). The 3D structure of LL-37 determined by 3D NMR reveals a long amphipathic helix (residues 2–31) and a C-terminal tail (residues 32–37) (Fig. 12.2). To validate the structure, we measured the ps-ns peptide backbone dynamics using a ^{15}N-labeled LL-37 sample. The dynamics data indicate the helical region is rigid, while the C-terminal tail is mobile, fully consistent with the structure (Wang 2008). There are also other structures for LL-37 and its fragments. Additional information can be found in other articles (Wang 2008; Wang et al. 2014b).

12.3.3 Structural Basis of Bacterial Membrane Binding of LL-37

We also asked whether the high-resolution structure determined in SDS micelles could be applied to bacterial membranes. First, we used dioctanoyl phosphatidylglycerol (D8PG) for NMR studies because it has the same head group as the major anionic lipid in bacteria (Wang et al. 2004; Keifer et al. 2004). Based on all the NMR data, the structure of LL-37 in complex with D8PG is the same as that bound to SDS micelles, implying that the detergent/lipid head groups play little role (Wang 2008). Because D8PG is not deuterated, it enabled us to observe intermolecular NOE cross peaks between peptide and lipid as well. In particular, aromatic F5, F6, F17, and F27 as well as basic R23 of LL-37 show direct interactions with the phosphatidylglycerol, confirming their important role in bacterial membrane binding (Wang 2008). The significance of R23 in interaction with bacteria was initially observed

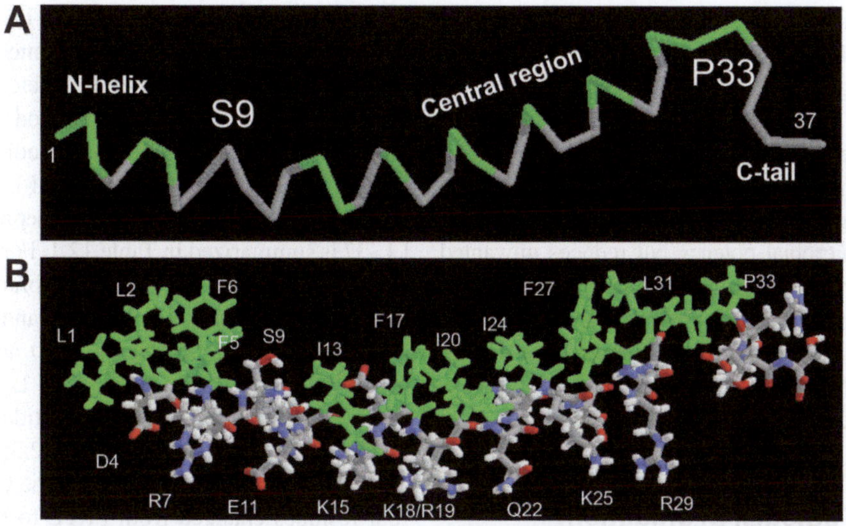

Fig. 12.2 3D structure of human innate immune peptide LL-37 bound to bacteria membrane-mimetic micelles determined by 3D NMR (PDB ID: 2K6O). (**a**) A backbone view of the long helix corresponding to residues 2-31 of LL-37 with the C-terminal tail disordered. (**b**) A stick view of the LL-37 structure with hydrophobic and hydrophilic residues selectively labeled. In both views, hydrophobic amino acids are in green. A discontinuation of the hydrophobic surface at S9, rather than the helix, is thus evident (Wang 2008)

using the central fragment of LL-37 (Wang 2007) and later confirmed by alanine scan (Wang et al. 2012b) as well as the lysine/arginine swap (Wang et al. 2018). Second, we also studied the interaction of LL-37 with *E. coli* LPS. While the signals for the helical region are absent, the signals for the C-terminal tail are evident. Thus, the disordered tail of LL-37 observed in micelles recurs when bound to LPS. We conclude that human LL-37 utilizes the hydrophobic surface of a long amphipathic helix (residues 2–31) to interact with bacterial membranes for antimicrobial killing without the need of the C-terminal tail.

Because the SDS-bound structure reflects the bacterial membrane-bound form, the 3D NMR-derived structure can be used to explain peptide LPS-binding and activity data. The synergistic binding to LPS by an ovine cathelicidin SMAP-29 can be explained by a helix break caused by a proline (Tack et al. 2002). This is not the case for LL-37 (Turner et al. 1998) because of a lack of proline in the middle and the continuity of the helix. However, S9 separates the hydrophobic surface of the long helix of LL-37 into two domains (Fig. 12.2), providing a novel mechanism for synergistic binding of human LL-37 to LPS (Wang 2008). In addition, S9 is also important for antibacterial activity since an effort to render the hydrophobic surface continuous reduced the peptide activity (Wang et al. 2012a). The 3D structure determined by 3D NMR also laid a foundation for discovering active fragments within LL-37.

12.4 Discovery of Antibacterial Regions Within LL-37

This section highlights LL-37 fragments generated under natural conditions and in laboratories. The natural conditions include protease processing from both the host and pathogens. Peptides are also chemically synthesized in laboratories to understand sequence-activity relationship.

12.4.1 Skin Processing of Human LL-37

The Gallo lab studied the protease cleavage of LL-37 in human skin by kallikreins. They found multiple fragments of LL-37, including both active and inactive forms. Some examples are KR-20, LL-23, KS-30, and RK-31 (Table 12.1). Interestingly, KS-30 and RK-31 are more active than their parent LL-37 in bacterial killing, indicating the N-terminal region is less important (Murakami et al. 2004). This processing may be a natural regulatory mechanism that enhances peptide antimicrobial potency but reduces unwanted immune responses via the intact molecule. This mechanism further enriches the molecule reservoir of human cathelicidin-based defense line. Future studies may elucidate the details of this skin regulatory mechanism of human cathelicidin.

12.4.2 Pathogen Degradation of LL-37

As a resistance mechanism, there are multiple pathogen proteases that can degrade human defense peptide LL-37. However, only in a few cases were the resultant fragments documented. Sieprawska-Lupa et al. (2004) studied the digested products of *S. aureus* proteases such as aureolysin and V8 proteases. Aureolysin cleaved LL-37 between R19-I20, R23-I24, and L31-V32 peptide bonds, leading to the instant inactivation of LL-37. In contrast, the V8 protease cleaves the E16-F17 peptide bond of LL-37 (Fig. 12.3), leading to the accumulation and isolation of FK-21 (Table 12.1), a fragment corresponding to residues 17-37 of LL-37. Interestingly, FK-21 is more potent than LL-37 in killing *E. coli, B. subtilis, P. aeruginosa*, and *E. faecalis*. This is understandable since FK-21 contains the major antimicrobial region FK-16 discovered by NMR (Li et al. 2006a). In another study, Rapala-Kozik et al. (2015) isolated an intermediate peptide LL-25 (Table 12.1) from the *C. albicans* cleavage of LL-37. LL-25 appears to have a different immune modulating role compared to LL-37 (see below).

12.4.3 Identification of Active Regions Within Human LL-37 in Laboratories

One of the major barriers toward peptide therapeutics is production cost. LL-37 is relatively long with 37 amino acids. To reduce the synthesis cost, it is necessary to trim unwanted regions. Two major methods are used to locate the active regions of LL-37. The first method is library screening. There are numerous studies with a goal of locating the active region of LL-37. The sequence relationship of these peptides with LL-37 is summarized in Table 12.1. For example, Braff et al. (2005) made a peptide library to identify the antibacterial, antifungal, and antiviral regions. Two long peptides, KS-30 and RK-31 (Table 12.1), are more potent than LL-37. Nell et al. (2006) identified a 24mer peptide P60 (i.e., residues 13-36 of LL-37 in Fig. 12.3; IG-24 in Table 12.1). A peptide P60.4 with the C-terminal four-residues changed from PRTE to RPLR has antimicrobial activity, LPS and LTA neutralization ability comparable to LL-37. Kanthawong et al. (2010) found IG-19, RK-25, and LL-31 (Table 12.1) against Gram-negative *Burkholderia pseudomallei* both in planktonic and biofilm forms. Among these peptides, LL-31 is most potent.

The second method is structure-based. Based on the helix-forming propensity of each amino acid along the sequence, Sigurdardottir et al. (2006) found GKE-21 with reduced cytotoxicity than LL-37. Li et al. (2006a) identified multiple LL-37 peptides based on NMR structural studies. In that study, LL-37 was initially split into LL-12 and IG-25 for structural studies (Table 12.1). While LL-12 is inactive, IG-25 is antibacterial. The study of a micelle-bound form of IG-25 led to the discovery of the major antimicrobial peptide FK-16 of LL-37 by using the NMR-trim technology that removes nonessential membrane-binding regions, such as the disordered C-terminal tail (Li et al. 2006a). Compared to the library approach that requires the synthesis of at least dozens of peptides, this NMR technology is efficient since it arrived at the major antimicrobial region using only a couple of synthetic peptides. In particular, IG-25 is only one residue longer than IG-24 found above via library screening (Nell et al. 2006). An N-terminally glycine appended version of FK-16 is called GF-17. FK-16 and GF-17 have very similar antibacterial activity (Table 12.2). A comparison of the antibacterial activity of multiple peptides described

Table 12.2 Comparison of antibacterial activity of the LL-37-derived antimicrobial peptides FK-16, GF-17, and 17BIPHE2

	MIC (μM)			
Peptide	*S. aureus* USA300	*E. coli* ATCC 25922	*K. pneumonia* ATCC 13883	*P. aeruginosa* PAO1
FK-16	3.1	≥3.1	3.1	25
GF-17	3.1	6.25	3.1	25
17BIPHE2	3.1	6.25	3.1	6.25

Peptide sequences in Table 12.1. 17BIPHE2 is a peptide antibiotic designed based on GF-17 (Wang et al. 2014a)

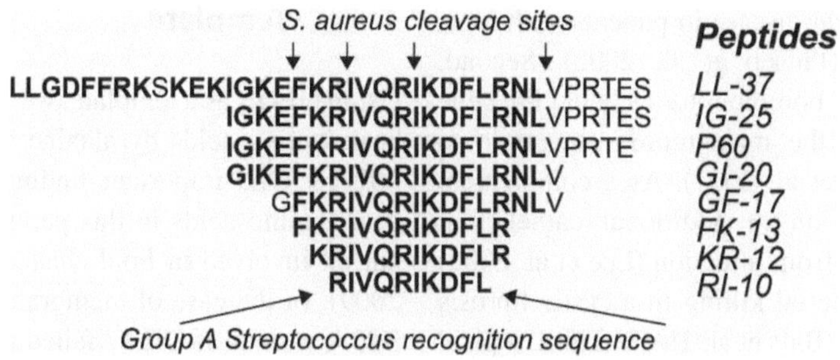

Fig. 12.3 Host-pathogen interaction at the peptide level. Listed are select LL-37 peptides discovered from structure and library approaches (Li et al. 2006a, b; Nell et al. 2006; Wang 2008; Wang et al. 2008). It is amazing that the core antimicrobial region of human LL-37 is also recognized by bacteria for degradation or to release virulence factors (see the text)

above confirmed GF-17 is most potent against methicillin-resistant *Staphylococcus aureus* (MRSA) USA300 (Wang et al. 2014a). This result may be due to a sufficient hydrophobicity of the major antimicrobial peptide of LL-37 (Table 12.1). In addition, a balance in charged and hydrophobic amino acids in the sequence led to a near zero GRAVY value in Table 12.1. Our NMR studies also led to the identification of FK-13, the core antibacterial region of LL-37 corresponding to residues 17–29 (Li et al. 2006a). Using a series of shorter peptides, it was found that the antibacterial region of FK-13 could be further shortened to 12-residue KR-12, the smallest antibacterial peptide of LL-37 (Wang, 2008). KR-12 is active against Gram-negative *E. coli* K12 (MIC 40 μM) but not Gram-positive *S. aureus* USA300. In 2018, Jessen lab made several 12-residue peptides along the LL-37 sequence. Their study showed no activity (>100 μg/mL) for N-terminal fragments LL-12, SK-12, C-terminal fragments VQ-12, IK-12, and DF-12 (net charge = 0, Table 12.1). This is not surprising since most of these fragments do not have the minimal hydrophobicity set by KR-12, the minimal antibacterial peptide we found (Wang 2008). In contrast, weak antibacterial activity (MIC 50 μg/mL) was observed for central fragments KR-12 and FK-12 against *S. epidermidis*. Note that both KR-12 and FK-12 (Saporito et al. 2018) were derived from FK-13 (Table 12.1), further validating the antimicrobial core region (Fig. 12.3) of LL-37 (Li et al. 2006a).

12.5 Peptide Design Based on Select LL-37 Peptides

The ESKAPE pathogens, including *Enterococcus faecium*, *Staphylococcus aureus*, *Klebsiella pneumoniae*, *Acinetobacter baumannii*, *Pseudomonas aeruginosa*, and *Enterobacter* species, are responsible for the majority of hospital-acquired infections (Boucher et al. 2009). For

MRSA alone, the estimated total deaths are already comparable to those caused by HIV-1/AIDS (Klevens et al. 2007). Therefore, new antimicrobials are needed. There is also a high interest in developing AMPs into therapeutic molecules (Zasloff 2002). Being linear, human cathelicidin LL-37 is an important template. The therapeutic implication of LL-37 is evident. First, a lack of LL-37 in neutrophils may be responsible for periodontal disease in patients with morbus Kostmann (Pütsep et al. 2002). Second, knockout of the homologous cathelicidin gene CRAMP makes the mice more susceptible to infection (Nizet et al. 2001). As a complement strategy, expression of additional cathelicidin protects the skin from infection (Lee et al. 2005) and restores bacterial killing in a cystic fibrosis xenograft model (Bals et al. 1999). Third, topical treatment of chronic wounds such as venous leg ulcers with LL-37 (0.5 or 1.6 mg/mL) markedly reduces the mean ulcer area without safety concerns (Grönberg et al. 2014). Furthermore, LL-37 can inhibit biofilm formation probably by opsonization that enhances bacterial clearance (Overhage et al. 2008; Sol et al. 2013). In addition, LL-37 boosts immune response to clear pathogens (Overhage et al., 2014). Fourth, LL-37 may also be induced to help eliminate pathogens (van der Does et al. 2012; Jiang et al. 2013; Schögler et al. 2016). These observations laid a solid basis for peptide design based on LL-37 and its derived fragments.

12.5.1 IG-24 Derived P60.4

Serum/plasma binding is believed to be a key factor that might have limited the systemic use of cationic AMPs. Based on the P60.4 template originally identified from a library screen (Nell et al. 2006), OP-145 or P60.4Ac (Table 12.1) was initially obtained via further amino acid modification, including N-terminal acetylation. OP-145 is effective against *S. aureus* clinical strains, but its activity can be reduced in the presence of plasma (de Breij et al. 2016). Recently, de Breij et al. (2018) made additional peptide mutants to search for candidates with reduced binding to plasma. SAAP-148, with such a property, is found to have antimicrobial and antibiofilm ability against the ESKAPE pathogens, including persisters, which cannot be killed by traditional antibiotics.

12.5.2 The FK-16-Derived GF-17 Template

Using GF-17 as a template, we studied the role of basic amino acids by alanine scan (Wang et al. 2012b). One important finding is that the five basic amino acids in this peptide are not equal, but all involved in lipid clustering (Epand et al. 2009). In the case of membrane permeation, the R23A variant of GF-17 failed to cross either the outer or inner membranes of *E. coli*, indicative of an essential role of R23 for membrane damage (Wang et al. 2012b), consistent with the observation of intermolecular NOE cross peaks between R23 and D8PG by NMR spectroscopy (Wang 2007). As a conservative change, we also swapped the positions between a pair of lysine and arginine in GF-17. Interestingly, the charged swapped peptides showed reduced killing efficiency and increased cytotoxicity to human cells, revealing the evolutional significance of the native sequence of the peptide (Wang et al. 2017).

Cytotoxicity also limits the applications of cationic AMPs. It is commonly observed that peptide cytotoxicity results from high hydrophobicity (Zasloff 2002; Boman 2003). In the case of GF-17, the major antimicrobial region of LL-37 (Fig. 12.3), two approaches were utilized to reduce the peptide hydrophobicity. The first method is peptide truncation that led to the discovery of KR-12 (Wang 2008). The second method is to incorporate D-amino acids. The structural basis for this has been elucidated for GF-17d3, which contains D-amino acids at positions 20, 24, and 28 of GF-17 (Table 12.1). The D-amino acids distorted the backbone of GF-17 and caused non-coherent packing of side chains, leading to hydrophobic defects (Li et al. 2006a). A hydrophobic gap exists on the hydrophobic surface of LL-23, explaining its poor antibacterial activity (Wang et al. 2012a).

Another hurdle in peptide development is protease stability. Any peptide that is cleaved prior to bacterial killing is not useful. We used a library screen method to identify a peptide template with stability to chymotrypsin. This was initially accomplished during antibacterial assays in duplicated wells containing proteases; a stable peptide remains bacterial inhibition even in the presence of proteases. In this experiment, GF-17d3, with 3 D-amino acids incorporated, is active against *E. coli* K12, but not GF-17d1 (one D-amino acid at position 20) or GF-17d2 (two D-amino acids at positions 20 and 24) containing 1-2 D-amino acids. This chymotrypsin-resistant template, GF-17d3, was used to design peptide analogs to combat resistant pathogens. One of the peptides, 17BIPHE2, with both F17 and F27 replaced with biphenylalanines, is potent, selective, and stable (Wang et al. 2014a). This is the first LL-37-derived peptide illustrated to kill the ESKAPE pathogens (MIC 3.1–6.2 μM) and can inhibit biofilm formation in vivo. 17BIPHE2 showed a 50% hemolytic concentration (HL_{50}) between 150 and 220 μM depending on the types of blood cells used. Our further investigation of 17BIPHE2 confirmed its stability to chymotrypsin digestion. In addition, it is not degraded after digestion with pathogen protease *S. aureus* V8 or fungal protease K, but can be cleaved by trypsin. In contrast, our recently designed Trp-rich peptide TetraF2W-RK with eight amino acids is inherently stable to trypsin and *S. aureus* V8 but can be cleaved by chymotrypsin (Mishra et al. 2017a). However, TetraF2W-RK, synthesized in D-amino acids, is stable to both trypsin and chymotrypsin. Peptide stability to such proteases from the digestive system can be important for future engineering peptides as oral drugs.

With the idea of personalized medicine, it may be necessary to design peptides with a desired activity spectrum. Recently, we found it possible to convert the broad-spectrum GF-17 into narrow-spectrum AMPs. First, by partially incorporating D-amino acids, we obtained GF-17d3 (Li et al. 2006a), which is only active against Gram-negative *E. coli* ATCC 25922 and *A. baumannii* B28-16, but not Gram-positive pathogens we tested (Wang et al. 2017). Alternatively, hydrophobic truncation of GF-17 led to KR-12 that is only active against *E. coli* K12. Second, 17BIPHE2-3RA, where three arginines are changed to alanines, is only active against Gram-positive staphylococcal strains such as *S. aureus* USA300 and *S. epidermidis* 1457. 17BIPHE2-3RA, with a net charge of +2 and hydrophobic content of 64%, resembles the database designed anti-MRSA peptide DFTamP1 that kills only Gram-positive pathogens (Mishra and Wang 2012). These results underscore the importance of amino acid composition in determining the peptide activity spectrum (Wang et al. 2018).

12.5.3 FK-13 and KR-12 Templates

NMR studies established a helical structure for KR-12 in complex with D8PG (Wang 2008). Understanding the role of each residue in KR-12 is important for peptide design. We investigated the effect of a single alanine substitution of basic amino acids on KR-12 activity (Mishra et al. 2013). Consistent with our finding for GF-17 that R23 and K25 are important for antimicrobial action, KR-12R23A and KR-12K25A are less effective in binding to anionic D8PG (Mishra et al. 2013). We did not replace hydrophobic amino acids because KR-12 is already the smallest, and such substitutions would lead to inactive peptides. This is indeed the case when we substituted I20, I24, or L28 of KR-12 into an alanine (Wang 2010). Recently, Gunasekera et al. (2018) made a systematic amino acid substitution within KR-12. It seems the culture media and bacterial strains used for antimicrobial assays can substantially influence the MIC values and make the results incomparable with the previously published results (Mishra et al. 2013).

Because of its low cytotoxicity and short length, KR-12 (Fig. 12.3) becomes an attractive template for peptide engineering. Jacob et al. (2013) generated several variants of KR-12 by increasing basic and hydrophobic residues at positions 27, 26, 22, 23, 25, and 18. KR-12-a1 to KR-12-a6 contains 1–6 changes of the listed residues of KR-12. KR-12a1 (a F27W variant) showed MIC values in the range of 1–8 μM using six Gram-positive and negative bacteria. Of note, additional substitutions did not increase peptide activity substantially against *S. aureus*, *S. typhimurium*, *B. subtilis*, and *S. epidermidis*, and in the case of KR-12-a6 with all six changes (K18L, Q22K, R23L, K25L, D26K, and F27W), antibacterial activity of the peptide against *E. coli* and *P. aeruginosa* actually reduced. Meanwhile, KR-12-a5 and KR-12-a6 became more hemolytic, indicating the K25L change is not favorable. Of note, a topical use of KR-12-a2 to treat MRSA otorrhea found no hearing loss or cochlear damage in guinea pigs (Sung et al. 2017), while KR-12-a5 is potent against oral pathogens (Caiaffa et al. 2017). A similar mutational study was also conducted using FK-13 (Rajasekaran et al. 2017). Da Silva et al. (2017) also utilized KR-12 as a template to design antibiofilm peptides against oral pathogens such as *Streptococcus mutans*. Substitution of I20 with W and appending KAEK at the C-terminus of KR-12 led to a more potent peptide against biofilms. These studies indicate a great potential of KR-12 as a template or a sequence unit for peptide design.

12.6 Potential Applications of LL-37 Peptides

12.6.1 Antibiofilm Peptides

Many pathogens are able to form biofilms, where a surface-anchored bacterial community shares a tower-like structure decorated with polymers (e.g., polysaccharides and DNA). Such biofilms, once formed on medical devices, are very challenging to remove. Therefore, much research effort of AMPs was oriented toward the development of potent antibiofilm agents and infection-free implants to combat these pathogens. Biofilm inhibition by human LL-37 at a low concentration was initially observed by the Hancock Group (Overhage et al. 2008). As one advantage of peptide design, however, both GF-17 and 17BIPHE2 possess antibiofilm capability superior to LL-37 (Mishra et al. 2016). In addition, we have demonstrated the antibiofilm ability of the engineered peptide 17BIPHE2 in a mouse catheter model (Wang et al. 2014a). As an independent path, a European group demonstrated antibiofilm ability of OP-145 and SAAP-148 (de Breij et al. 2016, 2018). However, combination treatment may be necessary and a natural choice for preformed biofilms that are difficult to remove. Our recent antipseudomonal biofilm results are summarized here: (1) Antibiotics are not effective, while AMPs are active, underscoring the importance of such peptides as a new antibiofilm agent. (2) Combined use of 17BIPHE2 with antibiotics is more effective to remove the *P. aeruginosa* biofilms. (3) Early treatment prior to biofilm maturation (<10 h) is advantageous since monotherapy of either antibiotic or peptide is effective (Mishra and Wang 2017). Based on these observations, there is now a growing interest in developing biofilm prevention approaches.

12.6.2 Antibiofilm Coating of LL-37 Peptides

An attractive antimicrobial strategy is to coat human cathelicidin LL-37 to the surfaces of medical devices to prevent infection, especially when biofilms form. In particular, peptide surface immobilization confers some advantages such as a local high peptide concentration, reduced cytotoxicity, and increased host cell adhesion. This section describes covalent coupling of LL-37 and its peptides (Table 12.3). There are multiple factors that affect the antimicrobial activity of immobilized peptides. These include the types of materials (metals or polymers), the linker or spacer, and the site for peptide coupling (terminus vs. side chain). There is no consensus as to the nature of linkers (short vs. long).

The full-length LL-37 molecule, with a cysteine added at the N-terminus, was first immobilized on a titanium surface via the maleimide chemistry. Gabriel et al. (2006) found that site-specific coupling of LL-37 with a flexible hydrophilic poly(ethylene glycol) (PEG) spacer is better than randomized coating. The LL-37-coated surface shows antimicrobial activity

Table 12.3 Covalent immobilization of LL-37 and its fragments onto various substrates

Study	Peptide	Substrate/ surface	Chemistry	Spacer	Activity organism	Ref
1	LL-37	Titanium	Surface hydroxylation, APTES amine generation, and peptide coupling via maleimide chemistry	PEG	*EC* (K12)	Gabriel et al. (2006)
2	LL-37	Nickel NPs	Dielectric barrier discharge glow plasma fluidized bed (GPFB) for deposition of nanolayer polyacrylic acid (PAA) on NPs. And peptide reaction using EDC/NHS	Polyacrylic acid	*EC* (DH5α)	Chen et al. (2009)
3	LL-37	Etafilcon A contact lens (pHEMA)	Carbodiimide reaction linking surface acidic group to the peptide amino group	Carboxylic acid	*PA*	Dutta et al. (2016)
4	LL-37	Amino saline coated magnetic nanoparticle	LL-37 was reacted with APTMS-coated NPs via amidation	APTMS	*CA, CG and CT*	Niemirowicz et al. (2017)
5	IG-25	Fluorous slide and Fluoroperm 60 contact lens	Perfluorocarbon chains with alkynyl group reacted with peptide-containing azido group via Cu-catalyzed azide-alkyne cycloaddition reaction (click chemistry)	Perfluorocarbon chains (C_8F_{17})	*PA* (PAO1)	Santos et al. (2013)
6	FK-16	Titanium	Similar to LL-37/Ti	6-Maleimidohexanoic acid	ESKAPE pathogens	Mishra et al. (2017a)
7	KR-12	Silk fibroin (SF) nanofiber	Carboxylic acid on the SF nanofiber reacted with aminoethyl maleimide via the EDC/NHS reaction. Peptide was conjugated with the maleimide-Cys reaction	Aminoethyl maleimide	*SA, SE, EC, and PA*	Song et al. (2016)
8	KR-12	Titanium	The hydroxylated surface was amine-functionalized and reacted with Asp of peptide for coupling	3-(2-aminoethylamino) propyltrimethoxysilane	*SE*	Nie et al. (2016)

EC Escherichia coli, SA Staphylococcus aureus, PA Pseudomonas aeruginosa, SE Staphylococcus epidermidis, CA C. albicans, CG C. glabrata, CT C. tropicalis

against *E. coli*. Recently, LL-37 has also been immobilized on other types of surfaces, including nanoparticles (Gustafsson et al. 2010; Cassin et al. 2016; Niemirowicz et al. 2017; Comune et al. 2017). The chemistries are summarized in Table 12.3. These studies illustrated (1) antifungal activity of LL-37 against *C. albicans*, (2) anti-adhesion properties, (3) LPS binding, (4) no toxicity to mammalian cells, (5) immune modulatory activity such cell migration, and (6) treatment of wounds in a mouse model.

Several fragments (IG-25, FK-16, and KR-12) initially reported by us (Li et al. 2006a; Wang 2008) have also been surface immobilized. Santos et al. (2013) immobilized IG-25 (Table 12.1) on fluorous thin films and fluorosilicone contact lens with activity against an ocular pathogen *P. aeruginosa* (Table 12.3). KR-12 was also covalently immobilized via a site-specific maleimide chemistry on electrospun silk fibroin nanofiber. The surface could inhibit *S. aureus*, *S. epidermidis*, *E. coli*, and *P. aeruginosa*. Compared to the free peptide described above, this surface-coated KR-12 appeared to have gained antibacterial activity. In addition, the immobilized peptide suppressed the LPS-induced TNF-α expression of monocytes (RAW264.7), helped the proliferation of fibroblasts and keratinocytes, and promoted the differentiation of keratinocytes with enhanced cell-cell attachment (Song et al. 2016). Likewise, KR-12 was also immobilized on the titanium surface (Nie et al. 2016), which shows antimicrobial and antibiofilm activities against *S. epidermidis*. Importantly, the peptide-coated surface is able to increase the adhesion and proliferation of human bone marrow mesenchymal stem cells (Nie et al. 2016). To obtain a surface with broad-spectrum activity, we recently immobilized FK-16, the major antimicrobial peptide of LL-37, on the titanium surface using a similar maleimide chemistry (Mishra and Wang 2017). FK-16, with a higher coating density than LL-37 on the titanium surface, shows potent activity against the ESKAPE pathogens but is nontoxic to human erythrocytes and epidermal keratinocytes HaCaT cells. Significantly, the FK-16-coated surface is able to inhibit biofilm formation of *S. aureus* USA300 (initial inoculation at 10^3 CFU) for up to 72 h. A lower CFU (10^3) used here should be medically more relevant considering the sterile condition of the surgery room. This study provides a proof-of-concept example for generating a broad-spectrum antimicrobial surface to prevent bacterial adhesion and biofilm formation on medical implants.

12.6.3 Antiviral Peptides

Viral infection is life threatening and has raised concerns from the public. This ranges from the well-known human immunodeficiency virus type 1 (HIV-1) to the recent Zika and Ebola viruses in the news. Often, we do not have therapeutic molecules to treat such outbreaks. Infected patients can only be sent to a containment facility and wait for miracle medicine. The development of new vaccines is slow. Therefore, alternative antiviral strategies are clearly needed. Current rise in HIV patients with coinfections has required researchers to revisit AMPs with a hope to overcome ever-increasing cases of antibiotic resistance and to increase the immune response status. It is interesting to explore how nature has devised AMPs for protection against various viruses. As of June 2018, the APD registered 182 antiviral peptides (Wang et al. 2016). While a systematic review on HIV inhibitory AMPs (Wang 2012) can be found online, this section focuses on antiviral effects of human cathelicidin LL-37. Up to date, LL-37 has been demonstrated to have inhibitory effects against at least ten types of viruses, including herpes simplex virus (HSV; Yasin et al. 2000), smallpox vaccinia virus (Howell et al. 2004), HIV-1 (Bergman et al. 2007), respiratory syncytial virus (RSV; Tian et al. 2011), varicella zoster virus (VZV; Crack et al. 2012), human adenovirus (Uchio et al. 2013), influenza A virus (IAV; Tripathi et al. 2013), dengue virus type 2 (DENV-2; Alagarasu et al. 2017), human rhinovirus (HRV; Sousa et al. 2017), and Zika virus (He et al. 2018). To facilitate our discussion, these LL-37 inhibited viruses are classified in Table 12.4 based on the type of nucleic acids (DNA/RNA) and whether they are enveloped. Thus, LL-37 has an inhibitory effect on both enveloped and non-enveloped viruses.

It is natural to ask whether the antibacterial segments derived from LL-37 also effectively inhibit viruses. While fragments from either the N-terminus or C-terminus are inactive, a central fragment of LL-37 is active against HIV-1 (Wang et al. 2008). Different from the anti-MRSA case, GI-20 (Fig. 12.3) has the highest therapeutic index, indicating that the central region of LL-37 is also important to inhibit viruses. In collaboration with ImQuest Biosciences, we also identified the minimal anti-HIV peptide of LL-37. Different from the bacteria case, the LL-37 core peptide FK-13 retains anti-HIV activity but not KR-12. This fact implies a significant role of F17 in inhibiting HIV-1. There are also other dif-

Table 12.4 Classification of LL-37-inhibited viruses

RNA viruses		DNA viruses	
Naked	Enveloped	Naked	Enveloped
Rhinoviruses	Human immunodeficiency virus type 1 (HIV-1)	Adenovirus	Herpes simplex virus (HSV)
	Influenza A virus (IAV)		Vaccinia virus (smallpox)
	Respiratory syncytial virus (RSV)		Varicella zoster virus (VZV)
	Zika virus (ZIKV)		
	Dengue virus type 2 (DENV-2)		

ferences from the antibacterial case. In particular, after reversal of the FK-13 sequence or incorporation of D-amino acids into GF-17, the peptides remain active against bacteria but not the virus, implying a different molecular target. It appears that LL-37 and its peptides IG-25 and FK-16 are able to inhibit HIV-1 reverse transcriptase in vitro at IC_{50} of 15, 7, and 70 μM, respectively (Wong et al. 2011). The weak enzyme inhibition activity of these peptides observed here (20–30% inhibition at 100 μM) does not sufficiently explain anti-HIV activity observed below 1 μM. Therefore, additional mechanistic studies are needed to better understand peptide activity. It is clear that the sequence requirement for inhibiting HIV-1 differs from that for bacterial inhibition. Other cathelicidins can also be useful. While BMAP-27 is toxic, a C-terminal truncated peptide is not (Skerlavaj et al. 1996). BMAP-18 is also demonstrated to be inhibitory to HIV-1 (Wang et al. 2008). Future studies may further define the mechanism of action and evaluate the therapeutic potential of these anti-HIV peptides in animal models.

Considering other benefits of LL-37 such as spermicidal effects and selective killing of invading pathogens without damaging commensal bacteria (Tanphaichitr et al. 2016), these LL-37 peptides may be promising candidates as topical spermicides/microbicides. However, further studies are required to resolve the complication from coinfected viruses such as HSV-2 that appears to alter the defense role of LL-37 into an offense role. The mechanistic studies revealed that LL-37 produced by HSV-2-infected epithelial cells upregulates HIV receptors (CD4 and CCR5) on the surface of monocyte-derived Langerhans cells (the mucosal epithelium resident dendritic cells), thereby enhancing HIV infection (Ogawa et al. 2013).

LL-37 peptides also have inhibitory effects on other RNA viruses (Table 12.4). LL-37 can directly disrupt enveloped influenza A virus (Tripathi et al. 2013). In the case of the pandemic H1N1 strain of 2009 (A/California/04/09/H1N1 or "Cal09"), LL-37 is inactive. However, GI-20, a central LL-37 fragment, retains anti-IAV activity against this strain (Tripathi et al. 2015b). This observation indicates the advantage of LL-37 reengineering.

Mosquito-borne Zika is another enveloped RNA virus first isolated in Uganda in 1947 near the Zika forest. After its outbreak in 2007, it became a rapidly emerging public health threat. Although clinical infection is frequently mild, significant neurological manifestations have been demonstrated in infants born to Zika virus (ZIKV)-infected mothers (McArthur 2017). Currently, there is no drug to treat ZIKV infection. Although vaccines are under active development, other effective countermeasures may also be considered. Recently, the efficacy of LL-37 and its derived peptides (e.g., GI-20 and GF-17) against ZIKA has been demonstrated in vitro, whereas RI-10 (Fig. 12.3) is ineffective (He et al. 2018). Further characterization reveals that GF-17 (Fig. 12.3) can directly inactivate this virus and work via the interferon pathway. In addition, 17BIPHE2, an engineered version of GF-17, also inhibits ZIKA effectively. However, its antiviral effect is reduced when R23 is altered to ornithine, indicating the important role of this arginine in inhibiting ZIKA.

Other labs also searched active antiviral regions of LL-37 against respiratory syncytial virus (RSV), which is responsible for lower respiratory tract infections of children. A low level of human cathelicidin is directly correlated with human RSV infection. Harcourt et al. (2016) found a better anti-RSV effect when LL-37 is used prophylactically (treat before infection) than therapeutically (treat after infection). Currie et al. (2016) also compared the effect of treatment time. They found that co-administration of LL-37 with virus clearly protects animals better against infection either when the peptide is treated before or after viral colonization. It is established that LL-37 can directly damage the viral envelope, disrupt virus particles, and inhibit infection of epithelial cells in vitro (Currie et al. 2016). The direct anti-RSV effect of LL-37 stimulates the interest in identifying the active region of LL-37. As we observed in the antibacterial study (Li et al. 2006a), Tian et al. (2011) found that the N-terminal fragment LL-12 is inactive while the C-terminal fragment IG-25 is. Moreover, among the four 22mer LL-37 peptides corresponding to residues 13–34 (IG-22), 14–35 (GK-22), 15–36 (KE-22), and 16–37 (EF-22), EF-22 shows an anti-RSV effect similar to LL-37. Currie et al. (2013) also showed that, while neither the N-terminal fragment LL-22 nor the C-terminal fragment EF-22 is inhibitory, a central peptide KI-22 (Table 12.1) is effective against viral particles. Despite the conflicting results between Tian et al. (2011) and Currie et al. (2013) regarding EF-22, these studies also point at the important antiviral role of the central fragment of LL-37 previously found based on bacteria (Li et al. 2006a; Nell et al. 2006) and later demonstrated against HIV-1 (Wang et al. 2008). It seems that the central region of LL-37 plays a general protective role against both bacterial and viral infection (Fig. 12.3).

Because peptides have a relatively short half-life, potential cytotoxicity, and non-specific interactions with cells, one possible way to improve the treatment outcome is to combine LL-37 with nanoparticles, which can be internalized, increasing the uptake of the peptide. LL-37 containing liposomes (size 106.8 ± 10.1 nm, shelf-life stability >1 year) are found to be superior to its free form in protecting keratinocytes from RSV infection without displaying cytotoxicity even at 400 μM (Ron-Doitch et al. 2016). In future studies, the central fragment of LL-37 may be formulated in the same manner. Other preventative measures may include the administration of vitamin D and increased exposure to sunlight to boost the expression of LL-37. However, the interaction of LL-37 with combustion-derived carbon nanoparticles (especially in winter) can compromise peptide antiviral and antibacterial activity, making immunocompromised people more susceptible to infection in highly polluted environments (Findlay et al. 2017).

12.6.4 Antifungal Peptides

High mortality and morbidity rates due to invasive mycosis have been increasing over the last 20 years. Medically significant pathogenic fungi (~300 species) are almost always molds. Opportunistic fungal infections create therapeutic challenges, particularly in high-risk immunocompromised patients with AIDS, cancer, and those undergoing transplantation. In light of growing resistance to antifungal drugs, novel medicine and treatment approaches are required.

The current APD registered 1067 antifungal peptides (Wang et al. 2016). Cathelicidin α-helical peptides have shown activity (BMAP-27 and BMAP-28 from cows) against Candida spp. and *C. neoformans*, but they are less active against filamentous fungi. Both LL-37 and a close mouse analog mCRAMP have a similar MIC range (15–20 μM) against *C. albicans*. In one study, mCRAMP was induced by *C. albicans* at the skin surface in a mouse model, demonstrating that these peptides provide a natural barrier to fungal infection. In addition, LL-37 can also regulate the immune response to better clear fungal infection. Gallo and colleagues tested antifungal activity of LL-37 and its fragments against *C. albicans* and found that KS-30 and RK-31 are more active (Braff et al. 2005).

Using fluorescein-labeled peptides, den Hertog et al. (2006) investigated the cell location of LL-37-derived peptides. While LL-37 and its C-terminally truncated peptide LL-31 (Table 12.1) are found at the perimeter of *C. albicans*, the N-terminally truncated peptide RK-31 (Table 12.1) enters the cytoplasm within 30 min. LL-25 can enter cells faster than RK-31. It seems that both the N-terminal helix and the central helix play a role in determining the peptide location on the cell perimeter since all the phenylalanines are important bacterial membrane anchors (Wang 2008). In addition, these peptides can all induce lipid phase separation in fungal membranes.

12.6.5 Immune Modulating Peptides

Human LL-37 also plays an important role in regulating the immune response (Scott et al. 2002; Choi et al. 2012; Kahlenberg and Kapplan 2013). This section highlights some differences between LL-37 and its derived peptides. The LPS neutralization property of LL-37 forms the basis for its use to treat sepsis. Arginines are important in this neutralization as citrullination of LL-37 makes it inactive in a sepsis mouse model (Koziel et al. 2014). Using a designed peptide 17BIPHE2, it is shown that an alteration of R23 to ornithine slightly reduces LPS neutralization (Wang et al. 2018). In addition, synergistic LPS binding of LL-37 requires two domains (See Fig. 12.2). This explains the reduced LPS-binding ability of LL-37 peptides without the N-terminal domain, including the 24mer peptide P60 found by Nell et al. (2006) and IG-19 (Nan et al. 2012) (sequences in Table 12.1). Increasing basic/hydrophobic amino acids and changing F17 and F27 to W, however, enhance the LPS-binding ability of IG-19. By binding to LPS, LL-37 suppresses the LPS binding to receptors such as CD14 and TLR-4, thereby reducing the apoptosis of liver endothelial cells (Suzuki et al., 2011). In addition, LL-37 also plays a role in LPS clearance. In this process, LL-37 enhances LPS uptake by liver cells via endocytosis (Suzuki et al. 2016).

Chemotaxis is a recognized role of LL-37. There is no correlation between immune modulation (e.g., IL-8 release from keratinocytes) and antimicrobial activity (Braff et al. 2005). Nell et al. (2006) also found that, although P60.4 has antimicrobial activity comparable to LL-37, it loses its chemotactic ability, while P60 and LL-37 are nearly equivalent in inducing neutrophil migration. Thus, an alteration of the follow-up sequence behind the central helix (Fig. 12.3) regulates chemotaxis. Chemotaxis of LL-37 is important for host defense against pathogen invasion. As a counteracting strategy, fungi can cleave LL-37 using aspartic proteases. LL-25 is identified as an intermediate peptide that shows a lower chemotactic activity to neutrophils than LL-37, reducing the recruitment of neutrophils to the infection sites (Rapala-Kozik et al. 2015).

Interestingly, LL-37 can also be cleaved into LL-23 in human skin (Murakami et al. 2004). There is a clear difference between LL-37 and its N-terminal fragment LL-23 in immune regulation (Wang et al. 2012a). Immune modulation by LL-37 also plays a role in viral control. For example, LL-37 can inhibit the release of IL-8 from neutrophils induced by IAV (Tripathi et al. 2014). Of note, the central peptide GI-20 is equally effective in reducing IL-8 (Tripathi et al. 2015a). LL-37 can also stimulate immune response by binding to nucleic acids. In this process, LL-37 forms oligomers and serves as a carrier. A comparison of LL-37 with its fragments reveals that both the N- and C-terminal regions are required in this process (Singh et al. 2014). This sequence requirement agrees with the NMR study of free LL-37 at pH 7 that the entire region of LL-37 is involved in the oligomerization into tetramers (Wang 2017).

It is interesting to note that there is also a sequence difference required for antimicrobial action and pathogen response. While FK-13 (Table 12.1) is the minimal peptide to inhibit HIV-1, KR-12 (obtained by deleting the N-terminal F17 of FK-13) retains antibacterial activity against *E. coli* (Wang 2008). Interestingly, RI-10 (Table 12.1), obtained by deleting one residue from the N- and C-termini of KR-12, lost antimicrobial activity (Wang 2008; He et al. 2018). However, RI-10 retains the minimal sequence information of LL-37 that triggers the bacterial two-component system. A direct interaction between LL-37 and the CSrRS receptor increases the virulence factor expression of group A streptococcus (GAS) (Velarde et al. 2014). With the expansion in our investigation, other functional sequence motifs of human LL-37 may emerge, further enriching our understanding of this innate immune peptide.

Taken together, there are different sequence requirements for antimicrobial action and cell response from both the host and pathogen sides. Such sequence differences are determined by different molecular targets: usually bacterial membranes for antimicrobial activity but cell receptors for host immune stimulation and pathogen response.

12.6.6 Anticancer Peptides

Anticancer activity of AMPs was demonstrated rather early (Ohsaki et al. 1992). AMPs with anticancer activity are discussed elsewhere in this book. An updated list of anticancer peptides can be found in the APD (Wang et al. 2016). Human cathelicidin LL-37 is linked to cancers in different manners. Its level is increased in ovarian, breast, and lung cancers, but LL-37 suppresses colon and gastric cancers (Wu et al. 2010b). Whether and how LL-37 promotes cancer and metastasis deserves further studies. This section discusses anticancer activity of LL-37 as a basis for developing alternative cancer treatment approaches. Such methods can be important for cancers that are resistant to existing anticancer drugs. However, the poor cell selectivity between normal and malignant cells may make it challenging to put them into practical use directly (Ohsaki et al. 1992; Li et al. 2006a; Mishra et al. 2018). Advanced engineering strategies discussed elsewhere may be helpful (Mishra et al. 2017b).

12.6.6.1 Colon Cancers

Human cathelicidin LL-37 is expressed strongly in normal colon mucosa but downregulated in colon cancer tissues. Kuroda et al. (2012) showed that LL-37 and a peptide analog FF/CAP18 suppresses colon cancer cell (HCT116) proliferation. FF/CAP18 corresponds to residues 5–32 of LL-37 with residues E16 and K25 changed to F. The peptide works by depolarization of the mitochondrial membrane independent of the p53 pathway. Ren et al. (2012) showed that treatment of colon cancer cells with LL-37 induces anionic phosphatidylserine (PS) exposure and DNA fragmentation, indicative of apoptosis. Previously, FK-16 is found to be the major antimicrobial and anticancer peptide of LL-37 (Li et al. 2006a). It is interesting that FK-16 shows a similar anti-colon cancer effect independent of caspase activation (Ren et al. 2013). Mechanistically, FK-16 causes the upregulation of Bax and downregulation of Bcl-2 by activating p53. As an alternative anticancer strategy, Cheng et al. (2014) observed tumor size shrinking when cathelicidin-expressing adeno-associated virus was administered intravenously into HT-29-derived subcutaneous tumors in nude mice.

12.6.6.2 Gastric Cancer

Helicobacter pylori is linked with gastric cancers (Li and Perez Perez 2018). Hase et al. (2003) noticed that the level of LL-37 in various types of gastric cancers is significantly reduced. During *Helicobacter pylori* infection, the level of LL-37/hCAP-18 secreted into gastric juice is increased. This is understandable since the *H. pylori*-induced expression of LL-37 exerts bactericidal effects. Wu et al. (2010a) showed that LL-37 suppresses gastric cancer by increasing the tumor-suppressing bone morphogenetic protein signaling via inhibiting proteasome. Schauber et al. (2004) detected upregulation of LL-37 in both colonic and gastric cells when histone-deacetylase (HDAC) inhibitors (e.g., butyrate and trichostatin A) were administered. Induction of LL-37 appears to be promising also for treatment of colon cancer.

12.7 Concluding Remarks

Human cathelicidin LL-37 is an interesting moonlighting peptide with multiple functional roles and involvement in numerous diseases. The wide functions of this peptide provide a scientific basis for developing its potential applications. The high interest in the therapeutic potential of antimicrobial peptides originates from their potency against drug-resistant bacteria, RNA viruses, and cancer. The design of AMPs based on LL-37 starts from the identification of its active regions (Table 12.1 and Fig. 12.3). Both library screen and structure-based approaches are utilized. It is also possible to combine the library screen with structure-based design (e.g., Wang et al. 2004). What we have learned to date from LL-37 peptide design can be summarized below:

1. Through multiple studies, the benefits of LL-37 engineering emerge. While the activity of full-length LL-37 is media dependent, the designer peptides such as GF-17 and 17BIPHE2 are not (Li et al. 2006a; Wang et al. 2014, 2018; Mishra et al. 2016). While LL-37 is poor in preformed biofilm disruption, both GF-17 and 17BIPHE2 work well (Mishra et al. 2016). While LL-37 is inactive against a seasonal flu virus, GI-20 remains active (Tripathi et al. 2015a, b). While the effect of LL-37 on cancer is controversial, FK-16 is anticancer and works superior to LL-37 (Li et al. 2006a; Ren et al. 2013). These examples underscore the medical significance of LL-37 fragments as well as peptide engineering.
2. There is a consensus that the antimicrobial action of LL-37 is achieved primarily via its central region. This central antimicrobial region (residues 13–32) becomes remarkably evident in the 3D structure of LL-37 (Fig. 12.2) because it is sandwiched between two punctuation signals of the LL-37 sequence: hydrophilic S9 (which splits the hydrophobic surface into two domains) and P33 (which ends the helical region) (Wang 2008). Peptides KR-12, FK-13, GF-17, and GI-20 are all derived from this region (Fig. 12.3). The sequence of IG-24 (Nell et al. 2006), the template for SAAP-148, however, extends beyond this central region by including four residues from the non-membrane-targeting C-terminal tail of LL-37.
3. An alteration of the amino acid composition affects the peptide activity spectrum. The wide-spectrum GF-17 (i.e., killing both Gram-positive and Gram-negative bacteria) has been converted to narrow-spectrum peptides that kill only either Gram-positive or Gram-negative bacteria. These results underscore the importance of basic amino acids for antimicrobial activity against Gram-negative bacteria and hydrophobic residues for killing Gram-positive pathogens, consistent with the database findings (Wang et al. 2018).
4. Consistent with the classic view, hydrophobic amino acids of LL-37 peptides are important for membrane anchoring. The protruding aromatic rings of F17 and F27 imply their interdigitating membranes. While all basic amino acids participate in lipid clustering (Epand et al. 2009), they are not equal. R23 of LL-37 is essential for pathogen recognition, LPS neutralization, membrane permeation, pathogen killing, and antibiofilm effects (Wang 2007; Wang et al. 2012b, 2017, 2018; He et al. 2018). Even a lysine-arginine positional swap can affect such properties of LL-37 peptides, revealing the evolutional significance of the native sequence (Wang et al. 2017). As an unwanted effect, such an interfacial arginine also contributes to hemolysis and an change to ornithine improves peptide selectivity (Wang et al. 2018).
5. To reduce peptide production cost, there has been a desire to identify the minimally active regions of LL-37. Interestingly, the shortest active peptide varies with biological activity (i.e., RI-10 for receptor binding, KR-12 against bacteria, FK-13 against HIV-1, and GF-17 against MRSA).

6. While NMR studies have accurately mapped the core antimicrobial region of LL-37 (Li et al. 2006a), bacteria also counteract on this region. *S. aureus* can secrete aureolysin to cut this region of LL-37 (Sieprawska-Lupa et al. 2004). In addition, another Gram-positive pathogen GAS can recognize RI-10 via the CsrRS receptor (Velarde et al. 2014) to increase the expression of virulence factors (Fig. 12.3). Such bacterial recognition regions shed new light on the significance of the core antimicrobial region of LL-37 as well as host-pathogen interactions at the molecular level.
7. While an increase of excessive basic amino acids can make the peptide more toxic (e.g., Jacob et al. 2013), a general and classic wisdom to improve cell selectivity of peptides is to decrease hydrophobicity. Sequence mutation, deletion, and truncation are routinely utilized for this purpose. Partial incorporation of D-amino acids can alter peptide conformation and generate incoherent side chain packing, reducing hydrophobicity (Li et al. 2006a). The incoherent packing of side chains leads to hydrophobic defects in the structure, the basis for cell selectivity. The importance of peptide hydrophobicity for targeting bacterial membranes sets a limit on the degree of hydrophobic reduction for cell selectivity.
8. Usually, peptides made of D-amino acids are more resistant to proteases (Boman 2003). Protease-stable peptides can also be screened from a peptide library followed by structure-guided peptide design to enhance activity against the ESKAPE pathogens. 17BIPHE2 can kill bacteria in the presence of host chymotrypsin, *S. aureus* V8 protease, or fungal proteinase K (Wang et al. 2014).
9. Likewise, plasma binding to SAAP-148 has been minimized (de Breij et al. 2018). This is regarded as a main reason for the failure of AMPs in vivo. Whether this is a general requirement remains to be validated.
10. To date, the therapeutic use of LL-37 is limited to topical treatment due to poor bioavailability, production cost, and potential cytotoxicity. The template for engineering 17BIPHE2 is obtained based on structure, whereas the template for SAAP-148 originates from peptide library screen (Wang et al. 2014a; de Breij et al. 2018). Both peptides show effects in vitro and in vivo on resistant pathogens and biofilms.
11. It is preferred to prevent biofilm formation because preformed biofilms (e.g., *P. aeruginosa*) are notoriously difficult to get rid of. Combined treatment using 17BIPHE2 and antibiotics give better results. This practical approach not only potentiates the effect of traditional antibiotics but also reduces the peptide needed, reducing cost and potential cytotoxicity (Wang 2017).
12. Antimicrobial activity and immune modulation are the two faces of the same coin. These innate immune peptides use different molecular targets (e.g., bacterial membranes for antimicrobial effects and host receptors for signal transduction and immune modulation). In a catheter-associated mouse biofilm model, 17BIPHE2 is not only antibacterial but also chemotactic to monocytes (Wang et al. 2014). GF-17 can directly inactivate Zika virus. It also acts via the interferon pathway (He et al. 2018). Through peptide design, it is possible to retain antibacterial activity and tune host immune response (Nell et al. 2006).

We anticipate that research interest in human cathelicidin will continue to expand, and new functions of LL-37 may be discovered. Both prevention and treatment strategies are under active development. To avoid infection, we anticipate continued research interest in covalent immobilization of AMPs on medical implants. The basis for this is that peptide injection or non-covalent coating of LL-37 peptides can reduce *S. aureus* infection (Wang et al. 2014a; de Breij et al. 2016). Efforts will also continue to find the optimal approach for LL-37 induction at a right time and location. In winter, sunlight and vitamin D supplement can be beneficial. Our results indicate the importance of early treatment. In addition, the anti-infective potential of LL-37-derived peptides may be expanded by combining them with traditional antibiotics or other approved simple compounds. As a milestone, topical treatment of infections using LL-37-derived peptides has been demonstrated in animal models (e.g., Wang et al. 2014a; de Breij et al. 2018), and OP-145 was used to treat chronic otitis media in a clinical phase 2 trial in 2009 (de Breij et al. 2018). These achievements will reignite the hope to develop the peptide into a new systemic antibiotic, the holy grail of future LL-37 engineering.

Acknowledgments This study is supported by the NIAID/NIH grant R01 AI105147 and AI128230 to GW. This chapter reflects the point of view of the authors and may not represent the funding agency.

References

Agerberth B, Gunne H, Odeberg J, Kogner P, Boman HG, Gudmundsson GH (1995) FALL-39, a putative human peptide antibiotic, is cysteine-free and expressed in bone marrow and testis. Proc Natl Acad Sci U S A 92(1):195–199

Alagarasu K, Patil PS, Shil P, Seervi M, Kakade MB, Tillu H, Salunke A (2017) In-vitro effect of human cathelicidin antimicrobial peptide LL-37 on dengue virus type 2. Peptides 92:23–30

Bals R, Weiner DJ, Meegalla RL, Wilson JM (1999) Transfer of a cathelicidin peptide antibiotic gene restores bacterial killing in a cystic fibrosis xenograft model. J Clin Invest 103:1113–1117

Bergman P, Walter-Jallow L, Broliden K, Agerberth B, Söderlund J (2007) The antimicrobial peptide LL-37 inhibits HIV-1 replication. Curr HIV Res 5(4):410–415

Boman HG (2003) Antibacterial peptides: basic facts and emerging concepts. J Intern Med 254(3):197–215

Boucher HW, Talbot GH, Bradley JS, Edwards JE, Gilbert D, Rice LB, Scheld M, Spellberg B, Bartlett J (2009) Bad bugs, no drugs: no ESKAPE! An update from the Infectious Diseases Society of America. Clin Infect Dis 48:1–12

Braff MH, Hawkins MA, Di Nardo A, Lopez-Garcia B, Howell MD, Wong C, Lin K, Streib JE, Dorschner R, Leung DY, Gallo RL (2005) Structure-function relationships among human cathelicidin peptides: disso ciation of antimicrobial properties from host immunostimulatory activities. J Immunol 174(7):4271–4278

Caiaffa KS, Massunari L, Danelon M, Abuna GF, Bedran TBL, Santos-Filho NA, Spolidorio DMP, Vizoto NL, Cilli EM, Duque C (2017) KR-12-a5 is a non-cytotoxic agent with potent antimicrobial effects against oral pathogens. Biofouling 33(10):807–818

Cassin ME, Ford AJ, Orbach SM, Saverot SE, Rajagopalan P (2016) The design of antimicrobial LL37-modified collagen-hyaluronic acid detachable multilayers. Acta Biomater 40:119–129

Chen G, Zhou M, Chen S, Lv G, Yao J (2009) Nanolayer biofilm coated on magnetic nanoparticles by using a dielectric barrier discharge glow plasma fluidized bed for immobilizing an antimicrobial peptide. Nanotechnology 20(46):465706

Cheng M, Ho S, Yoo JH, Tran DH, Bakirtzi K, Su B, Tran DH, Kubota Y, Ichikawa R, Koon HW (2014) Cathelicidin suppresses colon cancer development by inhibition of cancer associated fibroblasts. Clin Exp Gastroenterol 8:13–29

Choi KY, Chow LN, Mookherjee N (2012) Cationic host defence peptides: multifaceted role in immune modulation and inflammation. J Innate Immun 4(4):361–370

Comune M, Rai A, Chereddy KK, Pinto S, Aday S, Ferreira AF, Zonari A, Blersch J, Cunha R, Rodrigues R, Lerma J, Simões PN, Préat V, Ferreira L (2017) Antimicrobial peptide-gold nanoscale therapeutic formulation with high skin regenerative potential. J Control Release 262:58–71

Cowland JB, Johnsen AH, Borregaard N (1995) hCAP-18, a cathelin/probactenecin-like protein of human neutrophil specific granules. FEBS Lett 368(1):173–176

Crack LR, Jones L, Malavige GN, Patel V, Ogg GS (2012) Human antimicrobial peptides LL-37 and human β-defensin-2 reduce viral replication in keratinocytes infected with varicella zoster virus. Clin Exp Dermatol 37(5):534–543

Currie SM, Findlay EG, McHugh BJ, Mackellar A, Man T, Macmillan D, Wang H, Fitch PM, Schwarze J, Davidson DJ (2013) The human cathelicidin LL-37 has antiviral activity against respiratory syncytial virus. PLoS One 8(8):e73659

Currie SM, Gwyer Findlay E, McFarlane AJ, Fitch PM, Böttcher B, Colegrave N, Paras A, Jozwik A, Chiu C, Schwarze J, Davidson DJ (2016) Cathelicidins have direct antiviral activity against respiratory syncytial virus in vitro and protective function in vivo in mice and humans. J Immunol 196(6):2699–2710

da Silva BR, Conrado AJS, Pereira AL, Evaristo FFV, Arruda FVS, Vasconcelos MA, Lorenzón EN, Cilli EM, Teixeira EH (2017) Antibacterial activity of a novel antimicrobial peptide [W7]KR12-KAEK derived from KR-12 against Streptococcus mutans planktonic cells and biofilms. Biofouling 33(10):835–846

de Breij A, Riool M, Kwakman PH, de Boer L, Cordfunke RA, Drijfhout JW, Cohen O, Emanuel N, Zaat SA, Nibbering PH, Moriarty TF (2016) Prevention of *Staphylococcus aureus* biomaterial-associated infections using a polymer-lipid coating containing the antimicrobial peptide OP-145. J Control Release 222:1–8

de Breij A, Riool M, Cordfunke RA, Malanovic N, de Boer L, Koning RI, Ravensbergen E, Franken M, van der Heijde T, Boekema BK, Kwakman PHS, Kamp N,

El Ghalbzouri A, Lohner K, Zaat SAJ, Drijfhout JW, Nibbering PH. (2018) Sci Transl Med. 10(423). pii: eaan4044

den Hertog AL, van Marle J, Veerman EC, Valentijn-Benz M, Nazmi K, Kalay H, Grün CH, Van't Hof W, Bolscher JG, Nieuw Amerongen AV (2006) The human cathelicidin peptide LL-37 and truncated variants induce segregation of lipids and proteins in the plasma membrane of *Candida albicans*. Biol Chem 387(10–11):1495–1502

Dürr UH, Sudheendra US, Ramamoorthy A (2006) Membrane fragmentation by an amyloidogenic fragment of human islet amyloid polypeptide detected by solid-state NMR spectroscopy of membrane nanotubes. Biochim Biophys Acta 1768(9):2026–2029

Dutta D, Kumar N, D P Willcox M (2016) Antimicrobial activity of four cationic peptides immobilised to polyhydroxyethylmethacrylate. Biofouling 32(4):429–438

Epand RF, Wang G, Berno B, Epand RM (2009) Lipid segregation explains selective toxicity of a series of fragments derived from the human cathelicidin LL-37. Antimicrob Agents Chemother 53(9):3705–3714

Findlay F, Pohl J, Svoboda P, Shakamuri P, McLean K, Inglis NF, Proudfoot L, Barlow PG (2017) Carbon nanoparticles inhibit the antimicrobial activities of the human cathelicidin LL-37 through structural alteration. J Immunol 199(7):2483–2490

Frohm M, Agerberth B, Ahangari G, Stâhle-Bäckdahl M, Lidén S, Wigzell H, Gudmundsson GH (1997) The expression of the gene coding for the antibacterial peptide LL-37 is induced in human keratinocytes during inflammatory disorders. J Biol Chem 272(24):15258–15263

Gabriel M, Nazmi K, Veerman EC, Nieuw Amerongen AV, Zentner A (2006) Preparation of LL-37-grafted titanium surfaces with bactericidal activity. Bioconjug Chem 17(2):548–550

Grönberg A, Mahlapuu M, Ståhle M, Whately-Smith C, Rollman O (2014) Treatment with LL-37 is safe and effective in enhancing healing of hard-to-heal venous leg ulcers: a randomized, placebo-controlled clinical trial. Wound Repair Regen 22:613–621

Gudmundsson GH, Agerberth B, Odeberg J, Bergman T, Olsson B, Salcedo R (1996) The human gene FALL39 and processing of the cathelin precursor to the antibacterial peptide LL-37 in granulocytes. Eur J Biochem 238:325–332

Gunasekera S, Muhammad T, Strömstedt AA, Rosengren KJ, Göransson U (2018) Alanine and lysine scans of the LL-37-derived peptide fragment KR-12 reveal key residues for antimicrobial activity. Chembiochem 19(9):931–939

Gustafsson A, Olin AI, Ljunggren L (2010) LPS interactions with immobilized and soluble antimicrobial peptides. Scand J Clin Lab Invest 70(3):194–200

Harcourt JL, McDonald M, Svoboda P, Pohl J, Tatti K, Haynes LM (2016) Human cathelicidin, LL-37, inhibits respiratory syncytial virus infection in polarized airway epithelial cells. BMC Res Notes 9:11

Hase K, Murakami M, Iimura M, Cole SP, Horibe Y, Ohtake T, Obonyo M, Gallo RL, Eckmann L, Kagnoff MF (2003) Expression of LL-37 by human gastric epithelial cells as a potential host defense mechanism against Helicobacter pylori. Gastroenterology 125(6):1613–1625

He M, Zhang H, Li Y, Wang G, Tang B, Zhao J, Huang Y, Zheng J (2018) Cathelicidin-derived antimicrobial peptides inhibit Zika virus through direct inactivation and interferon pathway. Front Immunol 9:722

Henzler Wildman KA, Lee DK, Ramamoorthy A (2003) Mechanism of lipid bilayer disruption by the human antimicrobial peptide, LL-37. Biochemistry 42(21):6545–6558

Howell MD, Jones JF, Kisich KO, Streib JE, Gallo RL, Leung DY (2004) Selective killing of vaccinia virus by LL-37: implications for eczema vaccinatum. J Immunol 172(3):1763–1767

Jacob B, Park IS, Bang JK, Shin SY (2013) Short KR-12 analogs designed from human cathelicidin LL-37 possessing both antimicrobial and antiendotoxic activities without mammalian cell toxicity. J Pept Sci 19(11):700–707

Jayaram LN, Chen JY (2015) Antimicrobial peptides: possible anti-infective agents. Peptides 72:88–94

Jiang W, Sunkara LT, Zeng X, Deng Z, Myers SM, Zhang G (2013) Differential regulation of human cathelicidin LL-37 by free fatty acids and their analogs. Peptides 50:129–138

Kahlenberg JM, Kaplan MJ (2013) Little peptide, big effects: the role of LL-37 in inflammation and autoimmune disease. J Immunol 191(10):4895–4901

Kanthawong S, Bolscher JG, Veerman EC, van Marle J, Nazmi K, Wongratanacheewin S, Taweechaisupapong S (2010) Antimicrobial activities of LL-37 and its truncated variants against Burkholderia thailandensis. Int J Antimicrob Agents 36(5):447–452

Kay LE, Ikura M, Tschudin R, Bax A (2011) Three-dimensional triple-resonance NMR Spectroscopy of isotopically enriched proteins. 1990. J Magn Reson 213(2):423–441

Keifer PA, Peterkofsky A, Wang G (2004) Effects of detergent alkyl chain length and chemical structure on the properties of a micelle-bound bacterial membrane targeting peptide. Anal Biochem 331:33–39

Klevens RM, Morrison MA, Nadle J et al (2007) Invasive methicillin-resistant *Staphylococcus aureus* infections in the United States. JAMA 298:1763–1771

Koziel J, Bryzek D, Sroka A, Maresz K, Glowczyk I, Bielecka E, Kantyka T, Pyrć K, Svoboda P, Pohl J, Potempa J (2014) Citrullination alters immunomodulatory function of LL-37 essential for prevention of endotoxin-induced sepsis. J Immunol 192(11):5363–5372

Kuroda K, Fukuda T, Yoneyama H, Katayama M, Isogai H, Okumura K, Isogai E (2012) Anti-proliferative effect of an analogue of the LL-37 peptide in the colon cancer derived cell line HCT116 p53+/+ and p53−/−. Oncol Rep 28(3):829–834

Kyte J, Doolittle RF (1982) A simple method for displaying the hydropathic character of a protein. J Mol Biol 157:105–132

Larrick JW, Hirata M, Balint RF, Lee J, Zhong J, Wright SC (1995) Human CAP18: a novel antimicrobial lipopolysaccharide-binding protein. Infect Immun 63(4):1291–1297

Lee PH, Ohtake T, Zaiou M, Murakami M, Rudisill JA, Lin KH, Gallo RL (2005) Expression of an additional cathelicidin antimicrobial peptide protects against bacterial skin infection. Proc Natl Acad Sci U S A 102:3750–3755

Lee CC, Sun Y, Qian S, Huang HW (2011) Transmembrane pores formed by human antimicrobial peptide LL-37. Biophys J 100(7):1688–1696

Lehrer RI, Lu W (2012) α-Defensins in human innate immunity. Immunol Rev 245(1):84–112

Li J, Perez Perez GI. (2018) Is there a role for the non-helicobacter pylori bacteria in the risk of developing gastric cancer? Int J Mol Sci 19(5) pii: E1353

Li X, Li Y, Han H, Miller DW, Wang G (2006a) Solution structures of human LL-37 fragments and NMR-based identification of a minimal membrane-targeting antimicrobial and anticancer region. J Am Chem Soc 128:5776–5785

Li Y, Li X, Wang G (2006b) Cloning, expression, isotope labeling, and purification of human antimicrobial peptide LL-37 in *Escherichia coli* for NMR studies. Protein Expr Purif 47:498–505

Li Y, Li X, Li H, Lockridge O, Wang G (2007) A novel method for purifying recombinant human host defense cathelicidin LL-37 by utilizing its inherent property of aggregation. Protein Expr Purif 54:157–165

McArthur MA (2017) Zika virus: Recent advances toward the development of vaccines and therapeutics. Viruses 9(6) pii: E143

Meister M, Lemaitre B, Hoffmann JA (1997) Antimicrobial peptide defense in Drosophila. BioEssays 19(11):1019–1026

Mishra B, Wang G (2012) *Ab initio* design of potent anti-MRSA peptides based on database filtering technology. J Am Chem Soc 134:12426–12429

Mishra B, Wang G (2017) Individual and combined effects of engineered peptides and antibiotics on the Pseudomonas aeruginosa biofilms. Pharmaceuticals 10(3):58

Mishra B, Epand RF, Epand RM, Wang G (2013) Structural location determines functional roles of the basic amino acids of KR-12, the smallest antimicrobial peptide from human cathelicidin LL-37. RSC Adv 3:19560–19571

Mishra B, Golla R, Lau K, Lushnikova T, Wang G (2016) Anti-Staphylococcal biofilm effects of human cathelicidin peptides. ACS Med Chem Lett 7:117–121

Mishra B, Lushnikova T, Golla RM, Wang X, Wang G (2017a) Design and surface immobilization of short anti-biofilm peptides. Acta Biomater 49:316–328

Mishra B, Reiling S, Zarena D, Wang G (2017b) Host defense antimicrobial peptide as antibiotics: design and application strategies. Curr Opin Chem Biol 38:87–96

Mishra B, Wang X, Lushnikova T, Zhang Y, Golla RM, Narayana JL, Wang C, McGuire TR, Wang G (2018) Antibacterial, antifungal, anticancer activities and structural bioinformatics analysis of six naturally occurring temporins. Peptides 106:9–20

Murakami M, Lopez-Garcia B, Braff M, Dorschner RA, Gallo RL (2004) Postsecretory processing generates multiple cathelicidins for enhanced topical antimicrobial defense. J Immunol 172:3070–3077

Murakami M, Kameda K, Tsumoto H, Tsuda T, Masuda K, Utsunomiya R, Mori H, Miura Y, Sayama K (2017) TLN-58, an additional hCAP18 processing form, found in the lesion vesicle of palmoplantar pustulosis in the skin. J Invest Dermatol 137(2):322–331

Nan YH, Bang JK, Jacob B, Park IS, Shin SY (2012) Prokaryotic selectivity and LPS-neutralizing activity of short antimicrobial peptides designed from the human antimicrobial peptide LL-37. Peptides 35(2):239–247

Nell MJ, Tjabringa GS, Wafelman AR, Verrijk R, Hiemstra PS, Drijfhout JW, Grote JJ (2006) Development of novel LL-37 derived antimicrobial peptides with LPS and LTA neutralizing and antimicrobial activities for therapeutic application. Peptides 27(4):649–660

Nie B, Ao H, Chen C, Xie K, Zhou J, Long T, Tang T, Yue B (2016) Covalent immobilization of KR-12 peptide onto a titanium surface for decreasing infection and promoting osteogenic differentiation. RSC Adv 6(52):46733–46743

Niemirowicz K, Durnaś B, Tokajuk G, Piktel E, Michalak G, Gu X, Kułakowska A, Savage PB, Bucki R (2017) Formulation and candidacidal activity of magnetic nanoparticles coated with cathelicidin LL-37 and ceragenin CSA-13. Sci Rep 7(1):4610

Nizet V, Ohtake T, Lauth X, Trowbridge J, Rudisill J, Dorschner RA, Pestonjamasp V, Piraino J, Huttner K, Gallo RL (2001) Innate antimicrobial peptide protects the skin from invasive bacterial infection. Nature 414(6862):454–457

Ogawa Y, Kawamura T, Matsuzawa T, Aoki R, Gee P, Yamashita A, Moriishi K, Yamasaki K, Koyanagi Y, Blauvelt A, Shimada S (2013) Antimicrobial peptide LL-37 produced by HSV-2-infected keratinocytes enhances HIV infection of Langerhans cells. Cell Host Microbe 13(1):77–86

Ohsaki Y, Gazdar AF, Chen HC, Johnson BE (1992) Antitumor activity of magainin analogues against human lung cancer cell lines. Cancer Res 52(13):3534–3538

Oren Z, Lerman JC, Gudmundsson GH, Agerberth B, Shai Y (1999) Structure and organization of the human antimicrobial peptide LL-37 in phospholipid membranes: relevance to the molecular basis for its non-cell-selective activity. Biochem J 341(Pt 3):501–513

Overhage J, Campisano A, Bains M, Torfs EC, Rehm BH, Hancock RE (2008) Human host defense peptide

LL-37 prevents bacterial biofilm formation. Infect Immun 76(9):4176–4182
Overhage PE, McHugh B, Gwyer Findlay E, Mackellar A, Mackenzie KJ, Gallo RL, Govan JR, Simpson AJ, Davidson DJ (2014) Cathelicidin host defence peptide augments clearance of pulmonary Pseudomonas aeruginosa infection by its influence on neutrophil function in vivo. PLoS One 9:e99029
Peschel A, Sahl HG (2006) The co-evolution of host cationic antimicrobial peptides and microbial resistance. Nat Rev Microbiol 4:529–536
Pütsep K, Carlsson G, Boman HG, Andersson M (2002) Deficiency of antibacterial peptides in patients with morbus kostmann: an observation study. Lancet 360:1144–1149
Rajasekaran G, Kim EY, Shin SY (2017) LL-37-derived membrane-active FK-13 analogs possessing cell selectivity, anti-biofilm activity and synergy with chloramphenicol and anti-inflammatory activity. Biochim Biophys Acta 1859(5):722–733
Rapala-Kozik M, Bochenska O, Zawrotniak M, Wolak N, Trebacz G, Gogol M, Ostrowska D, Aoki W, Ueda M, Kozik A (2015) Inactivation of the antifungal and immunomodulatory properties of human cathelicidin LL-37 by aspartic proteases produced by the pathogenic yeast *Candida albicans*. Infect Immun 83(6):2518–2530
Ren SX, Cheng AS, To KF, Tong JH, Li MS, Shen J, Wong CC, Zhang L, Chan RL, Wang XJ, Ng SS, Chiu LC, Marquez VE, Gallo RL, Chan FK, Yu J, Sung JJ, Wu WK, Cho CH (2012) Host immune defense peptide LL-37 activates caspase-independent apoptosis and suppresses colon cancer. Cancer Res 72(24):6512–6523
Ren SX, Shen J, Cheng AS, Lu L, Chan RL, Li ZJ, Wang XJ, Wong CC, Zhang L, Ng SS, Chan FL, Chan FK, Yu J, Sung JJ, Wu WK, Cho CH (2013) FK-16 derived from the anticancer peptide LL-37 induces caspase-independent apoptosis and autophagic cell death in colon cancer cells. PLoS One 8(5):e63641
Ron-Doitch S, Sawodny B, Kühbacher A, David MMN, Samanta A, Phopase J, Burger-Kentischer A, Griffith M, Golomb G, Rupp S (2016) Reduced cytotoxicity and enhanced bioactivity of cationic antimicrobial peptides liposomes in cell cultures and 3D epidermis model against HSV. J Control Release 229:163–171
Saar-Dover R, Bitler A, Nezer R, Shmuel-Galia L, Firon A, Shimoni E, Trieu-Cuot P, Shai Y (2012) D-alanylation of lipoteichoic acids confers resistance to cationic peptides in group B streptococcus by increasing the cell wall density. PLoS Pathog 8:e1002891
Santos CM, Kumar A, Kolar SS, Contreras-Caceres R, McDermott A, Cai C (2013) Immobilization of antimicrobial peptide IG-25 onto fluoropolymers via fluorous interactions and click chemistry. ACS Appl Mater Interfaces 5(24):12789–12793
Saporito P, Vang Mouritzen M, Løbner-Olesen A, Jenssen H (2018) LL-37 fragments have antimicrobial activity against *Staphylococcus epidermidis* biofilms and wound healing potential in HaCaT cell line. J Pept Sci 8:e3080
Schauber J, Iffland K, Frisch S, Kudlich T, Schmausser B, Eck M, Menzel T, Gostner A, Lührs H, Scheppach W (2004) Histone-deacetylase inhibitors induce the cathelicidin LL-37 in gastrointestinal cells. Mol Immunol 41(9):847–854
Schögler A, Muster RJ, Kieninger E, Casaulta C, Tapparel C, Jung A, Moeller A, Geiser T, Regamey N, Alves MP (2016) Vitamin D represses rhinovirus replication in cystic fibrosis cells by inducing LL-37. Eur Respir J 47:520–530
Scott MG, Davidson DJ, Gold MR, Bowdish D, Hancock RE (2002) The human antimicrobial peptide LL-37 is a multifunctional modulator of innate immune responses. J Immunol 169(7):3883–3891
Sevcsik E, Pabst G, Richter W, Danner S, Amenitsch H, Lohner K (2008) Interaction of LL-37 with model membrane systems of different complexity: influence of the lipid matrix. Biophys J 94:4688–4699
Shah NR, Hancock RE, Fernandez RC (2014) Bordetella pertussis lipid A glucosamine modification confers resistance to cationic antimicrobial peptides and increases resistance to outer membrane perturbation. Antimicrob Agents Chemother 58(8):4931–4934
Sieprawska-Lupa M, Mydel P, Krawczyk K, Wójcik K, Puklo M, Lupa B, Suder P, Silberring J, Reed M, Pohl J, Shafer W, McAleese F, Foster T, Travis J, Potempa J (2004) Degradation of human antimicrobial peptide LL-37 by *Staphylococcus aureus*-derived proteinases. Antimicrob Agents Chemother 48(12):4673–4679
Sigurdardottir T, Andersson P, Davoudi M, Malmsten M, Schmidtchen A, Bodelsson M (2006) In silico identification and biological evaluation of antimicrobial peptides based on human cathelicidin LL-37. Antimicrob Agents Chemother 50(9):2983–2989
Singh D, Vaughan R, Kao CC (2014) LL-37 peptide enhancement of signal transduction by Toll-like receptor 3 is regulated by pH: identification of a peptide antagonist of LL-37. J Biol Chem 289(40):27614–27624
Skerlavaj B, Gennaro R, Bagella L, Merluzzi L, Risso A, Zanetti M (1996) Biological characterization of two novel cathelicidin-derived peptides and identification of structural requirements for their antimicrobial and cell lytic activities. J Biol Chem 271(45):28375–28381
Sol A, Ginesin O, Chaushu S, Karra L, Coppenhagen-Glazer S, Ginsburg I, Bachrach G (2013) LL-37 opsonizes and inhibits biofilm formation of Aggregatibacter actinomycetemcomitans at subbactericidal concentrations. Infect Immun 81:3577–3585
Song DW, Kim SH, Kim HH, Lee KH, Ki CS, Park YH (2016) Multi-biofunction of antimicrobial peptide-immobilized silk fibroin nanofiber membrane: implications for wound healing. Acta Biomater 39:146–155
Sorensen OE, Follin P, Johnsen AH, Calafat J, Tjabringa GS, Hiemstra PS et al (2001) Human cathelicidin, hCAP-18, is processed to the antimicrobial peptide

LL-37 by extracellular cleavage with proteinase 3. Blood 97(12):3951–3959

Sørensen OE, Gram L, Johnsen AH, Andersson E, Bangsbøll S, Tjabringa GS, Hiemstra PS, Malm J, Egesten A, Borregaard N (2003) Processing of seminal plasma hCAP-18 to ALL-38 by gastricsin: a novel mechanism of generating antimicrobial peptides in vagina. J Biol Chem 278(31):28540–28546

Sousa FH, Casanova V, Findlay F, Stevens C, Svoboda P, Pohl J, Proudfoot L, Barlow PG (2017) Cathelicidins display conserved direct antiviral activity towards rhinovirus. Peptides 95:76–83

Spencer JD, Schwaderer AL, Becknell B, Watson J, Hains DS (2014) The innate immune response during urinary tract infection and pyelonephritis. Pediatr Nephrol 29(7):1139–1149

Sung CM, Kim HC, Cho YB, Shin SY, Jang CH (2017) Evaluating the ototoxicity of an anti-MRSA peptide KR-12-a2. Braz J Otorhinolaryngol pii: S1808-8694(17)30078-2

Suzuki K, Murakami T, Kuwahara-Arai K, Tamura H, Hiramatsu K, Nagaoka I (2011) Human anti-microbial cathelicidin peptide LL-37 suppresses the LPS-induced apoptosis of endothelial cells. Int Immunol 23(3):185–193

Suzuki K, Murakami T, Hu Z, Tamura H, Kuwahara-Arai K, Iba T, Nagaoka I (2016) Human host defense cathelicidin peptide LL-37 enhances the lipopolysaccharide uptake by liver sinusoidal endothelial cells without cell activation. J Immunol 196(3):1338–1347

Tack BF, Sawai MV, Kearney WR, Robertson AD, Sherman MA, Wang W, Hong T, Boo LM, Wu H, Waring AJ, Lehrer RI (2002) SMAP-29 has two LPS-binding sites and a central hinge. Eur J Biochem 269(4):1181–1189

Takeuchi O, Hoshino K, Kawai T, Sanjo H, Takada H, Ogawa T, Takeda K, Akira S (1999) Differential roles of TLR2 and TLR4 in recognition of gram-negative and gram-positive bacterial cell wall components. Immunity 11(4):443–451

Tanphaichitr N, Srakaew N, Alonzi R, Kiattiburut W, Kongmanas K, Zhi R, Li W, Baker M, Wang G, Hickling D (2016) Potential use of antimicrobial peptides as vaginal spermicides/microbicides. Pharmaceuticals 9:13

Tian M, Zhao DY, Wang HW (2011) Studies on the core functional region of antimicrobial peptide LL-37 for inhibition of RSV replication. Zhonghua Shi Yan He Lin Chuang Bing Du Xue Za Zhi 25(5):355–357. In Chinese

Tripathi S, Tecle T, Verma A, Crouch E, White M, Hartshorn KL (2013) The human cathelicidin LL-37 inhibits influenza A viruses through a mechanism distinct from that of surfactant protein D or defensins. J Gen Virol 94(Pt 1):40–49

Tripathi S, Verma A, Kim EJ, White MR, Hartshorn KL (2014) LL-37 modulates human neutrophil responses to influenza A virus. J Leukoc Biol 96(5):931–938

Tripathi S, Wang G, White M, Rynkiewicz M, Seaton B, Hartshorn K (2015a) Identifying the critical domain of LL-37 involved in mediating neutrophil activation in the presence of influenza virus: functional and structural analysis. PLoS One 10(8):e0133454

Tripathi S, Wang G, White M, Qi L, Taubenberger J, Hartshorn KL (2015b) Antiviral activity of the human cathelicidin, LL-37, and derived peptides on seasonal and pandemic influenza A viruses. PLoS One 10:e0124706

Turner J, Cho Y, Dinh NN, Waring AJ, Lehrer RI (1998) Activities of LL-37, a cathelin-associated antimicrobial peptide of human neutrophils. Antimicrob Agents Chemother 42(9):2206–2214

Uchio E, Inoue H, Kadonosono K (2013) Anti-adenoviral effects of human cationic antimicrobial protein-18/LL-37, an antimicrobial peptide, by quantitative polymerase chain reaction. Korean J Ophthalmol 27(3):199–203

van der Does AM, Bergman P, Agerberth B, Lindbom L (2012) Induction of the human cathelicidin LL-37 as a novel treatment against bacterial infections. J Leukoc Biol 92(4):735–742

Vandamme D, Landuyt B, Luyten W, Schoofs L (2012) A comprehensive summary of LL-37, the factotum human cathelicidin peptide. Cell Immunol 280(1):22–35

Velarde JJ, Ashbaugh M, Wessels MR (2014) The human antimicrobial peptide LL-37 binds directly to CsrS, a sensor histidine kinase of group A Streptococcus, to activate expression of virulence factors. J Biol Chem 289(52):36315–36324

Wang G (2006) Structural biology of antimicrobial peptides by NMR spectroscopy. Curr Org Chem 10:569–581

Wang G (2007) Determination of solution structure and lipid micelle location of an engineered membrane peptide by using one NMR experiment and one sample. Biochim Biophys Acta 1768:3271–3281

Wang G (2008) Structures of human host defense cathelicidin LL-37 and its smallest antimicrobial peptide KR-12 in lipid micelles. J Biol Chem 283:32637–32643

Wang G (2010) Structure, dynamics and mapping of membrane-binding residues of micelle-bound antimicrobial peptides by natural abundance ^{13}C NMR spectroscopy. Biochim Biophys Acta 1798:114–121

Wang G (2012) Natural antimicrobial peptides as promising anti-HIV candidates. Curr Top Peptide Protein Res 13:93–110

Wang G (2014) Human antimicrobial peptides and proteins. Pharmaceuticals 7:545–594

Wang G (2017) In: Wang G (ed) Antimicrobial peptides. CABI, UK, pp 1–19. and 169–187

Wang Z, Wang G (2004) APD: the antimicrobial peptide database. Nucleic Acids Res 32:D590–D592

Wang G, Keifer PA, Peterkofsky A (2004) Short-chain diacyl phosphatidylglycerols: which one to choose for NMR structural determination of a membrane-associated peptide from *Escherichia coli?* Spectroscopy 18:257–264

Wang G, Waston K, Buckheit R Jr (2008) Anti-human immunodeficiency virus type 1 (HIV-1) activities of antimicrobial peptides derived from human and bovine cathelicidins. Antimicrob Agents Chemother 52:3438–3440

Wang G, Li X, Wang Z (2009) APD2: the updated antimicrobial peptide database and its application in peptide design. Nucleic Acids Res 37:D933–D937

Wang G, Elliott M, Cogen AL, Ezell EL, Gallo RL, Hancock REW (2012a) Structure, dynamics, antimicrobial and immune modulatory activities of human LL-23 and its single residue variants mutated based on homologous primate cathelicidins. Biochemistry 51:653–664

Wang G, Epand RF, Mishra B, Lushnikova T, Thomas VC, Bayles KW, Epand R (2012b) Decoding the functional roles of cationic side chains of the major antimicrobial region of human cathelicidin LL-37. Antimicrob Agents Chemother 56:845–856

Wang G, Hanke ML, Mishra B, Lushnikova T, Heim CE, Chittezham Thomas V, Bayles KW, Kielian T (2014a) Transformation of human cathelicidin LL-37 into selective, stable, and potent antimicrobial compounds. ACS Chem Biol 9:1997–2002

Wang G, Mishra B, Epand RF, Epand RM (2014b) High-quality 3D Structure shine light on antibacterial, anti-biofilm and antiviral activities of human cathelicidin LL-37 and its fragments. Biochim Biophys Acta 1838:2160–2172

Wang G, Mishra B, Lau K, Lushnikova T, Golla R, Wang X (2015) Antimicrobial peptides in 2014. Pharmaceuticals 8:123–150

Wang G, Li X, Wang Z (2016) APD3: the antimicrobial peptide database as a tool for research and education. Nucleic Acids Res 44:D1087–D1093

Wang X, Junior JCB, Mishra B, Lushnikova T, Epand RM, Wang G (2017) Arginine-lysine positional swap of the LL-37 peptides reveals evolutional advantages of the native sequence and leads to bacterial probes. Biochim Biophys Acta 1859:1350–1361

Wang X, Mishra B, Lushinikova T, Narayana JL, Wang G (2018) Amino acid composition determines peptide activity spectrum and hot spot-based design of merecidin. Adv Biosyst 2:1700259

Wong JH, Legowska A, Rolka K, Ng TB, Hui M, Cho CH, Lam WW, Au SW, Gu OW, Wan DC (2011) Effects of cathelicidin and its fragments on three key enzymes of HIV-1. Peptides 32(6):1117–1122

Wu WK, Sung JJ, To KF, Yu L, Li HT, Li ZJ, Chu KM, Yu J, Cho CH (2010a) The host defense peptide LL-37 activates the tumor-suppressing bone morphogenetic protein signaling via inhibition of proteasome in gastric cancer cells. J Cell Physiol 223(1):178–186

Wu WK, Wang G, Coffelt SB, Betancourt AM, Lee CW, Fan D, Wu K, Yu J, Sung JJ, Cho CH (2010b) Emerging roles of the host defense peptide LL-37 in human cancer and its potential therapeutic applications. Int J Cancer 127(8):1741–1747

Yasin B, Pang M, Turner JS, Cho Y, Dinh NN, Waring AJ, Lehrer RI, Wagar EA (2000) Evaluation of the inactivation of infectious Herpes simplex virus by host-defense peptides. Eur J Clin Microbiol Infect Dis 19(3):187–194

Zasloff M (1987) Magainins, a class of antimicrobial peptides from Xenopus skin: isolation, characterization of two active forms, and partial cDNA sequence of a precursor. Proc Natl Acad Sci U S A 84(15):5449–5453

Zasloff M (2002) Antimicrobial peptides of multicellular organisms. Nature 415(6870):389–395

Zhang LJ, Guerrero-Juarez CF, Hata T, Bapat SP, Ramos R, Plikus MV, Gallo RL (2015) Innate immunity. Dermal adipocytes protect against invasive *Staphylococcus aureus* skin infection. Science 347:67–71

Application of Synthetic Molecular Evolution to the Discovery of Antimicrobial Peptides

13

William C. Wimley

Abstract
Despite long-standing promise and many known examples, antimicrobial peptides (AMPs) have failed, with few exceptions, to significantly impact human medicine. Impediments to the systemic activity of AMPs include proteolysis, host cell interactions, and serum protein binding, factors that are not often considered in the early stages of AMP development. Here we discuss how synthetic molecular evolution, iterative cycles of library design, and physiologically relevant screening can be used to evolve AMPs that do not have these impediments.

13.1 Introduction

Antimicrobial peptides were first described in the 1980s, having been found in insect hemolymph (Steiner et al. 1981; Okada and Natori 1983), mammalian neutrophil granules (Patterson-Delafield et al. 1981), and frog skin secretions (Zasloff, 1987). Subsequently, more than a thousand natural AMPs have been found in many tissues of many different species (Fjell et al. 2007; Wang et al. 2016). Many more synthetic AMPs have been also created, often by mimicking natural sequences in combination with trial and error experimentation (Fjell et al. 2007; Wang et al. 2016) and sometimes by screening or computer-aided design (Rathinakumar and Wimley 2010; Rathinakumar et al. 2009; Moy et al. 2009; Kulagina et al. 2006, 2007; Hilpert et al. 2005). Since the beginning, AMPs have been promoted as novel antibiotics that might improve human health and well-being. Yet, at the time of their initial discovery, there was little urgency to the translational applications of AMPs. It was not known at the time that drug-resistant bacterial infections would grow over the next 30 years to become a global health crisis in morbidity and mortality (Arias and Murray 2009; Boucher et al. 2009; Otto 2012). We are now urgently and ever increasingly in need of novel antibiotic treatment options against drug-resistant bacteria. While some AMPs have been developed into potential topical drugs, and some are nearing clinical trials, or are in clinical trials (Fox 2013), AMPs have not succeeded, recently or in past decades (Gordon et al. 2005), to have any real impact on the systemic treatment options for drug-resistant bacteria.

Many of the known AMPs have good antibiotic activity in the culture tube, microwell plate, and petri dish, i.e., under standard laboratory conditions. Many known AMPs have potent, sterilizing activity at low μM concentrations against

W. C. Wimley (✉)
Department of Biochemistry and Molecular Biology, Tulane University School of Medicine, New Orleans, LA, USA
e-mail: wwimley@tulane.edu

K. Matsuzaki (ed.), *Antimicrobial Peptides*, Advances in Experimental Medicine and Biology 1117,
https://doi.org/10.1007/978-981-13-3588-4_13

multiple species of bacteria (Easton et al. 2009; Hamill et al. 2008), often including both Gram-positive and Gram-negative species. AMP activity is observed at the same concentration range, 1–10 μM, at which many conventional antibiotics are active under laboratory conditions. AMPs often have equally potent activity against drug-susceptible, drug-resistant, and multidrug-resistant bacteria (de Breij et al. 2018; Schlusselhuber et al. 2014; Mechkarska et al. 2013; Park et al. 2011) showing that the conventional mechanisms of drug resistance do not apply to AMPs. AMPs can also act against biofilms (de Breij et al. 2018; Wolfmeier et al. 2017), and they can create antibacterial surfaces by covalent tethering or physical adsorption (Kazemzadeh-Narbat et al. 2010). Importantly, AMPs may be less likely to induce resistance than conventional antibiotics (Dobson et al. 2013; Pollard et al. 2012; Fedders et al. 2010) although resistance to AMPs does occur (Perron et al. 2006). Despite their potent, broad-spectrum activity in the laboratory at low concentration, AMPs have not reached the ultimate goal: development into novel antibiotics that can be used systemically to prevent or treat drug-resistant bacterial infections.

This dearth of systemically active AMPs has many causes but may be due in part to impediments to bioavailability and dosing such as host cell inhibition (Starr et al. 2016), serum inhibition (de Breij et al. 2018), residual toxicity (Yeaman and Yount 2003), and proteolytic degradation (Starr and Wimley 2017). Systemically active AMPs are almost certainly possible, as none of these impediments seems to be unsurmountable on its own. In fact, a systemically active, lifesaving, anti-infective peptide drug already exists. The anti-HIV peptide enfuvirtide, which has been approved for human use in the USA and Europe since 2003 (Poveda et al. 2005), is a linear 36-residue peptide drug with over $1 billion in net sales (Poveda et al. 2005; Joly et al. 2010; LaBonte et al. 2003). This peptide is administered subcutaneously in 90 mg doses and has extended the lives of many patients infected with HIV that had become resistant to other drugs. Enfuvirtide has a long half-life in the human body (Joly et al. 2010) despite the lack of specific modifications to increase bioavailability or decrease proteolytic sensitivity. The peptide is an amphipathic α-helical peptide (Rapaport et al. 1995) that likely binds to cells and serum proteins, and this may help it to remain intact and in circulation. Despite this, enfuvirtide has low toxicity and is able to effectively inhibit the fusion of HIV viruses with cell membranes in vivo, probably by interfering with the structure-function relationships of the GP41 fusion protein, from which enfuvirtide was obtained (Qureshi et al. 1990).

If there are no insurmountable barriers to systemically active AMPs and we have thousands of known AMPs with activity in the laboratory, it is reasonable to ask: Why are there no systemically active AMP drugs? Why does there seem to be few in the development pipeline? In this chapter we hypothesize that the number of potential systemically active AMP drugs in the drug development pipeline is small because (i) the thousands of AMPs known have not evolved to be systemic AMPs or have not been discovered under the most relevant conditions and because (ii) rational engineering of AMP properties is not possible due to the fact that we do not have a sufficient knowledge of sequence-function relationships for any of these impediments or for antibacterial activity. Below, we discuss an approach that may be especially well suited in this situation: discovery of novel AMPs by iteratively screening small peptide libraries under experimental conditions that are increasingly relevant to physiological conditions. We have referred to this approach as "synthetic molecular evolution" (Krauson et al. 2013; Kauffman et al. 2018; Li et al. 2018). By these means we suggest approaching the most clinically relevant conditions in a stepwise manner and doing so as a first stage in the preclinical identification of peptide antibiotic drugs to feed a larger number of relevant candidates into the development pipeline. Below we detail some of the major impediments to systemic activity of AMPs and then describe how screening for AMP activity can be done under conditions that much more closely mimic in vivo conditions to identify peptides without these impediments.

13.2 Impediments to Systemic Activity

Compared to conventional antibiotic drugs, AMPs have a fundamentally different mode of action on bacteria. Most conventional antibiotics inhibit a critical biochemical process by targeting one molecule (e.g., an enzyme or ribosome). The number of functional molecules decreases until the microbe loses so much of that essential function that it cannot replicate or survive. The rate or degree to which critical activity is lost depends on local drug concentration. AMPs on the other hand kill microbes in a cooperative, saturation-like process in which the peptides must massively accumulate on bacteria to levels that essentially saturate the cell in order to kill them through the effect of their interfacial activity (Wimley 2010) on membrane integrity. Various measurements in the literature (Starr et al. 2016; Savini et al. 2018, 2017; Tran et al. 2002; Steiner et al. 1988) have shown that the number of bound AMPs required to kill a bacterial cell is extremely high, from 10^7 to 10^8 peptides *per cell.* Since an *E. coli* cell can be expected to have perhaps 2×10^7 total lipids, this means that killing does not occur until there is around one peptide bound *for every bacterial lipid*, which is more peptide than can possibly bind to the cytoplasmic membrane alone. This lethal amount of peptide is also equivalent to roughly one peptide per DNA base. If we assume that 10^8 peptides are evenly distributed in the volume of the cell, the local concentration of AMP is about 80 mM (Starr et al. 2016). In reality, AMPs will specifically accumulate on anionic structures such as cell wall, LPS, cytoplasmic membrane, and DNA and could have local concentrations that approach molar. As a result, the mode of action of AMPs is a successive attack on the cell architecture. Membranes are permeabilized within minutes (Rathinakumar and Wimley 2010; Rathinakumar et al. 2009), followed by leakage of macromolecules, including DNA. Within 60 min of treatment of bacteria with AMPs, the entire cell architecture is compromised, and individual cells are sometimes not discernable. It is thus reasonable that resistance would be more difficult to evolve, given that mechanism of action of AMPs may involve the entire cell architecture. When resistance is observed, it is usually due to changes in LPS or cell wall components (Peschel and Sahl 2006; Peschel 2002), making these structures less anionic so they do not accumulate large amounts of cationic AMPs.

13.3 Serum and Host Cell Inhibition

The need for accumulation of peptide means that systemic activity of AMPs faces different challenges than conventional drugs (Fig. 13.1) and will exhibit threshold behavior. There may be only a narrow window between saturation/killing of bacteria and survivable accumulation of an AMP (Fig. 13.2). Thus, any factor that competes for bacterial binding has the capacity to decrease accumulation on bacteria. Contrary to the commonly stated belief, external eukaryotic membranes are highly anionic overall, due to the large amount of anionic glycoconjugates attached to lipids and proteins. Thus, cationic AMPs bind to eukaryotic cells and tissues, at least moderately. Even weak competition can be problematic because host cells and tissue will always be orders of magnitude more abundant than pathogens. Using a set of 12 natural and synthetic AMPs, we have shown that even a few minutes of preincubation of AMPs with human erythrocytes strongly reduced the activity of most of them (Starr et al. 2016) (see Fig. 13.3) through a combination of host cell binding and proteolysis by the cytosolic proteases of RBCs (Starr and Wimley 2017). This effect is not always observed, indicating that it is a surmountable impediment. For example, the insect peptide cecropin A was not affected in our study by human RBCs (Starr et al. 2016). Similarly Stella and colleagues carefully examined the effect of RBCs on the activity of an AMP and found little inhibition (Savini et al. 2017).

Serum proteins, especially serum albumin, can also bind cationic AMPs and are also highly concentrated in the body (35–50 mg/mL in blood), further potentially reducing the effective concentration of peptide available to bind to

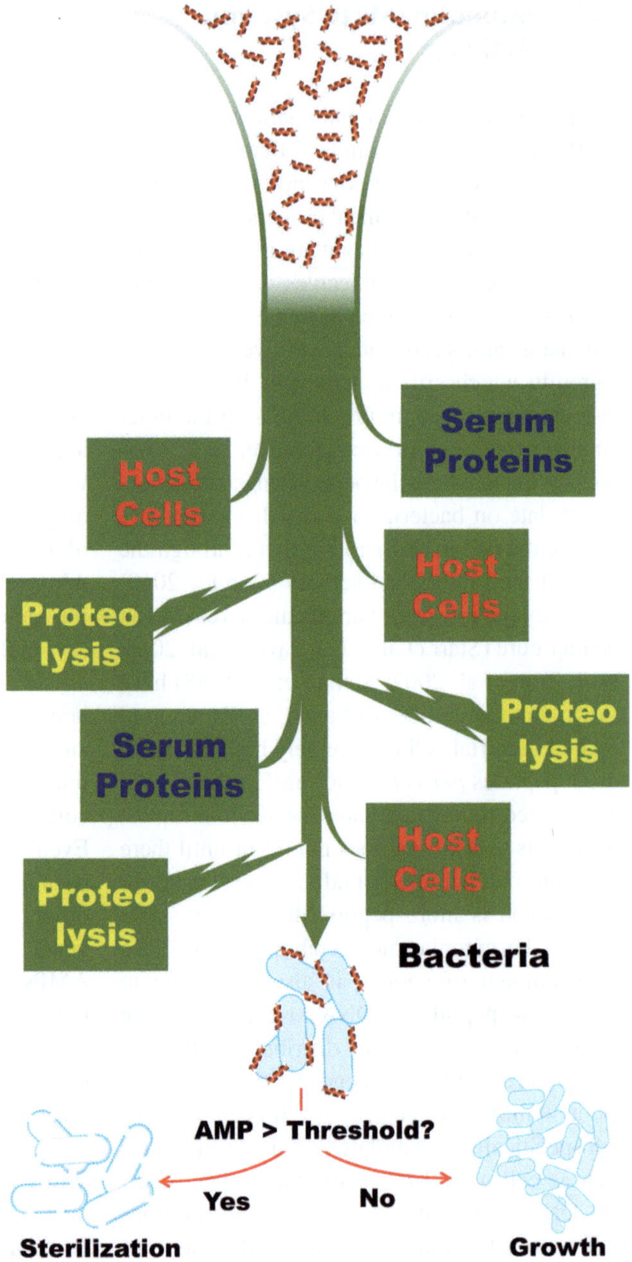

Fig. 13.1 Some impediments to the bioavailability and systemic activity of antimicrobial peptides. Antimicrobial peptides must accumulate significantly on bacteria to have bactericidal activity. In the body, interactions with host cells and tissue, interactions with serum proteins, and proteolytic degradation can decrease accumulation and decrease activity

bacteria (de Breij et al. 2018). As we discuss in detail below, host cell and serum protein binding are rarely considered in the early stages of novel AMP discovery or design. If they are tested at all, it is determined how much these factors interfere with AMP activity only very late in the preclinical development pipeline. Here we are proposing that these factors be included during initial discovery of AMPs by screening. This will presumably give rise to a large number of relevant AMPs that can enter the pipeline. Loading the front of the development pipeline with better candidates, in turn, will increase the probability of finding a few that can be developed into systemic drugs.

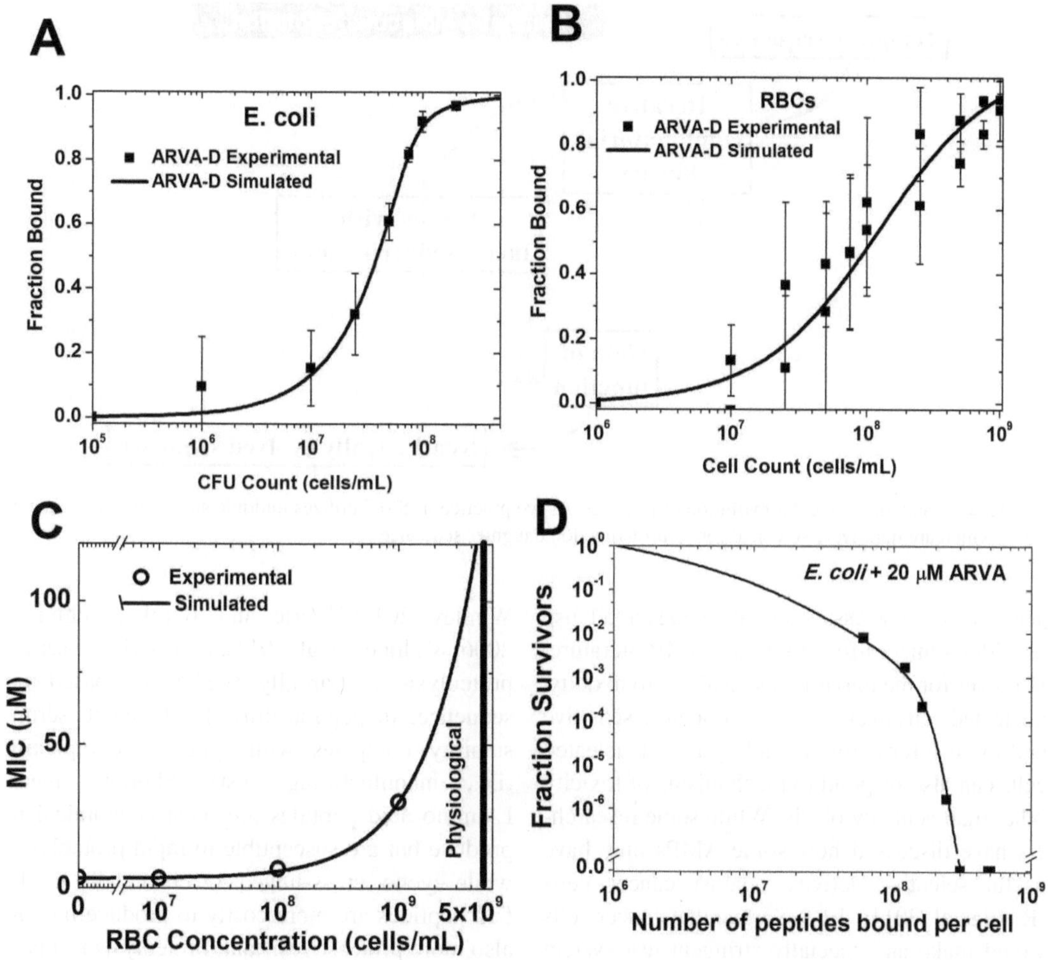

Fig. 13.2 Saturation-dependent activity of an antimicrobial peptide and its inhibition by host cells. The protease-resistant AMP D-ARVA (rrgwalrlvlay-NH_2) was used in these experiments (Starr et al. 2016; Rathinakumar and Wimley 2010; Rathinakumar et al. 2009). (**a**) Measured binding of D-ARVA to *E. coli* cells. (**b**) Measured binding of D-ARVA to human RBCs. (**c**) Experimental measurements compared to simulation of a mixed experiment assuming simple competition between *E. coli* and RBC show that the measured binding accounts for the loss of activity when host cells are present. (**d**) Survival of different innocula of *E. coli* incubated with 20 μM D-ARVA in conjunction with the binding curve in panel A enables comparison of peptide lethality and the number of peptides bound to each bacterial cell. More than 2×10^8 peptides bound per cell are required for sterilization

13.4 Toxicity Against Mammalian Cells

AMPs bind to anionic mammalian cells through electrostatic interactions (Starr et al. 2016; Riedl et al. 2011a; Selsted et al. 1985; Agawa et al. 1991) and have interfacial activity (Wimley 2010). As a result, many have at least some acute toxicity due to permeabilization of the plasma membrane of the host cells. Toxicities vary significantly among known AMPs, yet those with relatively little toxicity still have relatively poor therapeutic indices compared to conventional antibiotics. Some AMPs may become less toxic in the presence of serum or when host cells are highly concentrated, although this effect is rarely tested.

In the abundant literature on AMPs, researchers often measure lysis of erythrocytes (hemolysis) as a surrogate for eukaryotic cell toxicity.

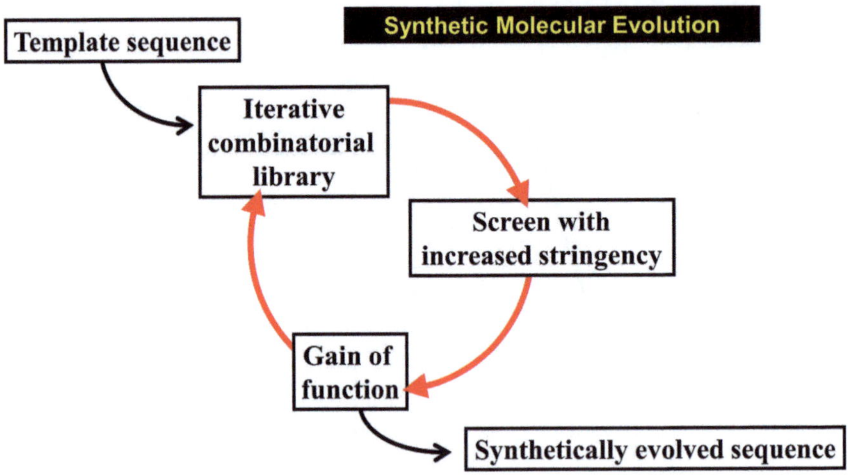

Fig. 13.3 Synthetic molecular evolution of peptides. As we practice it, SME utilizes multiple small libraries (generations) which are iteratively screened for gain-of-function daughter sequences

This is an easy assay and its widespread use provides some uniformity in the AMP literature. However for the maximum sensitivity to toxicity, nucleated cells in culture may be a more sensitive and more informative model system. Nucleated cells can also respond to mechanisms of toxicity other than acute cytolysis. While some researchers have discussed how some AMPs may have useful selective activity against cancer cells (Riedl et al. 2011a, b), we argue that cancer cells would make an especially stringent test system for selecting *against* toxicity because they are especially sensitive to AMPs. In other words, a synthetically evolved AMP that has potent antibacterial activity under relevant conditions and *no toxicity against cultured mammalian cancer cells*, such as HeLa cells, would seem to be an ideal candidate for development into a systemic drug. We discuss how this can be done in a screen below.

13.5 Proteolytic Degradation

Chemical stability of peptides, i.e., resistance to proteolysis, which does not come into play in laboratory assays, is also a critical consideration for systemic activity of peptides. Some peptides are degraded rapidly by serum exopeptidases, dipeptidases, and other proteases (Starr and Wimley 2017; Werle and Bernkop-Schnurch 2006; Molhoek et al. 2011). Sensitivity to serum proteolysis is partially predictable based on sequence. In peptide drug development, serum stability competes with synthetic complexity (i.e., manufacturing cost). Shorter, linear, L-amino acid peptides are most economical to produce but are susceptible to rapid proteolysis, while cyclic, cross-linked, or chemically modified peptides are more costly to produce but are also more protease resistant. Proteolysis is a pervasive threat to AMPs. For example, we have shown that washed human RBCs contain a very high concentration of multiple proteases in their cytosol (Starr and Wimley 2017). Incubation of a set of natural and synthetic linear AMPs with dilute RBCs leads to rapid degradation of peptide if there is even a small amount of hemolysis, which is almost always true. Cytosolic amino- and carboxy-exopeptidases removed amino acids one or two at a time from both termini (Starr and Wimley 2017). For this reason, even standard hemolysis assays in the laboratory may be strongly affected by the proteolytic sensitivity of peptides.

It is likely that host cell binding and serum protein binding will decrease susceptibility to degradation, but as stated above they may also interfere with activity. Cyclization or cross-linking, as found in many natural AMPs, will reduce proteolysis. For linear AMPs chemical modification of the termini (Werle and Bernkop-Schnurch 2006; Starr et al. 2018; Nguyen et al. 2010) with nonnatural terminal amino acids or selective substitution with D-amino acids can increase stability (Starr et al. 2018). Perhaps the simplest approach is to replace all residues with D-amino acids, as this will provide complete resistance to proteolysis while not changing activity (Starr et al. 2016; Rathinakumar and Wimley 2010; Savini et al. 2018)

13.6 Synthetic Molecular Evolution

To create an AMP that could have useful, systemic (in vivo) antibacterial activity, the factors described above will need to be simultaneously optimized. Specifically it will be necessary to (i) maximize selectivity for binding to bacteria over serum proteins and host cells, (ii) maximize bactericidal activity of bound peptide, (iii) minimize susceptibility to proteolytic degradation, (iv) minimize residual cytotoxicity, and (v) maximize solubility under physiological conditions. Thus, the design of a systemically active AMP is like a puzzle in which each of these coupled factors must be simultaneously minimized or maximized without negatively affecting the others. Yet, other than proteolytic susceptibility, the sequence-structure-function relationships for none of these factors are understood well enough to make useful predictions or to enable rational engineering. This is why most new AMPs described in the literature are either identified from natural sources or are discovered in the laboratory by simple trial and error under standard conditions.

How can one simultaneously optimize these various factors when they are incompletely coupled and when the molecular mechanisms are not understood in enough molecular detail to enable rational design? In this chapter, we discuss how this can be done using synthetic molecular evolution (SME). By this we mean iterative screening of rationally designed peptide libraries that are based on known AMPs and are designed using known physical principles to choose rational variations in library members. SME is especially useful for the development of AMPs when screening is done under conditions that mimic the environment in which a systemically active peptide must function. We call it "evolution" because it is most economical to screen iteratively such that each "generation" of gain-of-function AMP is selected from a library built around an active sequence from the previous generation, and each iterative screen further refines the selected sequences to have the properties that are sought. In this case we seek bactericidal activity at low μM concentration in the presence of concentrated host cells and serum, without toxicity.

13.7 Design and Synthesis of Combinatorial Peptide Libraries

Although there are many ways to synthesize and screen peptide libraries (Hilpert et al. 2005; Lam et al. 1991; Dooley et al. 1994; Chen et al. 1996; Frank 2002; Humet et al. 2003; Rathinakumar and Wimley 2008; Deuss et al. 2013; Wiedman et al. 2016), we focus here on the approach we have taken recently to identify membrane-active peptides with specific properties (Rathinakumar and Wimley 2010; Rathinakumar et al. 2009; Krauson et al. 2013; Kauffman et al. 2018; Li et al. 2018; Wiedman et al. 2016; Krauson et al. 2012, 2015; Marks et al. 2011; Rausch et al. 2005). We design small, iterative libraries of 10–30,000 members that are based on a template sequence with known activity. Using a library of this level of diversity means that we can design a library synthesis scheme that provides a relatively large amount of each library member to work with. This, in turn, enables us to use much more complex screens, and it allows us to screen the same library member in multiple parallel assays, which is needed to screen for bactericidal activity against multiple microbes as well as toxicity against host cells.

Our approach to library synthesis and quality control is well described in many papers (Rathinakumar and Wimley 2008, 2010;

Rathinakumar et al. 2009; Krauson et al. 2012, 2013, 2015; Wiedman et al. 2016; Marks et al. 2011; Rausch et al. 2005; He et al. 2011, 2013). In short, a photocleavable linker is added to Tentagel-NH_2 Megabeads, followed by library construction using the split and recombine approach (Chen et al. 1996). Quality control is assured with HPLC, mass spec, and sequencing performed on multiple individual beads. Each bead contains about 1 nmol of peptide which is released by UV light providing 100 μL of a 10 μM solution. This is sufficient peptide to perform multiple parallel assays on each library member. Screening of such small iterative libraries has long been routine in the laboratory (Rathinakumar and Wimley 2008, 2010; Rathinakumar et al. 2009; Krauson et al. 2012, 2013, 2015; Wiedman et al. 2016; Marks et al. 2011; Rausch et al. 2005; He et al. 2011, 2013) and requires no special robotic instrumentation.

13.8 High-Throughput Screening for Antibacterial Activity

As we envision the discovery of systemically active AMPs, screening must be done under conditions that most closely mimic the conditions experienced by a peptide antibiotic in vivo. At a minimum, there must be a high concentration of host cells and a high concentration of serum that has been heat inactivated to eliminate activated complement proteins. RBCs also contain a very high concentration of proteases that are somewhat different than serum proteases (Starr and Wimley 2017), so any screen with RBCs and/or serum will contain many realistic proteases. We have been experimenting with using human red blood cells (RBCs) as a host tissue analog. This is advantageous because it is easy to procure large amounts of fresh, concentrated human RBCs (Fig. 13.4).

In Fig. 13.5 we show the effect of preincubation of 1×10^9 RBC/mL on the antimicrobial activity of a set of natural and synthetic AMPs. In many cases, but not all, RBCs inhibit the antimicrobial activity when AMPs are preincubated for a few minutes with RBCs. We subsequently showed that this inhibition is due to two factors: (i) cytosolic RBC proteases, released by background autolysis or by a small amount of direct hemolysis, and (ii) direct interactions of AMPs with the host cells that reduce the pool of available AMP by competition. Some peptides are more susceptible to the former and some are more susceptible to the latter (Starr et al. 2016). Any AMP that will maintain antibacterial activity in vivo will need to be resistant to both proteolysis and direct host cell inhibition.

13.8.1 Radial Diffusion

There are a number of antimicrobial assays that can be used in a high-throughput screen format. Here we will discuss the strengths and weaknesses of some of them in the context of SME. Radial diffusion is an assay in which a thin bacteria-seeded agar layer is overlaid with a sterile, nutrient-rich agar enabling a lawn of bacteria to grow between the layers, except where growth is inhibited in a zone around a locally applied antibiotic. For peptides to be tested in the presence of host cells and serum, a small hole can be made in the lower agar layer and a small volume of peptide mixture can be introduced. After allowing some time for peptide diffusion into the agar, the nutrient overlay is added, and the plate is allowed to grow overnight. The following day, the lawn of bacteria is visible, and the zones of inhibition are readily observed and quantitated, as shown in Fig. 13.5. Radial diffusion has the advantage that it is readily adapted to high throughput and that it is quantitative: A larger zone of inhibition is related to better activity. However, the size of the zone of inhibition may also be strongly affected by the ability of the antibiotic to diffuse in agar rather than by its inherent antibacterial activity. MIC-based quantitation by radial diffusion using serial dilution circumvents this problem, but it cannot be accomplished in a screen. Further, radial diffusion reports on inhibition of bacterial growth, which does not necessarily indicate sterilization. In Fig. 13.5 we show some examples of radial diffusion-based screens of a peptide library tested simultaneously against the Gram-negative *Escherichia coli* and against

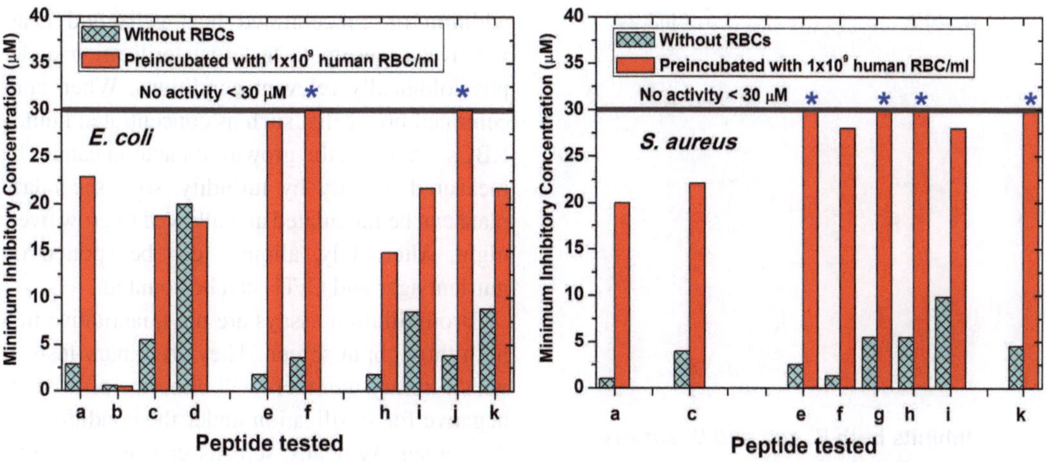

	Natural Peptides	
a	Melittin	*Apis mellifera* (Honey Bee)
b	Cecropin A	*Hyalophora cecropia* (Moth)
c	Indolicidin	*Bos Taurus* (Mow)
d	LL37	*Homo sapiens* (Human)

	Synthetic Peptides	
e	ARVA-L	Library Screen
f	ARVA-D	Library Screen
g	VVRG	Library Screen
h	NATT	Library Screen
i	MSI-78	Magainin analog
j	MelP5	Melittin analog
k	WLBU2	LL37 derived

Fig. 13.4 Human red blood cells (RBCs) inhibit antimicrobial peptides. As described elsewhere (Starr et al. 2016), preincubation of natural and synthetic AMPs with 1×10^9 human RBC/mL (2% of physiological concentration) causes inhibition of most. We have shown that such host cell inhibition is the result of both direct RBC binding and also proteolysis of the AMP by the cytosolic proteases found in human RBCs (Starr and Wimley 2017)

the Gram-positive *Staphylococcus aureus*. Some library members inhibit only one or the other species, while some inhibit both.

Incorporation of concentrated host cells and serum is easily accomplished in radial diffusion, as they can be mixed with peptide prior to introduction of the whole mixture into the well in the agar layer. However the environment in the gel mimics physiological conditions less well because peptides can diffuse into the agar/agarose, while the host cells cannot. On the other hand, strong binding of the peptide to the host cell will prevent diffusion of peptide and will inhibit activity. In Fig. 13.6 we show the effect of concentrated human RBCs on radial diffusion against *S. aureus*. While powerful and quantitative, we have found that the most significant fault of radial diffusion as a screening method is that it does not necessarily select for peptides with sterilizing activity. To overcome this barrier, we also have developed screens based on broth sterilization, which we describe next.

13.8.2 Broth Sterilization

Broth sterilization is an unambiguous, all-or-none assay for *sterilization*. Broth assays are done in liquid media inoculated with bacteria and treated with antibiotic. After overnight incubation, cultures are assayed for absence or presence of live bacteria. Survival of any bacteria generally means that they will grow to a high density after overnight incubation, while sterilized cultures will remain sterile. These two outcomes can easily be measured with optical density, and sterility can be verified by plating the sterile culture on nutrient agar and noting the presence or absence of colony-forming units (CFUs). Broth sterilization assays can be modified by the

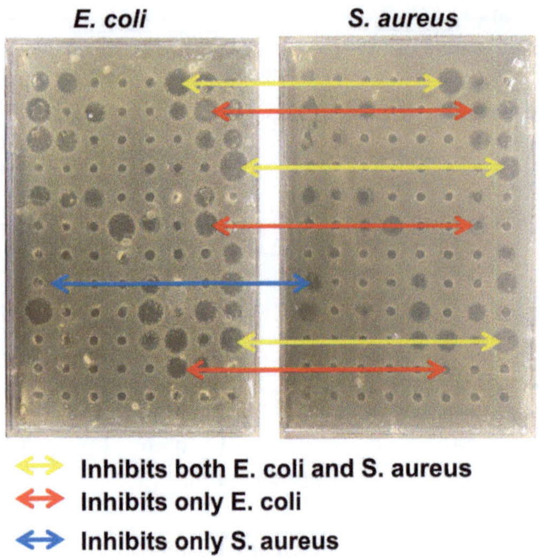

Fig. 13.5 Example radial diffusion-based screening of members of a peptide library in parallel against two organisms using radial diffusion. The same set of peptide library members were screened against the Gram-negative *E. coli* and the Gram-positive *S. aureus* with radial diffusion. Zones of inhibition are observed for some library members. Some active library members inhibit both bacteria (yellow), while others inhibit only one of the two

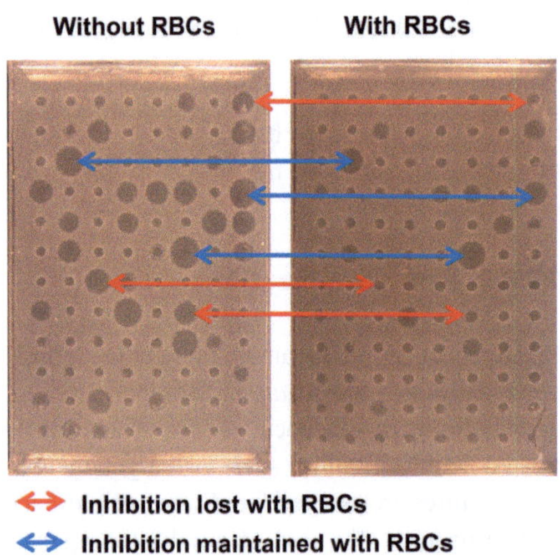

Fig. 13.6 The effect of human RBCs on the activity of peptide library members against S. aureus using radial diffusion. Example screening of a peptide library *S. aureus* with radial diffusion. Both plates were screened with the same library members. The samples added to the right plate had been incubated with 1x10^9 human RBC/mL prior to use. Most of the active library members are inhibited by RBCs. A few library members are not inhibited by RBCs

addition of concentrated host cells and heat-inactivated serum to test for sterilization under physiologically relevant conditions. When concentrated host cells, such as concentrated human RBCs, are used, the growth of bacteria cannot be measured directly by turbidity, so a secondary plate can be inoculated and allowed to grow overnight. Alternately, aliquots can be spotted on nutrient agar and CFUs can be counted.

Broth dilution assays are not quantitative in a high-throughput screen. They are binary tests in which library members will either be positive or negative for sterilization under the conditions of the screen. Typically screens can only be done under one condition for each library member tested. We propose testing antibiotic activity against multiple species simultaneously limiting the amount of each library member available. Thus it is critical to adjust the stringency such that a small number of leads are identified. The stringency of a broth sterilization assay can be modified by adjusting the inoculum size, antibiotic concentration, incubation time, or other factors. A possible scheme for a broth dilution screen is shown in Fig. 13.7, along with the results of screening members of an AMP library using broth dilution.

13.8.3 Reduction of Colony-Forming Units

CFU reduction is a hybrid assay that enables counting of live bacteria remaining in a solution after antibacterial treatment. It essentially reports on the same phenomenon as broth dilution, yet can be done with less labor and in less time. In this assay, which is readily adapted to high throughput, bacteria and antibiotic, with host cells and/or serum, are incubated together for an amount of time that enables killing and then are spotted on a nutrient agar plate at high nominal CFU counts and grown overnight. In the absence of bactericidal activity, a dense mat of bacteria will grow. However, when only a small fraction of bacteria survive, or none at all, a countable number of colonies will grow, a quantitative result that can be used to rank order AMPs in a screen. CFU

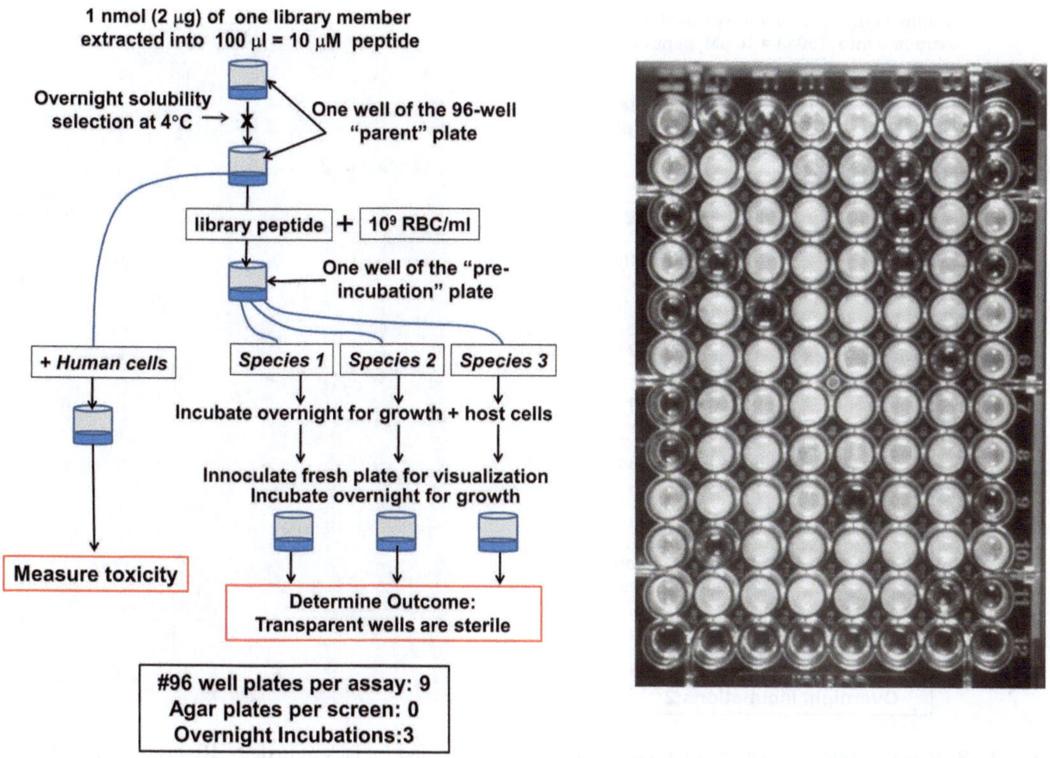

Fig. 13.7 Peptide library screening using broth sterilization. **Left:** One possible scheme for screening a peptide library for solubility, for physiologically relevant broad-spectrum bactericidal activity, and for lack of toxicity. **Right:** Example 96-well plate after screening a library for sterilizing activity against *E. coli*. Wells are either transparent or opaque. Transparent wells have no colony-forming units on nutrient agar, confirming sterility. Bottom row is for controls

reduction assays can readily be modified by the addition of host cells and serum, as above. In Fig. 13.8 we show a possible scheme for SME using CFU counts along with the results of a test screen of AMPs from a library. Note that positive, sterilizing sequences can readily be identified by the absence of CFUs, which amounts to more than four logs of CFU reduction.

13.8.4 Cytotoxicity and Hemolysis

In a screen for antimicrobial assays under physiological conditions, toxicity must also be measured simultaneously. This can be accomplished in assays that are done in the presence of RBCs as host cells by also measuring hemolysis. However, hemolysis may not be sensitive enough to be useful, especially in the presence of concentrated serum which can be protective. Here we suggest that toxicity be measured in parallel using sensitive human cancer cell lines, such as HeLa cells such that peptides with low toxicity can be identified during the screen.

13.9 ESKAPE Pathogens

While many bacteria can infect humans and harbor drug resistance, there are a small set that account for the majority of morbidity and mortality (Boucher et al. 2009; Centers For Disease Control 2014). These include *Clostridium difficile*, often associated with gastrointestinal infections, and the *ESKAPE* pathogens, whose acronym indicates ***E****nterococcus faecalis*, ***S****taphylococcus aureus*, ***K****lebsiella pneumonia*, ***A****cinetobacter baumannii*, ***P****seudomonas aeruginosa*, and

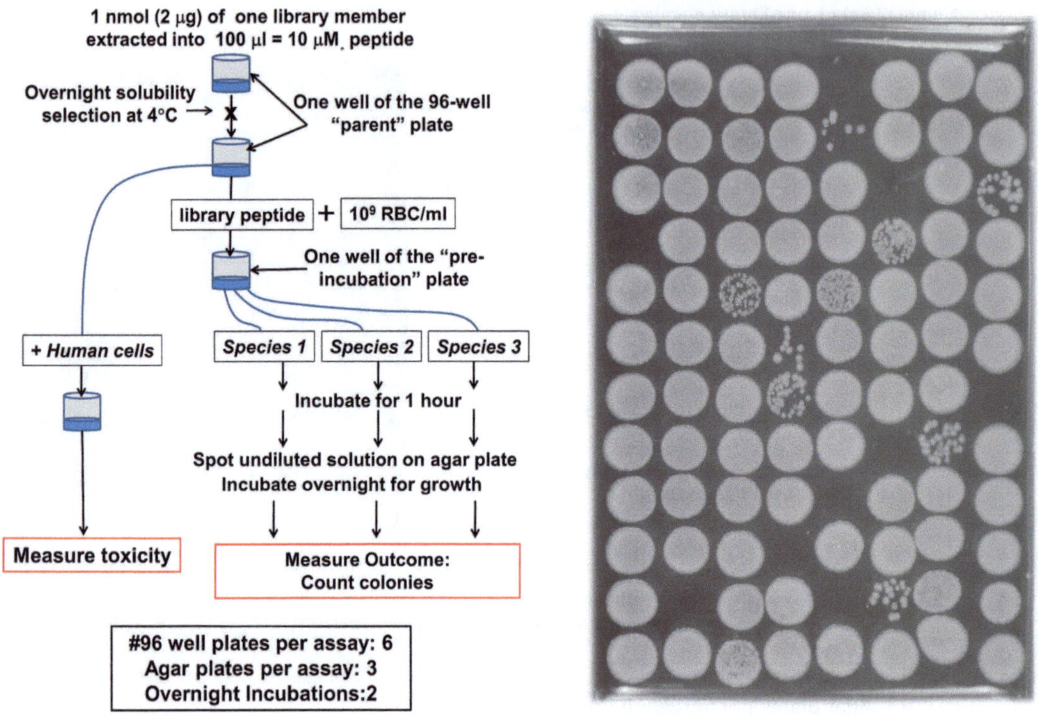

Fig. 13.8 Peptide library screening using the reduction in colony-forming units (CFU). **Left:** One possible scheme for screening a peptide library for solubility, for physiologically relevant broad-spectrum bactericidal activity, and for lack of toxicity. **Right:** Example nutrient agar plate after spotting library members mixed with bacteria for 1 h. Clear spots with no colonies have been sterilized

Enterobacteriaceae, which includes *Escherichia*, *Salmonella*, *Vibrio*, and *Shigella* species, among others. Since some AMPs have variable potencies against these different organisms, screening for the broadest activity must be done against multiple species simultaneously. We previously screened against two ESKAPE bacteria, *E. coli* and *S. aureus*, and a fungus, *Cryptococcus neoformans*, simultaneously and found very low overlap in activities. This enabled the identification of the rare peptides with broad-spectrum activity (Rathinakumar and Wimley 2010). We suggest screening in parallel against *S. aureus* two Gram-negative ESKAPE pathogens, *P. aeruginosa* and either *K. pneumonia* or *A. baumannii*.

13.10 Future Prospects

The physical chemistry-based action of AMPs on bacteria leads to broad-spectrum activity and more difficulty in evolving resistance, which accounts for some of the appeal of AMPs as potential drugs. Yet, these same properties also drive nonspecific interactions with serum protein and host cells that reduce the effectiveness of AMPs. In this chapter, we have presented the concept of synthetic molecular evolution as a tool to enable the discovery of AMPs, early in the development pipeline, that are less affected by host cell and serum protein binding. We remain hopeful that this new approach will finally enable

AMP researchers to bridge the gap between the laboratory bench and the clinic.

References

Agawa Y et al (1991) Interaction with phospholipid bilayers, ion channel formation, and antimicrobial activity of basic amphipathic a-helical model peptides of various chain lengths. J Biol Chem 266:20218–20222

Arias CA, Murray BE (2009) Antibiotic-resistant bugs in the 21st century – a clinical super-challenge. N Engl J Med 360:439–443

Boucher HW et al (2009) Bad bugs, no drugs: no ESKAPE! An update from the Infectious Diseases Society of America. Clin Infect Dis 48:1–12

Centers For Disease Control (2014) Antibiotic resistance threats in the United States, 2013. http://www.cdc.gov/drugresistance/pdf/ar-threats-2013-508.pdf.

Chen CL, Strop P, Lebl M, Lam KS (1996) One bead-one compound combinatorial peptide library: different types of screening. Methods Enzymol 267:211–219

de Breij A et al (2018) The antimicrobial peptide SAAP-148 combats drug-resistant bacteria and biofilms. Sci Transl Med 10(423):eaan4044

Deuss PJ, Arzumanov A, Williams DL, Gait MJ (2013) Parallel synthesis and splicing redirection activity of cell-penetrating peptide conjugate libraries of a PNA cargo. Org Biomol Chem 11:7621–7630

Dobson AJ, Purves J, Kamysz W, Rolff J (2013) Comparing selection on *S. aureus* between antimicrobial peptides and common antibiotics. PLoS One 8:e76521

Dooley CT et al (1994) An all D-amino acid opioid peptide with central analgesic activity from a combinatorial library. Science 266:2019–2022

Easton DM, Nijnik A, Mayer ML, Hancock RE (2009) Potential of immunomodulatory host defense peptides as novel anti-infectives. Trends Biotechnol 27:582–590

Fedders H, Podschun R, Leippe M (2010) The antimicrobial peptide Ci-MAM-A24 is highly active against multidrug-resistant and anaerobic bacteria pathogenic for humans. Int J Antimicrob Agents 36:264–266

Fjell CD, Hancock RE, Cherkasov A (2007) AMPer: a database and an automated discovery tool for antimicrobial peptides. Bioinformatics 23:1148–1155

Fox JL (2013) Antimicrobial peptides stage a comeback. Nat Biotechnol 31:379–382

Frank R (2002) The SPOT-synthesis technique. synthetic peptide arrays on membrane supports – principles and applications. J Immunol Methods 267:13–26

Gordon YJ, Romanowski EG, McDermott AM (2005) A review of antimicrobial peptides and their therapeutic potential as anti-infective drugs. Curr Eye Res 30:505–515

Hamill P, Brown K, Jenssen H, Hancock RE (2008) Novel anti-infectives: is host defence the answer? Curr Opin Biotechnol 19:628–636

He L, Hoffmann AR, Serrano C, Hristova K, Wimley WC (2011) High-throughput selection of transmembrane sequences that enhance receptor tyrosine kinase activation. J Mol Biol 412:43–54

He J et al (2013) Direct cytosolic delivery of polar cargo to cells by spontaneous membrane-translocating peptides. J Biol Chem 288:29974–29986

Hilpert K, Volkmer-Engert R, Walter T, Hancock RE (2005) High-throughput generation of small antibacterial peptides with improved activity. Nat Biotechnol 23:1008–1012

Humet M et al (2003) A positional scanning combinatorial library of peptoids as a source of biological active molecules: identification of antimicrobials. J Comb Chem 5:597–605

Joly V, Jidar K, Tatay M, Yeni P (2010) Enfuvirtide: from basic investigations to current clinical use. Expert Opin Pharmacother 11:2701–2713

Kauffman WB, Guha S, Wimley WC (2018) Synthetic molecular evolution of hybrid cell penetrating peptides. Nat Commun 9:2568

Kazemzadeh-Narbat M et al (2010) Antimicrobial peptides on calcium phosphate-coated titanium for the prevention of implant-associated infections. Biomaterials 31:9519–9526

Krauson AJ, He J, Wimley WC (2012) Gain-of-function analogues of the pore-forming peptide melittin selected by orthogonal high-throughput screening. J Am Chem Soc 134:12732–12741

Krauson AJ, He J, Hoffmann AR, Wimley AW, Wimley WC (2013) Synthetic molecular evolution of pore-forming peptides by iterative combinatorial library screening. ACS Chem Biol 8:823–831

Krauson AJ et al (2015) Conformational fine-tuning of pore-forming peptide potency and selectivity. J Am Chem Soc 137:16144–16152

Kulagina NV, Shaffer KM, Anderson GP, Ligler FS, Taitt CR (2006) Antimicrobial peptide-based array for Escherichia coli and Salmonella screening. Anal Chim Acta 575:9–15

Kulagina NV, Shaffer KM, Ligler FS, Taitt CR (2007) Antimicrobial peptides as new recognition molecules for screening challenging species. Sensors Actuators B Chem 121:150–157

LaBonte J, Lebbos J, Kirkpatrick P (2003) Enfuvirtide. Nat Rev Drug Discov 2:345–346

Lam KS et al (1991) A new type of synthetic peptide library for identifying ligand-binding activity. Nature (London) 354:82–84

Li S et al (2018) Potent macromolecule-sized poration of lipid bilayers by the Macrolittins, a synthetically

evolved family of pore-forming peptides. J Am Chem Soc 140:6441–6447
Marks JR, Placone J, Hristova K, Wimley WC (2011) Spontaneous membrane-translocating peptides by orthogonal high-throughput screening. J Am Chem Soc 133:8995–9004
Mechkarska M et al (2013) An analog of the host-defense peptide hymenochirin-1B with potent broad-spectrum activity against multidrug-resistant bacteria and immunomodulatory properties. Peptides 50:153–159
Molhoek EM, van Dijk A, Veldhuizen EJ, Haagsman HP, Bikker FJ (2011) Improved proteolytic stability of chicken cathelicidin-2 derived peptides by D-amino acid substitutions and cyclization. Peptides 32:875–880
Moy TI et al (2009) High-throughput screen for novel antimicrobials using a whole animal infection model. ACS Chem Biol 4:527–533
Nguyen LT et al (2010) Serum stabilities of short tryptophan- and arginine-rich antimicrobial peptide analogs. PLoS One 5:e12684
Okada M, Natori S (1983) Purification and characterization of an antibacterial protein from haemolymph of Sarcophaga peregrina (flesh-fly) larvae. Biochem J 211:727–734
Otto M (2012) MRSA virulence and spread. Cell Microbiol 14:1513–1521
Park SC et al (2011) Synthetic diastereomeric-antimicrobial peptide: antibacterial activity against multiple drug-resistant clinical isolates. Biopolymers 96:130–136
Patterson-Delafield J, Szklarek D, Martinez RJ, Lehrer RI (1981) Microbicidal cationic proteins of rabbit alveolar macrophages: amino acid composition and functional attributes. Infect Immun 31:723–731
Perron GG, Zasloff M, Bell G (2006) Experimental evolution of resistance to an antimicrobial peptide. Proc Biol Sci 273:251–256
Peschel A (2002) How do bacteria resist human antimicrobial peptides? Trends Microbiol 10:179–186
Peschel A, Sahl HG (2006) The co-evolution of host cationic antimicrobial peptides and microbial resistance. Nat Rev Microbiol 4:529–536
Pollard JE et al (2012) In vitro evaluation of the potential for resistance development to ceragenin CSA-13. J Antimicrob Chemother 67:2665–2672
Poveda E, Briz V, Soriano V (2005) Enfuvirtide, the first fusion inhibitor to treat HIV infection. AIDS Rev 7:139–147
Qureshi NM, Coy DH, Garry RF, Henderson LA (1990) Characterization of a putative cellular receptor for HIV-1 transmembrane glycoprotein using synthetic peptides. AIDS 4:553–558
Rapaport D, Ovadia M, Shai Y (1995) A synthetic peptide corresponding to a conserved heptad repeat domain is a potent inhibitor of Sendai virus-cell fusion: an emerging similarity with functional domains of other viruses. EMBO J 14:5524–5531
Rathinakumar R, Wimley WC (2008) Biomolecular engineering by combinatorial design and high-throughput screening: small, soluble peptides that permeabilize membranes. J Am Chem Soc 130:9849–9858
Rathinakumar R, Wimley WC (2010) High-throughput discovery of broad-spectrum peptide antibiotics. FASEB J 24:3232–3238
Rathinakumar R, Walkenhorst WF, Wimley WC (2009) Broad-spectrum antimicrobial peptides by rational combinatorial design and high-throughput screening: the importance of interfacial activity. J Am Chem Soc 131:7609–7617
Rausch JM, Marks JR, Wimley WC (2005) Rational combinatorial design of pore-forming beta-sheet peptides. Proc Natl Acad Sci U S A 102:10511–10515
Riedl S, Zweytick D, Lohner K (2011a) Membrane-active host defense peptides – challenges and perspectives for the development of novel anticancer drugs. Chem Phys Lipids 164:766–781
Riedl S et al (2011b) In search of a novel target – phosphatidylserine exposed by non-apoptotic tumor cells and metastases of malignancies with poor treatment efficacy. Biochim Biophys Acta 1808:2638–2645
Savini F et al (2017) Cell-density dependence of host-defense peptide activity and selectivity in the presence of host cells. ACS Chem Biol 12:52–56
Savini F, Bobone S, Roversi D, Mangoni M, Stella L (2018) From liposomes to cells: filling the gap between physicochemical and microbiological studies of the activity and selectivity of host-defense peptides. Pept Sci 110:e24041
Schlusselhuber M et al (2014) In vitro effectiveness of the antimicrobial peptide eCATH1 against antibiotic-resistant bacterial pathogens of horses. FEMS Microbiol Lett 350:216–222
Selsted ME, Brown DM, DeLange RJ, Harwig SSL, Lehrer RI (1985) Primary structures of six antimicrobial peptides of rabbit peritoneal neutrophils. J Biol Chem 260:4579–4584
Starr CG, Wimley WC (2017) Antimicrobial peptides are degraded by the cytosolic proteases of human erythrocytes. Biochim Biophys Acta 1859:2319–2326
Starr CG, He J, Wimley WC (2016) Host cell interactions are a significant barrier to the clinical utility of peptide antibiotics. ACS Chem Biol 11:3391–3399
Starr CG, Maderdrut JL, He J, Coy DH, Wimley WC (2018) Pituitary adenylate cyclase-activating polypeptide is a potent broad-spectrum antimicrobial peptide: structure-activity relationships. Peptides 104:35–40
Steiner H, Hultmark D, Engstrom A, Bennich H, Boman HG (1981) Sequence and specificity of two antibacterial proteins involved in insect immunity. Nature (London) 292:246–248
Steiner H, Andreu D, Merrifield RB (1988) Binding and action of cecropin and cecropin analogues: antibacterial peptides from insects. Biochim Biophys Acta 939:260–266

Tran D et al (2002) Homodimeric theta-defensins from rhesus macaque leukocytes: isolation, synthesis, antimicrobial activities, and bacterial binding properties of the cyclic peptides. J Biol Chem 277:3079–3084

Wang G, Li X, Wang Z (2016) APD3: the antimicrobial peptide database as a tool for research and education. Nucleic Acids Res 44:D1087–D1093

Werle M, Bernkop-Schnurch A (2006) Strategies to improve plasma half life time of peptide and protein drugs. Amino Acids 30:351–367

Wiedman G, Kim SY, Zapata-Mercado E, Wimley WC, Hristova K (2016) PH-triggered, macromolecule-sized poration of lipid bilayers by synthetically evolved peptides. J Am Chem Soc 139:937–945

Wimley WC (2010) Describing the mechanism of antimicrobial peptide action with the interfacial activity model. ACS Chem Biol 5:905–917

Wolfmeier H, Pletzer D, Mansour SC, Hancock REW (2017) New perspectives in biofilm eradication. ACS Infect Dis 4:93–106

Yeaman MR, Yount NY (2003) Mechanisms of antimicrobial peptide action and resistance. Pharmacol Rev 55:27–55

Zasloff M (1987) Magainins, a class of antimicrobial peptides from Xenopus skin: isolation, characterization of two active forms, and partial cDNA sequence of a precursor. Proc Natl Acad Sci U S A 84:5449–5453

AMPs as Anti-biofilm Agents for Human Therapy and Prophylaxis

14

Hawraa Shahrour, Raquel Ferrer-Espada, Israa Dandache, Sergio Bárcena-Varela, Susana Sánchez-Gómez, Ali Chokr, and Guillermo Martinez-de-Tejada

Abstract

Microbial cells show a strong natural tendency to adhere to surfaces and to colonize them by forming complex communities called biofilms. In this growth mode, biofilm-forming cells encase themselves inside a dense matrix which efficiently protects them against antimicrobial agents and effectors of the immune system. Moreover, at the physiological level, biofilms contain a very heterogeneous cell population including metabolically inactive organisms and persisters, which are highly tolerant to antibiotics. The majority of human infectious diseases are caused by biofilm-forming microorganisms which are responsible for pathologies such as cystic fibrosis, infective endocarditis, pneumonia, wound infections, dental caries, infections of indwelling devices, etc. AMPs are well suited to combat biofilms because of their potent bactericidal activity of broad spectrum (including resting cells and persisters) and their ability to first penetrate and then to disorganize these structures. In addition, AMPs frequently synergize with antimicrobial compounds and were recently reported to repress the molecular pathways leading to biofilm formation. Finally, there is a very active research to develop AMP-containing coatings that can prevent biofilm formation by killing microbial cells on contact or by locally releasing their active principle. In this chapter we will

H. Shahrour
Department of Microbiology and Parasitology, University of Navarra, Pamplona, Spain

Laboratory of Microbiology, Department of Life & Earth Sciences, Faculty of Sciences I, Lebanese University, Hadat campus, Beirut, Lebanon

Platform of Research and Analysis in Environmental Sciences (PRASE), Doctoral School of Sciences and Technologies, Lebanese University, Hadat Campus, Beirut, Lebanon

R. Ferrer-Espada
Department of Microbiology and Parasitology, University of Navarra, Pamplona, Spain

Wellman Center for Photomedicine, Massachusetts General Hospital, Harvard Medical School, Boston, MA, USA

I. Dandache · A. Chokr
Laboratory of Microbiology, Department of Life & Earth Sciences, Faculty of Sciences I, Lebanese University, Hadat campus, Beirut, Lebanon

Platform of Research and Analysis in Environmental Sciences (PRASE), Doctoral School of Sciences and Technologies, Lebanese University, Hadat Campus, Beirut, Lebanon

S. Bárcena-Varela · G. Martinez-de-Tejada (✉)
Department of Microbiology and Parasitology, University of Navarra, Pamplona, Spain
e-mail: gmartinez@unav.es

S. Sánchez-Gómez
Bionanoplus S.L. Polígono Mocholí, Noain, Navarra, Spain

K. Matsuzaki (ed.), *Antimicrobial Peptides*, Advances in Experimental Medicine and Biology 1117,
https://doi.org/10.1007/978-981-13-3588-4_14

describe these strategies and discuss the perspectives of the use of AMPs as anti-biofilm agents for human therapy and prophylaxis.

Keywords
Biofilm · Antimicrobial peptide · Host-defense peptide · Antibiotic lock therapy · Medical implant

14.1 Introduction to Microbial Biofilms

Biofilms are complex aggregates made up of cells adhered to each other and to a solid surface via an extracellular matrix (Stanley and Lazazzera 2004; Mielich-Süss and Lopez 2015). Frequently, they originate from a single bacterial or fungal clone, although sometimes these structures contain mixtures of different organisms including bacteria, fungi, algae, and protozoa. Within the biofilm, cells are embedded in a self-produced highly hydrated extracellular matrix mediating cell-to-cell and cell-to-surface interactions (Fig. 14.1).

The matrix accounts for 50–90% of the total organic content of the biofilm mass and is composed of a complex mixture of extracellular polymeric substances (EPS) including exopolysaccharides, proteins, and extracellular DNA (Donlan 2002; Kostakioti et al. 2013; Flemming and Wingender 2010). EPS acts as a physical barrier against external threats and traps exogenous substances such as nucleic acids, proteins, minerals, nutrients, and cell wall components found in the local environment. Thus, biofilm formation is beneficial to microorganisms which have a general tendency to colonize all types of environments using this growth mode (Jefferson 2004).

The development of a biofilm is a multi-step process starting with an initial reversible attachment of planktonic cells to a surface followed by a maturation phase. Attachment can occur both on abiotic and biotic surfaces including living tissues (Palmer et al. 2007). Initial attachment is mediated by electrostatic or hydrophobic interactions between planktonic cells and a particular surface. Depending on conditions such as steric hindrance, temperature, and hydrodynamic forces at the site of attachment, this initial interaction can lead to irreversible adhesion to that surface (Fig. 14.1).

Following adhesion, a community of microbial cells (or "microcolony") develops as a result of clonal growth and stable cell-cell interactions. This phase is usually accompanied with secretion of EPS matrix components that stabilize the biofilm structure. Each microcolony is separated from the others by water channels that allow the diffusion of nutrients, oxygen, and other substances. Subsequent growth of microcolonies gives rise to a mature biofilm (or "macrocolony") that acquires the typical mushroom-like structure (Lehner et al. 2005). A final step involves the detachment or dispersal of single cells or cell clusters that have the potential to colonize surrounding sites either in planktonic form or by establishing new sessile communities (Landini et al. 2010) (Fig. 14.1).

Biofilm-forming cells are much better suited than their planktonic counterparts to colonize and persist inside a host (Jamal et al. 2018). In humans, the former cells cause serious medical

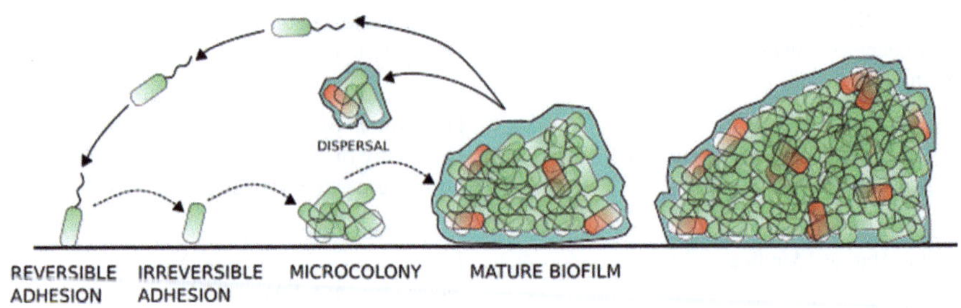

Fig. 14.1 Life cycle of biofilm-forming organisms (Based on Lebeaux et al. 2014). Persister cells are depicted in red

complications such as chronic infections that are very difficult to eradicate. Biofilms have been reported to be involved in 80% of all human infections including native valve endocarditis (Giamarellou 2002), osteomyelitis (Gbejuade et al. 2014), dental caries (Karygianni et al. 2016), middle ear infections (Akyıldız et al. 2013), ocular implant infections (Bispo et al. 2015), and chronic lung infections in cystic fibrosis patients (Høiby et al. 2010). The presence of inert surfaces like those of medical devices (e.g., catheters, prostheses, valves, pacemakers) greatly facilitates biofilm formation and the emergence of severe implant-associated infections (Davey and O'Toole 2000; Guggenbichler et al. 2011). Efficient clearance of these infections often requires implant removal which is associated with high morbidity and economic losses.

Biofilm-forming cells have markedly different gene expression profiles, physiology, and morphology compared to planktonic cells. In part, this explains the increased antibiotic resistance of the former (up to 1000 times, according to some reports) compared to the latter (Hall and Mah 2017). Moreover, it has been proposed that the EPS can slow down or completely block the penetration of some antimicrobials into the biofilm cells. Finally, a fraction of cells, named persisters or dormant cells, was shown to differentiate into a metabolically inactive state that renders them completely tolerant to antibiotics (Mah and O'Toole 2001; Stewart and Costerton 2001; Harms et al. 2016). Globally, the main changes occurring in biofilm-forming cells include induction of the general stress response, increasing expression of multiple drug resistance (MDR) pumps, activation of quorum sensing (QS) systems, and reorganization of outer membrane proteins (OMP) (Rabin et al. 2008).

The combination of the abovementioned mechanisms is the basis for the increased resistance of biofilms, not only to conventional antimicrobials but also to phagocytosis and other effectors of the immune system (Roilides et al. 2015). Moreover, horizontal transfer of resistance and virulence genes takes place with high efficiency in the dense cell population of the biofilm community. While these mechanisms are reported to play an important role in the persistence of infections, especially in immunocompromised patients, a more detailed understanding of these phenomena at the molecular level is necessary to design new agents that target biofilms.

The high doses of antimicrobials and lengthy regimes needed to treat biofilm-associated infections increase the risk of adverse reactions and facilitate the emergence of multidrug-resistant strains. In this context, it has been estimated that drug-resistant infections lead to the death of at least 25,000 people per year in Europe (ECDC 2009) and cause 23,000 deaths in the USA (CDC 2013). Incidence of infections caused by antimicrobial-resistant pathogens is in constant expansion, and it has been estimated that mortality due to these diseases will reach an annual toll of ten million deaths by 2050 if no breakthroughs in antimicrobial therapy occur (Hancock 2015). This situation represents a major medical challenge highlighting the need for new therapeutic options with specific anti-biofilm activity (Wu et al. 2015).

14.2 AMPs as Anti-biofilm Agents: Mechanisms of Action

In the last years, numerous biofilm control strategies have been proposed but no clinically useful therapy has yet emerged. The problem is particularly severe in hospitals, where patients are often immunocompromised and are confronted with increased chances of infection with antimicrobial-resistant pathogens. Numerous characteristics make AMPs very attractive candidates for the development of anti-biofilm therapies. Among other features, many of these compounds have broad-spectrum of microbicidal activity, reduced tendency to induce resistance, ability to kill metabolically inactive cells, and synergistic activity when combined with commonly used drugs (Pletzer et al. 2016). In addition, some AMPs display multiple mechanisms of action which enable them to interfere with various stages of biofilm formation. These include the ability to (i) inhibit the growth of planktonic cells, (ii) prevent the initial adhesion of microbial cells to surfaces,

(iii) disorganize mature biofilms, and (iv) kill biofilm-embedded organisms (Fig. 14.2). However, of all AMPs developed in the last decades, only a few exhibit a significant anti-biofilm activity below their minimal inhibitory concentration (MIC) (Pletzer and Hancock 2016).

Some prominent examples of AMPs displaying biofilm preventive activity include the human cathelicidin peptide LL-37, a compound produced by mucosal epithelial cells and several cells of the immune system (Table 14.1). LL-37 was reported to inhibit biofilm formation at concentrations much lower than its MIC (Overhage et al. 2008; De la Fuente-Núñez et al. 2012). In a different report, this activity was detectable against urinary tract isolates of *Staphylococcus aureus* and *Escherichia coli* at 1/32 to 1/2 of the peptide MIC (Luo et al. 2017). Other human peptides such as lactoferrin were shown to display potent biofilm preventive activity at 20 µg/mL (a sub-MIC value) against the opportunistic pathogen *Pseudomonas aeruginosa* (Singh et al. 2002) (Table 14.1). Notably, this activity seemed to be dependent on the ability of the peptide to activate twitching motility which triggers the daughter cells to move away from the point of parental cell division.

On the other hand, AMPs were shown by other authors to kill preformed biofilms at subinhibitory concentrations. For instance, De la Fuente-Nuñez and collaborators developed synthetic cathelicidin-derived peptides (DJK-5 and DJK-6) which eradicated biofilms established by multidrug-resistant (MDR) organisms (De la Fuente-Núñez et al. 2015) (Table 14.1). Notably, Anunthawan and collaborators reported that the two tryptophan-rich cationic amphipathic peptides KT2 and RT2 killed biofilm-forming cells of *E. coli* O157:H7 at sub-MBIC (minimal biofilm inhibition concentration) levels (Anunthawan et al. 2015). Other researchers found that 4 chimeric AMPs developed by them exerted potent antibacterial and anti-biofilm activity against 19

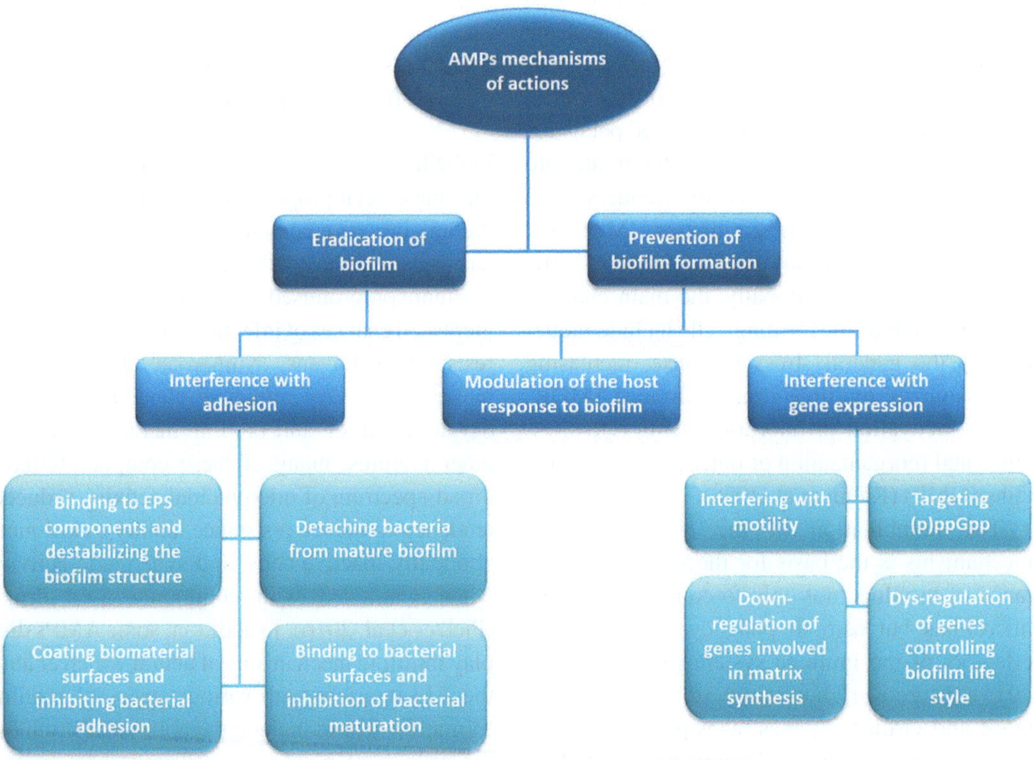

Fig. 14.2 Mechanisms of anti-biofilm action displayed by AMPs

Table 14.1 Selected examples of antimicrobial peptides with anti-biofilm activity

Peptide	Source	MIC	Active anti-biofilm concentration	Microbial species	Anti-biofilm mechanism	Reference
LL-37	Human	32 µg/mL	1–16 µg/mL	*P. aeruginosa; S. epidermidis; E. coli; S. aureus; Burkholderia* spp.; *L. monocytogenes; Candida* spp.	Decreased attachment of bacterial cells, stimulation of twitching motility, influences Las and Rhl systems	De la Fuente-Núñez et al. (2012), Kanthawong et al. (2012), Gabriel et al. (2006), Blower et al. (2015), Scarsini et al. (2015), and Dean et al. (2011)
Lactoferrin	Human	–	20 µg/mL	*S. mutans; S. gordonii; P. gingivalis; F. nucleatum*	Binding and sequestering iron in the environment	Singh et al. (2002) and Ammons and Copié (2013)
Oritavancin	Semisynthetic lipoglycopeptide	2–8 µg/ml	0.5–8 µg/ml	*S. aureus*	Affects membrane integrity, kills slow-growing cells	Belley et al. (2009)
DD13-RIP	Chimeric peptide	2 µg/ml	10–20 µg/ml	*S. aureus*, *S. epidermidis*	Interrupts QS mechanisms	Balaban et al. (2004)
(IDR-)1018	Derivative of bactenicin	8–128 µg/mL	2–10 µg/ml	*A. baumannii; E. coli; K. pneumonia; P. aeruginosa; S. enterica; S. aureus*	Degradation of (p)ppGpp	De la Fuente-Núñez et al. (2014)
DJK-5	Synthetic analog of active anti-biofilm peptides	1.6–16 µg/mL	0.8–4 µg/ml	*P. aeruginosa; E. coli; A. baumannii; K. pneumonia; S. enterica*	Degradation of (p)ppGpp	De la Fuente-Núñez et al. (2015)
DJK-6	Synthetic analog of active anti-biofilm peptides	1.6–16 µg/mL	0.5–8 µg/ml	*P. aeruginosa; E. coli; A. baumannii; K. pneumonia; S. enterica*	Degradation of (p)ppGpp	De la Fuente-Núñez et al. (2015)
1037	Derivative of LL-37	304 µg/mL	5–10 µg/mL	*P. aeruginosa; L. monocytogenes*	Inhibition of swimming and swarming motilities and stimulation of twitching motility	De la Fuente-Núñez et al. (2012)

MDR-resistant clinical *Acinetobacter baumannii* isolates. These peptides were also shown to synergize with conventional antibiotics and exhibited low cytotoxicity against human skin cells (Gopal et al. 2014). Finally, lactoferricin-derived peptides and lipopeptides were reported by Sánchez-Gómez and collaborators to cause a 10,000-fold reduction in the viability of *P. aeruginosa* biofilms after 1 h of treatment at 10 times the peptide MIC. In addition to this bactericidal activity, some of these compounds were shown to have a potent biofilm disorganizing and removing activity (Sánchez-Gómez et al. 2015).

Although the most prevalent mechanism of AMP activity against biofilms seems to involve their well-known ability to bind and disturb microbial membranes, these compounds can also exploit other modes of action. Thus, some AMPs have been reported to interfere with mechanisms necessary for proper biofilm formation including EPS biosynthesis, QS cell communication, and regulation of genes involved in motility, biofilm maturation, and persister cell generation (Kanthawong et al. 2012; De la Fuente-Núñez et al. 2014). Moreover, endogenous peptides are potent immune modulators which can control biofilms indirectly by enhancing the activities of immune cells against infecting microbes (Lai and Gallo 2009). Importantly, these alternative mechanisms are detectable at AMPs concentrations well below their MIC, and this is a clear indication that selection of candidates for anti-biofilm treatment should not be based only on the possession of a significant antimicrobial activity.

The mechanism of *P. aeruginosa* biofilm inhibition by the human cathelicidin LL-37 has been intensively studied. Using microarray technology, Overhage and collaborators demonstrated that the peptide was able to downregulate more than 50 QS-controlled genes at subinhibitory concentrations. These genes include *lasI* and *rhlR* which code for the QS autoinducer synthesis protein LasI and the QS regulator RhlR, respectively. Other genes were involved in flagella biosynthesis, a process interfering with the flagellar-mediated motility that the cells need prior to its attachment to a solid surface. Globally, these changes in gene expression decreased cell adhesion and stimulated twitching motility of *P. aeruginosa* leading to potent inhibition of biofilm development (Overhage et al. 2008). Other authors reported that LL-37 was effective against a slime-producing strain of *S. epidermidis* by inhibiting the initial attachment stage and subsequent biofilm formation (Dean et al. 2011).

Other peptides exploit totally different modes of action. For instance, the synthetic compound IDR-1018 prevented biofilm formation and eradicated preformed biofilms of clinically relevant bacterial species including *P. aeruginosa*, *E. coli*, *A. baumannii*, *Klebsiella pneumoniae*, *methicillin-resistant S. aureus (MRSA)*, *Salmonella typhimurium*, and *Burkholderia cenocepacia* at sub-MIC concentrations (De la Fuente-Núñez et al. 2014) (Table 14.1). Authors found that IDR-1018 binds to the second messenger ppGpp and stimulates its degradation inside the cell. Endogenous ppGpp is synthesized in response to environmental signals as part of the bacterial stringent stress response and seems to play an important role in biofilm formation and in regulating the formation of persister cells (De la Fuente-Núñez et al. 2014). On the other hand, Belley and coworkers reported that oritavancin, a semisynthetic lipoglycopeptide in clinical development for the treatment of serious Gram-positive infections, displays multiple mechanisms of anti-biofilm action (Table 14.1). This compound not only inhibited cell wall and RNA synthesis but also disrupted the membrane potential and increased the membrane permeability affecting exponential and stationary-phase *S. aureus* cells (Belley et al. 2009).

Another example of unique mechanisms of action is that of peptides hepcidin 20 and HBD3 which appear to interfere with the production of the EPS of *S. epidermidis* strains (Brancatisano et al. 2014; Zhu et al. 2013). Authors of these reports found that the peptides upregulated the expression of *icaR* (a transcriptional repressor of the *ica* operon), thereby inhibiting the expression of *icaA* and *icaD*, two genes of the *ica* operon responsible for the synthesis of the major extracellular polysaccharide PIA (polysaccharide-intercellular-adhesin). Additionally, hepcidin 20 was reported to destabilize biofilm structure by binding to negatively charged bacterial cells and extracellular DNA, thus hindering proper interactions between the components of the extracellular matrix. A summary of selected examples of AMPs exerting anti-biofilm properties in vitro or in vivo is reported in Table 14.1.

Up to the present time, more than 1200 AMPs have been isolated and tested for their biological activity against biofilms, and there is a specialized database gathering data from those compounds that display the highest potency in this regard (Biofilm Active Antimicrobial Peptide Database http://www.baamps.it/) (Di Luca et al. 2015).

14.3 Prevention of Biofilm Growth Using AMPs

An ideal strategy to combat infections due to biofilms involves prevention of microbial growth at the site of initial colonization. This implies the development of surfaces resistant to biofilm formation or able to kill microbes on contact. A surface of this type would be particularly important and necessary for medical devices that are implanted or inserted into patients (i.e., intravenous catheters, shunts, prostheses) (Rabin et al. 2008). Despite a very intensive research effort in this field, so far, a truly efficient biofilm-resistant surface has not been developed yet.

In 1987, Gristina proposed the concept of "race for the surface" suggesting that host cells and microbial cells compete for a spot on the implant surface. If microorganisms win this race, initial colonization will result in a biofilm-associated infection (Gristina 1987). This concept also highlights the risk of microbial colonization of the tissue surrounding the implant which constitutes another mechanism of infection (Riool et al. 2017). For the prevention of implant-associated infections, several strategies have been explored including the development of surfaces: (i) endowed with physicochemical modifications aimed at preventing biofilm adhesion or growth (Bryers and Ratner 2004), (ii) that incorporate immobilized antimicrobials, and (iii) which release antimicrobials to the surrounding area.

Microbial adhesion and subsequent proliferation can be prevented by modifying surface properties such as charge, hydrophobicity/hydrophilicity, or surface chemistry in a way that precludes or hinders microbial colonization. One example of this strategy is the use of hydrophilic polymer coatings like those based on immobilized polyethylene glycol (PEG). This approach has been reported to greatly reduce colonization of contact lenses, shunts, endotracheal tubes, and urinary catheters (Busscher et al. 2012; Banerjee et al. 2011). Other researchers successfully used surfaces functionalized with a dense layer of polymer chains to develop the so-called polymer brush coatings (Yu et al. 2017a).

Another strategy aimed at preventing implant colonization involves the development of surfaces that incorporate immobilized AMPs. Regardless of the method used to immobilize the peptide (Silva et al. 2016), it is of critical importance that the compound retains its activity after attachment (Costa et al. 2011). For this purpose, different parameters should be taken into consideration including length, flexibility, and the type of spacer connecting the peptide to the surface. In addition, the orientation of the immobilized peptide and the AMP surface density can affect the successful tethering of the molecule to the biomaterial (Li et al. 2015a).

A wide variety of AMPs, like GZ3.27 (De Zoysa and Sarojini 2017), GL13K (Chen et al. 2014), SESB2V (Tan et al. 2012), bacitracin (Nie et al. 2017), hLF1-11 (Godoy-Gallardo et al. 2014), chimeric peptides (Yazici et al. 2016), LL-37 (Gabriel et al. 2006), melamine (Willcox et al. 2008), lactoferricin (Yoshinari et al. 2010), and Mel-4 (Dutta et al. 2016), have been covalently coupled onto several surfaces, such as glass, silicon, and titanium using various immobilization strategies. Cleophas and collaborators designed a contact-killing hydrogel containing a covalently attached inverso-CysHHC10 peptide that showed a good in vitro antimicrobial activity against *S. aureus*, *S. epidermidis*, and *E. coli* (Cleophas et al. 2014). Other authors developed a brush coating polymer conjugated with the E6 peptide that reduced bacterial adhesion to catheters in a mouse model of urinary catheter infection (Yu et al. 2017b). Hoyos-Nogués and collaborators utilized a promising strategy based on combining the RGD cell-adhesive sequence with the lactoferrin-derived AMP LF1-11 and developed a multifunctional coating that inhibited bacterial colonization by *S. aureus* and *Streptococcus sanguinis* (Hoyos-Nogués et al. 2017). Similar antimicrobial efficacy was obtained when the unrelated peptide CWR11 was immobilized on commercial catheters (Lim et al. 2013, 2015). These devices are normally made of a widely used silicone-like material called polydimethylsiloxane (PDMS).

It should be noted that biomaterials containing immobilized peptides do not exert anti-biofilm preventive activity in the surrounding tissue but only kill microbes that are in direct contact with their surface. In addition, efficiency of these coatings is greatly compromised after insertion into the patient, due to the formation of a layer of host proteins covering the implanted device (Sánchez-Gómez and Martínez-de-Tejada 2017). The only strategy preventing both the colonization of the implant and that of the surrounding tissue involves the use of drug release systems. The advantage of this approach is that it allows achieving high concentrations of the antimicrobial in the vicinity of the implant. In turn, local delivery of the drug facilitates microbial killing while minimizing toxicity concerns associated with systemic administration of the antimicrobial.

In this respect, catheters coated with antibiotic-releasing layers are already approved for medical application although their use is still limited (Singha et al. 2017). This is probably due not only to their higher cost compared to regular catheters but also to the increased likelihood of the coated devices to promote antibiotic resistance. This is a realistic scenario in the case of drug-releasing biomaterials, because of the antimicrobial gradient generated and the possibility of reaching subinhibitory levels at a certain distance of the surface. In addition, the increasing proliferation of multidrug-resistant strains compromises more and more the efficiency of these devices. All these facts underscore the importance of developing release systems based on broad-spectrum antimicrobials unrelated to conventional antibiotics (Alt et al. 2011), and AMPs can be ideal drugs for this purpose.

Applications of AMPs in different types of release coatings have been described, including hydrogels, nanotubes, microporous calcium phosphate coatings, and polymer coatings. For instance, Cheng and collaborators developed a gelatin-based hydrogel on titanium surfaces leading to a controlled release of the AMP HHC36 that prevented biofilm formation by *S. aureus*, *S. epidermidis*, *E. coli*, and *P. aeruginosa* (Cheng et al. 2017). Similarly, Kazemzadeh-Narbat and collaborators reported high in vitro bactericidal activity of Tet213 loaded microporous calcium phosphate coatings applied on titanium against *S. aureus* and *P. aeruginosa* (Kazemzadeh-Narbat et al. 2010). Using a combination approach, Forbes and collaborators studied the anti-biofilm potential of the human apolipoprotein E peptide (apoEdp) and its tryptophan-rich analog (apoEdpL-W) along with other antimicrobials when incorporated onto hydrogels, polyethylene glycol, and nonporous polymers [polyurethane (PU) and PDMS]. It was observed that the coated surfaces could entirely eradicate *S. aureus* and *P. aeruginosa* planktonic and biofilm cells (Forbes et al. 2013).

Other authors preferred to encapsulate AMPs into drug delivery vehicles such as degradable hydrogels, nanoparticles, and liposomes that allow for a controlled and more progressive release of the active principle. A study by d'Angelo and collaborators evaluated the ability of engineered poly(lactide-co-glycolide) (PLGA) nanoparticles (NPs) carrying the cationic peptide polymyxin E (colistin) to eradicate preformed *P. aeruginosa* biofilms (d'Angelo et al. 2015). The authors showed that those NPs were able to release 50% of the encapsulated colistin in 6 h and sustain its release during ~15 days. Interestingly, an inhalable solution containing those NPs proved to have a good efficiency against *P. aeruginosa* biofilms in the airways of lung-infected patients. This strategy represents a promising alternative for inhaled-based treatment (d'Angelo et al. 2015).

14.4 Combination Strategies

One attractive approach to strengthen the anti-biofilm activity of AMPs is to combine these agents with other drugs that target biofilms at a different level. Some of the drugs that have been used successfully for this purpose include conventional antibiotics, compounds targeting the extracellular matrix, inhibitors of QS, and/or other signaling pathways implicated in biofilm formation or dispersal. In addition, AMPs have been administered concomitantly with other AMPs (or other peptide-based molecules) (Orlando et al. 2008) or even combined with physical methods such as ultrasounds. Alternatively other researchers developed hybrid peptides by fusing two different compounds or by combining functional domains with distinct function in the same molecule (Xu et al. 2014).

Some of these strategies showed that it is possible not only to complement the activities of the combination components but also to generate synergistic effects between them (Grassi et al. 2017a). For example, peptide IDR-1018 was capable of enhancing the activities of different classes of conventional antibiotics against biofilm-forming bacterial strains. Thus, at sub-MBIC (minimal biofilm inhibition concentration) levels, IDR-1018 caused a 64-fold decrease in the respective MBIC of several antibiotics. Combination of this peptide with ceftazidime completely eradicated mature biofilms formed by *A. baumannii* and was able to disrupt MRSA mature biofilms leading to cell death at low concentrations. Furthermore, IDR-1018 in combination with tobramycin led to killing of biofilm-forming cells of both *K. pneumoniae* and *E. coli* O157 (Reffuveille et al. 2014). Similar results were obtained with the D-enantiomeric peptides DJK-5 and DJK-6 which exhibited synergistic interactions with different classes of antibiotics to prevent formation or promote eradication of *P. aeruginosa*, *E. coli*, *A. baumannii*, *K. pneumonia*, and *S. enterica* biofilms (De la Fuente-Núñez et al. 2015).

The molecular basis of peptide-based potentiation appears to depend on the peptide ability to bind to microbial membranes even at subinhibitory concentrations. This leads to membrane destabilization and disruption of the cell permeability barrier. Besides combination with antibiotics, AMPs have been tested in combination with other AMPs (see previous section; Forbes et al. 2013) and as hybrid chimeric peptides. An example of a chimeric peptide is DD13-RIP which is composed of dermaseptin derivative (DD13) and an RNA III-inhibiting peptide (RIP) (Table 14.1). Whereas the dermaseptin moiety allows the peptide to target the bacterial cytoplasmic membranes, the RIP segment interrupts QS mechanisms of staphylococcal cells. This hybrid compound has been reported to prevent biofilm formation by *S. aureus* and *S. epidermidis* in rat graft models (Balaban et al. 2004).

An additional advantage of these types of therapies is that they allow reducing the total dose of drug administered because concentrations of components when associated are lower than those given in mono-dose. In turn, this favors a decrease in toxicity and lowers the chances of inducing resistance to treatments (Grassi et al. 2017a). In Table 14.2, we show selected examples of combination strategies published to date.

14.5 AMPs in Antibiotic Lock Therapy

Antibiotic lock therapy (ALT) was developed in the late 1980s as an alternative to catheter removal in patients with catheter-associated bacteremia in which withdrawal represented a serious threat (Messing et al. 1988). The technique consists of filling the lumen of the contaminated catheter with a solution containing antimicrobial agents at very high concentrations. Generally, antimicrobials are used at concentrations ranging from 100 to 1000 times their planktonic MICs in the presence of an anticoagulant such as heparin (Justo and Bookstaver 2014). The solution remains inside the catheter for the necessary time to achieve catheter sterilization, and this often requires adding freshly prepared solution to the catheter every 24 h for several days.

ALT is frequently applied to catheters implanted for long-term treatments (e.g., central

Table 14.2 Selected examples of combination strategies involving AMPs

Combined with	AMP	Combined compound	Bacterial specie(s)	Reference
Antibiotics	17BIPHE2 and DASamP2	Colistin, doripenem, tobramycin, tigecycline	*Pseudomonas aeruginosa*	Mishra and Wang (2017)
	AMP38	Imipenem	*Pseudomonas aeruginosa*	Rudilla et al. (2016)
	BMAP-28	Vancomycin	*Enterococcus faecalis*, *Staphylococcus aureus*, *Salmonella enterica* serovar Typhimurium	Orlando et al. (2008)
	Colistin	Doripenem	*Pseudomonas aeruginosa*	Lora-Tamayo et al. (2014)
	Colistin	Tobramycin	*Pseudomonas aeruginosa*	Herrmann et al. (2010)
	Colistin	Ciprofloxacin, cotrimoxazole, ceftazidime, azithromycin	*Pseudomonas aeruginosa*	Hill et al. (2005)
	Coprisin	Ampicillin, vancomycin, and chloramphenicol	*Enterococcus faecium, Staphylococcus aureus, Escherichia coli, Pseudomonas aeruginosa and Streptococcus mutans*	Hwang et al. (2013)
	CRAMP	Vancomycin	*Salmonella enterica* serovar Typhimurium	Mishra et al. (2015)
	DJK-5 and DJK-6	Ciprofloxacin, ceftazidime, tobramycin	*Acinetobacter baumannii, Klebsiella pneumoniae, Escherichia coli, Pseudomonas aeruginosa*	De La Fuente-Núñez et al. (2015)
	DJK-6	Imipenem, meropenem	*Klebsiella pneumoniae*	Carmona-Ribeiro (2000)
	FLIP7	Meropenem, amikacin, kanamycin, ampicillin, vancomycin, cefotaxime, clindamycin, erythromycin, chloramphenicol, oxacillin, tetracycline, ciprofloxacin, gentamicin, polymyxin B	*Staphylococcus aureus, Escherichia coli, Pseudomonas aeruginosa, Klebsiella pneumoniae*, and *Acinetobacter baumannii*	Chernysh et al. (2018)
	FK-13-a1 and FK-13-a7	Chloramphenicol	*Pseudomonas aeruginosa*	Rajasekaran et al. (2017)
	G10KHc	Tobramycin	*Pseudomonas aeruginosa*	Eckert et al. (2006)
	GL13NH2	Tobramycin	*Pseudomonas aeruginosa*	Hirt and Gorr (2013)
	HPMA	Ciprofloxacin	*Acinetobacter baumannii*	Gopal et al. (2014)
	IDR-1018	Ciprofloxacin	*Pseudomonas aeruginosa*	Reffuveille et al. (2014)
	IDR-1018	Ciprofloxacin, tobramycin, ceftazidime	*Pseudomonas aeruginosa, Escherichia coli, Klebsiella pneumoniae, Staphylococcus aureus, Salmonella enterica, Acinetobacter baumannii*	Mansour et al. (2015)
	Lactoferrin	Ciprofloxacin	*Porphyromonas gingivalis*	Wakabayashi et al. (2009)

(continued)

Table 14.2 (continued)

Combined with	AMP	Combined compound	Bacterial specie(s)	Reference
	Nisin	Penicillin	*Enterococcus faecalis*	Tong et al. (2014a)
	Nisin V or Nisin I4V	Penicillin, chloramphenicol	*Staphylococcus aureus* and *Staphylococcus pseudintermedius*	Field et al. (2016a)
	PMBN	Vancomycin	*Pseudomonas aeruginosa, Escherichia coli, Klebsiella pneumonia*	Ferrer-Espada and collaborators manuscript in preparation
	r(P)$ApoB_L$ and r(P)$ApoB_S$	Antibiotics, EDTA	*Staphylococcus aureus, Escherichia coli, Pseudomonas aeruginosa*	Gaglione et al. (2017)
	RNAIII-inhibiting peptide	Tigecycline	*Staphylococcus aureus*	Simonetti et al. (2016)
	UP-5	Levofloxacin, chloramphenicol, rifampicin, ampicillin, and erythromycin	*Staphylococcus aureus, Enterococcus faecalis, Escherichia coli*, and *Pseudomonas aeruginosa*	Almaaytah et al. (2018)
	WLBU2	Tobramycin, ciprofloxacin, ceftazidime, and meropenem	*Pseudomonas aeruginosa*	Lashua et al. (2016)
Antifungals	DS6	Amphotericin B, fluconazole	*Candida tropicalis*	Singh et al. (2017)
	HsLin06_18	Caspofungin	*Candida albicans*	Cools et al. (2017)
	β-peptides	Fluconazole or ketoconazole	*Candida albicans*	Mora-Navarro et al. (2015)
Chelators	Nisin	DHBA	*Staphylococcus aureus*	Ahire and Dicks (2014)
	Temporin 1 Tb	EDTA	*Staphylococcus epidermidis*	Maisetta et al. (2016)
	TB_KKG6A and TB-L1FK	EDTA	*Pseudomonas aeruginosa*	Grassi et al. (2017b)
	Lin-SB056–1	EDTA	*Pseudomonas aeruginosa*	Maisetta et al. (2017)
AMP	Citropil 1.1, temporin A, analog of tachyplesin I	Colistin	*Pseudomonas aeruginosa, Staphylococcus aureus*	Jorge et al. (2017)
	DD_{13}	RNA-inhibiting peptide	*Staphylococcus aureus, Staphylococcus epidermidis*	Balaban et al. (2004)
	Gramicidin S	PMB	*Pseudomonas aeruginosa*	Berditsch et al. (2015)
	Nisin	Colistin, PMB	*Pseudomonas aeruginosa*	Field et al. (2016b)
	Nisin	MTAD	*Enterococcus faecalis*	Tong et al. (2013)
	PPMO	PMBN	*Pseudomonas aeruginosa*	Howard et al. (2017)
	RNAIII-inhibiting peptide (RIP)	Bismuth ethanedithiol (BisEDT), rifampin	*Staphylococci*	Domenico et al. (2004)

(continued)

Table 14.2 (continued)

Combined with	AMP	Combined compound	Bacterial specie(s)	Reference
Disinfectant	β-defensin-3 (HBD3)	Calcium hydroxide (CH) and chlorhexidine digluconate (CHX)	*Streptococcus mutans*	Ahn et al. (2017)
	Peptide 1018	Chlorhexidine	Oral multispecies	Wang et al. (2015)
	DJK-5	Chlorhexidine	Oral multispecies, *Streptococcus mutans*, and *Enterococcus faecalis*	Zhang et al. (2016)
Amino acids	Nisin	D-cysteine (Cys), D- or L-aspartic acid (Asp), and D- or L-glutamic acid (Glu)	*Streptococcus mutans*	Tong et al. (2014b)
	Platelet-derived peptide (PD4)	Two arginine-tryptophan repeats (RW3 and RW4)	*Staphylococcus epidermidis*	Alabdullatif et al. (2018)
	Temporin 1 Tb	L-cysteine	*Staphylococcus epidermidis*	Maisetta et al. (2016)
Enzymes	Human β-defensin-3	DN-ase I	*Haemophilus influenzae*	Jones et al. (2013)
	TN-5	Alginate lyase	*Pseudomonas aeruginosa*	Bahar et al. (2015)
	KSL-W	Dispersin B	*Acinetobacter baumannii*, *Klebsiella pneumoniae*, *Staphylococcus aureus*, *Staphylococcus epidermidis*	Gawande et al. (2014)
Gas	Human β-defensin-2	Nitric oxide	*Pseudomonas aeruginosa*	Ren et al. (2016)
Ultrasound	Human β-defensin-3 (HBD-3)	Low-frequency ultrasound-targeted microbubble destruction (UTMD)	*Staphylococcus*	Li et al. (2015b)
Fatty acids	PMB and colistin	Polyunsaturated fatty acids	*Klebsiella pneumoniae*	Hobby et al. (2018)
Triple combinations	Colistin	Vancomycin and low-frequency ultrasound	*Acinetobacter baumannii*	Liu et al. (2016)
	CSA-13 and CSA-131	Core-shell magnetic nanoparticles (MNPs), vancomycin and colistin	*Pseudomonas aeruginosa* and *Staphylococcus aureus*	Niemirowicz et al. (2016)
	PMBN	Efflux pump inhibitors, ceftazidime, aztreonam, doxycycline, azithromycin, levofloxacin, piperacillin	*Pseudomonas aeruginosa*	Ferrer-Espada and collaborators unpublished data
	PMBN	β-lactams and β-lactamase inhibitors	*Pseudomonas aeruginosa*	Ferrer-Espada and collaborators unpublished data
	Tachyplesin III	Piperacillin/tazobactam	*Pseudomonas aeruginosa*	Minardi et al. (2007)
	Bacitracin	Octyl gallate	*Staphylococcus aureus*	Oh et al. (2018)
	GT peptide 10	Triethyl citrate (TEC)	*Propionibacterium acnes*	Eilers and Alexeyev (2016)

venous catheters), because they are more likely to get contaminated. Central venous catheters are used for hemodialysis, parenteral nutrition and for the administration of chemotherapy or other treatments. It has been estimated that patients may suffer an episode of bacteremia per 1000 days of catheter use (Dudeck et al. 2013). These episodes are the cause of early withdrawal of these devices in 25% of patients. However, catheter salvage is indicated in cases of reduced venous access or coagulation disorders (Lebeaux et al. 2014).

ALT has several advantages, including (i) the possibility of avoiding the removal of the device, thus deriving benefits both for the patient and for the economy, (ii) the reduction of the adverse effects associated with a systemic antimicrobial treatment and the possibility of administering a drug at levels that are not viable through other routes, and (iii) the fact that this therapy can be carried out on an outpatient basis. On the other hand, ALT may have some disadvantages such as (i) the possibility of promoting catheter blockade, especially if the ALT solution does not incorporate an anticoagulant, (ii) the risk of exposing the patient to high concentrations of antibiotics and anticoagulants if the procedure is not properly performed, and (iii) the possible delay in achieving infection clearance because of opting for a non-systemic treatment along with the risk of more severe complications if ALT fails (Justo and Bookstaver 2014; Fernández-Hidalgo and Almirante 2014). For these reasons, it is recommended to supplement the ALT treatment with a systemic antibiotic therapy (Fernández-Hidalgo and Almirante 2014).

The use of AMPs in ALT is being increasingly favored not only for the broad spectrum of microbicidal activity of these compounds but also for the biofilm disaggregating efficiency displayed by some of them (Di Luca et al. 2014).

Currently, there are several experimental lock solutions based on AMPs under study that include (i) the administration of a single compound, (ii) therapies combining peptides and chelators that destabilize the biofilm matrix, or (iii) synergistic combinations of AMPs with different antimicrobials (Table 14.3).

14.6 Challenges to the Use of AMPs

As mentioned above, AMPs possess many of the characteristics of an ideal antimicrobial including a broad spectrum of bactericidal activity, low tendency to promote antimicrobial resistance and ability to kill even metabolically inactive microorganisms (Batoni et al. 2011). Despite all these attractive characteristics and their potent anti-biofilm activity, AMPs have also drawbacks that constraint their use as drugs. On the one hand, many of these compounds show reduced activity in the presence of physiological concentrations of salts or other biological fluids. In addition, AMPs are often very sensitive to degradation by serum proteases, and this greatly diminishes their half-life in vivo. Finally, due to their unspecific mechanism of action, these compounds frequently have a low therapeutic index and exhibit certain level of cytotoxicity near their therapeutic concentration.

Regardless of their clinical performance, these molecules are still expensive to produce at industrial scale (see below). In addition, some AMPs display poor physical-chemical properties (e.g., tendency to aggregate or self-associate) that require further improvement (Shire et al. 2004). All these limitations probably explain why many promising AMPs have not passed clinical trials. In other instances, use of these compounds has been restricted to topical application (Heinbockel et al. 2014). However, in the last years, several strategies have been devised to overcome these downsides, and, as a consequence, the list of AMPs reaching clinical trials is growing very significantly.

Concerning safety, several studies have focused their efforts in determining the molecular level and the structural features that govern the interaction of AMPs with prokaryotic and eukaryotic membranes (Gupta et al. 2013). The modifications of some physicochemical properties such as their hydrophobicity, amphipathicity, and helicity successfully enhanced their selectivity (Chen et al. 2005).

Other chemical modifications applied to AMPs to extend their half-life once administered

Table 14.3 Experimental lock solutions based on antimicrobial peptides

AMP	Bacterial strain	Age of the biofilm	Time of exposure	Concentration	Result	Type of investigation	Reference
Temporin 1 Tb	*S. epidermidis*	24 h in 96-well microplate	24 h	50 μg/mL	Prophylaxis 3 log red	*In vitro* bioactivity study	Maisetta et al. (2016)
				100 μg/mL	Treatment 0.5–2 log red		
				100 μg/mL + EDTA 4 mg/mL	Treatment 3 log red		
CST	*P. aeruginosa*	24 h in 1-cm segments of 7-French, triple-lumen, CVCs	24, 48, 72 and 96 h	400 or 200 μg/mL	4 log red	*In vitro* bioactivity study	Ozbek and Mataraci-Kara (2016)
				200 mg/ml CLR	6 log red		
				1000 mg/mL HEP	3.5 log red		
CST	*A. baumannii*	24 h in 1-cm segments of 7-French, triple-lumen, CVCs	24, 48, 72 and 96 h	800 μg/mL	5–6 log red	*In vitro* bioactivity study	Ozbek and Mataraci (2013)
				800 μg/mL + 200 mg/ml CLR	5.5–6 log red		
				800 μg/mL + 1000 mg/mL HEP	3–6 log red		
				800 μg/mL + 200 mg/ml CLR + 1000 mg/mL HEP	5–6 log red		
CST	*K. pneumoniae*	48 h in 1-cm segments of 7-French, triple-lumen, CVCs	2, 4, 6 and 24 h	50 or 100 μg/mL	0.5–5 log red	*In vitro* bioactivity study	Mataraci Kara and Ozbek Celik (2018)
				50 or 100 μg/mL + 200 mg/ml CLR	1–4 log red		
				50 or 100 μg/mL + 0.25 mM ESO	2–3.5 log red		
Tachyplesin III	*P. aeruginosa*	Coated ureteral stent	120 h	10 μg/mL tachyplesin III-coat	Prophylaxis 3 log red	In vivo study	Minardi et al. (2007)
				120 mg/kg intraperitoneal TZP	Prophylaxis 3 log red		
				120 mg/kg intraperitoneal TZP + 10 μg/mL tachyplesin III-coat	Prophylaxis 5 log red		

RIP	*S. aureus*	Rat model of central venous catheter	30 min RIP prophylaxis +1 h Ab treatment	RIP 1 mg/mL	3 log red	*In vivo* study	Cirioni et al. (2006)
				RIP 1 mg/mL + 16 μg/mL vancomycin	3 log red		
				RIP 1 mg/mL + 1024 μg/mL vancomycin	5 log red		
				RIP 1 mg/mL + 16 μg/mL ciprofloxacin	3 log red		
				RIP 1 mg/mL + 1024 μg/mL ciprofloxacin	5 log red		
				RIP 1 mg/mL + 8 μg/mL imipenem	3 log red		
				RIP 1 mg/mL + 1024 μg/mL imipenem	5 log red		
TLF	*S. epidermidis* and *C. albicans* polymicrobial biofilm	24 h in 96-well microplate	0, 24, 48, and 72 h	TLF 8 mg/mL	1 log red *S. epidermidis*	*In vitro* bioactivity study	Venkatesh et al. (2009)
					0.5 log red *C. albicans*		
CST				0.1 mg/mL, vancomycin 0.1 mg/mL + heparin 100 units/mL	Compatibility and stability confirmed: 10% degradation at 60 days at 4 °C and 25 °C; vancomycin had ~25% decrease in concentrations after day 15 at room temperature	Stability study	Vincentelli et al. (1997)

CST colistin, *CLR* clarithromycin, *HEP* heparin, *ESO* esomeprazole, *TZP* piperacillin-tazobactam, *RIP* RNAIII-inhibiting peptide, *TLF* talactoferrin, *CVC* Central venous catheter include strategies to avoid their proteolytic degradation such as the substitution of standard amino acids with nonnatural amino acids (e.g., D-amino acids) (Miller et al. 1994; Strömstedt et al. 2009), peptide cyclization or terminal modification via amidation, alkylation or acetylation (Strömstedt et al. 2009; Nguyen et al. 2010), peptide mimetics, (Fjell et al. 2011), or end-tagging by hydrophobic oligo amino acid stretches (Malmsten et al. 2011), among others.

Alternatively, to increase their in vivo stability and lower their toxicity, AMPs can be either immobilized on solid surfaces to prevent biofilm colonization or encapsulated into natural or synthetic polymeric carriers and delivery systems (Kazemzadeh-Narbat et al. 2010; Mokkaphan et al. 2014) (see Sect. 14.3 of this chapter).

The large-scale production of peptides is still a challenge that needs to be faced. AMP manufacturing is difficult and expensive mainly because it requires complex extraction and purifications steps (Beckloff et al. 2007). Indeed, the production cost of a 5,000 Da molecular mass peptide is higher than that of a typical 500 Da molecule (at a comparable scale of operation) (Bray 2003). One possible solution to address this issue entails the production of AMPs in genetically manipulated microorganisms carrying expression vectors (Mygind et al. 2005; Cao et al. 2018). As the technology improves, the cost of large-scale peptide manufacturing is expected to decrease significantly in the next decade.

14.7 Conclusions and Future Directions

Currently, one of the major challenges in medicine is the treatment of biofilm-associated infections. We cannot rely solely on antibiotics, since efficiency of these drugs is being increasingly compromised by the emergence of multiresistant strains. In this context, AMPs appear as a very promising alternative, since these agents target highly conserved molecular motifs that are essential for the microbes and hence less likely to be modified by them. In addition, AMPs are potent bactericidal agents endowed with a broad spectrum of activity including biofilm-forming cells and persisters. Finally, AMPs display multiple mechanisms of action which enable them to interfere with various stages of biofilm formation. Importantly, some of these mechanisms are independent of the peptide antimicrobial activity (e.g., inhibition of extracellular matrix biosynthesis, blocking of QS cell communication, or downregulation of genes involved in biofilm formation).

Although AMPs have some clear drawbacks, many strategies are being actively pursued to speed up the incorporation of these molecules into the clinical practice. Simultaneously, other approaches are being explored, including the use AMPs as enhancing agents (Grassi et al. 2017a). Thus, co-administration of AMPs with antibiotics and anti-biofilm compounds offer an attractive strategy that allows killing a broad spectrum of microbial species at low antibiotic concentrations.

Medical device-associated infections are of great concern for public health, being one of the causes of recurrent surgeries, prolonged administration of antibiotics, and eventually patient death. The use of AMPs to coat the surfaces of implanted devices appears as a promising way to address this problem. However, a truly efficient biofilm-resistant surface has not been developed and more research is needed in this field. Specifically, efforts should be directed to design more efficient systems of peptide immobilization in biomaterials and to develop nano- and micro-systems able to sustain the release of AMPs at the infection sites.

Finally, a new arising field of study is the rational design of AMPs. This methodology is focused first on unraveling the complex interaction between AMPs and microbial cells and then to use this knowledge to design optimized molecules. This is a complex field of study, since many features of AMPs influence their activity such as size, sequence, charge, helicity, hydrophobicity, among others. New bioinformatic tools are being used and developed to compile the enormous information generated and to extract valid conclusions from these analyses. Current approaches aim at the screening and *in silico* modeling of novel AMPs to accelerate the discovery of new safe and effective compounds (Brancatisano et al. 2014; Jorge et al. 2012; Sharma et al. 2016).

Biofilm-associated infections are a critical health problem that needs to be addressed. Although further research is needed, AMPs have the potential to become efficient anti-biofilm agents alone or combined with other drugs.

Acknowledgments This work was supported by the Proyectos de Investigación Universidad de Navarra, Spain (PIUNA-P2011-17 and P2015-14 to G. M. T.). R. F. E and S.B.V. are recipients of doctoral fellowships granted by Gobierno Vasco, Spain (BFI-2011-9) and Asociación de Amigos de la Universidad de Navarra (Spain), respectively.

H.S. is grateful to financial support received from the Lebanese National Council for Scientific Research (CNRS) and by the Lebanese University.

References

Ahire JJ, Dicks LMT (2014) Nisin incorporated with 2,3-Dihydroxybenzoic acid in nanofibers inhibits biofilm formation by a methicillin-resistant strain of Staphylococcus aureus. Probiotics Antimicrob Proteins 7:52–59. https://doi.org/10.1007/s12602-014-9171-5

Ahn KB, Kim AR, Kum KY et al (2017) The synthetic human beta-defensin-3 C15 peptide exhibits antimicrobial activity against Streptococcus mutans, both alone and in combination with dental disinfectants. J Microbiol 55:830–836. https://doi.org/10.1007/s12275-017-7362-y

Akyıldız I, Take G, Uygur K et al (2013) Bacterial biofilm formation in the middle-ear mucosa of chronic otitis media patients. Indian J Otolaryngol Head Neck Surg 65:557–561. https://doi.org/10.1007/s12070-012-0513-x

Alabdullatif M, Atreya CD, Ramirez-Arcos S (2018) Antimicrobial peptides: an effective approach to prevent bacterial biofilm formation in platelet concentrates. Transfusion 00:1–9. https://doi.org/10.1111/trf.14646

Almaaytah A, Qaoud MT, Mohammed GK et al (2018) Antimicrobial and antibiofilm activity of UP-5, an ultrashort antimicrobial peptide designed using only arginine and biphenylalanine. Pharmaceuticals 11:1–18. https://doi.org/10.3390/ph11010003

Alt V, Bitschnau A, Böhner F et al (2011) Effects of gentamicin and gentamicin–RGD coatings on bone ingrowth and biocompatibility of cementless joint prostheses: an experimental study in rabbits. Acta Biomater 7:1274–1280. https://doi.org/10.1016/J.ACTBIO.2010.11.012

Ammons MC, Copié V (2013) Mini-review: lactoferrin: a bioinspired, anti-biofilm therapeutic. Biofouling 29:443–455. https://doi.org/10.1080/08927014.2013.773317

Anunthawan T, De La Fuente-Núñez C, Hancock REW, Klaynongsruang S (2015) Cationic amphipathic peptides KT2 and RT2 are taken up into bacterial cells and kill planktonic and biofilm bacteria. https://doi.org/10.1016/j.bbamem.2015.02.021

Bahar AA, Liu Z, Garafalo M et al (2015) Controlling persister and biofilm cells of gram-negative bacteria with a new 1,3,5-triazine derivative. Pharmaceuticals 8:696–710. https://doi.org/10.3390/ph8040696

Balaban N, Gov Y, Giacometti A et al (2004) A chimeric peptide composed of a dermaseptin derivative and an RNA III-inhibiting peptide prevents graft-associated infections by antibiotic-resistant staphylococci. Antimicrob Agents Chemother 48:2544–2550. https://doi.org/10.1128/AAC.48.7.2544–2550.2004

Banerjee I, Pangule RC, Kane RS (2011) Antifouling coatings: recent developments in the design of surfaces that prevent fouling by proteins, bacteria, and marine organisms. Adv Mater 23:690–718. https://doi.org/10.1002/adma.201001215

Batoni G, Maisetta G, Lisa Brancatisano F et al (2011) Use of antimicrobial peptides against microbial biofilms: advantages and limits. Curr Med Chem 18:256–279. https://doi.org/10.2174/092986711794088399

Beckloff N, Laube D, Castro T et al (2007) Activity of an antimicrobial peptide mimetic against planktonic and biofilm cultures of oral pathogens. Antimicrob Agents Chemother 51:4125–4132. https://doi.org/10.1128/AAC.00208-07

Belley A, Neesham-Grenon E, McKay G et al (2009) Oritavancin kills stationary-phase and biofilm Staphylococcus aureus cells in vitro. Antimicrob Agents Chemother 53:918–925. https://doi.org/10.1128/AAC.00766-08

Berditsch M, Jäger T, Strempel N et al (2015) Synergistic effect of membrane-active peptides polymyxin B and gramicidin s on multidrug-resistant strains and biofilms of Pseudomonas aeruginosa. Antimicrob Agents Chemother 59:5288–5296. https://doi.org/10.1128/AAC.00682-15

Bispo P, Haas W, Gilmore M (2015) Biofilms in infections of the eye. Pathogens 4:111–136. https://doi.org/10.3390/pathogens4010111

Blower RJ, Barksdale SM, van Hoek ML (2015) Snake cathelicidin NA-CATH and smaller helical antimicrobial peptides are effective against Burkholderia thailandensis. PLoS Negl Trop Dis 9:e0003862. https://doi.org/10.1371/journal.pntd.0003862

Brancatisano FL, Maisetta G, Di Luca M et al (2014) Inhibitory effect of the human liver-derived antimicrobial peptide hepcidin 20 on biofilms of polysaccharide intercellular adhesin (PIA)-positive and PIA-negative strains of *Staphylococcus epidermidis*. Biofouling 30:435–446. https://doi.org/10.1080/08927014.2014.888062

Bray BL (2003) Large-scale manufacture of peptide therapeutics by chemical synthesis. Nat Rev Drug Discov 2:587–593. https://doi.org/10.1038/nrd1133

Bryers JD, Ratner BD (2004) Bioinspired implant materials befuddle bacteria. ASM News 70:232–237. https://www.asm.org/index.php/fellows-info/fellows-elected-in-2016/5464-gerard-wong

Busscher HJ, van der Mei HC, Subbiahdoss G et al (2012) Biomaterial-associated infection: locating the finish line in the race for the surface. Sci Transl Med 4:153rv10. https://doi.org/10.1126/scitranslmed.3004528

Cao J, De La Fuente-Nunez C, Ou RW et al (2018) Yeast-based synthetic biology platform for antimicrobial peptide production. ACS Synth Biol 7:896–902. https://doi.org/10.1021/acssynbio.7b00396

Carmona-Ribeiro (2000) Interactions between cationic liposomes and drugs or biomolecules. An Acad Bras Cienc. 72(1):39–43. https://doi.org/10.1590/S0001-37652000000100005

CDC (2013) Antibiotic resistance threats in the United States, 2013. https://www.cdc.gov/drugresistance/pdf/ar-threats-2013-508.pdf

Chen Y, Mant CT, Farmer SW et al (2005) Rational design of alpha-helical antimicrobial peptides with enhanced activities and specificity/therapeutic index. J Biol Chem 280:12316–12329. https://doi.org/10.1074/jbc.M413406200

Chen X, Hirt H, Li Y et al (2014) Antimicrobial GL13K peptide coatings killed and ruptured the wall of Streptococcus gordonii and prevented formation and growth of biofilms. PLoS One 9:e111579. https://doi.org/10.1371/journal.pone.0111579

Cheng H, Yue K, Kazemzadeh-Narbat M et al (2017) Mussel-inspired multifunctional hydrogel coating for prevention of infections and enhanced osteogenesis. ACS Appl Mater Interfaces 9:11428–11439. https://doi.org/10.1021/acsami.6b16779

Chernysh S, Gordya N, Tulin D, Yakovlev A (2018) Biofilm infections between Scylla and Charybdis: interplay of host antimicrobial peptides and antibiotics. Infect Drug Resist 11:501–514. https://doi.org/10.2147/IDR.S157847

Cirioni O, Giacometti A, Ghiselli R et al (2006) RNAIII-inhibiting peptide significantly reduces bacterial load and enhances the effect of antibiotics in the treatment of central venous catheter-associated Staphylococcus aureus infections. J Infect Dis 193:180–186. https://doi.org/10.1086/498914

Cleophas RTC, Riool M, Quarles van Ufford H, Linda C et al (2014) Convenient preparation of bactericidal hydrogels by covalent attachment of stabilized antimicrobial peptides using thiol–ene click chemistry. ACS Macro Lett 3:477–480. https://doi.org/10.1021/mz5001465

Cools TL, Struyfs C, Drijfhout JW et al (2017) A linear 19-mer plant defensin-derived peptide acts synergistically with caspofungin against Candida albicans biofilms. Front Microbiol 8:1–14. https://doi.org/10.3389/fmicb.2017.02051

Costa F, Carvalho IF, Montelaro RC et al (2011) Covalent immobilization of antimicrobial peptides (AMPs) onto biomaterial surfaces. Acta Biomater 7:1431–1440. https://doi.org/10.1016/j.actbio.2010.11.005

d'Angelo I, Casciaro B, Miro A et al (2015) Overcoming barriers in Pseudomonas aeruginosa lung infections: engineered nanoparticles for local delivery of a cationic antimicrobial peptide. Colloids Surf B Biointerfaces 135:717–725. https://doi.org/10.1016/j.colsurfb.2015.08.027

Davey ME, O'Toole GA (2000) Microbial biofilms: from ecology to molecular genetics. Microbiol Mol Biol Rev 64:847–867. https://doi.org/10.1128/MMBR.64.4.847-867.2000

De la Fuente-Núñez C, Korolik V, Bains M et al (2012) Inhibition of bacterial biofilm formation and swarming motility by a small synthetic cationic peptide. Antimicrob Agents Chemother 56:2696–2704. https://doi.org/10.1128/AAC.00064-12

De la Fuente-Núñez C, Reffuveille F, Haney EF et al (2014) Broad-spectrum anti-biofilm peptide that targets a cellular stress response. PLoS Pathog 10:e1004152. https://doi.org/10.1371/journal.ppat.1004152

De La Fuente-Núñez C, Reffuveille F, Mansour SC et al (2015) D-enantiomeric peptides that eradicate wild-type and multidrug-resistant biofilms and protect against lethal Pseudomonas aeruginosa infections. Chem Biol 22:196–205. https://doi.org/10.1016/j.chembiol.2015.01.002

De Zoysa GH, Sarojini V (2017) Feasibility study exploring the potential of novel Battacin Lipopeptides as antimicrobial coatings. ACS Appl Mater Interfaces 9:1373–1383. https://doi.org/10.1021/acsami.6b15859

Dean SN, Bishop BM, van Hoek ML (2011) Natural and synthetic cathelicidin peptides with anti-microbial and anti-biofilm activity against Staphylococcus aureus. BMC Microbiol 11. https://doi.org/10.1186/1471-2180-11-114

Dutta D, Kumar N, Willcox MDP (2016) Antimicrobial activity of four cationic peptides immobilised to poly-hydroxyethylmethacrylate. Biofouling 32(4):429–438

Di Luca M, Maccari G, Nifosì R (2014) Treatment of microbial biofilms in the post-antibiotic era: prophylactic and therapeutic use of antimicrobial peptides and their design by bioinformatics tools. Pathog Dis 70:257–270. https://doi.org/10.1111/2049-632X.12151

Di Luca M, Maccari G, Maisetta G, Batoni G (2015) BaAMPs: the database of biofilm-active antimicrobial peptides. Biofouling 31:193–199. https://doi.org/10.1080/08927014.2015.1021340

Domenico P, Gurzenda E, Giacometti A et al (2004) BisEDT and RIP act in synergy to prevent graft infections by resistant staphylococci. Peptides 25:2047–2053. https://doi.org/10.1016/j.peptides.2004.08.005

Donlan RM (2002) Biofilms: microbial life on surfaces. Emerg Infect Dis 8:881–890. https://doi.org/10.3201/eid0809.020063

Dudeck MA, Weiner LM, Allen-Bridson K et al (2013) National Healthcare Safety Network (NHSN) report, data summary for 2012, device-associated module. Am J Infect Control 41:1148–1166. https://doi.org/10.1016/j.ajic.2013.09.002

ECDC (2009) The bacterial challenge: time to react. A call to narrow the gap between multidrug-resistant bacteria in the EU and the development of new antibacterial agents. https://www.ema.europa.eu/documents/report/bacterial-challenge-time-react_en.pdf

Eckert R, Brady KM, Greenberg EP et al (2006) Enhancement of antimicrobial activity against Pseudomonas aeruginosa by coadministration of G10KHc and tobramycin. Antimicrob Agents Chemother 50:3833–3838. https://doi.org/10.1128/AAC.00509-06

Eilers H, Alexeyev OA (2016) Effect of GT-peptide 10 and triethyl citrate on *P. acnes* biofilm formation, viability, and dispersion. J Drugs Dermatol 15:778–781. http://jddonline.com/articles/dermatology/S1545961616P0778X

Fernández-Hidalgo N, Almirante B (2014) Antibiotic-lock therapy: a clinical viewpoint. Expert Rev Anti-Infect Ther 12:117–129. https://doi.org/10.1586/14787210.2014.863148

Field D, O'Connor R, Cotter PD et al (2016a) In vitro activities of nisin and nisin derivatives alone and in combination with antibiotics against staphylococcus biofilms. Front Microbiol 7:1–11. https://doi.org/10.3389/fmicb.2016.00508

Field D, Seisling N, Cotter PD et al (2016b) Synergistic nisin-polymyxin combinations for the control of pseudomonas biofilm formation. Front Microbiol 7:1–7. https://doi.org/10.3389/fmicb.2016.01713

Fjell CD, Hiss JA, Hancock REW, Schneider G (2011) Designing antimicrobial peptides: form follows function. Nat Rev Drug Discov 11:37–51. https://doi.org/10.1038/nrd3591

Flemming H-C, Wingender J (2010) The biofilm matrix. Nat Rev Microbiol 8:623–633. https://doi.org/10.1038/nrmicro2415

Forbes S, McBain AJ, Felton-Smith S et al (2013) Comparative surface antimicrobial properties of synthetic biocides and novel human apolipoprotein E derived antimicrobial peptides. Biomaterials 34:5453–5464. https://doi.org/10.1016/J.BIOMATERIALS.2013.03.087

Gabriel M, Nazmi K, Veerman EC et al (2006) Preparation of LL-37-grafted titanium surfaces with bactericidal activity. Bioconjug Chem 17:548–550. https://doi.org/10.1021/bc050091v

Gaglione R, Dell'Olmo E, Bosso A et al (2017) Novel human bioactive peptides identified in apolipoprotein B: evaluation of their therapeutic potential. Biochem Pharmacol 130:34–50. https://doi.org/10.1016/j.bcp.2017.01.009

Gawande PV, Leung KP, Madhyastha S (2014) Antibiofilm and antimicrobial efficacy of Dispersinb®-KSL-w peptide-based wound gel against chronic wound infection associated bacteria. Curr Microbiol 68:635–641. https://doi.org/10.1007/s00284-014-0519-6

Gbejuade HO, Lovering AM, Webb JC (2014) The role of microbial biofilms in prosthetic joint infections. Acta Orthop 86:1–12. https://doi.org/10.3109/17453674.2014.966290

Giamarellou H (2002) Nosocomial cardiac infections. J Hosp Infect 50:91–105. https://doi.org/10.1053/jhin.2001.1144

Godoy-Gallardo M, Mas-Moruno C, Fernández-Calderón MC et al (2014) Covalent immobilization of hLf1-11 peptide on a titanium surface reduces bacterial adhesion and biofilm formation. Acta Biomater 10:3522–3534. https://doi.org/10.1016/j.actbio.2014.03.026

Gopal R, Kim YG, Lee JH et al (2014) Synergistic effects and antibiofilm properties of chimeric peptides against multidrug-resistant Acinetobacter baumannii strains. Antimicrob Agents Chemother 58:1622–1629. https://doi.org/10.1128/AAC.02473-13

Grassi L, Maisetta G, Esin S, Batoni G (2017a) Combination strategies to enhance the efficacy of antimicrobial peptides against bacterial biofilms. Front Microbiol 8:2409. https://doi.org/10.3389/fmicb.2017.02409

Grassi L, Maisetta G, Maccari G et al (2017b) Analogs of the frog-skin antimicrobial peptide temporin 1Tb exhibit a wider spectrum of activity and a stronger antibiofilm potential as compared to the parental peptide. Front Chem 5:1–13. https://doi.org/10.3389/fchem.2017.00024

Gristina AG (1987) Biomaterial-centered infection: microbial adhesion versus tissue integration. Science 237:1588–1595. https://doi.org/10.1126/science.3629258

Guggenbichler JP, Assadian O, Boeswald M, Kramer A (2011) Incidence and clinical implication of nosocomial infections associated with implantable biomaterials – catheters, ventilator-associated pneumonia, urinary tract infections. GMS Krankenhhyg Interdiszip 6:Doc18. https://doi.org/10.3205/dgkh000175

Gupta S, Kapoor P, Chaudhary K et al (2013) In silico approach for predicting toxicity of peptides and proteins. PLoS One 8:e73957. https://doi.org/10.1371/journal.pone.0073957

Hall CW, Mah T-F (2017) Molecular mechanisms of biofilm-based antibiotic resistance and tolerance in pathogenic bacteria. FEMS Microbiol Rev 41:276–301. https://doi.org/10.1093/femsre/fux010

Hancock REW (2015) Rethinking the antibiotic discovery paradigm. EBIOM 2:629–630. https://doi.org/10.1016/j.ebiom.2015.07.010

Harms A, Maisonneuve E, Gerdes K (2016) Mechanisms of bacterial persistence during stress and antibiotic exposure. Science 354:aaf4268. https://doi.org/10.1126/science.aaf4268

Heinbockel L, de Tejada G, Sánchez-Gómez S, et al (2014) Anti-infective polypeptides for combating bacterial and viral infections. In: Frontiers in clinical drug research: anti-infectives. Bentham Science Publishers, p 3–31. https://doi.org/10.2174/9781608058549114010003

Herrmann G, Yang L, Wu H et al (2010) Colistin-tobramycin combinations are superior to monotherapy concerning the killing of biofilm *Pseudomonas aeruginosa*. J Infect Dis 202:1585–1592. https://doi.org/10.1086/656788

Hill D, Hill D, Rose B et al (2005) Antibiotic susceptibility of Pseudomonas aeruginosa isolates derived from patients with cystic fibrosis under aerobic, anaerobic, and biofilm conditions. J Clin Microbiol 43:5085–5090. https://doi.org/10.1128/JCM.43.10.5085

Hirt H, Gorr SU (2013) Antimicrobial peptide GL13K is effective in reducing biofilms of Pseudomonas aeruginosa. Antimicrob Agents Chemother 57:4903–4910. https://doi.org/10.1128/AAC.00311-13

Hobby CR, Herndon JL, Morrow CA et al (2018) Exogenous fatty acids alter phospholipid composition, membrane permeability, capacity for biofilm formation, and antimicrobial peptide susceptibility in Klebsiella pneumoniae. Microbiology 635:1–11. https://doi.org/10.1002/mbo3.635

Høiby N, Ciofu O, Bjarnsholt T (2010) Pseudomonas aeruginosa biofilms in cystic fibrosis. Future Microbiol 5:1663–1674. https://doi.org/10.2217/fmb.10.125

Howard JJ, Sturge CR, Moustafa DA et al (2017) Inhibition of Pseudomonas aeruginosa by peptide-conjugated phosphorodiamidate morpholino oligomers. Antimicrob Agents Chemother 61:e01938–e01916. https://doi.org/10.1128/AAC.01938-16

Hoyos-Nogués M, Velasco F, Ginebra M-P et al (2017) Regenerating bone via multifunctional coatings: the blending of cell integration and bacterial inhibition properties on the surface of biomaterials. ACS Appl Mater Interfaces 9:21618–21630. https://doi.org/10.1021/acsami.7b03127

Hwang IS, Hwang JS, Hwang JH et al (2013) Synergistic effect and antibiofilm activity between the antimicrobial peptide coprisin and conventional antibiotics against opportunistic bacteria. Curr Microbiol 66:56–60. https://doi.org/10.1007/s00284-012-0239-8

Jamal M, Ahmad W, Andleeb S et al (2018) Bacterial biofilm and associated infections. J Chin Med Assoc 81:7–11. https://doi.org/10.1016/j.jcma.2017.07.012

Jefferson KK (2004) What drives bacteria to produce a biofilm? FEMS Microbiol Lett 236:163–173. https://doi.org/10.1016/j.femsle.2004.06.005

Jones EA, McGillivary G, Bakaletz LO (2013) Extracellular DNA within a nontypeable haemophilus influenzae-induced biofilm binds human beta defensin-3 and reduces its antimicrobial activity. J Innate Immun 5:24–38. https://doi.org/10.1159/000339961

Jorge P, Lourenço A, Pereira MO (2012) New trends in peptide based anti-biofilm strategies: a review of recent achievements and bioinformatic approaches. Biofouling 28:1033–1061. https://doi.org/10.1080/08927014.2012.728210

Jorge P, Grzywacz D, Kamysz W et al (2017) Searching for new strategies against biofilm infections: colistin-AMP combinations against pseudomonas aeruginosa and staphylococcus aureus single- and double-species biofilms. PLoS One 12:1–21. https://doi.org/10.1371/journal.pone.0174654

Justo JA, Bookstaver PB (2014) Antibiotic lock therapy: review of technique and logistical challenges. Infect Drug Resist 7:343–363. https://doi.org/10.2147/IDR.S51388

Kanthawong S, Bolscher JGM, Veerman ECI et al (2012) Antimicrobial and antibiofilm activity of LL-37 and its truncated variants against Burkholderia pseudomallei. Int J Antimicrob Agents 39:39–44. https://doi.org/10.1016/j.ijantimicag.2011.09.010

Karygianni L, Al-Ahmad A, Argyropoulou A et al (2016) Natural antimicrobials and oral microorganisms: a systematic review on herbal interventions for the eradication of multispecies oral biofilms. Front Microbiol 6:1–17. https://doi.org/10.3389/fmicb.2015.01529

Kazemzadeh-Narbat M, Kindrachuk J, Duan K et al (2010) Antimicrobial peptides on calcium phosphate-coated titanium for the prevention of implant-associated infections. Biomaterials 31:9519–9526. https://doi.org/10.1016/j.biomaterials.2010.08.035

Kostakioti M, Hadjifrangiskou M, Hultgren SJ (2013) Bacterial biofilms: development, dispersal, and therapeutic strategies in the dawn of the postantibiotic era. Cold Spring Harb Perspect Med 3:1–23. https://doi.org/10.1101/cshperspect.a010306

Lai Y, Gallo RL (2009) AMPed up immunity: how antimicrobial peptides have multiple roles in immune defense. Trends Immunol 30:131–141. https://doi.org/10.1016/j.it.2008.12.003

Landini P, Antoniani D, Burgess JG, Nijland R (2010) Molecular mechanisms of compounds affecting bacterial biofilm formation and dispersal. Appl Microbiol Biotechnol 86:813–823. https://doi.org/10.1007/s00253-010-2468-8

Lashua LP, Melvin JA, Deslouches B et al (2016) Engineered cationic antimicrobial peptide (eCAP) prevents Pseudomonas aeruginosa biofilm growth on airway epithelial cells. J Antimicrob Chemother 71:2200–2207. https://doi.org/10.1093/jac/dkw143

Lebeaux D, Ghigo J-M, Beloin C (2014) Biofilm-related infections: bridging the gap between clinical management and fundamental aspects of recalcitrance toward antibiotics. Microbiol Mol Biol Rev 78:510–543. https://doi.org/10.1128/MMBR.00013-14

Lehner A, Riedel K, Eberl L et al (2005) Biofilm formation, extracellular polysaccharide production, and cell-to-cell signaling in various Enterobacter sakazakii strains: aspects promoting environmental persistence. J Food Prot 68:2287–2294. https://www.researchgate.net/publication/7468730_Biofilm_Formation_Extracellular_Polysaccharide_Production_and_Cell-to-Cell_Signaling_in_Various_

Enterobacter_sakazakii_Strains_Aspects_Promoting_Environmental_Persistence

Li Y, Wei S, Wu J et al (2015a) Effects of peptide immobilization sites on the structure and activity of surface-tethered antimicrobial peptides. J Phys Chem C 119:7146–7155. https://doi.org/10.1021/jp5125487

Li S, Zhu C, Fang S et al (2015b) Ultrasound microbubbles enhance human β-defensin 3 against biofilms. J Surg Res 199:458–469. https://doi.org/10.1016/j.jss.2015.05.030

Lim K, Chua RRY, Saravanan R et al (2013) Immobilization studies of an engineered arginine–tryptophan-rich peptide on a silicone surface with antimicrobial and antibiofilm activity. ACS Appl Mater Interfaces 5:6412–6422. https://doi.org/10.1021/am401629p

Lim K, Chua RRY, Ho B et al (2015) Development of a catheter functionalized by a polydopamine peptide coating with antimicrobial and antibiofilm properties. Acta Biomater 15:127–138. https://doi.org/10.1016/j.actbio.2014.12.015

Liu X, Yin H, Weng CX, Cai Y (2016) Low-frequency ultrasound enhances antimicrobial activity of colistin-vancomycin combination against pan-resistant biofilm of acinetobacter baumannii. Ultrasound Med Biol 42:1968–1975. https://doi.org/10.1016/j.ultrasmedbio.2016.03.016

Lora-Tamayo J, Murillo O, Bergen PJ et al (2014) Activity of colistin combined with doripenem at clinically relevant concentrations against multidrug-resistant pseudomonas aeruginosa in an in vitro dynamic biofilm model. J Antimicrob Chemother 69:2434–2442. https://doi.org/10.1093/jac/dku151

Luo Y, McLean DTF, Linden GJ et al (2017) The naturally occurring host defense peptide, LL-37, and its truncated mimetics KE-18 and KR-12 have selected biocidal and antibiofilm activities against Candida albicans, Staphylococcus aureus, and Escherichia coli in vitro. Front Microbiol 8:544. https://doi.org/10.3389/fmicb.2017.00544

Mah TFC, O'Toole GA (2001) Mechanisms of biofilm resistance to antimicrobial agents. Trends Microbiol 9:34–39. https://doi.org/10.1016/S0966-842X(00)01913-2

Maisetta G, Grassi L, Di Luca M et al (2016) Anti-biofilm properties of the antimicrobial peptide temporin 1Tb and its ability, in combination with EDTA, to eradicate Staphylococcus epidermidis biofilms on silicone catheters. Biofouling 32:787–800. https://doi.org/10.1080/08927014.2016.1194401

Maisetta G, Grassi L, Esin S et al (2017) The semi-synthetic peptide Lin-SB056-1 in combination with EDTA exerts strong antimicrobial and antibiofilm activity against pseudomonas aeruginosa in conditions mimicking cystic fibrosis sputum. Int J Mol Sci 18:1–18. https://doi.org/10.3390/ijms18091994

Malmsten M, Kasetty G, Pasupuleti M et al (2011) Highly selective end-tagged antimicrobial peptides derived from PRELP. PLoS One 6:e16400. https://doi.org/10.1371/journal.pone.0016400

Mansour SC, De La Fuente-Núñez C, Hancock REW (2015) Peptide IDR-1018: modulating the immune system and targeting bacterial biofilms to treat antibiotic-resistant bacterial infections. J Pept Sci 21:323–329. https://doi.org/10.1002/psc.2708

Mataraci Kara E, Ozbek Celik B (2018) Investigation of the effects of various antibiotics against Klebsiella pneumoniae biofilms on in vitro catheter model. J Chemother 30:82–88. https://doi.org/10.1080/1120009X.2017.1390633

Messing B, Peitra-Cohen S, Debure A et al (1988) Antibiotic-lock technique: a new approach to optimal therapy for catheter-related sepsis in home-parenteral nutrition patients. J Parenter Enter Nutr 12:185–189. https://doi.org/10.1177/0148607188012002185

Mielich-Süss B, Lopez D (2015) Molecular mechanisms involved in Bacillus subtilis biofilm formation. Environ Microbiol 17:555–565. https://doi.org/10.1111/1462-2920.12527

Miller SM, Simon RJ, Ng S et al (1994) Proteolytic studies of homologous peptide and N-substituted glycine peptoid oligomers. Bioorg Med Chem Lett 4:2657–2662. https://doi.org/10.1016/S0960-894X(01)80691-0

Minardi D, Ghiselli R, Cirioni O et al (2007) The antimicrobial peptide Tachyplesin III coated alone and in combination with intraperitoneal piperacillin-tazobactam prevents ureteral stent pseudomonas infection in a rat subcutaneous pouch model. Peptides 28:2293–2298. https://doi.org/10.1016/j.peptides.2007.10.001

Mishra B, Wang G (2017) Individual and combined effects of engineered peptides and antibiotics on Pseudomonas aeruginosa biofilms. Pharmaceuticals 10. https://doi.org/10.3390/ph10030058

Mishra NM, Briers Y, Lamberigts C et al (2015) Evaluation of the antibacterial and antibiofilm activities of novel CRAMP–vancomycin conjugates with diverse linkers. Org Biomol Chem 13:7477–7486. https://doi.org/10.1039/C5OB00830A

Mokkaphan J, Banlunara W, Palaga T et al (2014) Silicone surface with drug nanodepots for medical devices. ACS Appl Mater Interfaces 6:20188–20196. https://doi.org/10.1021/am505566m

Mora-Navarro C, Caraballo-Leõn J, Torres-Lugo M, Ortiz-Bermúdez P (2015) Synthetic antimicrobial β-peptide in dual-treatment with fluconazole or ketoconazole enhances the in vitro inhibition of planktonic and biofilm Candida albicans. J Pept Sci 21:853–861. https://doi.org/10.1002/psc.2827

Mygind PH, Fischer RL, Schnorr KM et al (2005) Plectasin is a peptide antibiotic with therapeutic potential from a saprophytic fungus. Nature 437:975–980. https://doi.org/10.1038/nature04051

Nguyen LT, Chau JK, Perry NA et al (2010) Serum stabilities of short tryptophan- and arginine-rich antimicrobial peptide analogs. PLoS One 5:e12684. https://doi.org/10.1371/journal.pone.0012684

Nie B, Ao H, Long T et al (2017) Immobilizing bacitracin on titanium for prophylaxis of infections and for improving osteoinductivity: an in vivo study. Colloids Surf B Biointerfaces 150:183–191. https://doi.org/10.1016/j.colsurfb.2016.11.034

Niemirowicz K, Piktel E, Wilczewska AZ et al (2016) Core–shell magnetic nanoparticles display synergistic antibacterial effects against Pseudomonas aeruginosa and Staphylococcus aureus when combined with cathelicidin LL-37 or selected ceragenins. Int J Nanomedicine 11:5443–5455. https://doi.org/10.2147/IJN.S113706

Oh E, Bae J, Kumar A et al (2018) Antioxidant-based synergistic eradication of methicillin-resistant Staphylococcus aureus (MRSA) biofilms with bacitracin. Int J Antimicrob Agents. https://doi.org/10.1016/j.ijantimicag.2018.03.006

Orlando F, Ghiselli R, Cirioni O et al (2008) BMAP-28 improves the efficacy of vancomycin in rat models of gram-positive cocci ureteral stent infection. Peptides 29:1118–1123. https://doi.org/10.1016/j.peptides.2008.03.005

Overhage J, Campisano A, Bains M et al (2008) Human host defense peptide LL-37 prevents bacterial biofilm formation. Infect Immun 76:4176–4182. https://doi.org/10.1128/IAI.00318-08

Ozbek B, Mataraci E (2013) In vitro effectiveness of colistin, tigecycline and levofloxacin alone and combined with clarithromycin and/or heparin as lock solutions against embedded Acinetobacter baumannii strains. J Antimicrob Chemother 68:827–830. https://doi.org/10.1093/jac/dks472

Ozbek B, Mataraci-Kara E (2016) Comparative *in vitro* efficacies of various antipseudomonal antibiotics based catheter lock solutions on eradication of *Pseudomonas aeruginosa* biofilms. J Chemother 28:20–24. https://doi.org/10.1179/1973947814Y.0000000212

Palmer J, Flint S, Brooks J (2007) Bacterial cell attachment, the beginning of a biofilm. J Ind Microbiol Biotechnol 34:577–588. https://doi.org/10.1007/s10295-007-0234-4

Pletzer D, Hancock REW (2016) Antibiofilm peptides: potential as broad-spectrum agents. J Bacteriol 198:2572–2578. https://doi.org/10.1128/JB.00017-16

Pletzer D, Coleman SR, Hancock REW (2016) Anti-biofilm peptides as a new weapon in antimicrobial warfare. Curr Opin Microbiol 33:35–40. https://doi.org/10.1016/j.mib.2016.05.016

Rabin N, Zheng Y, Opoku-Temeng C et al (2008) Biofilm formation mechanisms and targets for developing anti-biofilm agents. Future Med Chem 7:493–512. https://doi.org/10.4155/FMC.15.6

Rajasekaran G, Kim EY, Shin SY (2017) LL-37-derived membrane-active FK-13 analogs possessing cell selectivity, anti-biofilm activity and synergy with chloramphenicol and anti-inflammatory activity. Biochim Biophys Acta Biomembr 1859:722–733. https://doi.org/10.1016/j.bbamem.2017.01.037

Reffuveille F, de la Fuente-Núñez C, Mansour S, Hancock REW (2014) A broad-spectrum antibiofilm peptide enhances antibiotic action against bacterial biofilms. Antimicrob Agents Chemother 58:5363–5371. https://doi.org/10.1128/AAC.03163-14

Ren H, Wu J, Colletta A et al (2016) Efficient eradication of mature Pseudomonas aeruginosa biofilm via controlled delivery of nitric oxide combined with antimicrobial peptide and antibiotics. Front Microbiol 7:1–8. https://doi.org/10.3389/fmicb.2016.01260

Riool M, de Breij A, Drijfhout JW et al (2017) Antimicrobial peptides in biomedical device manufacturing. Front Chem 5:63. https://doi.org/10.3389/fchem.2017.00063

Roilides E, Simitsopoulou M, Katragkou A, Walsh TJ (2015) How biofilms evade host defenses. Microbiol Spectr 3:287–300. https://doi.org/10.1128/microbiolspec.MB-0012-2014

Rudilla H, Fusté E, Cajal Y et al (2016) Synergistic antipseudomonal effects of synthetic peptide AMP38 and carbapenems. Molecules 21:1–12. https://doi.org/10.3390/molecules21091223

Sánchez-Gómez S, Martínez-de-Tejada G (2017) Antimicrobial peptides as anti-biofilm agents in medical implants. Curr Top Med Chem 17:590–603. https://doi.org/10.2174/1568026616666160713141439

Sánchez-Gómez S, Ferrer-Espada R, Stewart PS et al (2015) Antimicrobial activity of synthetic cationic peptides and lipopeptides derived from human lactoferricin against Pseudomonas aeruginosa planktonic cultures and biofilms. BMC Microbiol 15:137. https://doi.org/10.1186/s12866-015-0473-x

Scarsini M, Tomasinsig L, Arzese A et al (2015) Antifungal activity of cathelicidin peptides against planktonic and biofilm cultures of Candida species isolated from vaginal infections. Peptides 71:211–221. https://doi.org/10.1016/J.PEPTIDES.2015.07.023

Sharma A, Gupta P, Kumar R, Bhardwaj A (2016) dPABBs: a novel in silico approach for predicting and designing anti-biofilm peptides. Sci Rep 6:21839. https://doi.org/10.1038/srep21839

Shire SJ, Shahrokh Z, Liu J (2004) Challenges in the development of high protein concentration formulations. J Pharm Sci 93:1390–1402. https://doi.org/10.1002/jps.20079

Silva RR, Avelino KYPS, Ribeiro KL et al (2016) Chemical immobilization of antimicrobial peptides on biomaterial surfaces. Front Biosci (Schol Ed) 8:129–142. https://doi.org/10.2741/s453

Simonetti O, Cirioni O, Cacciatore I et al (2016) Efficacy of the quorum sensing inhibitor FS10 alone and in combination with tigecycline in an animal model of staphylococcal infected wound. PLoS One 11:1–12. https://doi.org/10.1371/journal.pone.0151956

Singh PK, Parsek MR, Greenberg EP, Welsh MJ (2002) A component of innate immunity prevents bacterial biofilm development. Nature 417:552–555. https://doi.org/10.1038/417552a

Singh K, Shekhar S, Yadav Y et al (2017) DS6: anticandidal, antibiofilm peptide against Candida tropicalis and exhibit synergy with commercial drug. J Pept Sci 23:228–235. https://doi.org/10.1002/psc.2973

Singha P, Locklin J, Handa H (2017) A review of the recent advances in antimicrobial coatings for urinary catheters. Acta Biomater 50:20–40. https://doi.org/10.1016/j.actbio.2016.11.070

Stanley NR, Lazazzera BA (2004) Environmental signals and regulatory pathways that influence biofilm formation. Mol Microbiol 52:917–924. https://doi.org/10.1111/j.1365-2958.2004.04036.x

Stewart PS, Costerton JW (2001) Antibiotic resistance of bacteria in biofilms. Lancet 358:135–138. https://doi.org/10.1016/S0140-6736(01)05321-1

Strömstedt AA, Pasupuleti M, Schmidtchen A, Malmsten M (2009) Evaluation of strategies for improving proteolytic resistance of antimicrobial peptides by using variants of EFK17, an internal segment of LL-37. Antimicrob Agents Chemother 53:593–602. https://doi.org/10.1128/AAC.00477-08

Tan XW, Lakshminarayanan R, Liu SP et al (2012) Dual functionalization of titanium with vascular endothelial growth factor and β-defensin analog for potential application in keratoprosthesis. J Biomed Mater Res B Appl Biomater 100B:2090–2100. https://doi.org/10.1002/jbm.b.32774

Tong Z, Ling J, Lin Z et al (2013) The effect of MTADN on 10 enterococcus faecalis isolates and biofilm: an in vitro study. J Endod 39:674–678. https://doi.org/10.1016/j.joen.2012.12.010

Tong Z, Zhang Y, Ling J, et al (2014a) An in vitro study on the effects of nisin on the antibacterial activities of 18 antibiotics against enterococcus faecalis. PLoS One 9. https://doi.org/10.1371/journal.pone.0089209

Tong Z, Zhang L, Ling J et al (2014b) An in vitro study on the effect of free amino acids alone or in combination with nisin on biofilms as well as on planktonic bacteria of Streptococcus mutans. PLoS One 9:3–10. https://doi.org/10.1371/journal.pone.0099513

Venkatesh M, Rong L, Raad I, Versalovic J (2009) Novel synergistic antibiofilm combinations for salvage of infected catheters. J Med Microbiol 58:936–944. https://doi.org/10.1099/jmm.0.009761-0

Vincentelli J, Braguer D, Guillet P et al (1997) Formulation of a flush solution of heparin, vancomycin, and colistin for implantable access systems in oncology. J Oncol Pharm Pract 3:18–23. https://doi.org/10.1177/107815529700300103

Wakabayashi H, Yamauchi K, Kobayashi T et al (2009) Inhibitory effects of lactoferrin on growth and biofilm formation of Porphyromonas gingivalis and Prevotella intermedia. Antimicrob Agents Chemother 53:3308–3316. https://doi.org/10.1128/AAC.01688-08

Wang Z, De La Fuente-Núñez C, Shen Y et al (2015) Treatment of oral multispecies biofilms by an antibiofilm peptide. PLoS One 10:1–16. https://doi.org/10.1371/journal.pone.0132512

Willcox MDP, Hume EBH, Aliwarga Y et al (2008) A novel cationic-peptide coating for the prevention of microbial colonization on contact lenses. J Appl Microbiol 105:1817–1825. https://doi.org/10.1111/j.1365-2672.2008.03942.x

Wu H, Moser C, Wang H-Z et al (2015) Strategies for combating bacterial biofilm infections. Int J Oral Sci 7:1–7. https://doi.org/10.1038/ijos.2014.65

Xu W, Zhu X, Tan T et al (2014) Design of embedded-hybrid antimicrobial peptides with enhanced cell selectivity and anti-biofilm activity. PLoS One 9:e98935. https://doi.org/10.1371/journal.pone.0098935

Yazici H, O'Neill MB, Kacar T et al (2016) Engineered chimeric peptides as antimicrobial surface coating agents toward infection-free implants. ACS Appl Mater Interfaces 8:5070–5081. https://doi.org/10.1021/acsami.5b03697

Yoshinari M, Kato T, Matsuzaka K et al (2010) Prevention of biofilm formation on titanium surfaces modified with conjugated molecules comprised of antimicrobial and titanium-binding peptides. Biofouling 26:103–110. https://doi.org/10.1080/08927010903216572

Yu L, Hou Y, Cheng C et al (2017a) High-antifouling polymer brush coatings on nonpolar surfaces via adsorption-cross-linking strategy. ACS Appl Mater Interfaces 9:44281–44292. https://doi.org/10.1021/acsami.7b13515

Yu K, Lo JCY, Yan M et al (2017b) Anti-adhesive antimicrobial peptide coating prevents catheter associated infection in a mouse urinary infection model. Biomaterials 116:69–81. https://doi.org/10.1016/j.biomaterials.2016.11.047

Zhang T, Wang Z, Hancock REW et al (2016) Treatment of oral biofilms by a D-enantiomeric peptide. PLoS One 11:1–16. https://doi.org/10.1371/journal.pone.0166997

Zhu C, Tan H, Cheng T et al (2013) Human β-defensin 3 inhibits antibiotic-resistant staphylococcus biofilm formation. J Surg Res 183:204–213. https://doi.org/10.1016/j.jss.2012.11.048

Clinical Application of AMPs

15

Fabíola Costa, Cátia Teixeira, Paula Gomes, and M. Cristina L. Martins

Abstract

Antimicrobial peptides (AMPs) have been described as one of the most promising compounds able to address one of the main health threats of the twenty-first century that is the continuous rise of multidrug-resistant microorganisms. However, despite the clear advantages of AMPs as a new class of antimicrobials, such as broad spectrum of activity, high selectivity, low toxicity and low propensity to induce resistance, only a small fraction of AMPs reported thus far have been able to successfully complete all phases of clinical trials and become accessible to patients. This is mainly related to the low bioavailability and still somewhat expensive production of AMP along with regulatory obstacles. This chapter offers an overview of selected AMPs that are currently in the market or under clinical trials. Strategies for assisting AMP industrial translation and major regulatory difficulties associated with AMP approval for clinical evaluation will be also discussed.

15.1 Introduction

The increasing reports on antibiotic resistance among pathogenic microorganisms have alerted authorities to what has been named as the 'post-antibiotic era' (WHO 2018). Indeed, there are only a limited number of chemical classes within the currently used antibiotics, which share strong similarities regarding both spectrum of activity and mode of action, altogether favouring the spread of cross-resistance between pathogens (Martinez 2014). As there are only a few new antibiotics in the pharmaceutical industry pipeline, it is imperative to find new therapeutic classes to tackle the growing menace of multidrug-resistant microorganisms (Coates et al. 2011; WHO 2018).

Antimicrobial peptides (AMPs) have been suggested as a new therapeutic class with potential to overcome some of the major disadvantages associated with classical antibiotics, namely,

F. Costa
i3S, Instituto de Investigação e Inovação em Saúde, Universidade do Porto, Porto, Portugal

INEB, Instituto de Engenharia Biomédica, Universidade do Porto, Porto, Portugal

C. Teixeira · P. Gomes
LAQV-REQUIMTE, Departamento de Química e Bioquímica, Faculdade de Ciências, Universidade do Porto, Porto, Portugal

M. C. L. Martins (✉)
i3S, Instituto de Investigação e Inovação em Saúde, Universidade do Porto, Porto, Portugal

INEB, Instituto de Engenharia Biomédica, Universidade do Porto, Porto, Portugal

Instituto de Ciências Biomédicas Abel Salazar, Universidade do Porto, Porto, Portugal
e-mail: cmartins@ineb.up.pt

K. Matsuzaki (ed.), *Antimicrobial Peptides*, Advances in Experimental Medicine and Biology 1117,
https://doi.org/10.1007/978-981-13-3588-4_15

high prevalence of resistant microorganisms emerging only a few years after market entry, narrow scope of activity and significant systemic toxicity (Zasloff 2002; Brooks and Brooks 2014). As already explained in previous chapters, AMPs are a distinct and diverse class of molecules, most of which share some fundamental similarities such as having cationic and hydrophobic domains and amphipathic behaviour (Zasloff 2002). Despite the structural heterogeneity found within AMPs, they share some very important functions that include direct and broad-spectrum microbial killing (including of antibiotic-resistant microorganisms), endotoxin neutralization, favourable immune modulation, enhancement of adaptive immune responses and, most importantly, very low propensity to promote microorganism resistance (Gordon et al. 2005; Wang et al. 2015). In fact, AMPs constitute a new hope in the difficult fight against the so-called ESKAPE bacteria, i.e. multidrug-resistant strains of *Enterococcus faecium*, *Staphylococcus aureus*, *Klebsiella pneumoniae*, *Acinetobacter baumannii*, *Pseudomonas aeruginosa* and *Enterobacter* species responsible for grave nosocomial infections (Giuliani et al. 2007; Santajit and Indrawattana 2016; Gomes et al. 2017). AMPs are also involved in the intracellular processes of angiogenesis, wound healing and cell signalling, which make them especially interesting for the development of novel drugs (Afacan et al. 2012). These remarkable properties have driven researchers to make considerable efforts towards the use of AMPs as commercially available drugs, which have been translated to an ever-increasing number of submissions of AMP-related patents. In PATENTSCOPE of World Intellectual Property Organization (WIPO), thousands of hits appear when searching for AMPs (WIPO 2018). Most of these patents are related to the description of newly found AMPs either from natural or synthetic origin, as well as AMP derivatives with increased activity, stability and/or lower toxicity. Interestingly, more recently submitted patents also include a focus on both an effective pharmaceutical composition and an optimal administration route for AMPs (Kosikowska and Lesner 2016; WIPO 2018). However, AMPs are still regarded as a 'no go' area for many pharmaceutical companies, given their poor pharmacokinetics (PK), which arises from, among other factors, low absorption and short half-lives in vivo due to metabolic transformations and/or fast clearance from the organism (Henninot et al. 2018). While this can be sometimes an advantage, as peptides do not tend to accumulate in the body and hence have reduced toxicity risks, it also constitutes the major issue that has to be dealt with when developing peptide therapeutics (Fernandes and Martens 2017; Henninot et al. 2018). In fact, as herein reviewed, shortcomings, such as inadequate in vitro to in vivo translation, high production costs and, mainly, poor oral bioavailability, have been major barriers in AMP progression into the clinics.

15.2 AMPs in Clinical Use

Given their high potency and selectivity, broad range of targets, potentially low toxicity, low accumulation in tissues and no propensity to trigger resistance, AMPs have been considered as promising anti-infection candidates especially for polymicrobial infections as in the case of skin and soft tissue infections (SSTI) (Mahlapuu et al. 2016; Pfalzgraff et al. 2018). Still, despite the myriad of AMP-related patents filed, only a very narrow proportion of the AMPs have already received the certification, either by US Food and Drug Administration (FDA) or EU European Medicines Agency (EMA), mandatory for entry into market and clinical use. FDA-approved AMPs are listed on Table 15.1.

15.3 AMPs and Analogues Under Clinical Trials

Clinical trials are aimed at testing potential treatments in human volunteers, to assess if they should be approved for wider use in the general population. This means that both efficacy and safety of the proposed treatment should surpass the existing drugs. Therefore, clinical trials are a key moment in the new drug development process, representing

Table 15.1 FDA-approved AMPs in clinical use (DrugDataBase n.d.; Usmani et al. 2017)

AMP[a]	Sequence[b]	Indication	Route of administration
Bacitracin	I*C*LeI[$^{\varepsilon}$KoIfHdN][c]	Skin and eye infections	Topical application
	(Mixture of side chain-to-tail cyclic peptides from *Bacillus subtilis* var Tracy; the sequence of bacitracin A is shown, which possesses a modification of the Cys residue – the side chain thiol of Cys has undergone an intramolecular condensation with the preceding peptide bond, forming a thiazolidine ring; this modified Cys is noted by ***C*** in the above sequence)	Parenteral (intramuscular) administration restricted to infants with staphylococcal pneumonia and empyema when due to organisms shown to be susceptible to bacitracin	Intramuscular
Polymyxin B	***6-mo-****Dab*T*DabDab*[$^{\gamma}$*Dab*fL*DabDab*T][d]	Last-line treatment option for multidrug-resistant gram-negative bacterial infections	Parenteral administration (intramuscular, intravenous, intrathecal ophthalmic)
	(Mixture of side chain-to-tail cyclic lipopeptides from *Bacillus polymyxa*; sequence shown for polymyxin B1 that is *N*-terminally acylated with 6-methyloctanoic acid, ***6-mo***)		
Polymyxin E (colistins)	***6-mh***-*Dab*T*Dab*[$^{\gamma}$*Dab*lL*DabDab*T][d]	Gastrointestinal tract infections caused by *Escherichia coli* and *Salmonella* spp.	Parenteral administration (intramuscular, intravenous)
	(Mixture of side chain-to-tail cyclic lipopeptides from *Bacillus colistinus*; sequence shown for colistin A, which is *N*-terminally acylated with 6-methylheptanoic acid, ***6-mh***)	Multidrug-resistant gram-negative bacteria (e.g. *Pseudomonas aeruginosa*; *Klebsiella pneumoniae*; *Acinetobacter*; unresponsive to other antibiotics)	
Tyrothricin	[fPFfNQYVOL]	Local treatment of infected skin and infected oropharyngeal mucous membranes	Topical application only, as it is very toxic when administrated parenterally
	(Mixture of head-to-tail cyclic polypeptides from *Bacillus brevis*; the sequence of tyrothricin A is shown)	Effective against gram-positive bacteria	
Gramicidin D (or, simply, gramicidin)	^{f}VGALAVVVWLWLWLW-***ea***	Skin lesions, surface wounds and eye infections	External use only
	(Mixture of linear pentadecapeptides from *Bacillus brevis* whose *N*-terminal residue (V or L) is *N*-formylated (^{f}V or ^{f}L) and whose *C*-terminal tryptophan (W) is forming an additional peptide bond to ethanolamine (***ea***); the sequence of gramicidin A is represented)		
Gramicidin S	[fPVOLfPVOL]	Potent against gram-negative and gram-positive bacteria and fungi restricted use as spermicide and to treat genital ulcers caused by STD	Topical application only, as it is highly haemolytic
	(Cyclic peptide biosynthesized from gramicidin in *Bacillus brevis*; comprises two identical pentapeptides coupled head to tail)		

(continued)

Table 15.1 (continued)

AMP[a]	Sequence[b]	Indication	Route of administration
Daptomycin (lipopeptide)	***da***-WnD[$^{\beta}$TGODaDGs^{3m}E***Kyn***][e] (side chain-to-tail cyclic lipopeptide from *Streptomyces roseosporus*, which is *N*-terminally acylated with decanoic acid (***da***) and comprises the unusual *C*-terminal amino acid kynurenine (***Kyn***, metabolite of Trp)	Complicated skin infections caused by susceptible strains of gram-positive microorganisms	Intravenous injection

[a]Other well-known antibiotics that include amino acid building blocks, and are thus considered as AMPs by many authors, e.g. teicoplanin or vancomycin, are outside the scope of this chapter, whose focus is on AMPs that, regardless of having or not modified amino acid residues, retain a typical peptide backbone

[b]Uppercase single-letter codes for standard L-α-amino acids have been used as defined by the IUPAC-IUB Joint Commission on Biochemical Nomenclature; lowercase indicates the corresponding D-enantiomers (e.g. if 'O' stands for L-ornithine, then 'o' stands for D-ornithine); amino acid residues inside brackets are those comprised in cyclic homodetic (lactam) structures (which may cover either a part or the full peptide sequence)

[c]$^{\varepsilon}$K stands for the Lys residue whose side chain amine ($^{\varepsilon}NH_2$) group is forming an intramolecular lactam (amide) bridge with the *C*-terminal amino acid

[d]*Dab* stands for L-α,γ-diaminobutanoic acid, and $^{\gamma}$*Dab* stands for the *Dab* residue whose side chain amine ($^{\gamma}NH_2$) group is forming an intramolecular lactam (amide) bridge with the *C*-terminal amino acid

[e]$^{\beta}$T stands for a Thr residue whose side chain hydroxyl ($^{\beta}$OH) is forming an intramolecular lactone (ester) bridge with the C-terminal amino acid; ^{3m}E stands for 3-methyl glutamic acid

the major struggle towards market entry, in terms of both time and costs (ClinicalTrials.gov n.d.; ClinicalTrialsPage n.d.; Greber and Dawgul 2017). Benefits may only come after competent authorities such as the FDA (US) or EMA (EU) approval, where the patent owner/sponsor earns the right of exclusive marketing of the drug for the period of patent protection (usually for 10–15 years).

Clinical trials are organized in four phases, always in strict obedience to the standards of Good Clinical Practice (GCP) (ClinicalTrialsPage n.d.). *Phase I trial* tests are focused in judging the drug safety and potential side-effects and in finding the correct drug dosage. Therefore, PK (absorption, distribution, metabolism, excretion and toxicity – ADMET) is the main focus of the trial, while drug pharmacodynamics (PD – pharmacological activity of the drug in the body) is also surveilled; often, a small group (50–100) of healthy individuals is involved at this stage. A larger group of people is involved in *Phase II trials* (up to 600), whose focus is on the effectiveness rather than safety, as this phase aims at obtaining preliminary data on whether the drug works in people affected by a certain disease or condition; comparative studies (use of a new drug compared to a standard treatment and placebo) are mainly conducted in this phase (ClinicalTrialsPage n.d.). A *Phase III trial* aims at gathering additional information on safety and effectiveness, by studying different populations and different dosages and using the drug in combination with other drugs (drug interactions studies); this phase mainly targets assessment of whether the benefits of the new medicine are higher than those of currently available therapeutics for the same indication. Therefore, the number of subjects involved at this stage ranges from hundreds to thousands (ClinicalTrialsPage n.d.). If the drug obtains a positive outcome at this stage, authorities (either FDA or EMA) will grant the certification required for its introduction in the market. Finally, *Phase IV trials* are conducted when the new drug is already in the market, by monitoring effectiveness and safety in a much larger scale, involving diverse populations (ClinicalTrialsPage n.d.).

Given the broad spectrum of activity of AMPs, it is possible to find the same peptide undergoing clinical trials for different pathological conditions or for use via different administration routes (ClinicalTrials.gov n.d.). This means that failure in a specific clinical trial of a given AMP does not necessarily rule out its potential approval, and consequent clinical use, in a different scenario. As outlined by Greber et al. (2017), most AMPs under clinical development are being tested for topical application, as major challenges for peptide administration by other routes have not yet been satisfactorily addressed, as further discussed in the next section (Greber and Dawgul 2017). As such, the number of AMPs that have successfully crossed the full clinical trial path is low, particularly when comparing with the enormous number of patented AMPs for potential therapeutic applications (WIPO 2018). Table 15.2 highlights the selected AMPs currently in the latest stages of clinical development. Omiganan (an indolicin derivative, i.e. belonging to the cathelicidin family) was one of the first recent AMPs able to complete Phase III trials' effectiveness tests for treatment of rosacea while being also under Phase II trials for evaluation of safety and efficacy of an omiganan-based topical gel on female subjects with moderate to severe inflammatory *Acne vulgaris* (Cutanea Life Sciences, Inc.) (ClinicalTrials.gov n.d.; Sader et al. 2004; Melo and Castanho 2007). Interestingly, there are several clinical trials involving peptides from the cathelicidin family, such as those addressing both the immunomodulatory effects of hCAP18 and LL37 peptides in improving infection outcomes and the direct bactericidal effects of some of their derivatives (ClinicalTrials.gov n.d.). For instance, LL37 has been tested for the treatment of hard-to-heal venous leg ulcers (Phase II), whereas its derivative OP-145 (also known as AMP60.4Ac) has been tested for the treatment of chronic suppurative otitis media with very positive Phase II results (OctoPlus BV) (ClinicalTrials.gov n.d.; Greber and Dawgul 2017). Notwithstanding, the best-known AMP reaching Phase III trials is pexiganan, an analogue of magainin 2. A 0.8% cream containing pexiganan was tested in Phase III clinical trials for treatment of infected diabetic foot ulcers (Dipexium Pharmaceuticals, Inc.) (ClinicalTrials.gov n.d.). Despite the positive outcomes, the FDA did not approve its marketing on the grounds that its efficacy was comparable to that of the standard oral antibiotic ofloxacin (Greber and Dawgul 2017). Another Phase III achiever is iseganan, a protegrin-1-derived AMP. Iseganan was submitted to clinical trials for treatment of oral mucositis in patients receiving radiation therapy for head and neck cancer but terminated at Phase III due to low effectiveness. The same AMP was then tested in the prevention of ventilator-associated pneumonia, having failed again at Phase II/III. Interestingly, an iseganan derivative, POL7080 (murepavadin), has started a new promising path, having completed two Phase II studies in patients with ventilator-associated bacterial pneumonia and non-cystic fibrosis-associated bronchiectasis (Polyphor, Lda) (WebPage n.d.; Kosikowska and Lesner 2016). One final example of an AMP that is a Phase III achiever is bovine lactoferrin that is currently recruiting to improve neonatal survival in low birth weight babies, by fighting neonatal sepsis and necrotizing enterocolitis (ClinicalTrials.gov n.d.). Actually, several attempts have been made over the years to put forward into the clinics different AMPs from the lactoferrin family. Lactoferrin has been tested as a dietary supplement both for HIV-1/AIDS infection and for nosocomial infections in critically ill patients, having completed Phase II (ClinicalTrials.gov n.d.). AM-Pharma developed the human lactoferrin-derived peptide (hLF1-11) for two potential indications: treatment of infectious complications among autologous haematopoietic stem cell transplant (HSCT) recipients and bacteraemia due to *S. epidermidis*. However, Phase I/II clinical trials with hLF1-11 for these two applications were withdrawn prior to enrolment. Nevertheless, AM-Pharma was able to take to conclusion Phase I/II studies on the safety of a single dose of 5 mg of hLF1-11 given to autologous HSCT recipients (Greber and Dawgul 2017).

Another promising AMP is PAC-113, an analogue of histatin 5 that occurs naturally in saliva, which has high activity against *Candida albicans*, including drug-resistant strains isolated from HIV-1 carriers. PAC-113 has passed Phase II clinical trials aimed at determining the optimal peptide doses for mouth rinsing towards treatment of oral candidiasis in HIV-1 carriers (ClinicalTrials.gov n.d.; Greber and Dawgul 2017).

Table 15.2 Selected AMP and AMP derivatives under clinical trial as of July 2018

Peptide	Origin	Evaluated condition	Clinical trial phase	Sponsor/collaborator
Omiganan (MBI 226 or MX 226)	Cathelicidin family	Treatment of rosacea	III (completed)	Cutanea Life Sciences
		Treatment of severe inflammatory acne vulgaris	II	Cutanea Life Sciences
		Preventing central venous catheter infection	III	Mallinacckrodt
LL37	Cathelicidin family (derived from hCAP18)	Treatment of hard-to-heal venous leg ulcers	II	Lipopeptide AB
OP-145 (AMP60.4Ac)	Cathelicidin family (derived from LL37)	Treatment of chronic suppurative otitis media	II	OctoPlus BV
Pexiganan	Analogue of magainin-2	Treatment of infected diabetic foot ulcers	III	Dipexium Pharmaceuticals (has merged with PLx Pharma)
Iseganan	Protegrin-1 derivative	Treatment of oral mucositis in patients receiving radiation therapy for head and neck cancer	III (failed)	IntraBiotics Pharmaceuticals
		Treatment for ventilator-associated pneumonia	II/III (failed)	
POL7080 (murepavadin)	Derivative of iseganan	Treatment of ventilator-associated pneumonia	II	Polyphor, Lda
		Treatment of non-cystic fibrosis-associated bronchiectasis	II	
PAC-113	Histatin analogue	Mouth rinse in HIV-seropositive individuals with oral candidiasis	II	Pacgen Biopharmaceuticals Corporation/Quintiles, Inc.
Lactoferrin		HIV infection	II (dietary supplement)	Jason Baker/Ventri Bioscience/Minneapolis Medical Research Foundation
Lactoferrin		Nosocomial infection in critically ill patients	II (dietary supplement)	Queen's University
Bovine lactoferrin		Neonatal sepsis	III	Aga Khan University/ University of Sydney/United States Agency for International Development
		Necrotizing enterocolitis	III	
hLF1-11	Derived from human lactoferrin	Treatment of infectious complications among HSCT recipients	I/II	AM-Pharma
		Treatment of bacteraemia due to *Staphylococcus epidermidis*	I/II (withdrawn)	
Talactoferrin		Oral solution for nosocomial infection in preterm infants	I/II	Agennix/NIH

15.4 Challenges in the Clinical Application of AMPs

The very high number of promising AMPs found in numerous patents and literature reports fail in the translation into clinical trials mostly because the preclinical studies are frequently performed in absence of physiological conditions. This means that the influence of, e.g. (i) physiological levels of relevant ions, (ii) high ionic strength (due to high-salt concentration) and (iii) presence of proteases and of physiological barriers is frequently overlooked (Fox 2013; Mahlapuu et al. 2016). This often leads to in vivo performances that are unexpectedly low, as compared to in vitro data. On the other hand, the opposite sometimes happens: high-throughput analysis of an AMP library has revealed AMP with poor in vitro performance to be very effective in vivo, even in amounts below their bactericidal concentrations, which was ascribed to their immunomodulatory properties promoting bacterial killing via stimulation of the host immune system (Hancock and Sahl 2006). Also, variability associated to different animal models used for in vivo testing of AMPs may create results with poor translation into humans.

Adding to the above, one of the greatest challenges towards effective clinical applications of AMPs is their low oral bioavailability. Peptides are unstable in the gastrointestinal tract (GIT) due to both the harsh gastric pH and proteolytic enzymes, which is further aggravated by their poor penetration across the intestinal epithelium (Di 2015; Lundquist and Artursson 2016; Lau and Dunn 2018). Hence, the benefits that AMPs represent as a potential new class of antimicrobial agents require that efficient strategies are found which enable them to step out of their current almost exclusive confinement to topical applications. Some of the approaches that are being devised to tackle these challenges are next addressed.

15.4.1 Chemical Approaches to Tackle Low Bioavailability

Given their generally poor absorption and distribution, along with fast metabolic degradation and excretion (ADME), which is further aggravated by their propensity to aggregate with plasma proteins, the vast majority of natural peptides has an oral bioavailability no higher than 1% (Di 2015). Still, as compared with protein-/antibody-based therapeutics, peptides have much better permeation into tissues and are considerably more stable, less prone to trigger immune/allergic reactions and significantly cheaper to produce, which underpins the extensive research that has been devoted to find strategies to improve their bioavailability (Otvos and Wade 2014; Di 2015) .

A number of chemical strategies have been developed to tackle this hurdle, mainly by increasing AMP (i) metabolic stability and (ii) lipophilicity, towards enhanced permeability (Di 2015; Molchanova et al. 2017; Qvit et al. 2017; Sierra et al. 2017; Henninot et al. 2018). Common approaches include:

- Cyclization to increase metabolic stability and/or lock bioactive conformations
- Replacement of genetically encoded amino acids by noncanonical residues, to deliver peptidomimetics of increased stability and permeability
- *N-/C*-Terminal modifications, to avoid proteolytic degradation and increase permeability

Backbone modifications, i.e. replacement of peptide bonds by enzyme-stable bioisosteres, have been also applied, but will not be addressed in this chapter.

15.4.1.1 Cyclization

Cyclization has been proven quite effective which is supported by the fact that most orally active peptides are cyclic, including well-known potent peptide antibiotics like bacitracin A, colistin, gramicidins and polymyxins B1 and B2, just to name a few emblematic examples (Joo 2012; Tapeinou et al. 2015; Falanga et al. 2017).

Chemical strategies towards peptide cyclization span from heterodetic (cyclization through a non-amide, e.g. a disulphide), or homodetic (cyclization through an amide, i.e. lactamization) cyclic peptides, to fully unnatural enzyme-stable all-hydrocarbon tethers (Fig. 15.1) (Lambert

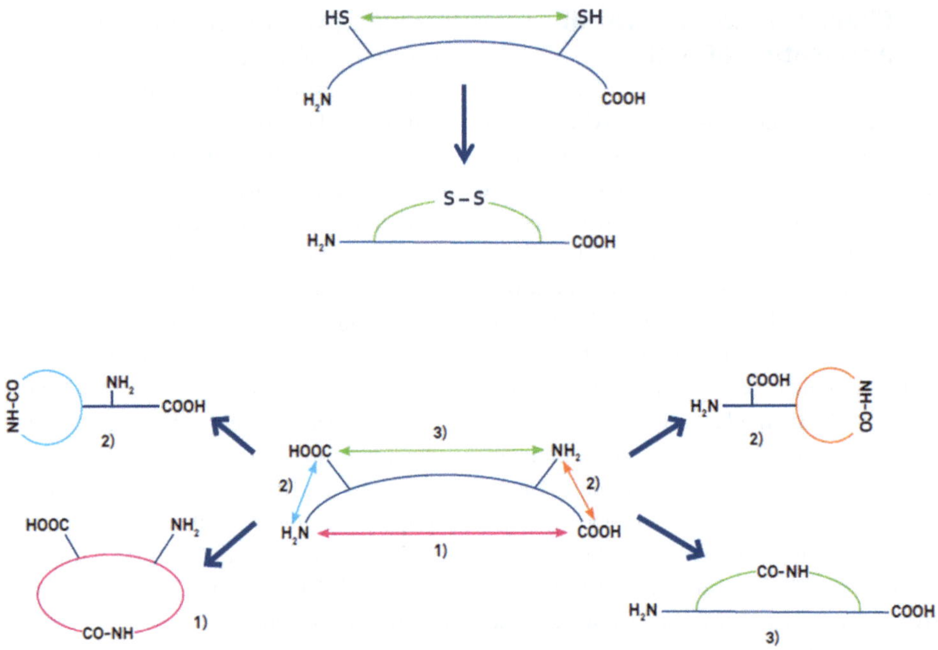

Fig. 15.1 Classic peptide cyclization approaches. *Top*: intramolecular oxidation of cysteine side chain thiols to produce cyclic disulphide (heterodetic) peptides; *bottom*: (1) head-to-tail, (2) side chain-to-head/tail and (3) side chain-to-side chain lactamization to give homodetic cyclic peptide (Adapted from http://www.bachem.com/service-support/newsletter/peptide-trends-march-2016/)

et al. 2001; Davies 2003; White and Yudin 2011; Chu et al. 2015; Marti-Centelles et al. 2015).

Heterodetic cyclization through formation of a single disulphide bridge via oxidation of the side chain thiols of a couple of cysteines (either belonging or added to the native peptide sequence) is quite straightforward from the chemical viewpoint (Fig. 15.1, top) and has delivered α-helical AMPs inspired in tenecin 1, an insect defensin peptide (Ahn et al. 2008). Still, given that peptides with a single disulphide bridge may be unstable in bioreducing environments (Yang et al. 2006), efforts have been focused on AMPs bearing multiple disulphide bridges, e.g. as those inspired in human defensins (Cheneval et al. 2014; Falanga et al. 2017), and on development of efficient chemical methods for the synthesis of cysteine-rich peptide (Cheneval et al. 2014). Another approach towards stable cyclic AMPs formed by means of 'natural' bonds has been lactamization (Fig. 15.1, bottom), i.e. formation of intramolecular peptide bonds to produce homodetic cyclic peptides (Monaim et al. 2018). Still, despite being more stable than their linear precursors, even against endopeptidases, cyclic homodetic peptides may remain somewhat susceptible to metabolic degradation (Bourne et al. 1996).

More recent approaches towards stable cyclic peptides are taking advantage of the wide panoply of commercial non-natural amino acids suitably functionalized to enable application of 'click'-type intramolecular reactions. These chemoselective cyclizations include, among many others, (i) Michael-type (thiol-ene) additions (Fig. 15.2a), (ii) Huisgen's 1,3-dipolar (azide-alkyne) cycloadditions (Fig. 15.2b) and (iii) ring-closure olefin metathesis (RCM), also known as Grubbs reaction (Fig. 15.2c) (Lambert et al. 2001; Davies 2003; White and Yudin 2011; Chu et al. 2015; Marti-Centelles et al. 2015; Lau and Dunn 2018).

Peptide cyclization via ring-closure metathesis (RCM), leading to the so-called stapled peptides, is actually gaining prominence towards development of synthetic AMPs and structurally related peptides, like cell-penetrating peptides

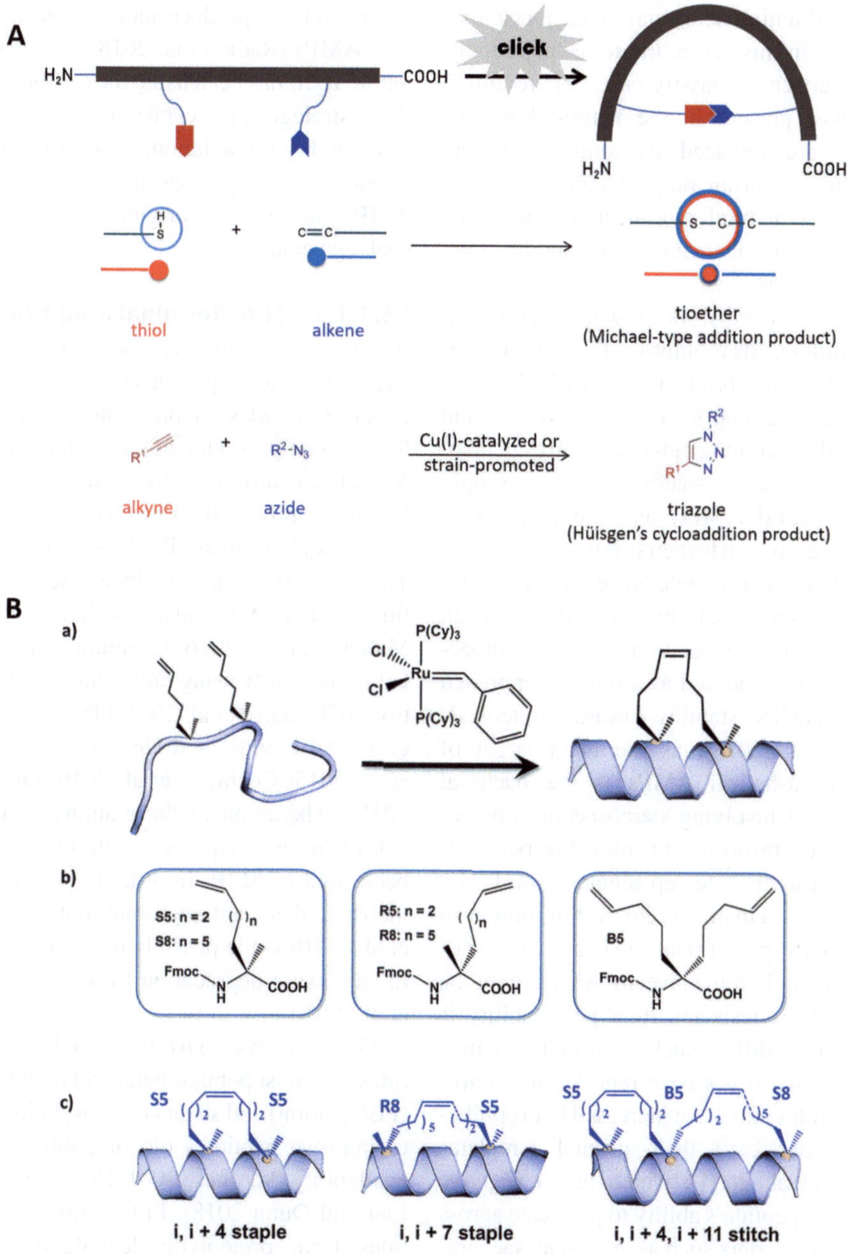

Fig. 15.2 Examples of chemoselective ('click'-type) cyclization reactions: (**a**) thiol-ene and azide-alkyne conjugations and (**b**) ring-closure metathesis (RCM) settling on use of (**a**) Grubbs catalyst and (**b**) non-natural amino acid building blocks, to yield (**c**) stapled peptides (Chu et al. 2015)

(CPPs), by presenting a restrained and metabolically stable helical conformation, widely known as relevant for the membrane interaction ability of most cationic amphipathic AMPs and CPPs (Walensky and Bird 2014; Lau and Dunn 2018; Migon et al. 2018).

15.4.1.2 Use of Noncanonical Amino Acids

Another popular approach to improve peptides' bioavailability has been the replacement of proteinogenic amino acids in the native sequence by non-coded ones, such as D-amino acids,

N-methylated amino acids and other noncanonical residues. In this sense, the term 'peptidomimetic' has arisen to classify cases where amino acid residues present in the natural bioactive sequence were replaced by similar building blocks, often differing only in chirality and/or other subtle structural modifications, with the main purpose of enhancing bioavailability (Wu and Gellman 2008).

Merrifield and co-workers were pioneers in using D-amino acids to improve the bioavailability of AMPs, by producing 'retro' (inverted sequence, all L-amino acids), 'enantio' (normal sequence, all D-amino acids) and 'retro-inverso' (or 'retro-enantio') analogues of cecropin A-melittin hybrid AMPs, and the latter were found to be equipotent to their original membrane-active AMP templates while presenting markedly higher proteolytic stability (Merrifield et al. 1995). Use of retro-inverso analogues of bioactive peptides has endured as a popular approach to increase AMPs' stability towards proteolysis (Cardoso et al. 2018), as the main target of amphipathic α-helical AMPs is the bacterial membrane, not involving stereospecific interactions. Hence, resorting to retro-inverso analogues, or just discrete replacements of L- by D-amino acids, remains an interesting option to improve AMPs' performance (Silva et al. 2014). However, though retro-inverso AMPs may be fully proteolysis resistant, their permeability is not expected to differ much from that of native AMPs. Actually, it has been reported that retro-inverso peptides may have increased hydrophilicity as compared to their normal templates (Alminana et al. 2004), which has a negative impact on the peptide's ability to permeate across physiological barriers such as, e.g. mucosae, epithelia and sub-epithelial tissues (Lundquist and Artursson 2016).

Given that orally available peptides often possess molecular weights no higher than 1000 Da and no more than six hydrogen bond donors (Doak et al. 2014), one way to improve AMP permeability is to reduce their number of hydrogen bond donors. This can be achieved by the use of *N*-methylated amino acids (Chatterjee et al. 2013); this strategy has been combined with cyclization, to produce more stable and permeable AMPs (Rader et al. 2018). Remarkably, the nature itself has been using the combination of all these strategies, i.e. cyclization, use of D-amino acids and *N*-methylation, to afford orally active peptides, like cyclosporine A (Fig. 15.3), or AMPs of marine origin, e.g. discodermins (Falanga et al. 2016).

15.4.1.3 N-/C-Terminal Modifications

Along with *N*-methylation, other reasonably straightforward approaches can be used to enhance peptides' stability and circulating half-lives, namely, modifications/conjugations at the *N*- and/or *C*-termini of the bioactive sequence or, less frequently, at the side chains of trifunctional amino acid residues like Cys, Ser, Thr or Lys. The simplest methods include *C*-terminal amidation and/or *N*-terminal acylation (Di 2015; Mahlapuu et al. 2016; Henninot et al. 2018), the latter eventually using fatty acids (peptide lipidation) (Chicharro et al. 2001; Oh et al. 2014; Trier et al. 2014; Vidal and Geffard 2014; Swiecicki et al. 2015; Cochrane et al. 2016; Di Pisa et al. 2015). Alkylation of the ε-amine group of Lys side chains is an equally simple approach that has been used in AMPs in order to preserve or even increase their antimicrobial potency (Teixeira et al. 2010) while possibly reducing their lability to enzymatic degradation by, e.g. trypsin (Arias et al. 2018).

Other moieties have been conjugated to peptides, the most popular being polyethylene glycol (PEGylation) and sugars (glycosylation), in order to improve peptide's biocompatibility and bioavailability (Veronese 2001; Henninot et al. 2018; Lau and Dunn 2018). PEGylation protects peptides from proteolytic degradation, decreases their immunogenicity by limiting their uptake via dendritic cells and, in some instances, increases their circulating half-life by over two orders of magnitude, partially due to reduced renal clearance (Veronese 2001; Henninot et al. 2018; Lau and Dunn 2018). PEGylated AMPs were found to have increased lung biocompatibility suitable for local inhaled therapy of respiratory infections (Morris et al. 2012). Chemoselective coupling of PEG has experienced some limitations in the

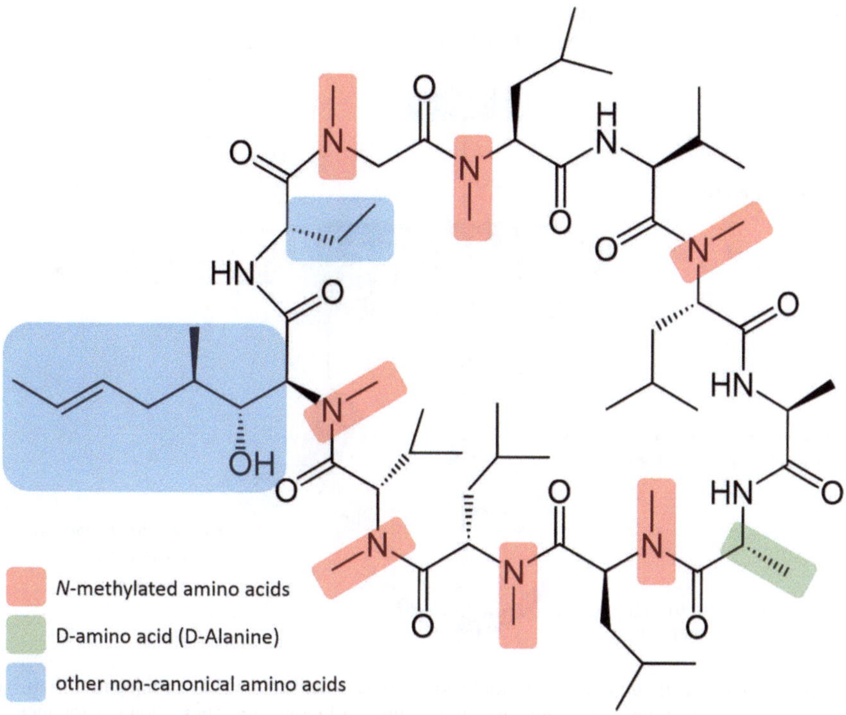

Fig. 15.3 Structure of cyclosporine A, an orally active peptide drug produced in nature by *Tolypocladium inflatum* fungi, which exhibits almost all of the features known to enhance peptides' oral bioavailability: *N*-methylated amino acids, namely, *N*-methyl-glycine (best known as sarcosine, Sar), *N*-methyl-L-valine (*N*-MeVal) and *N*-methyl-L-leucine (*N*-MeLeu); one D-amino acid (D-alanine); and two additional *N*-methylated noncanonical amino acids, *N*-methyl-2-aminobutanoic acid and *N*-methyl-2-amino-3-hydroxy-4-methyloct-6-enoic acid

early days of peptide PEGylation (Veronese 2001), including activity decrease for some AMPs (Guiotto et al. 2003; Imura et al. 2007). Suitable chemoselective methods became available in recent years (Roberts et al. 2002; Doherty et al. 2005; Hamley 2014), but the balance between improving PK and retaining good PD is a delicate one (Fishburn 2008); one recent approach to bypass this limitation is based on the use of chemoreversible PEGylation of arginine side chains via cleavable phenylglyoxal linkers (Fang et al. 2017); this produced prodrugs of arginine-rich AMPs, whose resistance to serum proteases was dramatically increased while allowing the parent bioactive peptide to be slowly released in a matter of hours to days (Gong et al. 2017). Actually, a temporary PEGylation strategy to increase stability, reduce immunogenicity and improve the PK of AMPs for intravenous administration had been previously described and patented by Hoffman et al. (2014). Remarkably, in this case, AMP (namely, apidaecin and oncocin) side chains were PEGylated using a PEG modified with a linker that includes a recognition sequence for trypsin-like serum proteases; in other words, a protease-sensitive linker was used to increase peptide stability and circulation time, as it allowed for a controlled release of the AMPs in their target tissues. Moreover, this strategy also contributed to delayed peptide renal excretion.

Glycosylation may also improve peptides' PK and PD (Moradi et al. 2016). Naturally glycosylated AMPs, namely, insect-derived Pro-rich AMPs (Lele et al. 2015), have inspired chemists and bioengineers to develop artificial site-specific glycosylation methods delivering glycopeptides with improved stability, bioactivity and specificity (Doores et al. 2006; Bednarska et al. 2017). Bioactive peptides have also been conjugated to selected moieties, including antibodies or albu-

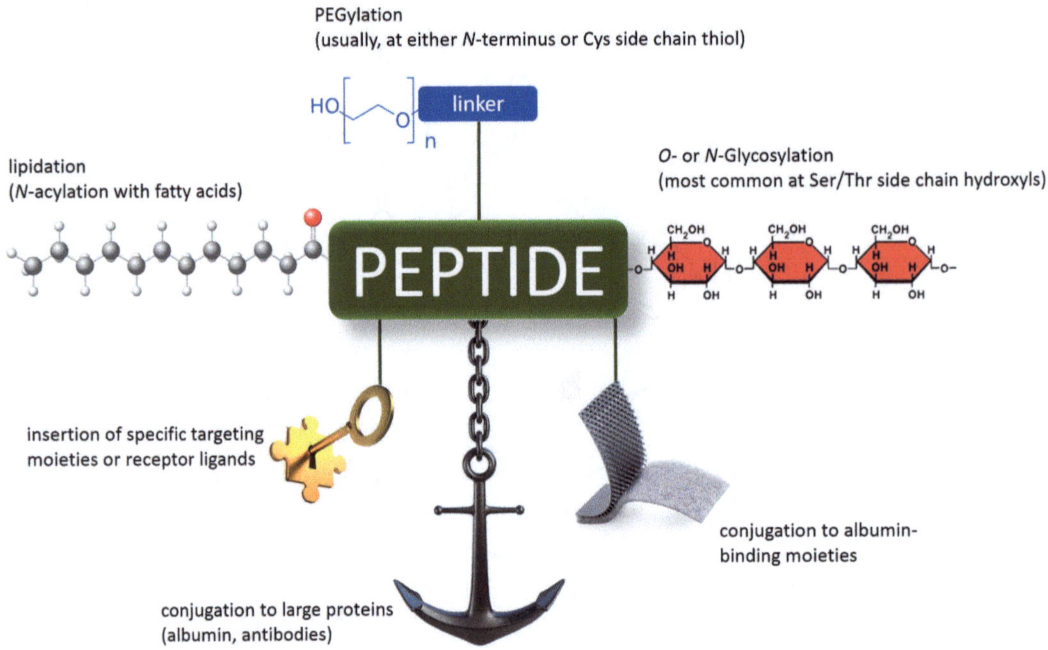

Fig. 15.4 Peptide modification/conjugation toolbox: from current approaches, like lipidation, glycosylation and PEGylation, to more sophisticated strategies, including insertion of specific peptide segments enabling targeted delivery, and conjugation to protein or protein-binding moieties allowing for increased peptide circulating half lives

min, to increase circulating time due to reduced renal clearance and rescue by the FcRn recycling pathway (Sockolosky and Szoka 2015; Schmidt et al. 2017), and albumin-binding moieties, leading to identical effects as when conjugated with large proteins (Henninot et al. 2018; Lau and Dunn 2018) (Fig. 15.4).

15.4.2 Bioengineering Strategies to Improve Pharmacokinetics and Pharmacodynamics

Bioengineering approaches, such as AMP encapsulation for local/targeted delivery, have been investigated to improve their stability and reduce systemic toxicity. Nanoparticles (NPs) prepared using biodegradable materials such as natural and synthetic polymers (e.g. chitosan NP (Almaaytah et al. 2017) and PLGA-NP (d'Angelo et al. 2015), respectively) and lipids (Carmona-Ribeiro and Carrasco 2014) have been used for AMP encapsulation due to their high degree of biocompatibility. Moreover, AMP-conjugated gold NPs have been also reported (Rai et al. 2016a, b; Reinhardt and Neundorf 2016; Zong et al. 2017). This strategy increases AMP serum stability and was able to improve their antimicrobial efficacy against a broad range of drug-resistant bacteria in comparison with its free form (Rajchakit and Sarojini 2017).

NPs can also offer the possibility to selectively target AMPs against bacteria (Reinhardt and Neundorf 2016) and to focus AMP delivery onto specific sites, minimizing side-effects and increasing efficacy (Eckert et al. 2014). Specifically targeted antimicrobial peptides (STAMP), containing a targeting peptide coupled to the AMP, were also proposed to inhibit the growing pathogenic bacteria in a selective way. STAMP developed by Eckert RH and co-workers (Eckert et al. 2014) demonstrated increased killing potency against targeted bacteria in mucosal infections, preserving the normal microflora on the affected region.

AMP attachment onto magnetic NPs was also proposed as a new nanotechnology strategy to increase AMP efficacy, since magnetic NPs can be directed to the infection site using an external magnetic field (Blin et al. 2011).

Antimicrobial dendrimer peptides, prepared by multiple peptide sequences linked to an inner branched core, represent a different approach to potentiate the antimicrobial effect of the AMP (decreasing their susceptibility to proteases due to steric hindrance) and decrease their haemolytic activity (Scorciapino et al. 2017). Antimicrobial dendrimeric peptides, designed by the attachment of two to eight copies of a tetrapeptide (RLYR) or an octapeptide (RLYRKVYG) around a core of lysine residues, have demonstrated very high antimicrobial efficacy against several bacteria and fungi in high- and low-salt conditions (Tam et al. 2002). Other types of antimicrobial dendrimeric peptides have been proposed by grafting unnatural peptides or using organic groups as the branching core (Scorciapino et al. 2017).

An innovative strategy was developed and patented by Collins et al. (2015) that uses engineered bacteriophages to express AMP and/or antimicrobial polypeptides (lytic enzymes) in a broad spectrum of host bacteria. Expressed AMP will be secreted from host bacteria or alternatively, released after bacteria lysis, improving bacterial infection treatment.

Targeted bioengineering approaches have been proposed to prevent implant-associated infections (IAI), namely, biofilm formation by AMP immobilization onto biomaterial coatings (Costa et al. 2011, 2015; Willcox et al. 2013; Rai et al. 2016a, b). These strategies imply covalent binding of the AMP directly, through PEG spacers, or even at polymeric brushes onto the surface of biomaterials. Surfaces with AMP-immobilized coatings were capable to decrease bacterial adhesion and/or affect viability of adherent bacteria. AMPs that were successfully covalently bound to a material surface and the chemical methods used for their immobilization were described in the following patent application (Willcox et al. 2013).

15.4.3 Dealing with Peptide Production Costs

One of the arguments disfavouring peptide-based drugs has been the alleged high cost of peptide synthesis, as well as difficulties faced in its scale-up. It is true that multi-gram production of peptides has been a challenge for many years, but the widely recognized benefits of peptide-based therapeutics have been contributing to a slow, but steady, paradigm shift in the way Big Pharma looks at peptides as drugs. Moreover, peptide synthesis costs represent a small fraction of the overall costs involved in the development of a novel active pharmaceutical ingredient (API) from bench to the clinics (Otvos and Wade 2014). As such, we have been 'watching peptide drugs go up' (Marx 2005), accompanied by considerable progress in chemical and biotechnological approaches towards peptide large-scale production (Marx 2005; Thayer 2011). Large-scale synthesis of peptides is now possible by means of both solid- (SPPS) and liquid-phase (LPPS) peptide synthesis methods (Fig. 15.5) (Marx 2005; Thayer 2011). The most popular example of a peptide-based drug produced in multi-tons is the T-20 peptide (Fuzeon, Roche), which is the first HIV fusion-inhibitor peptide produced at such scale by a combination of solid- and solution-phase methodologies. The decision of Roche to go fully synthetic, instead of doing this 26-mer peptide recombinantly, and the following massive investment, lowered the cost of all synthesis reagents, making industrial peptide synthesis more feasible and affordable.

One of the cheapest alternatives to synthetic peptides is production by recombinant expression methods using microorganisms for which the AMP produced is innocuous. For instance, plectasin, the first fungal defensin whose interesting therapeutic potential was unveiled, can be efficiently produced in high yields, via a fungal expression system that is currently used by Novozymes for industrial-scale production of proteins (Mygind et al. 2005). Still, one cannot underestimate the technical difficulties in producing large amounts of peptides by recombinant expression: at the current production efficiency, for each 100 kg of recombinant peptide, 1 million litres of fermentation mixture will be necessary. Nonetheless, it is foreseen that peptide production costs through this method can still be reduced by two orders of magnitude; hence, as with many protein production methods, large-scale produc-

Fig. 15.5 Modern peptide large-scale production facility (https://www.almacgroup.com/api-chemical-development/highly-potent-peptide-manufacture/)

tion of AMPs by recombinant means could be an economically viable and promising alternative.

In this connection, biotech companies are emerging which are investing on large-scale peptide production at highly competitive prices (Scott 2018).

15.4.4 Impact of Regulatory Impediments and Future Directions

Unlike the exuberance accompanying forays into AMP development a decade ago, current optimism is tempered by deeper recognition of the challenges in translating R&D leading to a new class of products (Fox 2013).

In the early years of AMP discovery, research focused on testing for activity in simplified in vitro settings with little or no concern about the legal prerequisites of an actual preclinical study. Therefore, many promising studies have proceeded possibly too fast, without a full pre-optimization of AMPs' performance in physiological conditions.

Researchers and early biotech companies were also caught by the intricate regulatory environment that encases the entry of AMPs into clinical trials. Even when this first obstacle is surpassed, it is important to highlight that each clinical trial must be precise on the application/pathological condition, focused population and route of administration. In other words, for a given AMP to be successful in completing the full clinical trial path, it has to undergo through a long and expensive iterative process. This explains why this is a difficult path that drains newly formed biotech firms, even when they manage to partner with established Big Pharma company. Fortunately, this paradigm seems to be reaching a point of renewal: the 'post-antibiotic era' is demanding for new solutions to enter the clinics in a near future; therefore, authorities are now more attentive and receptive to new therapeutics. The ever-evolving regulations are now being directed to create a clearer path for promoters. It is possible that, in the near future, cases such as that of pexiganan, which did not receive positive ruling by the FDA for the treatment of infected diabetic foot ulcers simply because it was not superior to oral ofloxacin, would have a different outcome in the present scenario. Moreover, it is also important to highlight that some AMPs have a synergistic effect when applied in combination with standard antibiotics; therefore, they could be used as part of multidrug approaches to enhance the rate of disease eradication and prevent the development of pathogen resistance to conventional therapy.

Altogether, the future of AMPs as a novel and viable class of antibiotics must be regarded with optimism.

References

Afacan NJ, Yeung AT, Pena OM, Hancock RE (2012) Therapeutic potential of host defense peptides in antibiotic-resistant infections. Curr Pharm Des 18(6):807–819

Ahn HS, Cho W, Kim JM, Joshi BP, Park JW, Lohani CR, Cho H, Lee KH (2008) Design and synthesis of cyclic disulfide-bonded antibacterial peptides on the basis of the alpha helical domain of Tenecin 1, an insect defensin. Bioorg Med Chem 16(7):4127–4137

Almaaytah A, Mohammed GK, Abualhaijaa A, Al-Balas Q (2017) Development of novel ultrashort antimicrobial peptide nanoparticles with potent antimicrobial and antibiofilm activities against multidrug-resistant bacteria. Drug Des Dev Ther 11:3159–3170

Alminana N, Alsina MA, Ortiz A, Reig F (2004) Comparative physicochemical study of SIKVAV peptide and its retro and retro-enantio analogues. Colloids Surf A: Physicochem Eng Asp 249(1–3):19–24

Arias M, Piga KB, Hyndman ME, Vogel HJ (2018) Improving the activity of Trp-rich antimicrobial peptides by Arg/Lys substitutions and changing the length of cationic residues. Biomolecules 8(2):1–17

Bednarska NG, Wren BW, Willcocks SJ (2017) The importance of the glycosylation of antimicrobial peptides: natural and synthetic approaches. Drug Discov Today 22(6):919–926

Blin T, Purohit V, Leprince J, Jouenne T, Glinel K (2011) Bactericidal microparticles decorated by an antimicrobial peptide for the easy disinfection of sensitive aqueous solutions. Biomacromolecules 12(4):1259–1264

Bourne DG, Jones GJ, Blakeley RL, Jones A, Negri AP, Riddles P (1996) Enzymatic pathway for the bacterial degradation of the cyanobacterial cyclic peptide toxin microcystin LR. Appl Environ Microbiol 62(11):4086–4094

Brooks BD, Brooks AE (2014) Therapeutic strategies to combat antibiotic resistance. Adv Drug Deliv Rev 78:14–27

Cardoso MH, Cândido ES, Oshiro KGN, Rezende SB, Franco OL (2018) Peptides containing D-amino acids and retro-inverso peptides: general applications and special focus on antimicrobial peptides. In: Koutsopoulos S (ed) Peptide applications in biomedicine, biotechnology and bioengineering. Woodhead Publishing, Elsevier Ltd, Kidlington, pp 131–155

Carmona-Ribeiro AM, Carrasco LDD (2014) Novel formulations for antimicrobial peptides. Int J Mol Sci 15(10):18040–18083

Chatterjee J, Rechenmacher F, Kessler H (2013) N-methylation of peptides and proteins: an important element for modulating biological functions. Angew Chem Int Ed 52(1):254–269

Cheneval O, Schroeder CI, Durek T, Walsh P, Huang YH, Liras S, Price DA, Craik DJ (2014) Fmoc-based synthesis of disulfide-rich cyclic peptides. J Org Chem 79(12):5538–5544

Chicharro C, Granata C, Lozano R, Andreu D, Rivas L (2001) N-terminal fatty acid substitution increases the leishmanicidal activity of CA(1-7)M(2-9), a cecropin-melittin hybrid peptide. Antimicrob Agents Chemother 45(9):2441–2449

Chu Q, Moellering RE, Hilinski GJ, Kim YW, Grossmann TN, Yeh JTH, Verdine GL (2015) Towards understanding cell penetration by stapled peptides. Medchemcomm 6(1):111–119

ClinicalTrials.gov (n.d.) https://www.clinicaltrials.gov/

ClinicalTrialsPage (n.d.) https://www.nia.nih.gov/health/what-are-clinical-trials-and-studies. Accessed Jul 2018.

Coates ARM, Halls G, Hu YM (2011) Novel classes of antibiotics or more of the same? Br J Pharmacol 163(1):184–194

Cochrane SA, Findlay B, Bakhtiary A, Acedo JZ, Rodriguez-Lopez EM, Mercier P, Vederas JC (2016) Antimicrobial lipopeptide tridecaptin A(1) selectively binds to gram-negative lipid II. Proc Natl Acad Sci U S A 113(41):11561–11566

Collins JJ, Koeris M, Lu TKT, et al. (2015) Bacteriophages expressing antimicrobial peptides and uses thereof. US 2015/0050717 A1

Costa F, Carvalho IF, Montelaro RC, Gomes P, Martins MCL (2011) Covalent immobilization of antimicrobial peptides (AMPs) onto biomaterial surfaces. Acta Biomater 7(4):1431–1440

Costa FMTA, Maia SR, Gomes PAC, Martins MCL (2015) Dhvar5 antimicrobial peptide (AMP) chemoselective covalent immobilization results on higher antiadherence effect than simple physical adsorption. Biomaterials 52:531–538

d'Angelo I, Casciaro B, Miro A, Quaglia F, Mangoni ML, Ungaro F (2015) Overcoming barriers in Pseudomonas aeruginosa lung infections: engineered nanoparticles for local delivery of a cationic antimicrobial peptide. Colloids Surf B:Biointerfaces 135:717–725

Davies JS (2003) The cyclization of peptides and depsipeptides. J Pept Sci 9(8):471–501
Di L (2015) Strategic approaches to optimizing peptide ADME properties. AAPS J 17(1):134–143
Di Pisa M, Chassaing G, Swiecicki JM (2015) When cationic cell-penetrating peptides meet hydrocarbons to enhance in-cell cargo delivery. J Pept Sci 21(5):356–369
Doak BC, Over B, Giordanetto F, Kihlberg J (2014) Oral druggable space beyond the rule of 5: insights from drugs and clinical candidates. Chem Biol 21(9):1115–1142
Doherty DH, Rosendahl MS, Smith DJ, Hughes JM, Chlipala EA, Cox GN (2005) Site-specific PEGylation of engineered cysteine analogues of recombinant human granulocyte-macrophage colony-stimulating factor. Bioconjug Chem 16(5):1291–1298
Doores KJ, Gamblin DP, Davis BG (2006) Exploring and exploiting the therapeutic potential of glycoconjugates. Chem Eur J 12(3):656–665
DrugDataBase (n.d.) http://www.drugsdb.eu/index.php?l=c. Accessed Jul 2018.
Eckert RH, Yarbrough D, Shi W, et al (2014) Selectively targeted antimicrobial peptides and the use thereof. EP2801368A1 2014
Falanga A, Lombardi L, Franci G, Vitiello M, Iovene MR, Morelli G, Galdiero M, Galdiero S (2016) Marine antimicrobial peptides: nature provides templates for the design of novel compounds against pathogenic bacteria. Int J Mol Sci 17(5):1–17
Falanga A, Nigro E, De Biasi MG, Daniele A, Morelli G, Galdiero S, Scudiero O (2017) Cyclic peptides as novel therapeutic microbicides: engineering of human defensin mimetics. Molecules 22(7):1–15
Fang Y, Xue JX, Gao S, Lu AQ, Yang DJ, Jiang H, He Y, Shi K (2017) Cleavable PEGylation: a strategy for overcoming the "PEG dilemma" in efficient drug delivery. Drug Deliv 24(2):22–32
Fernandes P, Martens E (2017) Antibiotics in late clinical development. Biochem Pharmacol 133:152–163
Fishburn CS (2008) The pharmacology of PEGylation: balancing PD with PK to generate novel therapeutics. J Pharm Sci 97(10):4167–4183
Fox JL (2013) Antimicrobial peptides stage a comeback(vol 31, pg 379, 2013). Nat Biotechnol 31(12):1066–1066
Giuliani A, Pirri G, Nicoletto SF (2007) Antimicrobial peptides: an overview of a promising class of therapeutics. Cent Eur J Biol 2(1):1–33
Gomes A, Teixeira C, Ferraz R, Prudencio C, Gomes P (2017) Wound-healing peptides for treatment of chronic diabetic foot ulcers and other infected skin injuries. Molecules 22(10):2–18
Gong Y, Andina D, Nahar S, Leroux JC, Gauthier MA (2017) Releasable and traceless PEGylation of arginine-rich antimicrobial peptides. Chem Sci 8(5):4082–4086
Gordon YJ, Romanowski EG, McDermott AM (2005) A review of antimicrobial peptides and their therapeutic potential as anti-infective drugs. Curr Eye Res 30(7):505–515
Greber KE, Dawgul M (2017) Antimicrobial peptides under clinical trials. Curr Top Med Chem 17(5):620–628
Guiotto A, Pozzobon M, Canevari M, Manganelli R, Scarin M, Veronese FM (2003) PEGylation of the antimicrobial peptide nisin a: problems and perspectives. Farmaco 58(1):45–50
Hamley IW (2014) PEG-peptide conjugates. Biomacromolecules 15(5):1543–1559
Hancock RE, Sahl HG (2006) Antimicrobial and host-defense peptides as new anti-infective therapeutic strategies. Nat Biotechnol 24(12):1551–1557
Henninot A, Collins JC, Nuss JM (2018) The current state of peptide drug discovery: back to the future? J Med Chem 61(4):1382–1414
Hoffmann R, Berthold N, Nollmann F (2014). Modified antibiotic peptides having variable systemic release. US20140309161 A1. 2014
Imura Y, Nishida M, Ogawa Y, Takakura Y, Matsuzaki K (2007) Action mechanism of tachyplesin I and effects of PEGylation. Biochim Biophys Acta-Biomembr 1768(5):1160–1169
Joo SH (2012) Cyclic peptides as therapeutic agents and biochemical tools. Biomol Ther 20(1):19–26
Kosikowska P, Lesner A (2016) Antimicrobial peptides (AMPs) as drug candidates: a patent review (2003-2015). Expert Opin Ther Pat 26(6):689–702
Lambert JN, Mitchell JP, Roberts KD (2001) The synthesis of cyclic peptides. J Chem Soc-Perkin Trans 1(5):471–484
Lau JL, Dunn MK (2018) Therapeutic peptides: historical perspectives, current development trends, and future directions. Bioorg Med Chem 26(10):2700–2707
Lele DS, Talat S, Kumari S, Srivastava N, Kaur KJ (2015) Understanding the importance of glycosylated threonine and stereospecific action of Drosocin, a proline rich antimicrobial peptide. Eur J Med Chem 92:637–647
Lundquist P, Artursson P (2016) Oral absorption of peptides and nanoparticles across the human intestine: opportunities, limitations and studies in human tissues. Adv Drug Deliv Rev 106:256–276
Mahlapuu M, Hakansson J, Ringstad L, Bjorn C (2016) Antimicrobial peptides: an emerging category of therapeutic agents. Front Cell Infect Microbiol 6:1–12
Marti-Centelles V, Pandey MD, Burguete MI, Luis SV (2015) Macrocyclization reactions: the importance of conformational, configurational, and template-induced preorganization. Chem Rev 115(16):8736–8834
Martinez JL (2014) General principles of antibiotic resistance in bacteria. Drug Discov Today Technol 11:33–39
Marx V (2005) Watching peptide drugs grow up. Chem Eng News 83(11):17–84
Melo MN, Castanho MARB (2007) Omiganan interaction with bacterial membranes and cell wall models. Assigning a biological role to saturation. Biochim Biophys Acta-Biomembr 1768(5):1277–1290
Merrifield RB, Juvvadi P, Andreu D, Ubach J, Boman A, Boman HG (1995) Retro and retroenantio analogs of

cecropin-melittin hybrids. Proc Natl Acad Sci U S A 92(8):3449–3453
Migon D, Neubauer D, Kamysz W (2018) Hydrocarbon stapled antimicrobial peptides. Protein J 37(1):2–12
Molchanova N, Hansen PR, Franzyk H (2017) Advances in development of antimicrobial peptidomimetics as potential drugs. Molecules 22(9):1–60
Monaim SAHA, Jad YE, El-Faham A, de la Torre BG, Albericio F (2018) Teixobactin as a scaffold for unlimited new antimicrobial peptides: SAR study. Bioorg Med Chem 26(10):2788–2796
Moradi SV, Hussein WM, Varamini P, Simerska P, Toth I (2016) Glycosylation, an effective synthetic strategy to improve the bioavailability of therapeutic peptides. Chem Sci 7(4):2492–2500
Morris CJ, Beck K, Fox MA, Ulaeto D, Clark GC, Gumbleton M (2012) Pegylation of antimicrobial peptides maintains the active peptide conformation, model membrane interactions, and antimicrobial activity while improving lung tissue biocompatibility following airway delivery. Antimicrob Agents Chemother 56(6):3298–3308
Mygind PH, Fischer RL, Schnorr KM, Hansen MT, Sonksen CP, Ludvigsen S, Raventos D, Buskov S, Christensen B, De Maria L, Taboureau O, Yaver D, Elvig-Jorgensen SG, Sorensen MV, Christensen BE, Kjaerulff S, Frimodt-Moller N, Lehrer RI, Zasloff M, Kristensen HH (2005) Plectasin is a peptide antibiotic with therapeutic potential from a saprophytic fungus. Nature 437(7061):975–980
Oh D, Shirazi AN, Northup K, Sullivan B, Tiwari RK, Bisoffi M, Parang K (2014) Enhanced cellular uptake of short polyarginine peptides through fatty acylation and cyclization. Mol Pharm 11(8):2845–2854
Otvos L, Wade JD (2014) Current challenges in peptide-based drug discovery. Front Chem 2:1–4
Pfalzgraff A, Brandenburg K, Weindl G (2018) Antimicrobial peptides and their therapeutic potential for bacterial skin infections and wounds. Front Pharmacol 9:1–23
Qvit N, Rubin SJS, Urban TJ, Mochly-Rosen D, Gross ER (2017) Peptidomimetic therapeutics: scientific approaches and opportunities. Drug Discov Today 22(2):454–462
Rader AFB, Reichart F, Weinmuller M, Kessler H (2018) Improving oral bioavailability of cyclic peptides by N-methylation. Bioorg Med Chem 26(10):2766–2773
Rai A, Pinto S, Evangelista MB, Gil H, Kallip S, Ferreira MGS, Ferreira L (2016a) High-density antimicrobial peptide coating with broad activity and low cytotoxicity against human cells. Acta Biomater 33:64–77
Rai A, Pinto S, Velho TR, Ferreira AF, Moita C, Trivedi U, Evangelista M, Comune M, Rumbaugh KP, Simoes PN, Moita L, Ferreira L (2016b) One-step synthesis of high-density peptide-conjugated gold nanoparticles with antimicrobial efficacy in a systemic infection model. Biomaterials 85:99–110
Rajchakit U, Sarojini V (2017) Recent developments in antimicrobial-peptide-conjugated gold nanoparticles. Bioconjug Chem 28(11):2673–2686
Reinhardt A, Neundorf I (2016) Design and application of antimicrobial peptide conjugates. Int J Mol Sci 17(5):1–21
Roberts MJ, Bentley MD, Harris JM (2002) Chemistry for peptide and protein PEGylation. Adv Drug Deliv Rev 54(4):459–476
Sader HS, Fedler KA, Rennie RP, Stevens S, Jones RN (2004) Omiganan pentahydrochloride (MBI 226), a topical 12-amino-acid cationic peptide: spectrum of antimicrobial activity and measurements of bactericidal activity. Antimicrob Agents Chemother 48(8):3112–3118
Santajit S, Indrawattana N (2016) Mechanisms of antimicrobial resistance in ESKAPE pathogens. Biomed Res Int 2016(2475067):1–8
Schmidt EGW, Hvam ML, Antunes F, Cameron J, Viuff D, Andersen B, Kristensen NN, Howard KA (2017) Direct demonstration of a neonatal Fc receptor (FcRn)-driven endosomal sorting pathway for cellular recycling of albumin. J Biol Chem 292(32):13312–13322
Scorciapino MA, Serra I, Manzo G, Rinaldi AC (2017) Antimicrobial dendrimeric peptides: structure, activity and new therapeutic applications. Int J Mol Sci 18(3):1–13
Scott A (2018) Peptide makers face threat from biotechnology how a 1-year-old German start-up could be about to shake up a $1 billion market. Chem Eng News 96(5):4–4
Sierra JM, Fuste E, Rabanal F, Vinuesa T, Vinas M (2017) An overview of antimicrobial peptides and the latest advances in their development. Expert Opin Biol Ther 17(6):663–676
Silva T, Magalhaes B, Maia S, Gomes P, Nazmi K, Bolscher JGM, Rodrigues PN, Bastos M, Gomes MS (2014) Killing of mycobacterium avium by lactoferricin peptides: improved activity of arginine- and D-amino-acid-containing molecules. Antimicrob Agents Chemother 58(6):3461–3467
Sockolosky JT, Szoka FC (2015) The neonatal Fc receptor, FcRn, as a target for drug delivery and therapy. Adv Drug Deliv Rev 91:109–124
Swiecicki JM, Di Pisa M, Lippi F, Chwetzoff S, Mansuy C, Trugnan G, Chassaing G, Lavielle S, Burlina F (2015) Unsaturated acyl chains dramatically enhanced cellular uptake by direct translocation of a minimalist oligo-arginine lipopeptide. Chem Commun 51(78):14656–14659
Tam JP, Lu YA, Yang JL (2002) Antimicrobial dendrimeric peptides. Eur J Biochem 269(3):923–932
Tapeinou A, Matsoukas MT, Simal C, Tselios T (2015) Cyclic peptides on a merry-go-round; towards drug design. Biopolymers 104(5):453–461
Teixeira V, Feio MJ, Rivas L, De la Torre BG, Andreu D, Coutinho A, Bastos M (2010) Influence of lysine N-epsilon-trimethylation and lipid composition on the membrane activity of the cecropin A-melittin hybrid peptide CA(1-7)M(2-9). J Phys Chem B 114(49):16198–16208
Thayer AM (2011) Making peptides at large scale. Chem Eng News 89(22):21–25

Trier S, Linderoth L, Bjerregaard S, Andresen TL, Rahbek UL (2014) Acylation of glucagon-like peptide-2: interaction with lipid membranes and in vitro intestinal permeability. PLoS One 9(10):1–10

Usmani SS, Bedi G, Samuel JS, Singh S, Kalra S, Kumar P, Ahuja AA, Sharma M, Gautam A, Raghava GPS (2017) THPdb: database of FDA-approved peptide and protein therapeutics. PLoS One 12(7):1–12

Veronese FM (2001) Peptide and protein PEGylation: a review of problems and solutions. Biomaterials 22(5):405–417

Vidal L, Geffard M (2014) Lauryl-poly-L-lysine: a new antimicrobial agent? FEBS J 281:328–328

Walensky LD, Bird GH (2014) Hydrocarbon-stapled peptides: principles, practice, and progress. J Med Chem 57(15):6275–6288

Wang G, Mishra B, Lau K, Lushnikova T, Golla R, Wang X (2015) Antimicrobial peptides in 2014. Pharmaceuticals (Basel) 8(1):123–150

WebPage, P (n.d.) http://www.polyphor.com/pol7080/. Accessed Jul 2018.

White CJ, Yudin AK (2011) Contemporary strategies for peptide macrocyclization. Nat Chem 3(7):509–524

WHO (2018) http://www.who.int/antimicrobial-resistance/global-action-plan/en/.

Willcox MDP, Kumar N, Cole N, et al. (2013) Antimicrobial peptides and uses thereof. WO2013076666 A1. 2013

WIPO (2018) https://patentscope.wipo.int/search/pt/search.jsf.

Wu YD, Gellman S (2008) Peptidomimetics. Acc Chem Res 41(10):1231–1232

Yang J, Chen H, Vlahov IR, Cheng JX, Low PS (2006) Evaluation of disulfide reduction during receptor-mediated endocytosis by using FRET imaging. Proc Natl Acad Sci U S A 103(37):13872–13877

Zasloff M (2002) Antimicrobial peptides of multicellular organisms. Nature 415(6870):389–395

Zong JY, Cobb SL, Cameron NR (2017) Peptide-functionalized gold nanoparticles: versatile biomaterials for diagnostic and therapeutic applications. Biomater Sci 5(5):872–886

Index

K. Matsuzaki (ed.), *Antimicrobial Peptides*, Advances in Experimental Medicine and Biology 1117,
https://doi.org/10.1007/978-981-13-3588-4

MIX
Papier aus verantwortungsvollen Quellen
Paper from responsible sources
FSC® C105338

If you have any concerns about our products,
you can contact us on
ProductSafety@springernature.com

In case Publisher is established outside the EU,
the EU authorized representative is:
Springer Nature Customer Service Center GmbH
Europaplatz 3, 69115 Heidelberg, Germany

Printed by Libri Plureos GmbH
in Hamburg, Germany